APPLIED FREQUENCY-DOMAIN ELECTROMAGNETICS

APPLIED
FREQUENCY-DOMAIN
ELECTROMAGNETICS

APPLIED FREQUENCY-DOMAIN ELECTROMAGNETICS

Robert Paknys

Concordia University, Montreal, Canada

WILEY

IEEE PRESS

Library of Congress Cataloging-in-Publication Data

Names: Paknys, Robert, author.
Title: Applied frequency-domain electromagnetics / Robert Paknys.
Description: Hoboken, New Jersey : John Wiley & Sons, 2016. | Includes
    bibliographical references and index.
Identifiers: LCCN 2016003736 (print) | LCCN 2016009152 (ebook) | ISBN
    9781118940563 (cloth) | ISBN 9781118940556 (pdf) | ISBN 9781118940549
    (epub)
Subjects: LCSH: Electromagnetic waves. | Electromagnetism–Mathematics.
Classification: LCC QC670 .P35 2016 (print) | LCC QC670 (ebook) | DDC
    537–dc23
LC record available at http://lccn.loc.gov/2016003736

A catalogue record for this book is available from the British Library.

Set in 9/11pt, TimesLTStd by SPi Global, Chennai, India.

1   2016

To Marilyn and Michael

Contents

Contents

About the Author

Robert Paknys was born in Montreal, Canada. He received the BEng degree from McGill University in 1979, and the MSc and PhD degrees from Ohio State University in 1982 and 1985, respectively, all in electrical engineering.

He was an assistant professor at Clarkson University during 1985-1987 and an engineer at MPB Technologies during 1987-1989. He joined Concordia University in 1989 as a faculty member in electrical and computer engineering, and is a professor. He has served as a consultant for the government and industry.

He was a visiting professor at the University of Auckland in 1996, the University of Houston in 2004 and the Ecole Polytechnique de Montreal in 2010.

Professor Paknys is a registered professional engineer, a member of CNC-URSI Commission B, a senior member of the IEEE, and a past associate editor for the *IEEE Transactions on Antennas and Propagation*.

Preface

The technologies related to electromagnetic waves go back to Hertz, Marconi and the radar systems of World War II. The knowledge gained during those eras propelled the subsequent development of microwave and satellite communications and the ubiquitous wireless technology of today. Understanding electromagnetic scattering is pivotal in the applications of radar target identification, underground geophysical probing as well as security applications such as airport scanners and seeing through walls. Computational electromagnetic modelling is a key element in the design of commercial and military aircraft, and navy ships, where the placement of dozens of collocated antennas must be carefully considered, so that intersystem interference can be mitigated.

Researchers behind these and other advances in technology need to understand both the classical theory of electromagnetics and modern techniques for solving Maxwell's equations. To this end, this book provides a graduate-level treatment of selected topics. Chapters 1 and 2 present background material on Maxwell's equations, plane waves and rigorous and approximate boundary conditions. Chapter 3 develops solutions for rectangular, cylindrical and dielectric waveguides and resonators. In Chapter 4, some crucial theorems, principles and potential theory are explained in detail. Chapter 5 presents the solutions to some canonical problems that have an exact solution, such as the cylinder, wedge and sphere. Chapter 6 describes the method of moments. Chapter 7 covers the finite element method. Chapter 8 is about the uniform geometrical theory of diffraction, and Chapter 9 covers physical optics and the physical theory of diffraction. Chapters 10–12 are about Green's functions and their applications.

Analytical methods provide physical insights that are valuable in the design process and the invention of new devices. The separation of variables method is applied to waveguides, cylinders, wedges and other canonical shapes. Asymptotic methods address the evaluation of integrals, as well as diffraction theory. Green's function concepts are presented in the two-dimensional (2D) scalar and three-dimensional (3D) dyadic forms, and their interpretation is given in relation to the surface equivalence principle.

Numerical methods are indispensable as they allow us to solve highly arbitrary and realistic problems that the purely analytical techniques cannot. The method of moments and the finite element method are described in dedicated chapters. The level of presentation allows the reader to immediately begin applying the methods to some problems of moderate complexity. It also provides an explanation of the underlying theory so that its capabilities and limitations can be understood. This has value as it helps one make informed decisions when using modern CAD tools.

Often, in the preliminary stages of research, it is very useful to investigate field behaviour by using 2D problems. This way, it is often possible to greatly simplify the problem while still retaining the essential characteristics of the fields. It is also a good way to learn the subject, as it minimizes the mathematical complexity and makes the field solutions easier to physically interpret. The book emphasizes a 2D approach, however, where appropriate, 3D is also used.

The book is aimed at graduate students and engineers in industry and R&D labs. The minimum assumed background is an undergraduate course in waves and transmission lines. The first three chapters aim to put all readers on an equal footing – thereby readers with diverse backgrounds and levels of familiarity are accommodated. The coverage is intended to assist research students who are beginning to explore the current engineering literature, as well as more experienced researchers who need to learn about new topics.

The way people look for relevant literature has changed dramatically in the past 20 years. For this reason, no attempt has been made to compile a comprehensive list of references, which in any case would be prone to rapid obsolescence. Rather, each chapter contains a small list of references that should help readers proceed and find the key books and papers that address their specific interests. Many fine works have been omitted, and should any authors feel slighted, I offer my apologies in advance.

The topics are not necessarily arranged by the subject category, but in the order that they are most easily learned and applied. Some topics are revisited at a gradually increasing depth. For instance, waveguides are in Chapters 3 and 4, and the surface equivalence principle is in Chapters 4 and 10.

The homework problems have been developed with an intention to provide motivation and opportunities for practice, as well as revealing new concepts. There are problems for review purposes, for analytical development and for programming.

For both analytical and numerical techniques, it is a rewarding step to generate numerical results. The computer-oriented homework problems allow the reader to apply numerical techniques. Some of the problems involve minor modifications of existing programs instead of coding from scratch. Therefore, larger amounts of material and more ambitious problems can be covered in a given time. Many other problems involve little or no computer work, so instructors can choose to opt out of the computation-oriented format or else solve some of the problems with their own code.

The supporting code is written in Fortran 90, which is widely used in computational science and high-performance computing. Well-tested subroutines are provided for special functions, diffraction coefficients, root finding, numerical integration and matrix manipulations. The Netlib repository is extensively used. In this book, the object-oriented capability of Fortran 90 has been used to develop easy-to-use interfaces that hide the complexity of large subroutines.

Computing and plotting can be done with public-domain software that is available under Linux, Windows and Mac OS X. It is assumed that the reader has some prior experience with a programming or scripting language, but not necessarily Fortran 90. Appendix F summarizes the essentials, so that the reader can begin computational work with little difficulty.

R. Paknys
Montreal
September 2015

Acknowledgements

I would like to thank a few colleagues for their help, encouragement and friendship over the years: Dr Amy R. Pinchuk, Infield Scientific Inc.; Professors Chris Trueman, Concordia University; Ayhan Altintas, Bilkent University; Michael J. Neve, University of Auckland; David R. Jackson and Donald R. Wilton, University of Houston; Jean-Jacques Laurin, Ecole Polytechnique de Montreal; and Derek McNamara, University of Ottawa. Particular appreciation goes to Professor Jackson for sharing his knowledge of Riemann surfaces, leaky waves and periodic structures.

Going back to my early years, I would like to acknowledge some inspiring professors. Their imparted knowledge and wisdom have served me well to this day. At McGill: Professors G. L. d'Ombrain, G. W. Farnell, E. L. Adler, C. W. Bradley, J. E. Turner, R. Vermes, and G. Bach; at Ohio State: Professors C. H. (Buck) Walter, R. G. Kouyoumjian, W. D. Burnside, P. H. Pathak, N. Wang, R. J. Marhefka, J. H. Richmond, E. H. Newman, B. A. Munk, and (visiting professor) R. E. Collin. A few, regrettably, are deceased – nevertheless their ideas live on.

I would like to thank our students. Their helpful feedback led to many improvements in this book. Most importantly, without them, there would have been no course and no book.

Finally I would like to thank a few people who helped transform my ideas into a book; in particular, Anna Smart and Sandra Grayson of the Wiley editorial staff, and Lincy Priya, the project manager at SPi Global.

1

Background

This chapter provides a review of Maxwell's equations in integral and differential forms. The capacitor and inductor are used to demonstrate and interpret the integral forms. The Poynting theorem, Lorentz reciprocity theorem, Friis transmission formula and radar range equation are also described. Some of the properties of high-frequency asymptotic techniques are reviewed.

1.1 Field Laws

Maxwell's equations in integral form are

$$\oint_C \boldsymbol{\mathcal{E}} \cdot \mathbf{d\ell} = -\frac{d}{dt} \int_S \boldsymbol{\mathcal{B}} \cdot \mathbf{dS} \tag{1.1}$$

$$\oint_C \boldsymbol{\mathcal{H}} \cdot \mathbf{d\ell} = \int_S \boldsymbol{\mathcal{J}} \cdot \mathbf{dS} + \frac{d}{dt} \int_S \boldsymbol{\mathcal{D}} \cdot \mathbf{dS} \tag{1.2}$$

$$\oint_S \boldsymbol{\mathcal{D}} \cdot \mathbf{dS} = \int_V \varrho_v \, dV = \mathcal{Q} \tag{1.3}$$

$$\oint_S \boldsymbol{\mathcal{B}} \cdot \mathbf{dS} = 0. \tag{1.4}$$

We will use the MKS system of units. The Volt, Ampere, Coulomb, Weber, and Tesla are abbreviated as V, A, C, Wb, and T. The electric field $\boldsymbol{\mathcal{E}}$ is in V/m; the magnetic field $\boldsymbol{\mathcal{H}}$ is in A/m; the electric flux density $\boldsymbol{\mathcal{D}}$ is in C/m^2, and the magnetic flux density $\boldsymbol{\mathcal{B}}$ is in T (equivalent to Wb/m^2). The electric current density $\boldsymbol{\mathcal{J}}$ is in A/m^2; charge $\mathcal{Q}$ is in C, and charge density ϱ_v is in C/m^3.

The surface and volume integrals are associated with the mathematical surfaces shown in Figure 1.1. The first equation is Faraday's law. The second one is credited to Ampère and Maxwell, and the third one is Gauss's law. The fourth equation is called Gauss's law for magnetism. The group of four equations is usually referred to as Maxwell's equations.

If a region has fields $\boldsymbol{\mathcal{E}}, \boldsymbol{\mathcal{B}}$, and a charge $\mathcal{Q}$ is moving through those fields with a velocity $\mathbf{u}$, the charge will experience a force, in accordance with the *Lorentz force law*

$$\boldsymbol{\mathcal{F}} = \mathcal{Q}\boldsymbol{\mathcal{E}} + \mathcal{Q}\mathbf{u} \times \boldsymbol{\mathcal{B}}. \tag{1.5}$$

Applied Frequency-Domain Electromagnetics, First Edition. Robert Paknys.
© 2016 John Wiley & Sons, Ltd. Published 2016 by John Wiley & Sons, Ltd.
Companion Website: www.wiley.com/go/paknys9981

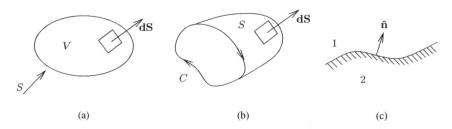

Figure 1.1 Mathematical surfaces associated with the field laws. (a) Closed surface and volume, (b) open surface and contour C, (c) boundary between regions 1 and 2.

Charge cannot be created or destroyed. Any increase or decrease of charge occurs because there is a current. This is stated mathematically as the *continuity equation*. In integral form, the outflux of current $\mathcal{J}$ across a closed surface S equals the time rate of decrease of the charge Q that is inside S

$$\oint_S \mathcal{J} \cdot d\mathbf{S} = -\frac{dQ}{dt}. \tag{1.6}$$

The point-form equivalent is

$$\nabla \cdot \mathcal{J} = -\frac{\partial \varrho_v}{\partial t}. \tag{1.7}$$

By integrating both sides of (1.7) over a volume V and applying the divergence theorem to the left-hand side, the integral form (1.6) is obtained.

From the electric field, the voltage is

$$V_{ab} = V_a - V_b = \int_a^b \mathcal{E} \cdot d\boldsymbol{\ell}. \tag{1.8}$$

An electric field having $\nabla \times \mathbf{E} = 0$ is said to be irrotational. This occurs in electrostatics and in the transverse cross section of a transmission line. In these cases, the line integral becomes path independent, and hence, the voltage is uniquely defined by the endpoints a and b.

From the magnetic field, the current is

$$I_t = \oint_C \mathcal{H} \cdot d\boldsymbol{\ell} \tag{1.9}$$

where I_t is in the direction of the right-hand thumb and C is a closed contour in the direction of the fingers. This relationship is strictly true for steady (DC) currents. It is still true in the AC case if there is no $\mathcal{E}$ component perpendicular to the surface bounded by C.

1.2 Properties of Materials

The electrical properties of materials are governed by their physical makeup. In this book, the physics and chemistry of these topics will not be covered, and the reader is referred to the references at the end of the chapter. It will be adequate for our purposes to describe the mathematical models that account for the presence of materials.

In free space, we have the *constitutive relations*

$$\mathcal{D} = \epsilon_0 \mathcal{E} \tag{1.10}$$

$$\mathcal{B} = \mu_0 \mathcal{H} \tag{1.11}$$

$$\epsilon_0 = 8.854187816 \times 10^{-12} \text{ F/m} \tag{1.12}$$

$$\mu_0 = 4\pi \times 10^{-7} \text{ H/m}. \tag{1.13}$$

Interestingly, in any system of units, the value of μ_0 or ϵ_0 can be arbitrarily chosen. However, $1/\sqrt{\mu_o \epsilon_0}$ must equal the speed of light. In the MKS system, μ_0 in (1.13) is chosen as an exact value. Then, ϵ_0 in (1.12) is determined.

In dielectric materials,

$$\mathcal{D} = \epsilon_0 \mathcal{E} + \mathcal{P}. \tag{1.14}$$

The term $\epsilon_0 \mathcal{E}$ is what we have in free space. If a dielectric is present, the applied electric field will push its atomic charges, positive towards one direction and negative in the opposite direction, forming dipoles. These dipoles contribute an additional electric flux density $\mathcal{P}$, the *polarization* in C/m^2. Generally, the relation between $\mathcal{D}$ and $\mathcal{P}$ can be complicated, that is, non-linear. In the special case of linear materials, $\mathcal{P}$ is linearly proportional to the applied field. More precisely, $\mathcal{P} = \epsilon_0 \chi_e \mathcal{E}$ where the constant of proportionality χ_e is called the *electric susceptibility*. In this case,

$$\mathcal{D} = \epsilon_0 \mathcal{E} + \mathcal{P} = \epsilon_0 (1 + \chi_e) \mathcal{E} = \epsilon \mathcal{E}. \tag{1.15}$$

Therefore, in linear materials, we can use the simple relation $\mathcal{D} = \epsilon \mathcal{E}$ where the permittivity is $\epsilon = \epsilon_0 (1 + \chi_e)$.

In magnetic materials,

$$\mathcal{B} = \mu_0 (\mathcal{H} + \mathcal{M}). \tag{1.16}$$

The term $\mu_0 \mathcal{H}$ is what we have in free space. If a magnetic material is present, the applied magnetic field will reorient the material's electronic orbits (which act as current loops) and contribute an additional magnetic flux density $\mu_0 \mathcal{M}$, the *magnetization* in T. In non-linear materials, the relation between $\mathcal{M}$ and $\mathcal{H}$ can be complicated. In the special case of linear materials, $\mathcal{M} = \chi_m \mathcal{H}$ where the constant of proportionality χ_m is called the *magnetic susceptibility*. In this case,

$$\mathcal{B} = \mu_0 (\mathcal{H} + \mathcal{M}) = \mu_0 (1 + \chi_m) \mathcal{H} = \mu \mathcal{H}. \tag{1.17}$$

Therefore, in linear materials, we can use the simple relation $\mathcal{B} = \mu \mathcal{H}$ where the permeability is $\mu = \mu_0 (1 + \chi_m)$.

In a good conductor, when an electric field is applied, the charges move immediately. Dipoles (as in a dielectric) do not have a chance to form. Therefore, $\mathcal{P} \approx 0$ and consequently $\epsilon \approx \epsilon_0$. In non-magnetic materials, $\mu \approx \mu_0$. Such approximations are good for non-magnetic conductors such as aluminium or copper.

1.3   Types of Currents

The *convection current* is associated with charges that are moving with a velocity **u**

$$\mathcal{J} = \varrho_v \mathbf{u}. \tag{1.18}$$

Such a 'stream' of charged particles occurs, for example, in a vacuum tube, a cathode ray tube or a scanning electron microscope.

Inside a conductor, an electric field will push on the charges and cause a *conduction current*

$$\boldsymbol{\mathcal{J}} = \sigma\boldsymbol{\mathcal{E}}. \tag{1.19}$$

The conductivity σ is in S/m (Siemens/m, or equvalently, mho/m). The main difference between a convection current and a conduction current is that the latter type occurs in an electrically neutral material. For example, in a wire, for every charge that enters at one end, a charge leaves at the other end. Therefore there is no net charge and $\varrho_v = 0$.

An *impressed current* is independent of the field around it, but the field around it depends on the impressed current. An example of an impressed current is a dipole antenna. An *induced current* comes from the interaction of a field with any surrounding media and/or boundaries. As an example, if a dipole antenna illuminates a metal body, it will cause surface currents to flow on the body; these are induced currents. The purpose of induced currents is that they adjust themselves in just the right way so that their field, when added to the impressed field, will give a total field that satisfies the boundary conditions, that is, $\boldsymbol{\mathcal{E}}_{tan} = 0$ on the metal. Inside dielectrics there are volume-equivalent induced currents; these are discussed in Chapter 4.

1.4   Capacitors, Inductors

To gain a better understanding of Maxwell's equations in the integral form, this section demonstrates their application to the fields inside capacitors and inductors.

First, the Ampere-Maxwell equation

$$\oint_C \boldsymbol{\mathcal{H}} \cdot d\boldsymbol{\ell} = \int_S \boldsymbol{\mathcal{J}} \cdot d\mathbf{S} + \frac{d}{dt}\int_S \boldsymbol{\mathcal{D}} \cdot d\mathbf{S} \tag{1.20}$$

will be applied to a capacitor, in Figure 1.2. The capacitor supports an electric field $\boldsymbol{\mathcal{E}} = \hat{\mathbf{z}}v(t)/d$ in the region $0 \leq z \leq d$. With $S = S_1$ in case (a), the current density $\boldsymbol{\mathcal{J}}$ pierces S_1. Because $\boldsymbol{\mathcal{E}}$ is zero outside the capacitor, $\boldsymbol{\mathcal{D}}$ will be zero on S_1, so that (1.20) becomes

$$\oint_C \boldsymbol{\mathcal{H}} \cdot d\boldsymbol{\ell} = \int_{S_1} \boldsymbol{\mathcal{J}} \cdot d\mathbf{S}. \tag{1.21}$$

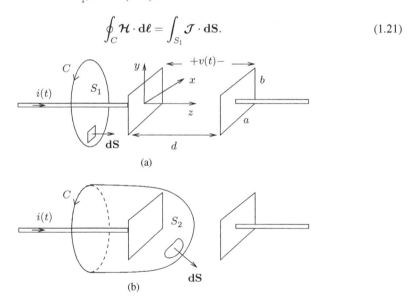

Figure 1.2   (a) Contour C bounds the disk S_1. (b) Contour C bounds the open surface S_2.

With $S = S_2$ in case (b), the current density $\mathcal{J}$ is zero on S_2 and $\mathcal{D} = \hat{z}\epsilon v(t)/d$ so that

$$\oint_C \mathcal{H} \cdot \mathbf{d\ell} = \frac{d}{dt} \int_{S_2} \hat{z}\epsilon v(t)/d \cdot \mathbf{dS}. \tag{1.22}$$

The right-hand side of (1.21) is the total current $i(t)$. Since C is the same in both cases, the left-hand side of (1.21) and (1.22) are equal. This leads to

$$i(t) = \oint_C \mathcal{H} \cdot \mathbf{d\ell} = \frac{d}{dt} \int_{S_2} \hat{z}\epsilon v(t)/d \cdot \mathbf{dS}$$

or

$$i(t) = \frac{\epsilon ab}{d} \frac{dv(t)}{dt}.$$

Recognizing the capacitance $C_0 = \epsilon ab/d$, we see that

$$i(t) = C_0 \frac{dv(t)}{dt}. \tag{1.23}$$

If S_2 is right at the surface of the $z = 0$ plate, then $D_z = \varrho_s$, and we can say that $\int \mathcal{D} \cdot \mathbf{dS} = \varrho_s ab = Q$ from which we obtain the well-known result $i(t) = dQ/dt$. Equating this with (1.23) implies that $C_0 dv = dQ$, which gives us the capacitance $C_0 = dQ/dv$.

Next, we apply Faraday's law to a wire loop and a toroidal inductor. Figure 1.3(a) shows a wire loop in the $x - y$ plane. The integration path C is tangent to the wire and crosses the gap at the terminals 1–2. Because $\mathcal{E}_{tan} = 0$ on the wire, the line integral is zero everywhere except at the gap, and

$$v_{12}(t) = \oint_C \mathcal{E} \cdot \mathbf{d\ell} = -\frac{d}{dt} \int_S \mathcal{B} \cdot \mathbf{dS}. \tag{1.24}$$

The direction of C implies that $\mathbf{dS} = -\hat{z} dS$. Let us denote the $+z$ flux through the loop as $\Phi = \int B_z dS$. If there are N turns, the flux is $N\Phi$, and Faraday's law (1.24) becomes

$$v_{12}(t) = N \frac{d\Phi}{dt}. \tag{1.25}$$

We can apply this result to the toroidal inductor in Figure 1.3(b). To better understand the relationships between v_{12}, i and Φ, it is helpful to consider what happens if a positive step of voltage $v_{12}(t) = v_0 U(t)$ is applied to the terminals. Equation (1.25) indicates that a ramp of flux $\Phi = (v_0/N)t$ will occur in the indicated direction. Also, by Ampere's law, if the right-hand thumb points in the direction of Φ, the fingers give the direction of i.

We can find the inductance L from

$$v_{12} = L \frac{di}{dt} = N \frac{d\Phi}{dt}.$$

It follows that $L di = N d\Phi$, so $L = N d\Phi/di$. For a toroidal core of length ℓ and cross-sectional area S, we also have that $Ni = H\ell$ and $\Phi = BS$, so that[1]

$$L = N \frac{d\Phi}{di} = \frac{N^2 S}{\ell} \frac{dB}{dH}.$$

---

[1] For a high-permeability toroid, the flux in the core is approximately constant. Ampere's law leads to $Ni \approx H\ell$, and the flux is $\Phi \approx BS$. More accurately, H for a toroid is not constant and is given by $H_\phi = Ni/2\pi\rho$ in cylindrical coordinates.

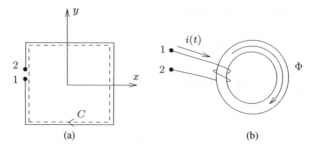

Figure 1.3   (a) Wire loop in the $x - y$ plane with terminals 1–2. The closed path C is tangential to the wire and crosses the terminal gap. (b) A toroidal inductor, showing the terminals 1–2 and the directions of i and Φ.

Interestingly, this even works for non-linear materials. For the special case of a linear material, $B = \mu H$ so that

$$L = \frac{\mu N^2 S}{\ell}.$$

In general, if a voltage is applied at the gap, a current will flow in the loop and a flux will be created (an inductor). If a time-varying magnetic flux is applied to the loop, then a voltage will be induced at the gap (a generator). In either case, (1.25) gives the relation between the time-varying flux and the terminal voltage.[2]

1.5   Differential Form

The Stokes and divergence theorems relate the integral form of the field laws (1.1)–(1.4) to their *differential*, or *point form*

$$\nabla \times \boldsymbol{\mathcal{E}} = -\frac{\partial \boldsymbol{\mathcal{B}}}{\partial t} \tag{1.26}$$

$$\nabla \times \boldsymbol{\mathcal{H}} = \boldsymbol{\mathcal{J}} + \frac{\partial \boldsymbol{\mathcal{D}}}{\partial t} \tag{1.27}$$

$$\nabla \cdot \boldsymbol{\mathcal{D}} = \varrho_v \tag{1.28}$$

$$\nabla \cdot \boldsymbol{\mathcal{B}} = 0. \tag{1.29}$$

The point form of the boundary conditions can be obtained by applying the integral form of the field laws (1.1)–(1.4) to infinitesimal volumes and surfaces that straddle the boundary surface in Figure 1.1(c). The result is

$$\hat{\mathbf{n}} \times (\boldsymbol{\mathcal{E}}_1 - \boldsymbol{\mathcal{E}}_2) = 0 \tag{1.30}$$

$$\hat{\mathbf{n}} \times (\boldsymbol{\mathcal{H}}_1 - \boldsymbol{\mathcal{H}}_2) = \boldsymbol{\mathcal{J}}_s \tag{1.31}$$

$$\hat{\mathbf{n}} \cdot (\boldsymbol{\mathcal{D}}_1 - \boldsymbol{\mathcal{D}}_2) = \varrho_s \tag{1.32}$$

$$\hat{\mathbf{n}} \cdot (\boldsymbol{\mathcal{B}}_1 - \boldsymbol{\mathcal{B}}_2) = 0. \tag{1.33}$$

---

[2] Many books on electromagnetics are a bit short on the practical details of how to make an inductor. The definitive book on this topic is by Grover (1962).

In words, (1.30) states that $\mathcal{E}_{tan}$ is continuous at the boundary. Equation (1.31) states that $\mathcal{H}_{tan}$ is continuous at the boundary, if there is no surface current $\mathcal{J}_s$ on the boundary. If a surface current is present, then $\mathcal{H}_{tan}$ has a step discontinuity across the sheet. Equation (1.32) states that the normal component of $\mathcal{D}$ is continuous at the boundary, if there is no surface charge ϱ_s on the boundary. If a surface charge is present, then the normal component of $\mathcal{D}$ has a step discontinuity across the sheet. Equation (1.33) states that the normal component of $\mathcal{B}$ is always continuous at the boundary. As a special case, perfect conductors support surface currents and have $\mathcal{E}_{tan} = 0$.

The integral version of the field laws is mathematically equivalent to the differential form plus the boundary conditions. A valid solution must satisfy (1.1)–(1.4). Alternatively, it must satisfy the differential forms (1.26)–(1.29) plus the boundary conditions (1.30)–(1.33).[3]

The following magnetostatic example illustrates the usage of integral-form and point-form field laws.

Example 1.1 (Thick Wire) A z-directed wire of radius a carries a uniform DC current I_0. Find the magnetic field from the integral form of the field laws, and demonstrate that the point-form field laws for magnetic fields are satisfied. The wire is regarded as 'thick', as we want to obtain the magnetic field both inside and outside the wire.

Solution: The current density is assumed to be uniformly spread out inside the wire and flowing in the z direction, so $\mathcal{J} = \hat{z}I_0/\pi a^2$. We use (1.2) with $d/dt = 0$ for the DC case. Noting that the surface integral of $\mathcal{J}$ is the current, we have

$$\oint_C \mathcal{H} \cdot d\boldsymbol{\ell} = \int_S \mathcal{J} \cdot d\mathbf{S} = I.$$

The path C is chosen to be a circle of radius ρ that bounds a disc S, so that $d\boldsymbol{\ell} = \hat{\phi}\rho d\phi$ and $d\mathbf{S} = \hat{z}\rho d\phi d\rho$. The left-hand side is equal to $2\pi\rho H_\phi$. When $\rho > a$, S captures all of the current, so the right-hand side is $I = I_0$. Therefore, $\mathcal{H} = \hat{\phi}I_0/2\pi\rho$ outside the wire. When $\rho < a$, S only captures part of the current, so that $I = I_0(\pi\rho^2)/(\pi a^2)$ and $\mathcal{H} = \hat{\phi}I_0\rho/2\pi a^2$ inside the wire. (From symmetry, one can show that H_ρ and H_z are zero and that H_ϕ does not depend on ϕ.)

The magnetic field should satisfy (1.27) and (1.29). Using the known $\mathcal{H}$, we obtain $\nabla \times \mathcal{H} = 0$ for $\rho > a$ and $\nabla \times \mathcal{H} = \hat{z}I_0/\pi a^2$ for $\rho < a$. This shows the 'point-form' nature of the field law (1.27). That is, the curl of $\mathcal{H}$ at some specific point (ρ, ϕ, z) in space equals the current density $\mathcal{J}$ *at that same point*. Similarly, for (1.29), it is straightforward to confirm that $\nabla \cdot \mu\mathcal{H} = 0$ both inside and outside the wire.

The boundary conditions (1.31) and (1.33) must hold at the surface $\rho = a$. If we choose an outward-pointing surface normal $\hat{n} = \hat{\rho}$, then region 1 is $\rho > a$ and region 2 is $\rho < a$. There are no surface currents, only a volume current, so $\mathcal{J}_s = 0$ and (1.31) becomes $H_\phi(a^+) - H_\phi(a^-) = 0$ where $a^\pm$ is the value of ρ just outside/inside the wire. Using the known H_ϕ, we obtain $H_\phi(a^+) - H_\phi(a^-) = I_0/2\pi a^+ - I_0a^-/2\pi a^2 = 0$, which proves that (1.31) is satisfied. Equation (1.33) requires that $\mu H_{1\rho} - \mu H_{2\rho} = 0$, which is satisfied, as $H_\rho = 0$ for this problem. ∎

1.6 Time-Harmonic Fields

AC steady-state (*a.k.a.* time-harmonic, or frequency-domain) analysis is efficiently done with phasors. A time-domain voltage $v(t) = V_0 \cos(\omega t + \phi)$ is represented by a phasor $V(\omega) = V_0 e^{j\phi}$, which is usually

---

[3] The integral form of Maxwell's equations is completely general, whereas the differential form is not – it is only valid away from boundaries. Therefore it is incorrect to 'derive' the integral form from the differential form via the Stokes and divergence theorems.

written in the Steinmetz notation as $V(\omega) = V_0 \underline{/\phi}$. A time-domain voltage is obtained from its phasor via the operation

$$v(t) = \mathrm{Re}(V(\omega)e^{j\omega t}).$$

Time-harmonic fields with an $e^{j\omega t}$ time dependence follow the same procedure. Using the notations $\mathcal{E}(x, y, z, t)$ and $\mathbf{E}(x, y, z, \omega)$ to distinguish between the time-domain and phasor quantities, we have

$$\mathcal{E}(x, y, z, t) = \mathrm{Re}(\mathbf{E}(x, y, z, \omega)e^{j\omega t}).$$

Time derivatives are replaced by $\partial/\partial t \rightarrow j\omega$, so the time-harmonic Maxwell's equations in integral form are

$$\oint_C \mathbf{E} \cdot d\boldsymbol{\ell} = -j\omega \int_S \mathbf{B} \cdot d\mathbf{S} \tag{1.34}$$

$$\oint_C \mathbf{H} \cdot d\boldsymbol{\ell} = \int_S \mathbf{J} \cdot d\mathbf{S} + j\omega \int_S \mathbf{D} \cdot d\mathbf{S} \tag{1.35}$$

$$\oint_S \mathbf{D} \cdot d\mathbf{S} = \int_V \rho_v \, dV = Q \tag{1.36}$$

$$\oint_S \mathbf{B} \cdot d\mathbf{S} = 0. \tag{1.37}$$

In point form, they are

$$\nabla \times \mathbf{E} = -j\omega \mathbf{B} \tag{1.38}$$

$$\nabla \times \mathbf{H} = \mathbf{J} + j\omega \mathbf{D} \tag{1.39}$$

$$\nabla \cdot \mathbf{D} = \rho_v \tag{1.40}$$

$$\nabla \cdot \mathbf{B} = 0 \tag{1.41}$$

and the boundary conditions are

$$(\mathbf{E}_1 - \mathbf{E}_2) \times \hat{\mathbf{n}} = 0 \tag{1.42}$$

$$\hat{\mathbf{n}} \times (\mathbf{H}_1 - \mathbf{H}_2) = \mathbf{J}_s \tag{1.43}$$

$$\hat{\mathbf{n}} \cdot (\mathbf{D}_1 - \mathbf{D}_2) = \rho_s \tag{1.44}$$

$$\hat{\mathbf{n}} \cdot (\mathbf{B}_1 - \mathbf{B}_2) = 0. \tag{1.45}$$

The constitutive relations in the frequency domain are

$$\mathbf{D} = \epsilon \mathbf{E} \tag{1.46}$$

$$\mathbf{B} = \mu \mathbf{H}. \tag{1.47}$$

As long as μ and ϵ are frequency independent, there is no difference between (1.46), (1.47) and the time domain forms (1.15), (1.17). If there is a frequency dependence, as in dispersive media, then the time-domain forms must be written as convolutions, that is, $\mathcal{D} = \epsilon \otimes \mathcal{E}$ and $\mathcal{B} = \mu \otimes \mathcal{H}$.

1.7 Sufficient Conditions

In general, fields must satisfy the four Maxwell equations (1.38)–(1.41) and the four boundary conditions (1.42)–(1.45). However, for the special case of a time-varying field, $j\omega \neq 0$, and it is sufficient to solve the two curl equations and meet the two tangential boundary conditions.

Table 1.1 Duality principle.

Electric sources $\longrightarrow$	**J**	**E**	**H**	ρ_v	μ	ϵ
Magnetic sources $\longrightarrow$	**M**	**H**	$-$**E**	ρ_{vm}	ϵ	μ

Taking the divergence of (1.38) and noting that $\nabla \cdot \nabla \times \mathbf{A} \equiv 0$ (where $\mathbf{A}$ is any vector), we obtain (1.41). Similarly, taking the divergence of (1.39) and using the continuity equation (1.7) in its time-harmonic form $\nabla \cdot \mathbf{J} = -j\omega\rho_v$ gives (1.40).

Similarly, it is sufficient to enforce just two of the four boundary conditions, the ones involving the continuity of $\mathbf{E}_{tan}$ and $\mathbf{H}_{tan}$, in (1.42) and (1.43). The normal boundary conditions on D_n and B_n in (1.44) and (1.45) will be automatically satisfied. For instance, the connection between $\mathbf{E}_{tan}$ in (1.42) and B_n in (1.45) can be shown in the following way. Suppose that $z = 0$ forms a boundary between two media. From (1.38), B_z obeys $-j\omega B_z = \partial E_y/\partial x - \partial E_x/\partial y$. If E_y is continuous across the $z = 0$ boundary for all (x, y), then $\partial E_y/\partial x$ is also continuous. The same argument holds for E_x and $\partial E_x/\partial y$. Therefore, the continuity of E_x and E_y at $z = 0$ ensures the continuity of B_z. A similar proof can be developed to show that the condition on $\mathbf{H}_{tan}$ in (1.43) ensures that (1.44) for D_n is met.

1.8 Magnetic Currents, Duality

The field laws in (1.38)–(1.45) express the fields in terms of the electric current $\mathbf{J}$ and electric charge ρ_v. In nature, there is no such thing as a magnetic current or magnetic charge. However, magnetic sources turn out to be very useful mathematical concepts that will be used extensively throughout this book. To this end, we now obtain the field laws for these types of sources.[4]

Let us denote the magnetic current and charge as $\mathbf{M}$ and ρ_{vm}. The basic idea is that if such sources existed, then Maxwell's equations would become symmetric. This symmetry is obtained when we apply the *principle of duality* to the field laws. This is accomplished by taking the equations for electric sources and replacing $\mathbf{J} \to \mathbf{M}, \mathbf{E} \to \mathbf{H}, \mathbf{H} \to -\mathbf{E}$ and so forth, as in Table 1.1. As a result, (1.38)–(1.45) become

$$\nabla \times \mathbf{H} = j\omega\mathbf{D} \tag{1.48}$$

$$\nabla \times \mathbf{E} = -\mathbf{M} - j\omega\mathbf{B} \tag{1.49}$$

$$\nabla \cdot \mathbf{B} = \rho_{vm} \tag{1.50}$$

$$\nabla \cdot \mathbf{D} = 0 \tag{1.51}$$

$$\hat{\mathbf{n}} \times (\mathbf{H}_1 - \mathbf{H}_2) = 0 \tag{1.52}$$

$$\hat{\mathbf{n}} \times (\mathbf{E}_1 - \mathbf{E}_2) = \mathbf{M}_s \tag{1.53}$$

$$\hat{\mathbf{n}} \cdot (\mathbf{B}_1 - \mathbf{B}_2) = \rho_{sm} \tag{1.54}$$

$$\hat{\mathbf{n}} \cdot (\mathbf{D}_1 - \mathbf{D}_2) = 0. \tag{1.55}$$

When both electric and magnetic sources are present, (1.38)–(1.45) can be combined with (1.48)–(1.55) to obtain

$$\nabla \times \mathbf{E} = -\mathbf{M} - j\omega\mathbf{B} \tag{1.56}$$

$$\nabla \times \mathbf{H} = \mathbf{J} + j\omega\mathbf{D} \tag{1.57}$$

$$\nabla \cdot \mathbf{D} = \rho_v \tag{1.58}$$

$$\nabla \cdot \mathbf{B} = \rho_{vm} \tag{1.59}$$

---

[4] To distinguish between the two types of currents, throughout this book, electric currents $\mathbf{J}$ will be shown as a black arrow $\longrightarrow$ and magnetic currents $\mathbf{M}$ will be shown as a white arrow $\longrightarrow\!\triangleright$.

$$(\mathbf{E}_1 - \mathbf{E}_2) \times \hat{\mathbf{n}} = \mathbf{M}_s \tag{1.60}$$

$$\hat{\mathbf{n}} \times (\mathbf{H}_1 - \mathbf{H}_2) = \mathbf{J}_s \tag{1.61}$$

$$\hat{\mathbf{n}} \cdot (\mathbf{D}_1 - \mathbf{D}_2) = \rho_s \tag{1.62}$$

$$\hat{\mathbf{n}} \cdot (\mathbf{B}_1 - \mathbf{B}_2) = \rho_{sm}. \tag{1.63}$$

In summary, Equations (1.56)–(1.63) give the most general and useful form of the field laws that we will need. Any static ($\omega = 0$) or time-harmonic field problem having electric and/or magnetic sources must satisfy these eight equations. In the time-harmonic case, it is sufficient to solve the curl equations (1.56) and (1.57) and impose the tangential boundary conditions (1.60) and (1.61).

1.9 Poynting's Theorem

Electromagnetic energy is stored in capacitors and inductors, dissipated in resistors, transported along wires and radiated by antennas. In 1884, J. H. Poynting developed a theorem for electromagnetic energy conservation that will now be derived.

A volume V is bounded by a closed surface S having an inward normal $\hat{\mathbf{n}}$ as in Figure 1.4. The material inside V is characterized by μ, ϵ and σ and are assumed to be real. The fields in V are $\mathbf{E}, \mathbf{H}$. Using a vector identity, we can write

$$\nabla \cdot (\mathbf{E} \times \mathbf{H}^*) = (\nabla \times \mathbf{E}) \cdot \mathbf{H}^* - (\nabla \times \mathbf{H}^*) \cdot \mathbf{E}. \tag{1.64}$$

With $\nabla \times \mathbf{E} = -j\omega\mathbf{B}$ and $\nabla \times \mathbf{H}^* = (\mathbf{J} + j\omega\mathbf{D})^*$, this becomes

$$\nabla \cdot (\mathbf{E} \times \mathbf{H}^*) = -j\omega\mathbf{B} \cdot \mathbf{H}^* + j\omega\mathbf{D}^* \cdot \mathbf{E} - \mathbf{J}^* \cdot \mathbf{E}. \tag{1.65}$$

Being mindful that $\hat{\mathbf{n}}dS = \mathbf{dS}$ points inwards, the divergence theorem is

$$-\oint_S \mathbf{F} \cdot \mathbf{dS} = \int_V \nabla \cdot \mathbf{F} \, dV.$$

Applying this to (1.65) gives

$$\oint_S (\mathbf{E} \times \mathbf{H}^*) \cdot \mathbf{dS} = -j\omega \int_V \mathbf{D}^* \cdot \mathbf{E} - \mathbf{B} \cdot \mathbf{H}^* \, dV + \int_V \mathbf{J}^* \cdot \mathbf{E} \, dV. \tag{1.66}$$

Multiplying (1.66) by 1/2 leads to Poynting's theorem in its usual form

$$\frac{1}{2}\oint_S (\mathbf{E} \times \mathbf{H}^*) \cdot \mathbf{dS} = -j2\omega \left[\frac{1}{4}\int_V \mathbf{D}^* \cdot \mathbf{E} \, dV - \frac{1}{4}\int_V \mathbf{B} \cdot \mathbf{H}^* \, dV \right] + \frac{1}{2}\int_V \mathbf{J}^* \cdot \mathbf{E} \, dV. \tag{1.67}$$

Since μ, ϵ and σ are real, using $\mathbf{D} = \epsilon\mathbf{E}$, $\mathbf{B} = \mu\mathbf{H}$ and $\mathbf{J} = \sigma\mathbf{E}$ in (1.67) gives

$$\frac{1}{2}\oint_S (\mathbf{E} \times \mathbf{H}^*) \cdot \mathbf{dS} = -j2\omega \left[\frac{1}{4}\int_V \epsilon|\mathbf{E}|^2 \, dV - \frac{1}{4}\int_V \mu|\mathbf{H}|^2 \, dV \right] + \frac{1}{2}\int_V \sigma|\mathbf{E}|^2 \, dV. \tag{1.68}$$

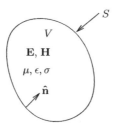

Figure 1.4   A volume V with material μ, ϵ, σ is bounded by a closed surface S having an inward normal $\hat{n}$.

The left-hand side of (1.68) is complex. On the right-hand side, the term is square brackets is real, and the last integral is real. Therefore,

$$\frac{1}{2} \operatorname{Re} \oint_S (\mathbf{E} \times \mathbf{H}^*) \cdot \mathbf{dS} = \frac{1}{2} \int_V \sigma |\mathbf{E}|^2 \, dV \tag{1.69}$$

and

$$\frac{1}{2} \operatorname{Im} \oint_S (\mathbf{E} \times \mathbf{H}^*) \cdot \mathbf{dS} = -2\omega \left[\frac{1}{4} \int_V \epsilon |\mathbf{E}|^2 \, dV - \frac{1}{4} \int_V \mu |\mathbf{H}|^2 \, dV \right]. \tag{1.70}$$

The left-hand side of (1.69) contains the *Poynting vector* $\mathbf{S}$, defined as

$$\mathbf{S} = \frac{1}{2} \operatorname{Re}(\mathbf{E} \times \mathbf{H}^*). \tag{1.71}$$

This very important vector represents the real power density in Watts per square metre. Integrated over the surface S, it gives the total power in Watts that flows into the volume V.

This same power can alternatively be found from the right-hand side of (1.69). If the conduction loss σ is known, then we can find the ohmic power loss P_L in Watts, from

$$P_L = \frac{1}{2} \int_V \sigma |\mathbf{E}|^2 \, dV. \tag{1.72}$$

Turning now to the imaginary parts in (1.70), there is a 'reactive power' that is associated with energy storage. At some point in an AC cycle, energy will flow into the volume V, but later on in the cycle, it must flow back out. This occurs, for instance, in capacitors and inductors. A standard way to write (1.70) is

$$\frac{1}{2} \operatorname{Im} \oint_S (\mathbf{E} \times \mathbf{H}^*) \cdot \mathbf{dS} = -2\omega(W_e - W_m) \tag{1.73}$$

where

$$W_e = \frac{1}{4} \int_V \epsilon |\mathbf{E}|^2 \, dV \tag{1.74}$$

and

$$W_m = \frac{1}{4} \int_V \mu |\mathbf{H}|^2 \, dV. \tag{1.75}$$

W_e and W_m are the average stored electric and magnetic energies in V.

The stored energy expressions may seem a bit abstract. It is helpful to consider a simple example such as a parallel-plate capacitor. Suppose the plates are at $x = 0$ and $x = d$ and are rectangular, with areas

$a \times b$. The applied voltage is $v(t) = V_0 \cos \omega t$, and its phasor is $V_0 \underline{/0°}$. The electric field is assumed to be approximately uniform between the plates and is therefore given by $E_x = V_0/d$. Evaluating (1.74) gives

$$W_e = \frac{1}{4} \int_V \epsilon |\mathbf{E}|^2 \; dV = \frac{\epsilon |\mathbf{E}|^2}{4} \int_V \; dV = \frac{\epsilon |V_0/d|^2 abd}{4} = \frac{\epsilon ab}{4d} \; |V_0|^2 = \frac{1}{4} C |V_0|^2$$

where we have identified the parallel-plate capacitance $C = \epsilon ab/d$. With the quasi-static assumption, ω is small and $\mathbf{H}$ is neglected, so (1.73) is approximately zero.

One thing left unexplained was the factor of 1/2 introduced in Equation (1.67). This was done so that it would represent the time average power. The 1/2 factor is a consequence of time averaging, and it also appears in circuits.[5]

One can also develop a time-domain version of Poynting's theorem. The derivation is omitted, and the result is given as follows:

$$\oint_S (\mathcal{E} \times \mathcal{H}) \cdot \mathbf{dS} = \int_V \mathcal{H} \cdot \frac{\partial \mathcal{B}}{\partial t} + \mathcal{E} \cdot \frac{\partial \mathcal{D}}{\partial t} \; dV + \int_V \mathcal{J} \cdot \mathcal{E} \; dV. \tag{1.76}$$

In (1.76), there is an instantaneous Poynting vector, given by

$$\mathcal{S} = \mathcal{E} \times \mathcal{H}. \tag{1.77}$$

It is also valid for DC. It can be used, for example, to find the power carried by the $\mathcal{E}, \mathcal{H}$ fields on a transmission line that has a steady current.

1.10   Lorentz Reciprocity Theorem

The *Lorentz reciprocity theorem* gives a useful relation between two sets of independent sources and their fields. Within a closed volume V, suppose that $\mathbf{J}_1, \mathbf{M}_1$ produce fields $\mathbf{E}_1, \mathbf{H}_1$. A second set of sources $\mathbf{J}_2, \mathbf{M}_2$ produce $\mathbf{E}_2, \mathbf{H}_2$. The theorem states that

$$\int_V \mathbf{E}_1 \cdot \mathbf{J}_2 - \mathbf{H}_1 \cdot \mathbf{M}_2 \; dV - \int_V \mathbf{E}_2 \cdot \mathbf{J}_1 - \mathbf{H}_2 \cdot \mathbf{M}_1 \; dV = -\oint_S (\mathbf{E}_1 \times \mathbf{H}_2 - \mathbf{E}_2 \times \mathbf{H}_1) \cdot \mathbf{dS} \tag{1.78}$$

where $\mathbf{dS}$ points out of V.

There are several cases where the surface integral vanishes. When this happens, (1.78) reduces to

$$\int_V \mathbf{E}_1 \cdot \mathbf{J}_2 - \mathbf{H}_1 \cdot \mathbf{M}_2 \; dV = \int_V \mathbf{E}_2 \cdot \mathbf{J}_1 - \mathbf{H}_2 \cdot \mathbf{M}_1 \; dV. \tag{1.79}$$

An important case where the surface integral vanishes is if S encloses all of the field sources $\mathbf{J}_{1,2}$, $\mathbf{M}_{1,2}$. Suppose that all the sources are inside V_1 and the bounding surface is S_1 with outward normal $\hat{\mathbf{n}}_1$. Then, (1.78) becomes

$$\int_{V_1} \mathbf{E}_1 \cdot \mathbf{J}_2 - \mathbf{H}_1 \cdot \mathbf{M}_2 \; dV - \int_{V_1} \mathbf{E}_2 \cdot \mathbf{J}_1 - \mathbf{H}_2 \cdot \mathbf{M}_1 \; dV$$

$$= -\oint_{S_1} (\mathbf{E}_1 \times \mathbf{H}_2 - \mathbf{E}_2 \times \mathbf{H}_1) \cdot \hat{\mathbf{n}}_1 dS. \tag{1.80}$$

---

[5] In an AC circuit having $v(t) = V_0 \cos(\omega t + \theta)$ and $i(t) = I_0 \cos(\omega t + \phi)$, the time average power is given by $P = \frac{1}{T} \int_T v(t)i(t)dt = \frac{1}{2} V_0 I_0 \cos(\theta - \phi)$. With phasors $V = V_0 \underline{/\theta}$ and $I = I_0 \underline{/\phi}$, the quantity $P = \frac{1}{2} \operatorname{Re}(VI^*)$ gives exactly the same result.

Now consider a second case, where the theorem is applied to V_2, the exterior of V_1. The bounding surface S_2 is the same as S_1 except that now, $\hat{\mathbf{n}}_2$ points out of V_2 (or into V_1) so that (1.78) becomes

$$\int_{V_2} \mathbf{E}_1 \cdot \mathbf{J}_2 - \mathbf{H}_1 \cdot \mathbf{M}_2 \, dV - \int_{V_2} \mathbf{E}_2 \cdot \mathbf{J}_1 - \mathbf{H}_2 \cdot \mathbf{M}_1 \, dV$$

$$= -\oint_{S_2} (\mathbf{E}_1 \times \mathbf{H}_2 - \mathbf{E}_2 \times \mathbf{H}_1) \cdot \hat{\mathbf{n}}_2 dS. \qquad (1.81)$$

There are no sources in V_2, so the left-hand side of (1.81) is zero. Using $\hat{\mathbf{n}}_1 = -\hat{\mathbf{n}}_2$ and $S_1 = S_2$ in (1.81) gives

$$\oint_{S_1} (\mathbf{E}_1 \times \mathbf{H}_2 - \mathbf{E}_2 \times \mathbf{H}_1) \cdot \hat{\mathbf{n}}_1 dS = 0.$$

Therefore, the right-hand side of (1.80) is also zero.

This proof covers most cases of practical interest. No boundary conditions were imposed on S, so (1.79) also applies to special cases, such as V being bounded by a perfect conductor, an impedance boundary or a sphere at infinity.

1.11   Friis and Radar Equations

Communication and radar systems involve antennas at large distances that, for all practical purposes, transmit and receive plane waves. An antenna sends power in a particular direction, at the expense of power in other directions. This effect is called the *directivity D*. If the antenna's power density pattern $\hat{\mathbf{r}}S(r,\theta,\phi)$ (the Poynting vector, in W/m^2) is known, then

$$D = \frac{S_{max}}{S_{iso}}; \quad S_{iso} = \frac{1}{4\pi}\int_\Omega S(r,\theta,\phi) \sin\theta \, d\theta d\phi \qquad (1.82)$$

where Ω is the surface of a sphere of radius r that encloses the antenna. S_{iso} is the power density that would be obtained if the antenna were to radiate its power isotropically. Equivalently, it is $S(r,\theta,\phi)$ averaged over the spherical surface Ω. S_{max} is the maximum value of $S(r,\theta,\phi)$.

From this definition, we see that D represents the increase in an antenna's power density as compared to an isotropic antenna. An isotropic antenna has $D=1$ and in general, $D \geq 1$. The isotropic antenna is an important and convenient concept, but it should be remembered that as a practical matter, it is not possible to build an antenna that sends its power equally in all directions.

Related to D is the *gain G*, which accounts for ohmic losses and is defined by $G = eD$. Here, $0 \leq e \leq 1$ is the efficiency. Microwave antennas and resonant dipoles have very high efficiencies, and usually, $e = 1$ is assumed. In contrast, a short dipole made with a good conductor might have an efficiency on the order of 50% or even less.

The effective aperture of a receive antenna, A_{er}, is a measure of the antenna's ability to collect the power incident upon it. It is in square metres. If the receive antenna gain G_r is known, then A_{er} can be found from

$$A_{er} = \frac{\lambda^2}{4\pi}G_r. \qquad (1.83)$$

It was easiest to explain the expressions (1.82) and (1.83) by considering (1.82) as a transmitting antenna and (1.83) as a receiver. However, it can be shown from reciprocity that the antenna gain is the same, for the transmit and receive modes. So, the distinction between a transmit and a receive antenna is not necessary but can be used when convenient.

Example 1.2 (Dipole Directivity) Find the directivity of a z-directed short dipole antenna. It has a Poynting vector

$$\mathbf{S}(r, \theta, \phi) = \hat{\mathbf{r}} \frac{K_0}{r^2} \sin^2 \theta$$

where K_0 is a constant.

Solution: The maximum power density is $S_{max} = K_0/r^2$ and

$$S_{iso} = \frac{K_0}{4\pi r^2} \int_{\theta=0}^{\pi} \int_{\phi=0}^{2\pi} \sin^2 \theta \sin \theta \, d\theta d\phi = \frac{K_0}{4\pi r^2} \frac{8\pi}{3}.$$

From this,

$$D = \frac{S_{max}}{S_{iso}} = \frac{K_0/r^2}{2K_0/3r^2} = 1.5.$$

Therefore, the power density produced by a short dipole is 1.5 times higher than that of an isotropic source.   ∎

Two useful equations for calculating the system performance are now described, the *Friis transmission formula* and the *radar range equation*.

The Friis transmission formula allows us to find the received power P_r in a communication link, in terms of the transmitter power P_t, the distance r between the antennas and the antenna gains G_t and G_r. The power density incident on a receiver at a distance r is

$$S^i = \frac{P_t}{4\pi r^2} G_t. \tag{1.84}$$

The received power is

$$P_r = S^i A_{er}. \tag{1.85}$$

From (1.83)–(1.85), we obtain the Friis transmission formula

$$P_r = \frac{P_t}{4\pi r^2} G_t \frac{\lambda^2}{4\pi} G_r. \tag{1.86}$$

The Friis formula also applies in two-dimensional problems. In this case, the directivity can be found from

$$D = \frac{S_{max}}{S_{iso}}; \quad S_{iso} = \frac{1}{2\pi} \int_{\Theta} S(\rho, \phi) \, d\phi, \tag{1.87}$$

where Θ is the surface of a cylinder of radius ρ that encloses the antenna. The effective aperture is $A_e = (\lambda/2\pi)G$, and the Friis formula becomes

$$P_r = \frac{P_t}{2\pi\rho} G_t \frac{\lambda}{2\pi} G_r. \tag{1.88}$$

Similar formulas exist for a radar system. The power density incident on a radar target at a distance r from the transmitter is still as in (1.84). If a portion of this power is collected by a fictitious aperture of area σ and scattered isotropically, then the scattered power density at a distance r from the target is

$$S^s = S^i \frac{\sigma}{4\pi r^2}. \tag{1.89}$$

From (1.83)–(1.85) and (1.89), it follows that

$$P_r = \frac{P_t}{4\pi r^2} G_t \frac{\sigma}{4\pi r^2} \frac{\lambda^2}{4\pi} G_r. \tag{1.90}$$

This is called the radar range equation, and σ is called the *scattering cross section* or *radar cross section* (RCS) and is in square metres.

Since the incident wave is a plane wave, from (1.89), it follows that

$$\frac{S^s}{S^i} = \frac{|\mathbf{E}^s|^2/2\eta_0}{|\mathbf{E}^i|^2/2\eta_0} = \frac{\sigma}{4\pi r^2}$$

from which

$$\sigma = \lim_{r \to \infty} 4\pi r^2 \frac{|\mathbf{E}^s|^2}{|\mathbf{E}^i|^2}. \tag{1.91}$$

In two dimensions, the radar cross section (sometimes called the *echo width*) is

$$\sigma = \lim_{\rho \to \infty} 2\pi\rho \frac{|\mathbf{E}^s|^2}{|\mathbf{E}^i|^2}. \tag{1.92}$$

Any real target does not scatter the power isotropically. Rather, power is scattered in a preferred direction at the expense of power in other directions. This is similar to the concept of antenna gain. So, in general, $\sigma = \sigma(\theta, \phi)$ varies with the aspect angles (θ, ϕ).

Usually, a radar transmitter and receiver are at the same position. This is called the *monostatic* case. Otherwise, when they are at different positions, it is the *bistatic* case. RCS is often given in dB with respect to a reference level of one square metre (dBsm).

1.12   Asymptotic Techniques

Asymptotic techniques are associated with the solution of 'high-frequency' problems. Their defining characteristic is that the radiating or scattering structure is large, in terms of the wavelength. Asymptotic techniques appear in physical optics, the geometrical theory of diffraction, the physical theory of diffraction, as well as the stationary-phase and saddle-point integration techniques. Common to all of these techniques is the characteristic that the higher the frequency, the more accurate they become. In any case, the field expressions appear as an 'asymptotic series'.

Under the right conditions, an asymptotic series is highly accurate. Most often in practice, only one term is used. This happens so frequently that in routine usage, it sometimes gets forgotten that the term came from a series. In some cases, we might talk about a *higher order asymptotic technique* – which means we are including the second and possibly third terms from the series.

To illustrate the point, a solution of Bessel's differential equation that gives an outward-travelling cylindrical wave is the Hankel function of the second kind of order zero, $H_0^{(2)}(kr)$; see Appendix C. When $kr = 2\pi r/\lambda$ is large, as is the case at high frequencies, it can be approximated by an asymptotic series (Felsen and Marcuvitz 1994, Appendix 6a). The first three terms are

$$H_0^{(2)}(kr) \sim \sqrt{\frac{2}{\pi kr}}\, e^{-j(kr - \pi/4)}$$

$$\times \left[1 - \frac{1}{4}\frac{1}{(j2kr)} + \frac{9}{32}\frac{1}{(j2kr)^2} + O\left(\frac{1}{(kr)^3}\right) + \cdots \right]. \tag{1.93}$$

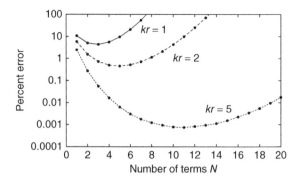

Figure 1.5   Percent error of $|H_0^{(2)}(kr)|$ versus number of terms, using the asymptotic formula (1.93).

Higher order terms can be easily obtained but are omitted here for brevity. The accuracy improves with increasing kr, as the higher order terms become negligible. The special symbol $\sim$ is used, which means 'asymptotically equal to'. That is, it becomes equal when the parameter kr is sufficiently large.

In general, an asymptotic series may or may not be convergent, and the series in (1.93) diverges as the number of terms tends to infinity! In spite of this seeming deficiency, it provides a very simple and accurate way to calculate $H_0^{(2)}(kr)$ when kr is large.

In contrast to an asymptotic series, the Hankel function also has a power series. From Appendix C, we have that

$$H_0^{(2)}(kr) \approx \left(1 + \frac{2}{j\pi}\left[\ln\left(kr/2\right) + 0.5772\right]\right)\left(1 - \frac{1}{4}(kr)^2 + \frac{1}{64}(kr)^4 - \cdots\right)$$
$$+ \frac{1}{j\pi}\left(\frac{1}{2}(kr)^2 - \frac{3}{64}(kr)^4 + \cdots\right). \tag{1.94}$$

The power series is convergent; however, in practice, a very large number of terms might be needed when kr is large. The power series uses the symbol $\approx$ which means 'approximately equal to' but otherwise does not carry any further implications.

The convergence behaviour of the asymptotic series is illustrated in Figure 1.5, which shows the percent error of $|H_0^{(2)}(kr)|$ versus the number of terms N, as calculated from (1.93). If N is fixed, the graphs show that as expected, accuracy improves with increasing kr. On the other hand, increasing N does not necessarily offer improvement. In fact, a larger number of terms might give a worse result, and this depends on the numerical value of kr. In practice, one has to ensure that for a given kr, adding more terms will not make the accuracy worse.

1.13   Further Reading

For a general background that complements the coverage of topics in this book, the following reading list is suggested. The list is by no means exhaustive.

Intermediate graduate-level electromagnetics is presented in the books by Harrington (2001), Balanis (2012), Ishimaru (1991) and Rothwell and Cloud (2008). More advanced topics are treated by Collin (1991), Felsen and Marcuvitz (1994) and Van Bladel (2007). The physics-oriented books by Stratton (1941) and Morse and Feshbach (1953) are older yet remain relevant in modern times.

A good general-purpose mathematics book that treats special functions, differential and integral equations, complex variables, Green's functions and many other topics used in electromagnetics is by

Arfken and Weber (2005). More specialized, Churchill and Brown (2013) covers complex variables. Gradshteyn and Ryzhik (1980) is an extraordinary compilation containing about 1200 pp. of formulas for integrals, series and products. Papoulis (1962) describes Fourier, Hilbert and Laplace transforms, Fourier series and related topics.

Properties of special functions and numerical evaluation techniques are compiled in Abramowitz and Stegun (1965) and more recently in Olver et al. (2010a,b). An engineering-oriented treatment of numerical methods is in Faires and Burden (2013).

References

Abramowitz M and Stegun I (1965) *Handbook of Mathematical Functions*. Dover Publications.
Arfken GB and Weber HJ (2005) *Mathematical Methods for Physicists*. Elsevier.
Balanis CA (2012) *Advanced Engineering Electromagnetics*. John Wiley & Sons, Inc.
Churchill RV and Brown JW (2013) *Complex Variables and Applications*. McGraw-Hill.
Collin RE (1991) *Field Theory of Guided Waves*. Oxford University Press.
Faires JD and Burden RL (2013) *Numerical Methods*. Brooks/Cole Publishing Company.
Felsen LB and Marcuvitz N (1994) *Radiation and Scattering of Waves*. IEEE Press.
Gradshteyn IS and Ryzhik IM (1980) *Table of Integrals, Series, and Products*. Academic Press.
Grover FW (1962) *Inductance Calculations Working Formulas and Tables*. Dover Publications.
Harrington RF (2001) *Time-Harmonic Electromagnetic Fields*. IEEE Press.
Ishimaru A (1991) *Electromagnetic Wave Propagation, Radiation and Scattering*. Prentice-Hall.
Morse PM and Feshbach H (1953) *Methods of Theoretical Physics*. McGraw-Hill.
Olver FWJ, Lozier DW, Boisvert RF and Clark CW (2010a) NIST digital library of mathematical functions, http://dlmf.nist.gov/. Accessed: 2015-02-04.
Olver FWJ, Lozier DW, Boisvert RF and Clark CW (2010b) *NIST Handbook of Mathematical Functions*. Cambridge University Press.
Papoulis A (1962) *The Fourier Integral and its Applications*. McGraw-Hill.
Rothwell EJ and Cloud MJ (2008) *Electromagnetics*. CRC Press.
Stratton JA (1941) *Electromagnetic Theory*. McGraw-Hill. (Reprinted, IEEE Press 2007).
Van Bladel J (2007) *Electromagnetic Fields*. John Wiley & Sons, Inc.

Problems

1.1 Point P is at $(x, y, z) = (1, 2, 5)$. Convert P to

    (a) cylindrical coordinates

    (b) spherical coordinates.

1.2 A vector goes from $(x, y, z) = (0, 0, 0)$ to $(x, y, z) = (1, 2, 5)$. Express the vector in

    (a) rectangular coordinates

    (b) spherical coordinates.

1.3 A vector at a point $P\,(x, y, z) = (1, 2, 5)$ is given in spherical coordinates by $\mathbf{A} = 3\,\hat{\boldsymbol{\theta}}$. Convert $\mathbf{A}$ to rectangular coordinates.

1.4 A vector field is given by $\mathbf{E} = 2x\,\hat{\mathbf{x}}$.

    (a) Sketch the field in the $x - y$ plane.

    (b) Find the divergence and curl.

1.5 A vector field is given by $\mathbf{E} = 3y\,\hat{\mathbf{x}}$.

    (a) Sketch the field in the $x - y$ plane.

    (b) Find the divergence and curl.

1.6 A parallel-plate capacitor with an air dielectric is charged up to 10 V. The voltage source is then disconnected. A dielectric slab having a permittivity of $\epsilon = 2\epsilon_0$ is slipped in between the plates. What is the new voltage? Hint: $\mathcal{D}$ does not change.

1.7 A rectangular wire loop having $0 \leq x \leq a$, $0 \leq y \leq b$ is in the $z = 0$ plane. A small gap with terminals AB is at $(x, y, z) = (a/2, 0, 0)$. A magnetic flux density $\mathcal{B} = \hat{z}B_0 \cos \omega t$ is applied to the loop.

(a) Find the terminal voltage V_{AB}.

(b) If a resistor R is placed at the gap, find the current and its direction.

1.8 A toroidal inductor with N turns is made of a linear material for which $B = \mu H$. The cross section in cylindrical coordinates is given by $a \leq \rho \leq b$ and $0 \leq z \leq h$. Use Ampere's law to obtain an expression for H that depends on the position ρ inside the core and use it to find the inductance.

1.9 Convert the following phasors to the time domain. The angular frequency ω is assumed to be known.

(a) A voltage $V = 3 + j4$ V.

(b) An electric field $\mathbf{E} = \hat{x}3$ V/m.

(c) An electric field $\mathbf{E} = \hat{x}(3 + j4)$ V/m.

1.10 A magnetic field phasor is given by $\mathbf{H} = 12\hat{x} \sin ay$ A/m, where a is a constant.

(a) Use $\nabla \times \mathbf{H}$ to find $\mathbf{E}$. Assume a non-conducting medium that has $\mathbf{J} = \sigma\mathbf{E} = 0$.

(b) From $\nabla \times \mathbf{E}$, find $\mathbf{H}$. Under what condition does this equal the original $\mathbf{H}$ that we started with?

(c) If $a = 3.48$ and the medium is Plexiglas with $\epsilon_r = 2.76$, find the frequency in MHz.

1.11 A current density phasor is $\mathbf{J} = \hat{x}J_0 \, e^{-ax}$ A/m^2 where J_0 and a are constants.

(a) Find the charge density phasor ρ_v.

(b) Find the time-domain current $\mathcal{J}(t)$ and charge density $\varrho_v(t)$.

1.12 Free charges placed inside a conductor will repel each other, and as they move apart, the charge density will decay exponentially. Use the point form of Gauss's law and the continuity equation to show that

$$\frac{\partial \varrho_v}{\partial t} = -\frac{\sigma}{\epsilon}\varrho_v$$

and hence $\varrho_v(t) = \varrho_v(0)e^{-\sigma t/\epsilon}$. Calculate the 'relaxation time' $\tau = \epsilon/\sigma$ for porcelain having $\sigma = 10^{-14}$ S/m and $\epsilon = 6\epsilon_0$ and for copper with $\sigma = 5.8 \times 10^7$ S/m. Note that $\epsilon \approx \epsilon_0$ in a metal.

1.13 By taking the divergence of the Ampere-Maxwell equation $\nabla \times \mathbf{H} = \mathbf{J} + j\omega\mathbf{D}$, and using the continuity equation, obtain Gauss's law $\nabla \cdot \mathbf{D} = \rho_v$.

1.14 The $z = 0$ plane is an interface between two media. The electric field at $z = 0^-$ is $\mathbf{E}_1$, and at $z = 0^+$, it is $\mathbf{E}_2$. Show that the boundary condition $\hat{n} \times (\mathbf{E}_1 - \mathbf{E}_2) = 0$ is equivalent to $E_{x1} = E_{x2}$; $E_{y1} = E_{y2}$.

1.15 A conductor in region 1 $z > 0$ has a conductivity σ_1. A second conductor in region 2 $z < 0$ has a conductivity σ_2. The two regions are connected, and both are good conductors with $\epsilon \approx \epsilon_0$. A uniform volume current $\mathbf{J} = \hat{z} J_0$ flows through both regions; J_0 is a constant. Find $\mathbf{E}$ in each region, and ρ_s at $z = 0$.

1.16 A boundary has medium 1 for $z > 0$ and medium 2 for $z < 0$. A small cylindrical 'pillbox' surface S_0 straddles the boundary and has flat endcaps at $z = \pm h$. The pillbox encloses surface charges ρ_s at $z = 0$, and volume charges ρ_v elsewhere. Shrink $h \to 0$ and develop a continuity equation in integral form, relating $\mathbf{J}_s$ and the total surface charge Q_s inside S_0. Show that the normal volume current J_z is continuous at $z = 0$.

1.17 An electric field in free space is given by $\boldsymbol{\mathcal{E}} = \hat{\mathbf{x}}E_0 \cos(\omega t - kz)$ and $\boldsymbol{\mathcal{J}} = \sigma\boldsymbol{\mathcal{E}} = 0$.

(a) Take the curl of $\boldsymbol{\mathcal{E}}$ and then integrate with respect to time, to get $\boldsymbol{\mathcal{H}}$. The constant of integration represents a magnetostatic solution; we can make it zero.

(b) Now take the curl of $\boldsymbol{\mathcal{H}}$ and integrate with respect to time to get $\boldsymbol{\mathcal{E}}$.

(c) What is the required relationship between ω and k so that after the two curl operations, we get back the original $\boldsymbol{\mathcal{E}}$?

1.18 Assuming $\boldsymbol{\mathcal{E}} = \hat{\mathbf{x}}E_0 \cos(\omega t - kz)$, find the phasors $\mathbf{E}$, $\mathbf{H}$ and the Poynting vector $\mathbf{S}$.

1.19 Assuming $\boldsymbol{\mathcal{E}} = \hat{\mathbf{x}}E_0 \cos(\omega t - kz)$, find $\boldsymbol{\mathcal{H}}$ and the time-domain Poynting vector $\boldsymbol{\mathcal{S}}$.

1.20 A coaxial cable has an inner radius a and an outer radius b. The space in between the conductors is air. A resistor R_L is connected at one end, and a battery at the other end supplies a DC voltage V_0. The cable carries a DC current I_0. The fields are of the form $\boldsymbol{\mathcal{E}} = \hat{\boldsymbol{\rho}}E_0/\rho$ and $\boldsymbol{\mathcal{H}} = \hat{\boldsymbol{\phi}}H_0/\rho$.

(a) Find the constants E_0 and H_0 in terms of V_0 and I_0.

(b) Find the time-domain Poynting vector $\boldsymbol{\mathcal{S}}$ and integrate over the coax cable's cross section, to find the power.

(c) Compare this to the power in R_L as obtained by circuit theory.

1.21 When the time derivatives are not zero, the four Maxwell equations become coupled and only two of them are needed. This is also true for the boundary conditions. For instance, if we enforce the continuity of $\mathbf{E}_{tan}$, then B_n is automatically continuous.

Prove this assertion by using the Maxwell curl equation, $\nabla \times \mathbf{E} = -j\omega\mathbf{B}$. Expand out the x component. Assume we have a planar boundary with medium 1 for $x < 0$ and medium 2 for $x > 0$. Let $\mathbf{E}_1 = \hat{\mathbf{y}}E_{y1} + \hat{\mathbf{z}}E_{z1}$ at $x = 0^-$, and $\mathbf{E}_2 = \hat{\mathbf{y}}E_{y2} + \hat{\mathbf{z}}E_{z2}$ at $x = 0^+$. If $E_{y1} = E_{y2}$ for all (y, z), their derivatives will also be equal. The same idea applies to E_{z1} and E_{z2}.

1.22 Similar to the previous problem, if we enforce the boundary condition on $\mathbf{H}_{tan}$, the condition for D_n is automatically satisfied. Prove this assertion.

Assume we have a planar boundary with medium 1 for $x < 0$ and medium 2 for $x > 0$. Expand $\hat{\mathbf{n}} \times (\mathbf{H}_1 - \mathbf{H}_2) = \mathbf{J}_s$ to obtain expressions for J_{sy} and J_{sz}. Then expand the x component of the Maxwell curl equation, $\nabla \times \mathbf{H} = \mathbf{J} + j\omega\mathbf{D}$.

Note that the x component of the volume current $\mathbf{J}$ is continuous at $x = 0$. In addition, you will need a surface form of the continuity equation, which relates $\mathbf{J}_s$ and ρ_s via

$$\partial J_{sy}/\partial y + \partial J_{sz}/\partial z = -j\omega\rho_s.$$

2

Transverse Electromagnetic Waves

This chapter describes *Transverse Electromagnetic* (TEM) waves. We will start with uniform plane waves, which have the electric and magnetic fields constant (uniform) over a plane. This is the simplest possible solution of the wave equation, so it is highly suitable for explaining many properties of electromagnetic waves. Plane waves will then be used to obtain the Fresnel reflection and transmission coefficients for an air–material interface. After this, the phenomenon of total internal reflection and non-uniform plane waves will be treated. Next, approximate impedance boundary conditions (IBCs) and their use will be explained. A recursive method for calculating the reflection by multi-layer materials will be discussed.

Transmission lines support TEM waves of finite transverse extent. Their characteristics will be discussed in terms of fields and also in terms of an equivalent LC-ladder circuit. Reflection and transmission of waves on transmission lines are considered next. Finally, the concepts of phase velocity, group velocity and dispersion are explained.

2.1 Introduction

We wish to solve the Maxwell curl equations

$$\nabla \times \mathbf{E} = -j\omega\mu\mathbf{H} \tag{2.1}$$

$$\nabla \times \mathbf{H} = \mathbf{J} + j\omega\epsilon\mathbf{E} \tag{2.2}$$

to obtain the wave equation, also known as the *Helmholtz equation*. The curl of (2.1) in (2.2) along with the double-curl identity $\nabla \times \nabla \times \mathbf{A} = -\nabla^2\mathbf{A} + \nabla\nabla \cdot \mathbf{A}$ from Appendix B gives the wave equation for $\mathbf{E}$

$$\nabla^2\mathbf{E} + k^2\mathbf{E} = j\omega\mu\mathbf{J} + \nabla(\nabla \cdot \mathbf{E}), \tag{2.3}$$

and similarly, the curl of (2.2) in (2.1) gives the wave equation for $\mathbf{H}$

$$\nabla^2\mathbf{H} + k^2\mathbf{H} = -\nabla \times \mathbf{J} + \nabla(\nabla \cdot \mathbf{H}) \tag{2.4}$$

where k is called the *wavenumber* and is defined by $k^2 = \omega^2\mu\epsilon$. For the moment, it is assumed that the medium is non-conducting, so that $\mathbf{J} = 0$. Since $\mathbf{J} = 0$ at all positions (x, y, z), its derivatives will also be

Applied Frequency-Domain Electromagnetics, First Edition. Robert Paknys.
© 2016 John Wiley & Sons, Ltd. Published 2016 by John Wiley & Sons, Ltd.
Companion Website: www.wiley.com/go/paknys9981

zero. From the continuity equation, $\nabla \cdot \mathbf{J} = -j\omega\rho_v = 0$, and then, Gauss's law gives $\nabla \cdot \mathbf{E} = \rho_v/\epsilon = 0$. Then, (2.3) becomes

$$\nabla^2 \mathbf{E} + k^2 \mathbf{E} = 0. \tag{2.5}$$

Similarly, $\nabla \times \mathbf{J} = 0$ and $\nabla \cdot \mu\mathbf{H} = 0$ so that (2.4) becomes

$$\nabla^2 \mathbf{H} + k^2 \mathbf{H} = 0. \tag{2.6}$$

2.2   Plane Waves

This section develops the most elementary solution for the wave equation, the uniform plane wave. The wave is constant over a plane, and the electric field vector $\mathbf{E}$ lies in that plane. Let us find a solution of the wave equation (2.5) that has these properties. We can begin by assuming that the coordinate axes have been oriented so that $\mathbf{E}$ points in the x direction. Then, the wave equation becomes

$$\nabla^2 E_x + k^2 E_x = 0.$$

If we assume that there is no variation in the $x - y$ plane (more generally, any plane containing the $\mathbf{E}$ vector would work), the Laplacian reduces to $\nabla^2 = d^2/dz^2$ so that

$$\frac{d^2 E_x}{dz^2} + k^2 E_x = 0 \tag{2.7}$$

which has a general solution[1]

$$E_x = K_1 e^{-jkz} + K_2 e^{jkz}$$

where K_1 and K_2 are arbitrary constants. By taking the curl of $\mathbf{E}$, the magnetic field is found to be

$$H_y = \frac{K_1}{\eta} e^{-jkz} - \frac{K_2}{\eta} e^{jkz}$$

where

$$\eta = \sqrt{\frac{\mu}{\epsilon}}$$

is the *intrinsic impedance* of the medium. In free space, $\eta \approx 377 \ \Omega$. Both $\mathbf{E}$ and $\mathbf{H}$ are in a plane transverse to the propagation direction z, so it is called a *transverse electromagnetic* (or TEM) wave.

Converting E_x and H_y to the time domain gives

$$\mathcal{E}_x = K_1 \cos(\omega t - kz) + K_2 \cos(\omega t + kz)$$

$$\mathcal{H}_y = \frac{K_1}{\eta} \cos(\omega t - kz) - \frac{K_2}{\eta} \cos(\omega t + kz).$$

By sketching $\cos(\omega t - kz)$ as a function of z at two instances of time, say $t = t_1$ and a slightly later time $t = t_1 + \delta t$, it can be seen that this is a wave that travels in the $+z$ direction; similarly, $\cos(\omega t + kz)$ travels in the $-z$ direction. Letting $K_2 = 0$, we observe that the $+z$ travelling wave has $\mathcal{E}_x/\mathcal{H}_y = \eta$, and

---

[1] With $\partial/\partial x = \partial/\partial y = 0$, solutions with $E_x(z)$ or $E_y(z)$ are possible. However, a wave solution $E_z(z) = e^{\pm jkz}$ is not possible because $\nabla \times \mathbf{E} = 0$, so there is no magnetic field and, hence, no wave.

the travel is in the same direction as $\mathcal{E} \times \mathcal{H}$. With $K_1 = 0$, the $-z$ travelling wave has $\mathcal{E}_x/\mathcal{H}_y = -\eta$, and the travel remains in the $\mathcal{E} \times \mathcal{H}$ direction. These ideas can be used to quickly construct plane waves with other travel directions and *polarizations*. The polarization of a wave is defined as the direction in which the wave's electric field vector is pointing.

Example 2.1 (Plane Wave Construction) In the frequency domain, construct a y-polarized plane wave that is travelling in the $+x$ direction.

Solution: $+x$ propagation requires a wave with e^{-jkx} dependence, so $E_y = K_1 e^{-jkx}$. To get **H**, we could take $\nabla \times \mathbf{E}$, but it is easier to just observe that $\mathbf{E} \times \mathbf{H}$ must point in the travel direction and that $|\mathbf{E}| = \eta|\mathbf{H}|$. Therefore, $H_z = (K_1/\eta)e^{-jkx}$. ∎

The wave's velocity can be defined as the speed of a point on the wave that has a constant phase. This can be found by considering a point where $\omega t - kz = $ const. Then, the velocity is $u = dz/dt = \omega/k$. Since $k = \omega\sqrt{\mu\epsilon}$, it follows that $u = 1/\sqrt{\mu\epsilon}$. Also, $u = \lambda f$ so that $k = 2\pi/\lambda$. If we substitute in the free-space numerical values $\mu = \mu_0$ and $\epsilon = \epsilon_0$, we obtain $u = 2.998 \times 10^8$ m/s, the speed of light.

The one-dimensional time-domain wave equation for an x-polarized, z-propagating wave can be found by noting that

$$-k^2 = (j\omega\sqrt{\mu\epsilon})^2 \rightarrow \mu\epsilon\frac{\partial^2}{\partial t^2}$$

so that (2.7) becomes

$$\frac{\partial^2 \mathcal{E}_x}{\partial z^2} - \frac{1}{u^2}\frac{\partial^2 \mathcal{E}_x}{\partial t^2} = 0$$

where $u = 1/\sqrt{\mu\epsilon}$. The time-domain wave equation solution is not limited to $\cos(\omega t \mp kz)$. By substitution into the wave equation, it can be confirmed that

$$\mathcal{E}_x = K_1\, f(t - z/u) + K_2\, g(t + z/u)$$

is a solution, where f and g are *any* twice-differentiable functions. This is known as *d'Alembert's formula*. Comparison of $f(t - z/u)$ with $\cos(\omega t - kz) = \cos\omega(t - z/u)$ shows that $\omega/u = k$ from which we obtain the velocity $u = 1/\sqrt{\mu\epsilon}$.

The idea that any function can be used is sensible, as an antenna or transmission line can send any kind of waveform, not just a sinusoidal signal.

Example 2.2 (Plane-Wave Generator) An electric current sheet $\mathbf{J}_s$ in the $z = 0$ plane in free space can generate outgoing plane waves $\mathbf{E} = \hat{\mathbf{x}}E_0 e^{\mp jkz}$ in the $\pm z$ directions. Find the surface current. (E_0 is assumed to be real.)

Solution: Denote $z < 0$ as region 1 and $z > 0$ as region 2. The postulated electric field is

$$\mathbf{E}_1 = \hat{\mathbf{x}}E_0 e^{jkz}; \quad z \leq 0$$

$$\mathbf{E}_2 = \hat{\mathbf{x}}E_0 e^{-jkz}; \quad z \geq 0.$$

Note that $\mathbf{E}_{tan}$ is continuous at $z = 0$. The corresponding magnetic fields are

$$\mathbf{H}_1 = -\hat{\mathbf{y}}\frac{E_0}{\eta}e^{jkz}; \quad z \le 0$$

$$\mathbf{H}_2 = \hat{\mathbf{y}}\frac{E_0}{\eta}e^{-jkz}; \quad z \ge 0.$$

There is a jump discontinuity in $\mathbf{H}_{tan}$ at $z = 0$, so there must be an electric current sheet. It is found from the boundary condition $\mathbf{J}_s = \hat{\mathbf{n}} \times (\mathbf{H}_1 - \mathbf{H}_2)$. The normal must point into region 1, so $\hat{\mathbf{n}} = -\hat{\mathbf{z}}$. This gives

$$\mathbf{J}_s = \hat{\mathbf{x}}\frac{2E_0}{\eta}.$$

This can be visualized in the time domain as

$$\boldsymbol{\mathcal{J}}_s(t) = \hat{\mathbf{x}}\frac{2E_0}{\eta}\cos\omega t.$$

As the current's direction alternates between $+x$ and $-x$, plane waves are generated.

This is only a mathematical example; the generator would be difficult to build. A metal sheet could be used to carry the current $\mathbf{J}_s$. However, the postulated $\mathbf{E}$ could not exist because $\mathbf{E}_{tan} = 0$ is required on the sheet. ∎

2.2.1 Lossy Medium

We now consider the wave equation in a source-free and lossy homogeneous medium. Since $\mathbf{J} = \sigma\mathbf{E}$, the wave equation (2.3) can be rewritten as

$$\nabla^2\mathbf{E} - \gamma^2\mathbf{E} = \nabla(\nabla \cdot \mathbf{E}) \tag{2.8}$$

where the complex propagation constant γ is defined by

$$\gamma^2 \equiv j\omega\mu(\sigma + j\omega\epsilon).$$

With $\mathbf{J} = \sigma\mathbf{E}$, the Ampere–Maxwell equation (2.2) becomes $\nabla \times \mathbf{H} = (\sigma + j\omega\epsilon)\mathbf{E}$. The divergence of the curl of any vector is identically zero, so that $\nabla \cdot \nabla \times \mathbf{H} = (\sigma + j\omega\epsilon)\nabla \cdot \mathbf{E} \equiv 0$ which implies that $\nabla \cdot \mathbf{E} = 0$ and $\nabla \cdot \mathbf{J} = 0$ as well. Equation (2.8) then reduces to

$$\nabla^2\mathbf{E} - \gamma^2\mathbf{E} = 0. \tag{2.9}$$

For the magnetic field, $\nabla \cdot \mu\mathbf{H} = 0$ and $\nabla \times \mathbf{J} = \nabla \times \sigma\mathbf{E} = -j\omega\mu\sigma\mathbf{H}$ so that (2.4) in the lossy case becomes

$$\nabla^2\mathbf{H} - \gamma^2\mathbf{H} = 0. \tag{2.10}$$

An alternate way to handle a lossy medium is to notice that (2.2) in the lossless case ($\mathbf{J} = 0$) is

$$\nabla \times \mathbf{H} = j\omega\epsilon\mathbf{E},$$

whereas in the lossy case ($\mathbf{J} = \sigma\mathbf{E}$)

$$\nabla \times \mathbf{H} = j\omega\left(\epsilon + \frac{\sigma}{j\omega}\right)\mathbf{E}.$$

Comparison of these two equations suggests that we can extend the lossless case to the lossy case by defining a new complex permittivity $\epsilon_c = \epsilon + (\sigma/j\omega)$ and replacing ϵ for the lossless case by ϵ_c. This affects k in the following way

$$jk = j\omega\sqrt{\mu\epsilon} \rightarrow j\omega\sqrt{\mu\left(\epsilon + \frac{\sigma}{j\omega}\right)} = \sqrt{j\omega\mu(\sigma + j\omega\epsilon)} = \gamma.$$

Since $\sqrt{-1} = \pm j$ there is always a sign to choose. Waves of the form $e^{\mp jkz}$ with a real propagation constant k become $e^{\mp\gamma z}$ with a complex propagation constant γ in the lossy case. Physical considerations have to be used to decide which sign is appropriate. We expect the wave to decay as $z \rightarrow \infty$, so we must choose the branch of the square root that gives $jk = +\gamma$.

It is customary to define the *attenuation constant* $\alpha = \mathrm{Re}\,\gamma$ and *phase constant* $\beta = \mathrm{Im}\,\gamma$. Then,[2]

$$\gamma = \alpha + j\beta = \sqrt{j\omega\mu(\sigma + j\omega\epsilon)}. \tag{2.11}$$

To understand the physical significance of α and β, it is helpful to consider an x-polarized wave travelling in the $\pm z$ directions

$$E_x = K_1 e^{-\gamma z} + K_2 e^{\gamma z}.$$

If we consider the $+z$ travelling wave

$$E_x = K_1 e^{-\alpha z} e^{-j\beta z},$$

we can see that α determines the exponential amplitude decay of the wave and β is associated with the phase. For the units, αz is in nepers (np) and βz is in radians (rad). Both αz and βz are dimensionless, so np and rad are fictitious units. Sometimes np/m and rad/m are simply given as 1/m. β is related to the wavelength through $\beta = 2\pi/\lambda$. The lossless case with $\sigma = 0$ reduces to $\gamma = jk = j\beta$, and $\alpha = 0$.

Materials are often characterized as being good dielectrics or good conductors. The difference between these two types can be understood by considering the current in a lossy capacitor. From Section 1.4, we know that there is a displacement current between the plates. However, with a lossy dielectric, there are two ways for the current to get from one plate to the other: by a conduction current, and by a displacement current. These currents obey

$$\nabla \times \mathbf{H} = \underbrace{\sigma\mathbf{E}}_{\mathbf{J}_c} + \underbrace{j\omega\epsilon\mathbf{E}}_{\mathbf{J}_d}.$$

In a good dielectric, the displacement current $\mathbf{J}_d$ dominates, which means that $\sigma \ll \omega\epsilon$. If $\sigma \approx \omega\epsilon$, we have what amounts to a capacitor and a resistor in parallel. If $\sigma \gg \omega\epsilon$, the conduction current $\mathbf{J}_c$ dominates and the device acts more like a resistor than a capacitor. Low-loss dielectrics are usually characterized by their *loss tangent*

$$\tan\delta = \frac{|\mathbf{J}_c|}{|\mathbf{J}_d|} = \frac{\sigma}{\omega\epsilon}. \tag{2.12}$$

The angle δ is interpreted as in Figure 2.1.

---

[2] Both β and k are commonly used to denote the phase constant. In the lossless case, they are the same. In the lossy case, $-j\gamma = k = k' - jk''$ is allowed to be complex which implies that $\beta = \mathrm{Im}\,k$ and $\alpha = \mathrm{Re}\,k$.

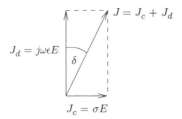

Figure 2.1   Phasor diagram showing the loss tangent $\tan \delta$.

A convenient way to deal with the presence of conductivity is to combine $\mathbf{J}_c$ and $\mathbf{J}_d$ into one term by defining a complex permittivity ϵ_c

$$\nabla \times \mathbf{H} = j\omega \underbrace{\left(\epsilon + \frac{\sigma}{j\omega} \right)}_{\epsilon_c} \mathbf{E}$$

that is,

$$\epsilon_c = \epsilon + \frac{\sigma}{j\omega} = \epsilon_0 \epsilon_r (1 - j \tan \delta). \tag{2.13}$$

Therefore, the lossless and lossy versions of the Ampere–Maxwell law look almost the same

$$\nabla \times \mathbf{H} = j\omega\epsilon\mathbf{E}; \text{ lossless} \qquad \nabla \times \mathbf{H} = j\omega\epsilon_c\mathbf{E}; \text{ lossy}$$

and we can account for loss by simply replacing $\epsilon \to \epsilon_c$ in any equation. In terms of real and imaginary parts, ϵ_c can be expressed as

$$\epsilon_c = \epsilon' - j\epsilon'' = \epsilon'(1 - j \tan \delta). \tag{2.14}$$

The complex propagation constant is

$$\gamma = \alpha + j\beta = \sqrt{j\omega\mu \, j\omega\epsilon'(1 - j \tan \delta)}.$$

The loss tangent is usually used in conjunction with lossy dielectrics where $\tan \delta$ is small. For this case, we can use the approximation

$$\sqrt{1+x} \approx 1 + \frac{1}{2}x + \cdots$$

to obtain

$$\alpha = \frac{\omega\sqrt{\mu\epsilon'}}{2} \tan \delta; \qquad \beta = \omega\sqrt{\mu\epsilon'}. \tag{2.15}$$

Let us now consider good conductors. The amplitude of a wave inside a good conductor decays rapidly as $e^{-\alpha z}$, and deep inside the conductor, the fields are essentially zero. Therefore, any volume current $\mathbf{J} = \sigma\mathbf{E}$ will be mostly near the surface. This concentration of current near the surface is called the *skin effect*. The point at which the field has decayed to $1/e$ of its maximum value is called the *skin depth*[3] δ and is $\delta = 1/\alpha$. Since a good conductor has $\sigma \gg \omega\epsilon$, we can say that

$$\alpha + j\beta = \sqrt{j\omega\mu(\sigma + j\omega\epsilon)} \approx \sqrt{j\omega\mu\sigma}$$

from which

$$\alpha = \beta = \sqrt{\omega\mu\sigma/2} \tag{2.16}$$

and, hence, $\delta \approx \sqrt{2/\omega\mu\sigma}$. At 1 MHz in copper, the skin depth is only $\delta = 0.066$ mm. An AWG 35 wire with a diameter of 0.143 mm would be able to carry most of the current. A $100\times$ drop in the frequency causes a $10\times$ increase in δ.

When considering a wide frequency range, it turns out that for many low-loss dielectrics, $\tan\delta$ tends to be fairly constant, whereas for good conductors, σ tends to be fairly constant. Therefore, conduction losses are usually given as $\tan\delta$ for dielectrics, and as σ for conductors.

2.2.2   Polarization

We have seen that the polarization of a wave is defined as the direction in which the wave's electric field vector is pointing. For instance, a z-travelling wave could be x-polarized or y-polarized. In either case, these situations are called *linearly polarized* waves. One could combine two waves as in

$$\mathbf{E}(z, \omega) = \hat{\mathbf{x}} E_0 e^{-j\beta z} + \hat{\mathbf{y}} E_0 e^{-j\beta z}.$$

In this case, the resultant field vector points along the direction $\hat{\mathbf{x}} + \hat{\mathbf{y}}$, but this is still linearly polarized.

Let us suppose now that the two components are combined as earlier, but with a $-90°$ phase shift on the y component

$$\mathbf{E}(z, \omega) = \hat{\mathbf{x}} E_0 e^{-j\beta z} - j\hat{\mathbf{y}} E_0 e^{-j\beta z}.$$

The consequences are easier to see in the time domain. Using $\mathcal{E}(z, t) = \text{Re}(\mathbf{E}e^{j\omega t})$ and assuming that E_0 is real,

$$\mathcal{E}(z, t) = \hat{\mathbf{x}} E_0 \cos(\omega t - \beta z) + \hat{\mathbf{y}} E_0 \sin(\omega t - \beta z).$$

To visualize this, let $z = 0$ so that

$$\mathcal{E}(0, t) = \hat{\mathbf{x}} E_0 \cos\omega t + \hat{\mathbf{y}} E_0 \sin\omega t.$$

At that position, when $\omega t = 0$, the wave is x-polarized. At a slightly later time $\omega t = \pi/2$, it becomes y-polarized. As time progresses, the tip of the electric field vector traces out a circle. This is called *circular polarization*.

There is a rule to describe the sense of rotation. The thumb should be oriented in the direction of propagation (in this case, $+z$). It is noticed that the $\mathcal{E}$ vector is rotating in the direction of the right-hand fingers. So, this is called a right-hand circularly polarized (RHCP) wave. With a different phasing, rotation in the other direction could be obtained, giving a left-hand circularly polarized (LHCP) wave. If the two components have different magnitudes, then *elliptical polarization* is obtained.

The electric field at a fixed instant in time can be visualized by letting $t = 0$. In this case,

$$\mathcal{E}(z, 0) = \hat{\mathbf{x}} E_0 \cos\beta z - \hat{\mathbf{y}} E_0 \sin\beta z.$$

The tip of the electric field vector traces out a helical path in space. Since $\beta = 2\pi/\lambda$, the helix repeats periodically with every λ of distance. With increasing z, the rotation follows the *left*-hand fingers.

---

[3] It is unfortunate that δ is used for both the skin depth and the loss tangent angle. Usually, the context makes it clear what is being used.

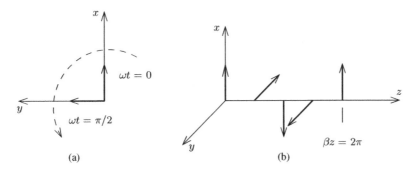

Figure 2.2 Time and space progression of a right-hand circularly polarized wave. (a) $\mathcal{E}(0,t)$ with increasing time; the electric field vector rotates counterclockwise. (b) $\mathcal{E}(z,0)$ as a function of z; the tip of the electric field vector traces out a helical path.

The time and space progression of the wave is shown in Figure 2.2. From the illustration, it can be seen that to reach the correct conclusion about the polarization sense, one should look at the time (not space) behaviour.

The space picture (b) can be better understood by considering that the 'early' part of the wave is at larger z whereas the 'later' part is at smaller z. Therefore, going forward in time in (b) corresponds to moving in the negative-z direction.

2.3   Oblique Plane Waves

Until now, we have considered plane waves that travel in the x, y or z direction. A so-called *oblique incidence* plane wave travels in an arbitrary direction. We will now generalize the previous plane-wave expressions to the oblique case.

We have seen that if we assume an electric field orientation of $\mathbf{E} = \hat{x}E_x$, the homogeneous wave equation becomes

$$\nabla^2 E_x + k^2 E_x = 0,$$

and if $\partial/\partial x = \partial/\partial y = 0$, then

$$\frac{d^2 E_x}{dz^2} + k^2 E_x = 0.$$

Assuming a lossless medium, the wavenumber $k = \omega\sqrt{\mu\epsilon} = 2\pi/\lambda$ is real. The general solution is

$$E_x = K_1 e^{-jkz} + K_2 e^{jkz}.$$

This result can be generalized to a plane wave travelling in an arbitrary direction. A wave e^{-jkz} has planes of constant phase where $kz = \text{const}$. Since the equation of a plane is $k_x x + k_y y + k_z z = \text{const.}$, a wave $e^{-j(k_x x + k_y y + k_z z)}$ will have planes of constant phase on those surfaces. Therefore, a plane wave can be written as

$$\mathbf{E}(x, y, z) = \hat{p}\, E_0\, e^{-j\mathbf{k}\cdot\mathbf{r}}$$

where we have defined a propagation vector $\mathbf{k} = \hat{x}k_x + \hat{y}k_y + \hat{z}k_z$ and a position vector $\mathbf{r} = x\hat{x} + y\hat{y} + z\hat{z}$. The propagation direction is given by the unit vector $\hat{\mathbf{k}} = \mathbf{k}/k$. The wavenumber k is

$$k = |\mathbf{k}| = \omega\sqrt{\mu\epsilon} = 2\pi/\lambda.$$

The $\mathbf{k}$ vector is perpendicular to the plane; this can be shown as follows. The normal to a surface $f(x, y, z) = $ const. is given by $\mathbf{n} = \nabla f$. Since the plane is $f(x, y, z) = k_x x + k_y y + k_z z = $ const., it follows that $\mathbf{n} = \nabla f = \hat{\mathbf{x}} k_x + \hat{\mathbf{y}} k_y + \hat{\mathbf{z}} k_z$ which is $\mathbf{k}$.

The unit vector $\hat{\mathbf{p}}$ is the polarization, that is, the direction of the electric field. The electric field is necessarily perpendicular to the propagation direction, so $\hat{\mathbf{p}} \cdot \mathbf{k} = 0$. We can obtain $\mathbf{H}$ from $\nabla \times \mathbf{E} = -j\omega\mu\mathbf{H}$. Recognizing that $\mathbf{E}$, $\mathbf{H}$ and $\mathbf{k}$ are mutually orthogonal leads to the convenient plane-wave relations

$$\mathbf{H} = \hat{\mathbf{k}} \times \mathbf{E}/\eta$$

$$\mathbf{E} = \eta\mathbf{H} \times \hat{\mathbf{k}}.$$

If an incident plane wave $\mathbf{E}^i$ undergoes reflection at a surface, then

$$\mathbf{E}^i = \hat{\mathbf{p}}_i \, E_0^i e^{-j\mathbf{k}_i \cdot \mathbf{r}}$$

$$\mathbf{E}^r = \hat{\mathbf{p}}_r \, \Gamma E_0^i e^{-j\mathbf{k}_r \cdot \mathbf{r}}.$$

The reflection coefficient Γ is defined in terms of the tangential electric fields at the reflecting surface S_0

$$\mathbf{E}^r_{tan}(S_0) = \Gamma \, \mathbf{E}^i_{tan}(S_0).$$

If the surface normal is $\hat{\mathbf{n}}$, this can also be written as

$$\hat{\mathbf{n}} \times \mathbf{E}^r(S_0) = \Gamma \, \hat{\mathbf{n}} \times \mathbf{E}^i(S_0).$$

Upon reflection, the normal parts of $\hat{\mathbf{p}}_i$ and $\mathbf{k}_i$ change sign, but the tangential parts do not, so that $\hat{\mathbf{n}} \cdot \hat{\mathbf{p}}_i = -\hat{\mathbf{n}} \cdot \hat{\mathbf{p}}_r, \hat{\mathbf{n}} \cdot \mathbf{k}_i = -\hat{\mathbf{n}} \cdot \mathbf{k}_r$ and $\hat{\mathbf{n}} \times \hat{\mathbf{p}}_i = \hat{\mathbf{n}} \times \hat{\mathbf{p}}_r, \hat{\mathbf{n}} \times \mathbf{k}_i = \hat{\mathbf{n}} \times \mathbf{k}_r$.

2.4 Plane-Wave Reflection and Transmission

Figure 2.3 shows a plane wave $\mathbf{E}^i$ that is incident on an interface between regions 1 and 2. Region 1 occupies $z < 0$ and is characterized by $\mu_1, \epsilon_1, \sigma_1$. Region 2 occupies $z > 0$ and is characterized by $\mu_2, \epsilon_2, \sigma_2$.

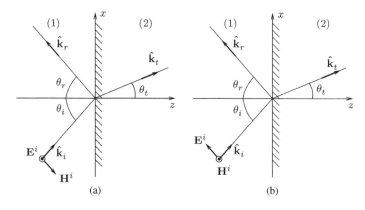

Figure 2.3 (a) Perpendicular polarization, $\mathbf{E}$ is $\perp$ to the paper and (b) parallel polarization, $\mathbf{E}$ is $\parallel$ to the paper. $z < 0$ is region 1 and $z > 0$ is region 2.

Part of the incident wave is reflected into region 1, and another part is transmitted into region 2. We want to develop expressions for the reflected and transmitted waves.

The incident, reflected and transmitted rays lie in the $x - y$ plane. Their directions are given by the unit vectors $\hat{\mathbf{k}}_i$, $\hat{\mathbf{k}}_r$ and $\hat{\mathbf{k}}_t$. The angles are all positive.

In the most general case, $\mathbf{E}^i$ can have a polarization in any direction that is perpendicular to $\hat{\mathbf{k}}_i$. Hence, $\mathbf{E}^i$ can always be resolved into perpendicular and parallel components, and it is sufficient to consider the two cases shown in Figure 2.3.

The incidence angle θ_i is assumed to be known. We need to find θ_r, θ_t and suitable reflection and transmission coefficients. To begin, we see from geometrical considerations in Figure 2.3 that we can construct the unit vectors

$$\hat{\mathbf{k}}_i = \hat{\mathbf{z}}\cos\theta_i + \hat{\mathbf{x}}\sin\theta_i \tag{2.17}$$

$$\hat{\mathbf{k}}_r = -\hat{\mathbf{z}}\cos\theta_r + \hat{\mathbf{x}}\sin\theta_r \tag{2.18}$$

$$\hat{\mathbf{k}}_t = \hat{\mathbf{z}}\cos\theta_t + \hat{\mathbf{x}}\sin\theta_t. \tag{2.19}$$

The position vector $\mathbf{r}$ for any point (x, y, z) is

$$\mathbf{r} = \hat{\mathbf{x}}x + \hat{\mathbf{y}}y + \hat{\mathbf{z}}z. \tag{2.20}$$

The complex propagation constant and impedance in each region are

$$jk_{1,2} = \gamma_{1,2} = \sqrt{j\omega\mu(\sigma + j\omega\epsilon)}|_{1,2} \tag{2.21}$$

$$\eta_{1,2} = \sqrt{\frac{j\omega\mu}{\sigma + j\omega\epsilon}}|_{1,2}. \tag{2.22}$$

The total fields for $z < 0$ are given by $\mathbf{E} = \mathbf{E}^i + \mathbf{E}^r$, $\mathbf{H} = \mathbf{H}^i + \mathbf{H}^r$. The total fields for $z > 0$ are given by $\mathbf{E} = \mathbf{E}^t$, $\mathbf{H} = \mathbf{H}^t$.

We now enforce the boundary conditions. $\mathbf{E}_{tan}$ and $\mathbf{H}_{tan}$ have to be continuous at $z = 0$. This is always true at an interface between two media. The only exception is when one of the media has an infinite conductivity. For example, if $z > 0$ is a perfect conductor, the skin depth becomes zero and the volume current $\mathbf{J} = \sigma\mathbf{E}$ in the conductor collapses down to a surface current $\mathbf{J}_s$. This in turn makes $\mathbf{H}_{tan}$ discontinuous at $z = 0$.

Enforcing the continuity of $\mathbf{E}_{tan}$ and $\mathbf{H}_{tan}$, we have

$$\hat{\mathbf{z}} \times (\mathbf{E}^i + \mathbf{E}^r)|_{z=0^-} = \hat{\mathbf{z}} \times \mathbf{E}^t|_{z=0^+} \tag{2.23}$$

$$\hat{\mathbf{z}} \times (\mathbf{H}^i + \mathbf{H}^r)|_{z=0^-} = \hat{\mathbf{z}} \times \mathbf{H}^t|_{z=0^+}. \tag{2.24}$$

2.4.1 *Perpendicular Polarization*

The incident, reflected and transmitted electric fields are

$$\mathbf{E}^i = \hat{\mathbf{y}}E_0 e^{-j\mathbf{k}_i \cdot \mathbf{r}} \tag{2.25}$$

$$\mathbf{E}^r = \hat{\mathbf{y}}\Gamma E_0 e^{-j\mathbf{k}_r \cdot \mathbf{r}} \tag{2.26}$$

$$\mathbf{E}^t = \hat{\mathbf{y}}T E_0 e^{-j\mathbf{k}_t \cdot \mathbf{r}}. \tag{2.27}$$

We need to solve for the unknowns Γ, T, $\hat{\mathbf{k}}_r$ and $\hat{\mathbf{k}}_t$. From (2.23), the continuity of $\mathbf{E}_{tan}$ at $z = 0$ is required. It must hold for all values of x and y. This is only possible if the fields on both sides have the same variation with x and y. This is often called the phase-matching condition. Therefore, we must have

$$j\mathbf{k}_i \cdot \mathbf{r} = j\mathbf{k}_r \cdot \mathbf{r} = j\mathbf{k}_t \cdot \mathbf{r}|_{z=0}. \tag{2.28}$$

Remembering that $\mathbf{k}_i$ and $\mathbf{k}_r$ are in medium 1 whereas $\mathbf{k}_t$ is in medium 2, we have that $\mathbf{k}_i = k_1\hat{\mathbf{k}}_i$, $\mathbf{k}_r = k_1\hat{\mathbf{k}}_r$ and $\mathbf{k}_t = k_2\hat{\mathbf{k}}_t$ so that (2.28) becomes

$$jk_1 x \sin\theta_i = jk_1 x \sin\theta_r = jk_2 x \sin\theta_t. \tag{2.29}$$

This leads to the law of reflection

$$\sin\theta_i = \sin\theta_r \tag{2.30}$$

and Snell's law

$$k_1 \sin\theta_i = k_2 \sin\theta_t. \tag{2.31}$$

As a special case, if medium 1 is air and medium 2 is a lossless dielectric, then $\mu_1 = \mu_2 = \mu_0$, $\sigma_1 = \sigma_2 = 0$ and $\epsilon_1 = \epsilon_0$. This leads to

$$\frac{jk_2}{jk_1} = \sqrt{\frac{\epsilon_2}{\epsilon_0}} = m,$$

where m is defined as the *index of refraction*. In this case, (2.31) becomes $\sin\theta_i = m \sin\theta_t$.

To determine the two unknowns Γ and T, we must use the two boundary conditions (2.23) and (2.24). Imposing (2.23) with $\mathbf{E}$ given by (2.25)–(2.27) and letting $z = 0$, we obtain

$$1 + \Gamma = T. \tag{2.32}$$

Now we need $\mathbf{H}$. We can get it from $\nabla \times \mathbf{E} = -j\omega\mu\mathbf{H}$. For a plane-wave field, it is easier to use the equivalent expression $\mathbf{H} = (\hat{\mathbf{k}} \times \mathbf{E})/\eta$. Applying this to (2.25)–(2.27) gives us

$$\mathbf{H}^i = \frac{E_0}{\eta_1}(-\hat{\mathbf{x}}\cos\theta_i + \hat{\mathbf{z}}\sin\theta_i)e^{-j\mathbf{k}_i \cdot \mathbf{r}} \tag{2.33}$$

$$\mathbf{H}^r = \frac{\Gamma E_0}{\eta_1}(\hat{\mathbf{x}}\cos\theta_r + \hat{\mathbf{z}}\sin\theta_r)e^{-j\mathbf{k}_r \cdot \mathbf{r}} \tag{2.34}$$

$$\mathbf{H}^t = \frac{T E_0}{\eta_2}(-\hat{\mathbf{x}}\cos\theta_t + \hat{\mathbf{z}}\sin\theta_t)e^{-j\mathbf{k}_t \cdot \mathbf{r}}. \tag{2.35}$$

Imposing (2.24) with $\mathbf{H}$ given by (2.33)–(2.35) and letting $z = 0$, we obtain

$$\frac{1}{\eta_1}(-\hat{\mathbf{y}}\cos\theta_i\, e^{-j\mathbf{k}_i \cdot \mathbf{r}} + \Gamma\hat{\mathbf{y}}\cos\theta_r\, e^{-jk_1\hat{\mathbf{k}}_r \cdot \mathbf{r}}) = \frac{T}{\eta_2}(-\hat{\mathbf{y}}\cos\theta_t\, e^{-j\mathbf{k}_t \cdot \mathbf{r}}).$$

Since $\theta_i = \theta_r$ and $jk_1\hat{\mathbf{k}}_i \cdot \mathbf{r} = jk_2\hat{\mathbf{k}}_r \cdot \mathbf{r} = jk_3\hat{\mathbf{k}}_t \cdot \mathbf{r}$ then

$$\frac{1}{\eta_1}\cos\theta_i(-1+\Gamma) = \frac{-T}{\eta_2}\cos\theta_t. \tag{2.36}$$

We can solve (2.32) and (2.36) for Γ and T with the result that

$$\Gamma_\perp = \frac{\eta_2 \cos \theta_i - \eta_1 \cos \theta_t}{\eta_2 \cos \theta_i + \eta_1 \cos \theta_t} \quad (2.37)$$

$$T_\perp = \frac{2\eta_2 \cos \theta_i}{\eta_2 \cos \theta_i + \eta_1 \cos \theta_t}. \quad (2.38)$$

These are known as the *Fresnel reflection and transmission coefficients*. Since they apply to the perpendicular polarization, the subscript $\perp$ has been added.

The Fresnel coefficients can be rewritten for the special case of an air-dielectric interface where $\epsilon_1 = \epsilon_0$ and $\epsilon_2 = \epsilon_r \epsilon_0$. Equation (2.31) can also be used to eliminate θ_t in (2.37), (2.38) with the result that

$$\Gamma_\perp = \frac{\cos \theta_i - \sqrt{\epsilon_r - \sin^2 \theta_i}}{\sqrt{\epsilon_r - \sin^2 \theta_i} + \cos \theta_i} \quad (2.39)$$

$$T_\perp = \frac{2 \cos \theta_i}{\sqrt{\epsilon_r - \sin^2 \theta_i} + \cos \theta_i}. \quad (2.40)$$

2.4.2 Parallel Polarization

The incident, reflected and transmitted electric fields are

$$\mathbf{E}^i = E_0(\hat{\mathbf{x}} \cos \theta_i - \hat{\mathbf{z}} \sin \theta_i)e^{-j\mathbf{k}_i \cdot \mathbf{r}} \quad (2.41)$$

$$\mathbf{E}^r = \Gamma E_0(\hat{\mathbf{x}} \cos \theta_r + \hat{\mathbf{z}} \sin \theta_r)e^{-j\mathbf{k}_r \cdot \mathbf{r}} \quad (2.42)$$

$$\mathbf{E}^t = T E_0(\hat{\mathbf{x}} \cos \theta_t - \hat{\mathbf{z}} \sin \theta_t)e^{-j\mathbf{k}_t \cdot \mathbf{r}}. \quad (2.43)$$

Using $\mathbf{H} = (\hat{\mathbf{k}} \times \mathbf{E})/\eta$, we obtain the magnetic fields

$$\mathbf{H}^i = \hat{\mathbf{y}} \frac{E_0}{\eta_1} e^{-j\mathbf{k}_i \cdot \mathbf{r}} \quad (2.44)$$

$$\mathbf{H}^r = -\hat{\mathbf{y}} \frac{\Gamma E_0}{\eta_1} e^{-j\mathbf{k}_r \cdot \mathbf{r}} \quad (2.45)$$

$$\mathbf{H}^t = \hat{\mathbf{y}} \frac{T E_0}{\eta_2} e^{-j\mathbf{k}_t \cdot \mathbf{r}}. \quad (2.46)$$

As before, we require that $j\mathbf{k}_i \cdot \mathbf{r} = j\mathbf{k}_r \cdot \mathbf{r} = j\mathbf{k}_t \cdot \mathbf{r}$. This leads to the same law of reflection and Snell's law as before. Therefore, these laws are *polarization independent*.

We enforce the continuity of $\mathbf{E}_{tan}$ at $z = 0$ by applying (2.23) to (2.41)–(2.43) with the result that

$$(1 + \Gamma) \cos \theta_i = T \cos \theta_t. \quad (2.47)$$

To enforce the continuity of $\mathbf{H}_{tan}$ at $z = 0$, we apply (2.24) to (2.44)–(2.46) to obtain

$$\frac{1 - \Gamma}{\eta_1} = \frac{T}{\eta_2}. \quad (2.48)$$

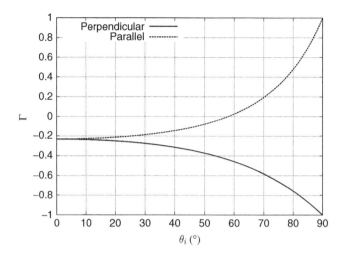

Figure 2.4 $\Gamma_\perp$ and $\Gamma_\parallel$ versus incidence angle θ_i for glass, $m = 1.6$.

Solving (2.47), (2.48) for Γ and T noting that this is the parallel case, we obtain

$$\Gamma_\parallel = \frac{\eta_2 \cos\theta_t - \eta_1 \cos\theta_i}{\eta_2 \cos\theta_t + \eta_1 \cos\theta_i} \tag{2.49}$$

$$T_\parallel = \frac{2\eta_2 \cos\theta_i}{\eta_2 \cos\theta_t + \eta_1 \cos\theta_i}. \tag{2.50}$$

Using (2.31) to eliminate θ_t and specializing the result to an air-dielectric interface,

$$\Gamma_\parallel = \frac{\sqrt{\epsilon_r - \sin^2\theta_i} - \epsilon_r \cos\theta_i}{\sqrt{\epsilon_r - \sin^2\theta_i} + \epsilon_r \cos\theta_i} \tag{2.51}$$

$$T_\parallel = \frac{2\sqrt{\epsilon_r} \cos\theta_i}{\sqrt{\epsilon_r - \sin^2\theta_i} + \epsilon_r \cos\theta_i}. \tag{2.52}$$

For the limiting case of a perfectly conducting medium 2, $\eta_2 = 0$ and $\Gamma_\perp = \Gamma_\parallel = -1$ for any angle of incidence. In this case, the incident and reflected fields add, as expected from image theory. For the limiting case of grazing incidence we see that if $\theta_i \to 90°$ then $\Gamma_\perp \to -1$ and $\Gamma_\parallel \to 1$. This holds for any finite η_2.

Figure 2.4 shows $\Gamma_\perp$ and $\Gamma_\parallel$ for glass. The index of refraction for light barium flint glass at optical frequencies is about $m = 1.6$.

2.4.3   The Brewster Angle

It is interesting to note that $\Gamma_\parallel$ can be zero. The energy of the incident wave then gets completely transmitted into medium 2. This is known as the *Brewster angle* $\theta_i = \theta_B$. It can be found by setting the numerator of (2.51) to zero,

$$\sqrt{\epsilon_r - \sin^2\theta_B} - \epsilon_r \cos\theta_B = 0$$

or

$$\sin \theta_B = \sqrt{\frac{\epsilon_r}{\epsilon_r + 1}}. \qquad (2.53)$$

It can be seen in Figure 2.4 that $\Gamma_\parallel = 0$ at the Brewster angle $\theta_B = 58°$. 'Brewster windows' are used in laser cavities to allow the light energy to exit.

2.4.4 Total Internal Reflection

Suppose that medium 1 is glass, and medium 2 is air. Snell's law (2.31) indicates that

$$\sqrt{\epsilon_1} \sin \theta_i = \sqrt{\epsilon_2} \sin \theta_t$$

and since $m = \sqrt{\epsilon_1/\epsilon_2}$,

$$m \sin \theta_i = \sin \theta_t.$$

The transmitted angle is $\theta_t = 90°$ when the incident angle θ_i equals the *critical angle* θ_c

$$\sin \theta_i = \sin \theta_c = \frac{1}{m}.$$

With $m = 1.6$ for glass, the critical angle is $\theta_c = 39°$. What happens if the incident angle exceeds θ_c? It turns out that our assumption that a plane wave is transmitted into medium 2 is no longer valid. A new solution, involving a *non-uniform plane wave* is needed. This is a type of plane wave that propagates along x but has an exponential amplitude decay along z. It is still a plane wave, because the phase is constant in an $x = $ const. plane. It is non-uniform because the amplitude varies in that plane. It is only with this kind of wave that the boundary conditions can be met. From this solution, it can be shown that when $\theta_i > \theta_c$, the reflection coefficient magnitude (for both polarizations) becomes $|\Gamma| = 1$. Hence, *total* reflection occurs. The transmitted wave is non-zero, but it decays exponentially, away from the surface.

Total internal reflection is often seen by people swimming under the water. Looking upwards, it is sometimes impossible to see out into the air. Total internal reflection is also used to contain a wave inside an optical fibre.

Let us derive the reflection coefficient. For the perpendicular polarization, we start by replacing (2.27) by the non-uniform plane wave so that

$$\mathbf{E}^i = \hat{\mathbf{y}} E_0 e^{-j\mathbf{k}_i \cdot \mathbf{r}} \qquad (2.54)$$

$$\mathbf{E}^r = \hat{\mathbf{y}} \Gamma E_0 e^{-j\mathbf{k}_r \cdot \mathbf{r}} \qquad (2.55)$$

$$\mathbf{E}^t = \hat{\mathbf{y}} T E_0 e^{-\alpha z - j\beta x}. \qquad (2.56)$$

As before, the tangential field continuity at $z = 0$ requires phase matching of the fields so that $k_1 x \sin \theta_i = k_1 x \sin \theta_r = \beta x$ which implies that

$$\sin \theta_i = \sin \theta_r \qquad (2.57)$$

and

$$\beta = k_1 \sin \theta_i. \qquad (2.58)$$

Using the wave equation in medium 2,

$$\nabla^2 E_y + k_2^2 E_y = 0$$

we can relate α and β

$$\beta^2 - \alpha^2 = k_2^2 = \omega^2 \mu_2 \epsilon_2. \tag{2.59}$$

The continuity of E_y at $z = 0$ requires

$$1 + \Gamma = T. \tag{2.60}$$

Taking $\nabla \times \mathbf{E}$ gives the magnetic field

$$\mathbf{H}^i = \frac{E_0}{\eta_1}(-\hat{\mathbf{x}}\cos\theta_i + \hat{\mathbf{z}}\sin\theta_i)e^{-j\mathbf{k}_i \cdot \mathbf{r}} \tag{2.61}$$

$$\mathbf{H}^r = \frac{\Gamma E_0}{\eta_1}(\hat{\mathbf{x}}\cos\theta_r + \hat{\mathbf{z}}\sin\theta_r)e^{-j\mathbf{k}_r \cdot \mathbf{r}} \tag{2.62}$$

$$\mathbf{H}^t = \frac{\hat{\mathbf{x}}\alpha - \hat{\mathbf{z}}j\beta}{-j\omega\mu_2}TE_0 e^{-\alpha z - j\beta x} \tag{2.63}$$

and continuity of H_x at $z = 0$ requires

$$\frac{-1}{\eta_1}\cos\theta_i + \frac{\Gamma}{\eta_1}\cos\theta_r = \frac{\alpha}{-j\omega\mu_2}T. \tag{2.64}$$

Solving for Γ gives

$$\Gamma_\perp = \frac{j\omega\mu_2\cos\theta_i - \alpha\eta_1}{j\omega\mu_2\cos\theta_i + \alpha\eta_1}. \tag{2.65}$$

From this expression, it can be seen that the magnitudes of the numerator and the denominator are the same, so that $|\Gamma_\perp| = 1$. (One has to remember that because of the assumed form of $\mathbf{E}^t$ in (2.56), the expression for Γ only holds for $\theta_i \geq \theta_c$.) The reflection coefficient has a phase, which causes a lateral (along x) phase shift of the reflected wave.

The reflection coefficient can be easily calculated. Using (2.58), the incident angle θ_i gives β and (2.59) gives α; then, (2.65) gives Γ. When $\theta_i = \theta_c$, there is no decay in the transmitted wave, so $\alpha = 0$. As θ_i is further increased, the transmitted wave begins to show some decay. As the incident angle approaches grazing, $\theta_i \to 90^\circ$ and the decay α becomes maximum.

A similar procedure applies for the parallel polarization. In this case,

$$\mathbf{E}^i = E_0(\hat{\mathbf{x}}\cos\theta_i - \hat{\mathbf{z}}\sin\theta_i)e^{-j\mathbf{k}_i \cdot \mathbf{r}} \tag{2.66}$$

$$\mathbf{E}^r = \Gamma E_0(\hat{\mathbf{x}}\cos\theta_r + \hat{\mathbf{z}}\sin\theta_r)e^{-j\mathbf{k}_r \cdot \mathbf{r}} \tag{2.67}$$

$$\mathbf{E}^t = \frac{(\hat{\mathbf{x}}\alpha - \hat{\mathbf{z}}j\beta)}{j\omega\epsilon_2}\frac{TE_0}{\eta_2}e^{-\alpha z - j\beta x} \tag{2.68}$$

and

$$\mathbf{H}^i = \hat{\mathbf{y}}\frac{E_0}{\eta_1}e^{-j\mathbf{k}_i \cdot \mathbf{r}} \tag{2.69}$$

$$\mathbf{H}^r = -\hat{\mathbf{y}}\frac{\Gamma E_0}{\eta_1}e^{-j\mathbf{k}_r \cdot \mathbf{r}} \tag{2.70}$$

$$\mathbf{H}^t = \hat{\mathbf{y}}\frac{TE_0}{\eta_2}e^{-\alpha z - j\beta x}. \tag{2.71}$$

Using phase matching and the wave equation, we find that the results (2.57)–(2.59) still apply. The continuity of E_x requires

$$\cos\theta_i + \Gamma\cos\theta_r = \frac{\alpha T}{j\omega\epsilon_2\eta_2}$$ (2.72)

and the continuity of H_y requires

$$\frac{1}{\eta_1}(1-\Gamma) = \frac{T}{\eta_2}.$$ (2.73)

Solving for Γ gives

$$\Gamma_\| = \frac{\alpha - j\omega\epsilon_2\eta_1\cos\theta_i}{\alpha + j\omega\epsilon_2\eta_1\cos\theta_i}.$$ (2.74)

It is seen that $|\Gamma_\|| = 1$ for this polarization as well.

2.5 Multilayer Slab

Figure 2.5 shows a multilayer material slab. The incident and reflected rays make an angle θ with respect to the $+z$ axis. There are N layers. The nth layer has a thickness d_n, permittivity ϵ_n and permeability μ_n.

To find the reflection and transmission coefficients, the recursive technique (Richmond 1965) will be used. Alternatively, a matrix multiplication solution (Collin 1991, Chapter 3) could also be used.

For the perpendicular polarization, the incident, reflected and transmitted waves are

$$E_y^i = E_0 e^{jk_0 x \sin\theta} e^{jk_0 z \cos\theta}$$

$$E_y^r = \Gamma_\perp E_0 e^{jk_0 x \sin\theta} e^{-jk_0 z \cos\theta}$$

$$E_y^t = T_\perp E_0 e^{jk_0 x \sin\theta} e^{jk_0 z \cos\theta}.$$

In the nth layer, we have a sum of right-travelling waves A_n and a sum of left-travelling waves B_n where

$$E_n = (A_n e^{\gamma_n z} + B_n e^{-\gamma_n z}) e^{jk_0 x \sin\theta}$$

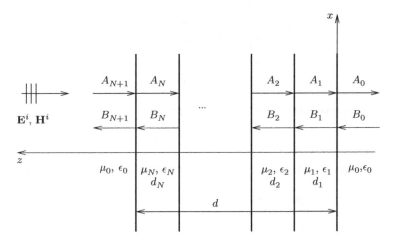

Figure 2.5 Multilayer slab, with N layers. In the nth layer, the sum of right-travelling waves is A_n and the sum of left-travelling waves is B_n.

and in the $n + 1$th layer

$$E_{n+1} = (A_{n+1}e^{\gamma_{n+1}z} + B_{n+1}e^{-\gamma_{n+1}z})e^{jk_0 x \sin\theta}.$$

By enforcing the continuity of E_y and H_x at the left side of layer n, one can obtain

$$A_{n+1} = P_n A_n + Q_n B_n \tag{2.75}$$

$$B_{n+1} = R_n A_n + S_n B_n \tag{2.76}$$

where

$$P_n = \frac{1}{2}(1 + \mu_{n+1}\gamma_n/\mu_n\gamma_{n+1})e^{(\gamma_n - \gamma_{n+1})z_n} \tag{2.77}$$

$$Q_n = \frac{1}{2}(1 - \mu_{n+1}\gamma_n/\mu_n\gamma_{n+1})e^{-(\gamma_n + \gamma_{n+1})z_n} \tag{2.78}$$

$$R_n = \frac{1}{2}(1 - \mu_{n+1}\gamma_n/\mu_n\gamma_{n+1})e^{(\gamma_n + \gamma_{n+1})z_n} \tag{2.79}$$

$$S_n = \frac{1}{2}(1 + \mu_{n+1}\gamma_n/\mu_n\gamma_{n+1})e^{-(\gamma_n - \gamma_{n+1})z_n}. \tag{2.80}$$

Here, $z_n = d_1 + d_2 + \cdots + d_n$. The wave equation is used to obtain the propagation constant

$$\gamma_n = j\sqrt{\omega^2 \mu_n \epsilon_n - k_0^2 \sin^2\theta} = jk_z.$$

We can now obtain the reflection and transmission coefficients. If the $z < 0$ region is air, we set

$$A_0 = 1; \quad B_0 = 0 \tag{2.81}$$

and then use the recursion equations to compute $A_1, B_1, A_2, B_2, \ldots A_N, B_N$ in that order. Then,

$$E_0 = A_{N+1} \tag{2.82}$$

$$\Gamma_\perp = \frac{E_y^r(0,0,0)}{E_y^i(0,0,0)} = \frac{B_{N+1}}{A_{N+1}} \tag{2.83}$$

$$T_\perp = \frac{E_y^t(0,0,0)}{E_y^i(0,0,0)} = \frac{A_0}{A_{N+1}} = \frac{1}{A_{N+1}}. \tag{2.84}$$

If a perfect electric conductor is situated at $z = 0$, then we set

$$A_1 = 1; \quad B_1 = -1 \tag{2.85}$$

and proceed to compute the A_n and B_n as before. Equations (2.83) and (2.84) are used to obtain $\Gamma_\perp$ and $T_\perp$.

The equations for parallel polarization are almost the same, except that the A_n and B_n now represent magnetic field intensities. In addition, we must modify (2.77)–(2.80) by replacing $\mu_n \to \epsilon_n$ and $\mu_{n+1} \to \epsilon_{n+1}$. It is also noted that B_{N+1}/A_{N+1} is the magnetic field reflection coefficient, which differs from the electric field reflection coefficient by a minus sign.

If $z < 0$ is air, we set

$$A_0 = 1; \quad B_0 = 0 \tag{2.86}$$

and use

$$H_0 = A_{N+1} \tag{2.87}$$

$$\Gamma_{\parallel} = -\frac{H_y^r(0,0,0)}{H_y^i(0,0,0)} = -\frac{B_{N+1}}{A_{N+1}} \tag{2.88}$$

$$T_{\parallel} = \frac{H_y^t(0,0,0)}{H_y^i(0,0,0)} = \frac{A_0}{A_{N+1}} = \frac{1}{A_{N+1}}. \tag{2.89}$$

If a perfect electric conductor is situated at $z = 0$, then

$$A_1 = 1; \quad B_1 = 1. \tag{2.90}$$

We then proceed to compute A_n and B_n as before. Equations (2.88) and (2.89) are used to obtain $\Gamma_{\parallel}$ and $T_{\parallel}$.

For both polarizations, Γ represents the ratio of the reflected/incident field at $(0,0,0)$. We can change the reference plane to the left interface $(0,0,d)$ where $d = d_1 + d_2 + \cdots + d_N$. Then, we can define a new reflection coefficient Γ', given by

$$\Gamma' = \Gamma e^{-j2k_0 d \cos\theta}.$$

2.6 Impedance Boundary Condition

Under certain conditions, the surface current and tangential electric field on a material can be described by an *impedance boundary condition* (IBC). It is defined by

$$Z_s \mathbf{J}_s = \mathbf{E}_{tan} \tag{2.91}$$

and Z_s is the surface impedance in ohms. The concept is usually used to describe thin material sheets that have reflection and transmission (a penetrable boundary) as well as boundaries having only reflection, such as a dielectric-coated conductor or a metal that is many skin depths thick (an impenetrable boundary). The terms *impedance sheet* and *impedance boundary* are sometimes used to distinguish between a penetrable and impenetrable boundary, respectively. To get an idea of how the surface impedance concept is used, it will be applied to a thin penetrable sheet and to a dielectric half-space.

2.6.1 Penetrable Boundary

Figure 2.6 shows a plane wave $\mathbf{E}^i$ that is incident on a thin planar material sheet. The surrounding regions 1 and 2 are air, with $k_0 = \omega\sqrt{\mu_0 \epsilon_0}$ and $\eta_0 = \sqrt{\mu_0/\epsilon_0}$. Under certain conditions, the sheet can be approximated by an IBC. The IBC equivalent has zero thickness. The actual thickness d should be small enough so that $\mathbf{E}_{tan}$ is the same on both faces of the sheet; in other words, the field inside should not vary appreciably with z. This situation occurs in a conductor, if the thickness d is much less than the skin depth, or in a dielectric, if $d \ll \lambda_0/\sqrt{\epsilon_r}$. The incidence angle can also have an effect.

The boundary conditions at $z = 0$ are

$$\hat{\mathbf{z}} \times (\mathbf{E}_2 - \mathbf{E}_1) = 0 \tag{2.92}$$

$$\hat{\mathbf{z}} \times (\mathbf{H}_2 - \mathbf{H}_1) = \mathbf{J}_s. \tag{2.93}$$

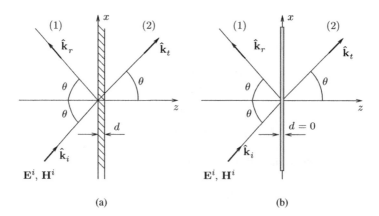

Figure 2.6 Impedance boundary, (a) thin sheet and (b) impedance boundary equivalent. Regions 1 and 2 are air.

Using (2.91) in (2.93), the IBC becomes

$$Z_s \hat{\mathbf{z}} \times (\mathbf{H}_2 - \mathbf{H}_1) = \mathbf{E}_{tan}.$$

The tangential part of $\mathbf{E}$ can be extracted by using $\mathbf{E}_{tan} = -\hat{\mathbf{z}} \times \hat{\mathbf{z}} \times \mathbf{E}$ so that

$$Z_s \hat{\mathbf{z}} \times (\mathbf{H}_2 - \mathbf{H}_1) = -\hat{\mathbf{z}} \times \hat{\mathbf{z}} \times \mathbf{E}.$$

We can obtain an alternate but equivalent form by taking $\hat{\mathbf{z}} \times$ again, so that

$$Z_s \hat{\mathbf{z}} \times \hat{\mathbf{z}} \times (\mathbf{H}_2 - \mathbf{H}_1) = -\hat{\mathbf{z}} \times \hat{\mathbf{z}} \times \hat{\mathbf{z}} \times \mathbf{E}.$$

The operation $-\hat{\mathbf{z}} \times \hat{\mathbf{z}}$ does not change the right-hand side, so we can write

$$Z_s \hat{\mathbf{z}} \times \hat{\mathbf{z}} \times (\mathbf{H}_2 - \mathbf{H}_1) = \hat{\mathbf{z}} \times \mathbf{E}. \tag{2.94}$$

Since $\mathbf{E}_{tan}$ is continuous, $\mathbf{E}$ can be either $\mathbf{E}_1$ or $\mathbf{E}_2$.

Returning to Figure 2.6, part of the incident wave is reflected into region 1 and another part is transmitted into region 2. The solution procedure is similar to that in Section 2.4 for a single interface. It turns out that the boundary conditions can be satisfied by waves having the same incident, reflected and transmitted angles, so that

$$\hat{\mathbf{k}}_i = \hat{\mathbf{z}} \cos \theta + \hat{\mathbf{x}} \sin \theta$$

$$\hat{\mathbf{k}}_r = -\hat{\mathbf{z}} \cos \theta + \hat{\mathbf{x}} \sin \theta$$

$$\hat{\mathbf{k}}_t = \hat{\mathbf{k}}_i.$$

Using (2.92) and (2.94),

$$\hat{\mathbf{z}} \times \mathbf{E}^t|_{z=0^+} = \hat{\mathbf{z}} \times (\mathbf{E}^i + \mathbf{E}^r)|_{z=0^-} \tag{2.95}$$

$$\hat{\mathbf{z}} \times \mathbf{H}^t|_{z=0^+} - \hat{\mathbf{z}} \times (\mathbf{H}^i + \mathbf{H}^r)|_{z=0^-} = \mathbf{J}_s = \frac{1}{Z_s} \hat{\mathbf{z}} \times \mathbf{E}^t|_{z=0^+}. \tag{2.96}$$

For the perpendicular polarization, the reflection and transmission coefficients can be found by using the **E** fields (2.25)–(2.27), **H** fields (2.33)–(2.35) and the boundary conditions (2.95) and (2.96). The result is

$$\Gamma_\perp = \frac{-\eta_0}{\eta_0 + 2Z_s \cos\theta} \tag{2.97}$$

$$T_\perp = 1 + \Gamma_\perp = \frac{2Z_s \cos\theta}{\eta_0 + 2Z_s \cos\theta}. \tag{2.98}$$

Following a similar procedure for the parallel polarization leads to

$$\Gamma_\parallel = \frac{-\eta_0 \cos\theta}{\eta_0 \cos\theta + 2Z_s} \tag{2.99}$$

$$T_\parallel = 1 + \Gamma_\parallel = \frac{2Z_s}{\eta_0 \cos\theta + 2Z_s}. \tag{2.100}$$

What is the relationship between Z_s and the actual material parameters? The conduction current is $\mathbf{J} = \sigma\mathbf{E}$. If the sheet is thin, the internal field will be uniform with respect to z, so that $\mathbf{J} \approx \mathbf{J}_s/d$. Since $\mathbf{E} = \mathbf{J}/\sigma$ and $\mathbf{J}$ in the thin sheet is assumed to have no z component, neither does $\mathbf{E}$ and hence $\mathbf{E} \approx \mathbf{E}_{tan}$. Using these conditions and (2.91), we get

$$\mathbf{E}_{tan} = \frac{\mathbf{J}}{\sigma} = \frac{\mathbf{J}_s/d}{\sigma} = \mathbf{J}_s Z_s$$

from which

$$Z_s = \frac{1}{\sigma d}. \tag{2.101}$$

For a thin dielectric sheet, the volume equivalent current $\mathbf{J} = j\omega(\epsilon - \epsilon_0)\mathbf{E}_{tan}$, is needed.[4] We should replace σ by $j\omega(\epsilon - \epsilon_0)$ with the result that

$$Z_s = \frac{1}{j\omega(\epsilon - \epsilon_0)d}. \tag{2.102}$$

When does the IBC fail? In the case of normal incidence, there will be a problem when d/λ is too large, as $\mathbf{E}_{tan}$ will no longer be the same on both sides of the sheet. For oblique incidence, we must also consider the validity of the approximation $\mathbf{E} \approx \mathbf{E}_{tan}$. In a thin conducting sheet, the volume current density $\mathbf{J} = \sigma\mathbf{E}$ flows mainly *along* the material, and not perpendicular to it, so this will give an electric field that is mostly tangential. As for the dielectric sheet, we note that rays inside the dielectric are refracted towards the normal. Therefore, if we avoid low ϵ_r and large incidence angles, the approximation will be good.

2.6.2 Impenetrable Boundary

An IBC can also be used for a very different situation, an exterior-only representation of an object. The impenetrable IBC is also called the *Leontovich* boundary condition, named after its inventor in 1940, who used it to examine radio-wave propagation on the finitely conducting spherical Earth.

As an example, in Figure 2.7, suppose that region 1 is air with a wavenumber $k_0 = \omega\sqrt{\mu_0\epsilon_0}$ and an impedance $\eta_0 = \sqrt{\mu_0/\epsilon_0}$. Region 2 is a dielectric with permittivity $\epsilon = \epsilon_0\epsilon_r$, a wavenumber $k_2 = k_0\sqrt{\epsilon_r}$ and an impedance $\eta_2 = \eta_0/\sqrt{\epsilon_r}$. We regard the dielectric as the object and seek an IBC equivalent in region 1. The interior fields in region 2 are assumed to be zero, so it is 'impenetrable'.

---

[4] The volume equivalent current replaces the dielectric with a current in free space. It is described in Section 4.11.

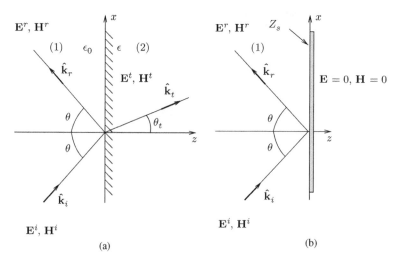

Figure 2.7   (a) Air–dielectric interface and (b) impenetrable impedance boundary equivalent.

With $\hat{\mathbf{n}} = -\hat{\mathbf{z}}$ and $\mathbf{J}_s = \hat{\mathbf{n}} \times \mathbf{H}_1$ at $z = 0^-$, (2.91) becomes

$$Z_s \hat{\mathbf{n}} \times \mathbf{H}_1 = \mathbf{E}_{tan}$$

and using $\mathbf{E}_{tan} = -\hat{\mathbf{n}} \times \hat{\mathbf{n}} \times \mathbf{E}_1$ gives

$$Z_s \hat{\mathbf{n}} \times \mathbf{H}_1 = -\hat{\mathbf{n}} \times \hat{\mathbf{n}} \times \mathbf{E}_1.$$

We can obtain an alternate but equivalent form by taking $\hat{\mathbf{n}} \times$ again, so that

$$Z_s \hat{\mathbf{n}} \times \hat{\mathbf{n}} \times \mathbf{H}_1 = -\hat{\mathbf{n}} \times \hat{\mathbf{n}} \times \hat{\mathbf{n}} \times \mathbf{E}_1.$$

The operation $-\hat{\mathbf{n}} \times \hat{\mathbf{n}}$ does not change the right-hand side so that

$$Z_s \hat{\mathbf{n}} \times \hat{\mathbf{n}} \times \mathbf{H}_1 = \hat{\mathbf{n}} \times \mathbf{E}_1. \tag{2.103}$$

For the perpendicular polarization, we can use the plane-wave expressions (2.25) and (2.26) for E_y and (2.33) and (2.34) for H_x. At $z = 0^-$, (2.103) becomes

$$(E_y^i + E_y^r) = -Z_s(H_x^i + H_x^r)$$

which gives

$$1 + \Gamma_\perp = -Z_s \frac{-1 + \Gamma_\perp}{\eta_0} \cos\theta$$

so that

$$\Gamma_\perp = \frac{\bar{Z}_s \cos\theta - 1}{\bar{Z}_s \cos\theta + 1}, \tag{2.104}$$

where $\bar{Z}_s = Z_s/\eta_0$. Conversely, we also have

$$\bar{Z}_s = \sec\theta \frac{1+\Gamma_\perp}{1-\Gamma_\perp}.$$

The tangential fields for the original problem are continuous at the interface; the transmitted E_y^t and H_x^t in (2.27) and (2.35) at $z = 0^+$ can be used to find the surface impedance

$$\bar{Z}_s = \frac{-(E_y^i + E_y^r)}{\eta_0(H_x^i + H_x^r)} = \frac{-E_y^t}{\eta_0 H_x^t} = \frac{\eta_2}{\eta_0 \cos\theta_t} = \frac{1}{\sqrt{\epsilon_r}\cos\theta_t}.$$

Using $\cos\theta_t = \sqrt{1 - \sin^2\theta_t}$ and Snell's law $\sin\theta = \sqrt{\epsilon_r}\sin\theta_t$,

$$\bar{Z}_s = \frac{1}{\sqrt{\epsilon_r - \sin^2\theta}}. \tag{2.105}$$

Using (2.105) in (2.104), we obtain

$$\Gamma_\perp = \frac{\cos\theta - \sqrt{\epsilon_r - \sin^2\theta}}{\cos\theta + \sqrt{\epsilon_r - \sin^2\theta}} \tag{2.106}$$

which is identical to the reflection coefficient that we originally had in (2.39).

A similar procedure can be used for the parallel polarization to obtain

$$\Gamma_\parallel = \frac{\bar{Z}_s - \cos\theta}{\bar{Z}_s + \cos\theta} \tag{2.107}$$

where

$$\bar{Z}_s = \frac{E_x^t}{\eta_0 H_y^t} = \frac{1}{\epsilon_r}\sqrt{\epsilon_r - \sin^2\theta}. \tag{2.108}$$

This gives

$$\Gamma_\parallel = \frac{\sqrt{\epsilon_r - \sin^2\theta} - \epsilon_r\cos\theta}{\sqrt{\epsilon_r - \sin^2\theta} + \epsilon_r\cos\theta} \tag{2.109}$$

which is the same as (2.51).

What has the IBC accomplished? We already knew the reflection coefficients. Furthermore, we needed to use the exact solution from Section 2.4 to find Z_s. Viewed this way, the IBC offers no advantages.

One use for the IBC is in the development of new, approximate solutions for possibly complex problems. For instance, suppose we had a curved metal body with several layers of material coatings. From a knowledge of the field behaviour at the surface, one might be able to obtain an *approximate* value of Z_s. Then, the problem could be solved for the scattered fields, and the more difficult part involving solving for the interior fields in the layers and on the metal could be avoided.

If the true surface fields obey (2.103) and Z_s is known, the IBC model will be exact. In practice, it still works well if Z_s is only approximately constant. This occurs for a good conductor with tangential fields that vary slowly along the surface, over distances on the order of λ in the conductor or the skin depth (Wait 1993).

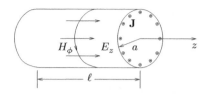

Figure 2.8   Illustration of the skin effect in a solid round wire of length ℓ and radius a.

A very good use for an IBC is to find the AC impedance of a wire. Figure 2.8 shows a solid round wire of length ℓ and radius a. Because of the skin effect, the volume current $\mathbf{J}$ inside the wire is tightly confined to the immediate inside of the outer surface and rapidly decays exponentially towards the inside of the wire. From the IBC in (2.103),

$$Z_s \hat{\mathbf{n}} \times \hat{\mathbf{n}} \times \mathbf{H} = \hat{\mathbf{n}} \times \mathbf{E}$$

$$Z_s \hat{\rho} \times \hat{\rho} \times \hat{\phi} H_\phi = \hat{\rho} \times \hat{z} E_z$$

$$-\hat{\phi} Z_s H_\phi = -\hat{\phi} E_z$$

so that $Z_s H_\phi = E_z$.

At the surface, E_z and H_ϕ are continuous and must be related by the conductor's intrinsic impedance η_c, so that $Z_s = \eta_c \approx \sqrt{j\omega\mu/\sigma}$. The total wire current in Amps is the surface current density J_{sz} times the surface width $2\pi a$, and $J_{sz} = H_\phi$ from the boundary condition at a current sheet. The voltage between the ends of the wire is $E_z \ell$ so that

$$Z = \frac{V}{I} = \frac{E_z \ell}{J_{sz} 2\pi a} = \frac{E_z \ell}{H_\phi 2\pi a} = Z_s \frac{\ell}{2\pi a}. \qquad (2.110)$$

If the wire has some other cross-section shape and if J_{sz} is uniform around the circumference w, we can replace $2\pi a$ by w.

Since this chapter is about plane waves and the wire field is cylindrical, the previous discussion may seem implausible. However, it turns out that a rigorous derivation using the wave equation in cylindrical coordinates (Silvester 1968) reaches the same conclusion. With this in hindsight, it is safe to say that the simple approach presented here captures all of the essential physics and provides the correct solution.

Example 2.3 (AC Wire Impedance) An AWG 35 copper wire has a length of 1 m. Find the AC impedance at 100 MHz and the DC resistance. Determine if the surface impedance approximation is valid.

Solution: AWG 35 corresponds to a wire diameter of 0.143 mm. The conductivity of copper is 5.8×10^7 S/m. Copper is non-magnetic, so $\mu = \mu_0$. With these numbers,

$$Z = Z_s \frac{\ell}{2\pi a} = \sqrt{\frac{j\omega\mu}{\sigma}} \frac{\ell}{2\pi a} = 5.81 + j5.81 \ \Omega.$$

The DC resistance is

$$R = \frac{\ell}{\sigma \pi a^2} = 1.07 \ \Omega.$$

Therefore, the DC resistance would be highly inaccurate at 100 MHz. The skin depth is

$$\delta = \sqrt{\frac{2}{\omega\mu\sigma}} = 6.61 \times 10^{-6} \text{ m.}$$

Since $\delta \ll a$, the assumption that most of the current is near the surface is a good one.   ■

2.7   Transmission Lines

TEM waves on z-directed transmission lines have $E_z = 0$ and $H_z = 0$. The fields can be obtained from a related electrostatic problem that is easier to solve. The method will now be described.

The TEM fields have only transverse components, so let us call them $\mathbf{E}_t, \mathbf{H}_t$. To simplify the discussion, rectangular coordinates can be assumed, so that

$$\mathbf{E}(x,y,z) = \hat{\mathbf{x}} E_x(x,y,z) + \hat{\mathbf{y}} E_y(x,y,z) = \mathbf{E}_t \qquad (2.111)$$

$$\mathbf{H}(x,y,z) = \hat{\mathbf{x}} H_x(x,y,z) + \hat{\mathbf{y}} H_y(x,y,z) = \mathbf{H}_t. \qquad (2.112)$$

Without loss of generality, we can also assume that we have a non-conducting medium with $\sigma = 0$. The fields have to satisfy Maxwell's equations

$$\nabla \times \mathbf{E}_t = -j\omega\mu\mathbf{H}_t \qquad (2.113)$$

$$\nabla \times \mathbf{H}_t = j\omega\epsilon\mathbf{E}_t. \qquad (2.114)$$

The ∇ operator can be written as a transverse part ∇_t and a z part

$$\nabla = \hat{\mathbf{x}}\frac{\partial}{\partial x} + \hat{\mathbf{y}}\frac{\partial}{\partial y} + \hat{\mathbf{z}}\frac{\partial}{\partial z} = \nabla_t + \hat{\mathbf{z}}\frac{\partial}{\partial z}.$$

Then, (2.113) and (2.114) become

$$\nabla \times \mathbf{E}_t = \nabla_t \times \mathbf{E}_t + \frac{\partial}{\partial z}\hat{\mathbf{z}} \times \mathbf{E}_t = -j\omega\mu\mathbf{H}_t \qquad (2.115)$$

$$\nabla \times \mathbf{H}_t = \nabla_t \times \mathbf{H}_t + \frac{\partial}{\partial z}\hat{\mathbf{z}} \times \mathbf{H}_t = j\omega\epsilon\mathbf{E}_t. \qquad (2.116)$$

Let us think about (2.115). The term $\nabla_t \times \mathbf{E}_t$ has only a $\hat{\mathbf{z}}$ component. However, on the right-hand side, $\mathbf{H}_t$ has no $\hat{\mathbf{z}}$ component. The vector on the left side of the equation must equal the vector on the right side of the equation, so we must have

$$\nabla_t \times \mathbf{E}_t = 0. \qquad (2.117)$$

The other part $(\partial/\partial z)\hat{\mathbf{z}} \times \mathbf{E}_t$ in (5) has only transverse components, and $\mathbf{H}_t$ can be found from $\mathbf{E}_t$ by using

$$\frac{\partial}{\partial z}\hat{\mathbf{z}} \times \mathbf{E}_t = -j\omega\mu\mathbf{H}_t. \qquad (2.118)$$

Similarly, we can show that the magnetic field obeys

$$\nabla_t \times \mathbf{H}_t = 0 \qquad (2.119)$$

and

$$\frac{\partial}{\partial z}\hat{\mathbf{z}} \times \mathbf{H}_t = j\omega\epsilon\mathbf{E}_t. \tag{2.120}$$

Operating with $\partial/\partial z$ and $\hat{\mathbf{z}}\times$ on (2.118), and using it in (2.120), along with the vector identity $\mathbf{A} \times (\mathbf{B} \times \mathbf{C}) = (\mathbf{A} \cdot \mathbf{C})\mathbf{B} - (\mathbf{A} \cdot \mathbf{B})\mathbf{C}$, we obtain

$$\frac{\partial^2 \mathbf{E}_t}{\partial z^2} + \beta^2 \mathbf{E}_t = 0 \tag{2.121}$$

where $\beta = \omega\sqrt{\mu\epsilon}$. This is the wave equation. The solution is

$$\mathbf{E}_t(x, y, z) = \mathbf{E}_s(x, y)e^{\pm j\beta z} \tag{2.122}$$

where $\mathbf{E}_s$ needs to be found. Notice that $\mathbf{E}_t$ and $\mathbf{E}_s$ only differ by the factor $e^{\pm j\beta z}$, so they are both transverse fields. In the following discussion, an $e^{-j\beta z}$ wave is assumed but the remarks apply just as well for an $e^{j\beta z}$ wave.

To find $\mathbf{E}_s(x, y)$, we proceed as follows. From (2.117) and (2.122),

$$\nabla_t \times \mathbf{E}_t(x, y, z) = \nabla_t \times (\mathbf{E}_s(x, y)e^{-j\beta z})$$

$$= e^{-j\beta z}\nabla_t \times \mathbf{E}_s(x, y) = e^{-j\beta z}\nabla \times \mathbf{E}_s(x, y) = 0$$

from which

$$\nabla \times \mathbf{E}_s(x, y) = 0. \tag{2.123}$$

Since there are no charges, $\nabla \cdot \mathbf{E} = \rho_v/\epsilon = 0$ so that

$$\nabla \cdot \mathbf{E} = \nabla \cdot \mathbf{E}_t = \nabla \cdot (\mathbf{E}_s(x, y)e^{-j\beta z}) = e^{-j\beta z}\nabla \cdot \mathbf{E}_s(x, y) = 0$$

which implies that

$$\nabla \cdot \mathbf{E}_s(x, y) = 0. \tag{2.124}$$

The results (2.123), (2.124) are exactly the conditions that a z-independent electrostatic field must obey. This leads to an important conclusion:

> The TEM field can always be found by taking a z-independent electrostatic solution $\mathbf{E}_s(x, y)$ of the transmission-line structure, then multiplying that solution by $e^{-j\beta z}$ as in (2.122).

The voltages and currents on a TEM transmission line are related to the electric and magnetic fields. The voltage at a particular position z is found from a line integral of the electric field between the two conductors, at a $z = $ const. plane so that

$$V_{ab}(z) = \int_a^b \mathbf{E}_t \cdot d\boldsymbol{\ell} = e^{-j\beta z}\int_a^b \mathbf{E}_s \cdot d\boldsymbol{\ell}.$$

Since $\mathbf{E}_s$ is an electrostatic field, the line integral does not depend on the shape of the path from a to b and the voltage is unique.

Ampere's law (1.35) allows us to find the current in a conductor from a line integral along a contour C that surrounds the conductor

$$\oint_C \mathbf{H} \cdot d\boldsymbol{\ell} = \int_S \mathbf{J} \cdot \hat{\mathbf{z}}\, dx\, dy + j\omega\epsilon \int_S \mathbf{E} \cdot \hat{\mathbf{z}}\, dx\, dy.$$

On the right-hand side, the integral involving $\mathbf{J}$ equals the z-directed current I. The integral involving $\mathbf{E}$ is zero because the TEM wave has $E_z = 0$. Therefore,

$$I = \oint_C \mathbf{H} \cdot d\boldsymbol{\ell}.$$

The right-hand thumb follows the current direction I and the fingers follow the contour C. An alternative way to find the current is to integrate the surface current density on the conductor.

In the following two examples, $V(z)$ and $I(z)$ are obtained for a parallel-plate line and a coaxial line.

Example 2.4 (Parallel-Plate Line) Parallel plates are at $x = 0$ and $x = d$. Each plate extends along $0 \leq y \leq w$. The wave propagation is along $+z$. Find the fields, voltage and current on this transmission line.

Solution: The electrostatic problem is a parallel-plate capacitor. We will use the approximate solution that neglects fringing effects, $\mathbf{E}_s = \hat{\mathbf{x}}E_0$ from which

$$\mathbf{E}_t = \hat{\mathbf{x}}E_0 e^{-j\beta z}.$$

If we define V_{ab} as $V(x = 0) - V(x = d)$, then

$$V_{ab} = \int_{x=0}^{d} E_x\, dx = \int_{x=0}^{d} E_0 e^{-j\beta z}\, dx = E_0 d\, e^{-j\beta z}.$$

Taking $\nabla \times \mathbf{E}$ and remembering that $\beta = \omega\sqrt{\mu\epsilon}$,

$$\mathbf{H}_t = \hat{\mathbf{y}}\,(E_0\beta/\omega\mu)\, e^{-j\beta z} = \hat{\mathbf{y}}\,(E_0/\eta)\, e^{-j\beta z}.$$

With C enclosing the bottom plate and noting that $\mathbf{H} = 0$ for $x < 0$, the $+z$-directed current is

$$I = \int_0^w H_y\, dx = (E_0 w/\eta) e^{-j\beta z},$$

and on the top plate, it is $-I$. ∎

Example 2.5 (Coaxial Line) Metal cylinders are at $\rho = a$ and $\rho = b$. Propagation is along $+z$. Find the fields, voltage and current on this transmission line.

Solution: The electrostatic problem has the solution

$$\mathbf{E}_s = \hat{\boldsymbol{\rho}}\frac{E_0}{\rho}$$

from which

$$\mathbf{E}_t = \hat{\rho}\frac{E_0}{\rho}e^{-j\beta z}.$$

If we define V_{ab} as $V(\rho = a) - V(\rho = b)$, then

$$V_{ab} = \int_{\rho=a}^{b} E_\rho \, d\rho = \int_{\rho=a}^{b} \frac{E_0}{\rho}e^{-j\beta z} \, d\rho = E_0 \ln{(b/a)}\, e^{-j\beta z}.$$

Taking $\nabla \times \mathbf{E}$ and remembering that $\beta = \omega\sqrt{\mu\epsilon}$,

$$\mathbf{H}_t = \hat{\phi}\frac{E_0}{\eta\rho}e^{-j\beta z}.$$

With C enclosing the inner cylinder, the $+z$-directed current is

$$I = \int_0^{2\pi} H_\phi \, a \, d\phi = \frac{2\pi E_0}{\eta}e^{-j\beta z},$$

and on the inside surface of the outer cylinder, the current is $-I$.   ■

2.7.1   Characteristic Impedance

All transmission lines have a *characteristic impedance* Z_0 in ohms that is defined as the ratio of the voltage and current of the travelling wave. The parallel-plate line of Example 2.4 has

$$Z_0 = \frac{V_{ab}}{I} = \frac{E_0 d\, e^{-j\beta z}}{(E_0 w/\eta)\, e^{-j\beta z}} = \frac{\eta d}{w}. \tag{2.125}$$

For a wave travelling in the $-z$ direction, we can find the relation between the voltage and current by replacing $\beta \rightarrow -\beta$. With $\pm\beta/\omega\mu = \pm 1/\eta$ we obtain

$$\frac{V_{ab}}{I} = \frac{E_0 d\, e^{j\beta z}}{(-E_0 w/\eta)\, e^{j\beta z}} = -\frac{\eta d}{w} = -Z_0. \tag{2.126}$$

Therefore, $V/I = \pm Z_0$ for $\pm z$-travelling waves.

Similarly, for the coaxial line in Example 2.5, the characteristic impedance is

$$Z_0 = \frac{V_{ab}}{I} = \frac{\eta}{2\pi}\ln{(b/a)}. \tag{2.127}$$

2.7.2   LC Ladder

A TEM transmission line can be modelled as an LC-ladder network, shown in Figure 2.9. Working in terms of L and C is a popular approach because it leads to general equations that can be used for any TEM transmission line, without regard to a specific geometry or fields. The quantities L, C are per unit length so that L is in H/m and C is in F/m. In a segment of length Δz, the total inductance and capacitance become $L\Delta z$ and $C\Delta z$, respectively.

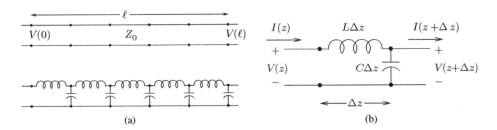

Figure 2.9 (a) Transmission line and LC-ladder equivalent. (b) One section of the ladder.

The ladder can be thought of as a sampled version of a transmission line. If each LC section is small in terms of wavelength, then the model will be reasonable. If we let the section size Δz shrink to zero, the mathematical model then becomes exact.

The governing network equations will now be derived. Referring to Figure 2.9(b), Kirchoff's voltage law applied to the inductor gives

$$V(z + \Delta z) - V(z) = -j\omega L \Delta z I.$$

Dividing by Δz and letting $\Delta z \to 0$ gives

$$\frac{dV}{dz} = -j\omega LI. \tag{2.128}$$

Similarly, Kirchoff's current law applied to the capacitor gives

$$I(z + \Delta z) - I(z) = -j\omega C \Delta z V.$$

Dividing by Δz and letting $\Delta z \to 0$ gives

$$\frac{dI}{dz} = -j\omega CV. \tag{2.129}$$

Differentiating (2.128) with respect to z and using (2.129) to eliminate I gives

$$\frac{d^2V}{dz^2} = \underbrace{(j\omega L)(j\omega C)}_{-\beta^2} V. \tag{2.130}$$

The solution is

$$V(z) = Ae^{-j\beta z} + Be^{j\beta z}. \tag{2.131}$$

From (2.128), the current is

$$I(z) = \frac{1}{Z_0}(Ae^{-j\beta z} - Be^{j\beta z}). \tag{2.132}$$

A and B are arbitrary constants giving the amplitudes of the incident and reflected waves. Since β^2 is defined in (2.130), the phase constant is

$$\beta = \omega\sqrt{LC} \tag{2.133}$$

and the characteristic impedance for a right-travelling wave can be found from (2.128), (2.131) and (2.132) with $B = 0$ to obtain V/I, with the result that

$$Z_0 = \sqrt{\frac{L}{C}}. \tag{2.134}$$

The wave velocity is $u = \omega/\beta = 1/\sqrt{LC}$. On a transmission line, the velocity is governed by the ambient medium and is $u = 1/\sqrt{\mu\epsilon}$. Equating the two velocities leads to an important connection between the LC ladder and a transmission line

$$LC = \mu\epsilon = \frac{1}{u^2}. \tag{2.135}$$

2.7.2.1   LC Ladder with Finite Δz

It is reasonable to wonder if an LC ladder with a finite but small Δz could be used to model a transmission line. A good way to get at least a partial answer is to try it out with a circuit simulator, using N sections. If the segments $\Delta z = \ell/N$ are small enough, then at the nodes of the ladder, we should be able to recover a sampled version of the true $V(z)$ and $I(z)$ that exist on a transmission line.

This will be illustrated using an RG58 coaxial line. It has a characteristic impedance of 50 Ω, a velocity factor $u/c = 0.667$, an inductance per unit length of $L = 250$ nH/m and a capacitance per unit length of $C = 100$ pF/m. The relative permittivity of the dielectric is related to the velocity factor and is $\epsilon_r = (c/u)^2 = 2.25$.

Let the transmission-line length be $\ell = 1$ m. An LC-ladder equivalent with $N = 5$ sections will have five 50 nH inductors and five 20 pF capacitors. We can assume a 50 Ω source and a 50 Ω load and calculate $V(\ell)$ at the load, $V(0)$ at the input and the voltage transfer function $V_t = V(\ell)/V(0)$. The ladder circuit results for $N = 1$ and 5 and the transmission-line results ($N = \infty$) are shown in Figure 2.10. It can be seen that the transmission line and the $N = 5$ ladder circuit have similar behaviour up to about 100 MHz. At this frequency, the wavelength in the dielectric is $\lambda_d = 2$ m and $\Delta z = 0.2$ m $= \lambda_d/10$. Therefore, an LC ladder can be a reasonable model for a transmission line when $\Delta z \leq \lambda_d/10$. For the transmission line, the constant amplitude and linear phase of V_t indicate that there is a time delay, but otherwise no distortion; this is further discussed in Section 2.10.

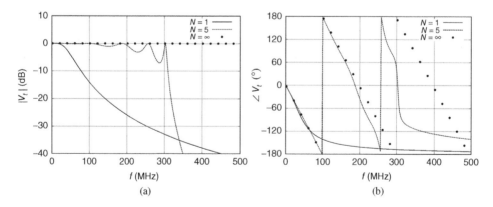

Figure 2.10   Voltage transfer function $V_t = V(\ell)/V(0)$ for an LC ladder with N sections, and a transmission line ($N = \infty$). (a) Magnitude, (b) phase.

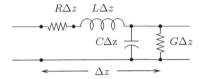

Figure 2.11   LC ladder with losses.

From Figure 2.10, we can see that the ladder is a low-pass filter having a frequency response that resembles a transmission line, if $f < 300$ MHz. Since there are no waves in a lumped-element circuit, the ladder's phase response can be associated with a time delay, but *not* a velocity.

2.7.3   *Small Losses*

Let us now consider a transmission line having some losses. Conductor loss involving the skin effect adds series resistance R, and dielectric loss creates a shunt conductance G. It is noted that the presence of some series resistance will cause a small E_z, which drives the conduction current. So, strictly speaking, the wave is no longer TEM. However, in practice, E_z is much smaller than the transverse electric field, so it is a very good approximation to retain the TEM assumption.

The modified ladder including losses is shown in Figure 2.11. It is not necessary to redo the circuit analysis. Comparing it with the lossless case in Figure 2.9(b) shows that simple replacements can be used

$$j\omega L \rightarrow R + j\omega L$$

$$j\omega C \rightarrow G + j\omega C.$$

Applying these substitutions leads to the voltage and current

$$V(z) = Ae^{-\gamma z} + Be^{\gamma z} \tag{2.136}$$

$$I(z) = \frac{1}{Z_0}\left(Ae^{-\gamma z} - Be^{\gamma z}\right) \tag{2.137}$$

where the complex propagation constant is

$$\gamma = \alpha + j\beta = \sqrt{(R + j\omega L)(G + j\omega C)} \tag{2.138}$$

and the characteristic impedance is

$$Z_0 = \sqrt{\frac{R + j\omega L}{G + j\omega C}}. \tag{2.139}$$

In the case of small losses, $R \ll \omega L$ and $G \ll \omega C$. Using the small-argument approximations

$$\sqrt{1 + x} \approx 1 + \frac{1}{2}x + \cdots$$

$$\frac{1}{\sqrt{1 + y}} \approx 1 - \frac{1}{2}y + \cdots$$

with (2.138) and (2.139) gives

$$\alpha \approx \frac{\sqrt{LC}}{2}\left(\frac{R}{L} + \frac{G}{C}\right); \quad \beta \approx \omega\sqrt{LC} \tag{2.140}$$

and

$$Z_0 \approx \sqrt{\frac{L}{C}}\left(1 + \frac{1}{j2\omega}\left(\frac{R}{L} - \frac{G}{C}\right)\right). \tag{2.141}$$

The TEM wave and the LC ladder should have the same wave velocity. Equating β in (2.15) with (2.140) implies that

$$LC = \mu\epsilon' = 1/u^2 \tag{2.142}$$

where $\epsilon' = \epsilon_0\epsilon_r$ is the real part of a complex permittivity.

As an application, let us find the losses for a coaxial line with a centre conductor of radius a and a shield with radius b. We can use (2.110) to find the conductor loss associated with the skin effect. Since R is a per-unit-length quantity, we let $\ell = 1$ m. Recognizing that there is loss for both the inner and outer conductors, we add them up, obtaining

$$Z = \frac{Z_s}{2\pi a} + \frac{Z_s}{2\pi b}.$$

The surface impedance for a conductor characterized by μ_c, σ_c is $Z_s = \sqrt{j\omega\mu_c/\sigma_c}$ so that Z due to conductor loss becomes

$$Z = R + jX = (1 + j)\sqrt{\frac{\omega\mu_c}{2\sigma_c}}\left(\frac{1}{2\pi a} + \frac{1}{2\pi b}\right).$$

The term jX is the self-inductance, associated with the current inside the conductor. It is normally neglected because it is much smaller than the series inductance $j\omega L$. Retaining only R leaves us with the series resistance

$$R = R_s\left(\frac{1}{2\pi a} + \frac{1}{2\pi b}\right)$$

where $R_s = \sqrt{\omega\mu_c/2\sigma_c}$.

$R = X$ is a characteristic of good conductors, and $X \ll \omega L$ implies that $R \ll \omega L$. In a practical transmission line, usually $G/\omega C \ll R/\omega L \ll 1$ so that $Z_0 \approx \sqrt{L/C}$.[5]

The next step is to find the shunt conductance G associated with dielectric loss. The electric field is

$$\mathbf{E} = \hat{\rho}\frac{E_0}{\rho}e^{-j\beta z},$$

and the current due to dielectric conductivity σ_d is $\mathbf{J} = \sigma_d\mathbf{E}$. The conductance per unit length is

$$G = \frac{I_t/\ell}{V} = \frac{\oint_0^{2\pi}\sigma_d\mathbf{E}\cdot\hat{\rho}\,\rho d\phi}{\int_a^b\mathbf{E}\cdot\hat{\rho}\,d\rho} = \frac{2\pi\sigma_d}{\ln(b/a)}.$$

---

[5] The higher order term in (2.141) contains useful information, and it is not always ignored. If $R/L = G/C$, then Z_0 will be frequency independent and the transmission line will be distortionless. This is known as the *Heaviside condition*. In the past, telephone companies accomplished this by placing loading coils on the poles every 6000 feet. Equalization to compensate for distortion is now done electronically with repeaters.

Here, I_t denotes the transverse conduction current which flows radially from one conductor to the other, through the lossy dielectric. The integrals for finding Z_0 and G are almost the same, and it turns out (see Problem 2.15) that

$$G = \eta \sigma_d / Z_0. \tag{2.143}$$

Example 2.6 (Coaxial Line Losses) Evaluate the losses for a 100 ft length (30.5 m) of RG58 line at 100, 200 and 400 MHz. The line has a characteristic impedance of 50 Ω. The polyethylene dielectric has $\epsilon_r = 2.25$, $\tan \delta = 2 \times 10^{-4}$ and a diameter of $2b = 0.116''$. The conductors are copper.

Solution: From Z_0 and $2b$, we can find out that $2a = 33.2$ mils. Copper has a conductivity of $\sigma = 5.8 \times 10^7$ S/m, and it is non-magnetic, so $\mu_c = \mu_0$. The dielectric's conductivity is $\sigma_d = \omega \epsilon_0 \epsilon_r \tan \delta$. By using (2.140), we can examine the copper and dielectric losses separately:

$$\alpha_c = \frac{\sqrt{LC}}{2} \frac{R}{L} = \frac{R}{2Z_0} = \frac{1}{2Z_0} \sqrt{\frac{\omega \mu_c}{2\sigma_c}} \left(\frac{1}{2\pi a} + \frac{1}{2\pi b} \right)$$

$$\alpha_d = \frac{\sqrt{LC}}{2} \frac{G}{C} = \frac{GZ_0}{2} = \frac{Z_0}{2} \frac{2\pi \sigma_d}{\ln(b/a)}.$$

The dB loss in a length ℓ is $L = -20 \log e^{-\alpha \ell} = 8.69 \, \alpha \ell$. The conduction and dielectric losses L_c, L_d for $\ell = 30.5$ m are tabulated as follows.

f (MHz)	L_c (dB)	L_d (dB)
100	3.4	0.1
200	4.7	0.2
400	6.7	0.3

Both types of losses increase with frequency, but conduction loss is the dominant effect. ∎

2.7.4 Transmission Line Parameters

Using the methods described in the previous section, Z_0 and R for some common transmission lines were obtained and are summarized in Table 2.1. The parallel-plate line neglects fringing at the edges and is an approximation. For the coaxial and two-wire lines, the results are exact.[6]

In Table 2.1, it is assumed that $\epsilon = \epsilon' - j\epsilon''$ is complex but that Z_0 is real; here, $\epsilon_r = \epsilon'/\epsilon_0$. Since $Z_0 = \sqrt{L/C}$ and $LC = \mu \epsilon'$, we can find L and C from Z_0 and ϵ' if needed. In fact, a good way to find Z_0 is from

$$Z_0 = \sqrt{\frac{L}{C}} = \sqrt{\frac{\mu \epsilon'/C}{C}} = \frac{1}{uC}.$$

Therefore, if we can solve the electrostatic problem for C, we will have L and Z_0, as well as G from (2.143). Finding R, however, does not come automatically and is a separate problem.

---

[6] An infinitely thin line charge above a ground plane has equipotentials that are circles. This boundary condition 'fits' a conducting circular cylinder and leads to closed-form capacitance formulas for a wire above a ground plane and a two-wire line. See, for example, Demarest (1997).

Table 2.1   Transmission-line parameters.

Geometry	Characteristic impedance	Series resistance
(parallel plates, width w, separation d)	$Z_0 \approx \eta \dfrac{d}{w}$	$R \approx \dfrac{2R_s}{w}$
(coaxial, inner radius a, outer radius b)	$Z_0 = \dfrac{\eta}{2\pi} \ln \dfrac{b}{a}$	$R = \dfrac{R_s}{2\pi}\left(\dfrac{1}{a} + \dfrac{1}{b}\right)$
(two-wire, separation D, diameter d)	$Z_0 = \dfrac{\eta}{\pi} \cosh^{-1} \dfrac{D}{d}$	$R = \dfrac{2R_s}{\pi d}\dfrac{D/d}{\sqrt{(D/d)^2 - 1}}$

Here, $\eta = \sqrt{\mu/\epsilon'}$ and $R_s = \sqrt{\omega \mu_c / 2\sigma_c}$.

A good way to find R, especially for more complicated geometries where the surface current J_{sz} may be non-uniform, is with the *Wheeler incremental inductance rule* (Pozar 2012, pp. 83–85). It is based on an observation that R and X are the same for a good conductor. The derivation leads to a simple result which states that

$$R = -\frac{R_s}{\eta}\frac{\partial Z_0(n)}{\partial n} \tag{2.144}$$

where n is a coordinate that is *outward* from the conductor surface. The application is best understood from an example.

Example 2.7 (Incremental Inductance Rule)   Use Wheeler's incremental inductance rule to find the AC resistance per unit length R for the parallel-plate transmission line shown in Table 2.1.

Solution: To evaluate the normal derivatives of Z_0, we can think of the plates as having a width w along x, with one plate at $y = a$ and the other at $y = b$. Then, $Z_0 = \eta d/w = \eta(b-a)/w$. For the bottom plate, the outward derivative is in the $+y$ direction and $\partial Z_0/\partial n = \partial Z_0/\partial a = -\eta/w$. The loss resistance is

$$R = \frac{-R_s}{\eta}\frac{\partial Z_0}{\partial n} = \frac{R_s}{w}.$$

For the top plate, the outward derivative is in the $-y$ direction and $\partial Z_0/\partial n = -\partial Z_0/\partial b = -\eta/w$. The loss resistance is

$$R = \frac{-R_s}{\eta}\frac{\partial Z_0}{\partial n} = \frac{R_s}{w}.$$

The total resistance is then

$$R = 2R_s/w. \qquad\blacksquare$$

2.7.5   Microstrip, Stripline and Coplanar Lines

Microstrip, stripline and coplanar transmission lines are widely used in microwave circuits and antennas. The electromagnetic analysis of these lines cannot be done in just a few pages and is not covered here.

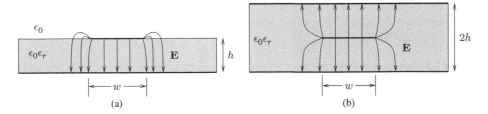

Figure 2.12   (a) Microstrip and (b) stripline transmission lines.

Rather, the following sections serve to summarize some final results and practical formulas (Collin 2001) for the wave velocity and characteristic impedance of these lines.

2.7.5.1   Microstrip

Figure 2.12(a) shows a microstrip line. The microstrip has a width w and is on top of a dielectric substrate with a relative permittivity ϵ_r and a ground plane. The wave is partially in the air and partially in the dielectric, so the wave velocity is necessarily different in the two regions. Strictly speaking, a TEM wave cannot meet the boundary conditions at the air–dielectric interface; the rigorous solution requires a TE/TM combination – a hybrid mode. However, *most* of the wave is within the dielectric, so it is possible to obtain a good approximate TEM solution.

Because the fields are partly in the air and partly in the dielectric, there will be a quasi-TEM wave with a velocity somewhere in the range

$$\frac{1}{\sqrt{\mu_0 \epsilon_0 \epsilon_r}} \leq u \leq \frac{1}{\sqrt{\mu_0 \epsilon_0}}.$$

This can be accounted for by using an *effective permittivity* ϵ_e such that $1 \leq \epsilon_e \leq \epsilon_r$.

If C is the scapacitance of the microstrip line, and C_0 is the capacitance of the same structure when there is no dielectric (i.e. $\epsilon_r = 1$), then the effective permittivity can be obtained from

$$\epsilon_e = \frac{C}{C_0}.$$

If solutions for C and C_0 can somehow be obtained from analytic or numerical techniques, then ϵ_e readily follows. In practice, solving the two problems can usually be avoided, as a convenient formula is available for ϵ_e

$$\epsilon_e = \frac{\epsilon_r + 1}{2} + \frac{\epsilon_r - 1}{2}\left(1 + 12\frac{h}{w}\right)^{-1/2} + f. \tag{2.145}$$

Here, $f = 0.02\,(\epsilon_r - 1)(1 - w/h)^2$ when $w/h \leq 1$, and $f = 0$ when $w/h > 1$.

For the quasi-TEM transmission line, the relationships $Z_0 = \sqrt{L/C}$ and $LC = \mu_0 \epsilon_0 \epsilon_e$ hold, so that Z_0 can be found from the capacitance of the microstrip. The final results are

$$Z_0 = \frac{\eta_0}{2\pi\sqrt{\epsilon_e}} \ln\left(\frac{8h}{w} + \frac{w}{4h}\right); \quad \frac{w}{h} \leq 1 \tag{2.146}$$

$$Z_0 = \frac{\eta_0}{\sqrt{\epsilon_e}}\left[\frac{w}{h} + 1.393 + 0.667\ln\left(\frac{w}{h} + 1.444\right)\right]^{-1}; \quad \frac{w}{h} > 1 \tag{2.147}$$

where $\eta_0 = \sqrt{\mu_0/\epsilon_0}$.

These approximate formulas are sufficiently accurate for most practical engineering purposes. The ϵ_e formula is accurate to 1% for $0.25 \leq w/h \leq 6$ and $\epsilon_r \leq 16$, and the Z_0 formulas are accurate to 1/4% for $0.05 \leq w/h \leq 10$.

Example 2.8 (Microstrip Calculation) A microstrip line has a substrate with a relative permittivity of 2.2, a substrate thickness of 31 mils, a strip width of 100 mils and a line length of $\ell = 1000$ mils. The frequency is 1 GHz. Find the effective permittivity, characteristic impedance and electrical length of the line.

Solution: Here, $w/h = 3.23$. From (2.145), we find that $\epsilon_e = 1.88$, and from (2.147), we obtain $Z_0 = 48.7 \,\Omega$. Note that the effective permittivity is in the range $1 < \epsilon_e < \epsilon_r$.

The line length is $\ell = 1000$ mils which is 1" or 2.54 cm. The electrical length is

$$\beta\ell = (\omega/u)\ell = (\omega\sqrt{\epsilon_e}/c)\ell = 0.729 \text{ rad} = 41.8^\circ.$$

∎

2.7.5.2 Stripline

Figure 2.12(b) shows a stripline. The centre conductor is a metal strip of width w that is surrounded by a dielectric having a relative permittivity ϵ_r. The line is symmetrically bounded by two ground planes that are spaced by a distance of $2h$. Since the field is entirely within the dielectric, the wave is purely TEM, having a velocity $u = 1/\sqrt{\mu_0\epsilon_0\epsilon_r} = 1/\sqrt{LC}$. This type of structure can be solved exactly in terms of elliptic integrals, using conformal mapping. The characteristic impedance is found to be

$$Z_0 = \frac{\eta_0}{4\pi\sqrt{\epsilon_r}} \ln\left(2\frac{1+\sqrt{k}}{1-\sqrt{k}}\right) ; \quad 0 < \frac{w}{h} \leq 1.14 \tag{2.148}$$

$$Z_0 = \frac{\pi\eta_0}{4\sqrt{\epsilon_r}} \left[\ln\left(2\frac{1+\sqrt{k'}}{1-\sqrt{k'}}\right)\right]^{-1} ; \quad 1.14 \leq \frac{w}{h} < \infty \tag{2.149}$$

where $k = \text{sech}(\pi w/4h)$ and $k' = \tanh(\pi w/4h)$; note that $k' = \sqrt{1-k^2}$. For narrow strips, (2.148) simplifies to

$$Z_0 = \frac{\eta_0}{2\pi\sqrt{\epsilon_r}} \ln\left(\frac{16h}{\pi w}\right) ; \quad \frac{w}{h} \leq 0.4 \tag{2.150}$$

and for wide strips, (2.149) simplifies to

$$Z_0 = \frac{\pi\eta_0}{8\sqrt{\epsilon_r}(\ln 2 + \pi w/4h)} ; \quad \frac{w}{h} \geq 2. \tag{2.151}$$

2.7.5.3 Coplanar Line

Figure 2.13 shows a coplanar line. This structure is also called a coplanar waveguide. The centre conductor is at a potential V_0, and the two outer conductors are ground planes. The conductors reside on a dielectric that is assumed to have a large thickness h. The entire structure can be enclosed in a rectangular shielding box, and if the box is large enough, its effect will be negligible. An advantage of the coplanar line is that the centre conductor and ground planes are on the same side of the dielectric, so it is good for attaching surface-mount components. In contrast, access to a microstrip ground is more difficult and requires drilling a via hole.

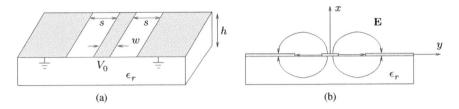

Figure 2.13 Coplanar line. (a) The gap widths are s, and the line width is w. (b) Electric field lines.

Since $\mathbf{E}_{tan}$ is continuous at $x = 0$ and $E_n = 0$ at $x = 0$, the boundary conditions at $x = 0^+$ and $x = 0^-$ will be the same. Consequently, the electrostatic field solutions in the air and dielectric regions will be the same. Therefore, the electric field is symmetric with respect to the $x = 0$ plane, as in Figure 2.13(b). The capacitance can be found via conformal mapping techniques and is in terms of elliptic integrals. The effective permittivity is

$$\epsilon_e = \frac{\epsilon_r + 1}{2} \tag{2.152}$$

and the characteristic impedance is

$$Z_0 = \frac{\eta_0}{4\pi\sqrt{\epsilon_e}} \ln\left(2\frac{1 + \sqrt{k'}}{1 - \sqrt{k'}}\right); \quad 0 < k \leq 0.7 \tag{2.153}$$

$$Z_0 = \frac{\pi\eta_0}{4\sqrt{\epsilon_e}}\left[\ln\left(2\frac{1 + \sqrt{k}}{1 - \sqrt{k}}\right)\right]^{-1}; \quad 0.7 \leq k \leq 1 \tag{2.154}$$

where $k = s/(s + 2w)$ and $k' = \sqrt{1 - k^2}$.

2.7.6 Reflection and Transmission on a Transmission Line

This section develops the expressions for voltages, currents and impedance on transmission lines. The lossless case will be assumed. If the lossy case is needed, it can be obtained by replacing β by $-j\gamma$ in all of the equations.

We will first consider a junction between two transmission lines as shown in Figure 2.14(a). The lines have characteristic impedances Z_{01} and Z_{02} and phase constants β_1 and β_2. There are incident plus reflected waves on the left and a transmitted wave on the right. The voltage reflection and transmission coefficients are defined by

$$\Gamma = \frac{V^r(z = 0^-)}{V^i(z = 0^-)} \qquad T = \frac{V^t(z = 0^+)}{V^i(z = 0^-)}. \tag{2.155}$$

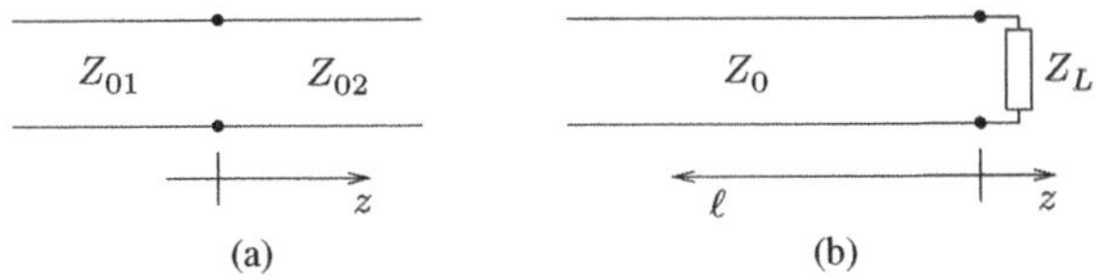

Figure 2.14 (a) Junction of two transmission lines. (b) Transmission line and a load Z_L.

$V/I = Z_0$ for the incident and transmitted waves, whereas $V/I = -Z_0$ for the reflected wave. Therefore, the voltages and currents on the two lines can be written as

$$z < 0 \qquad\qquad\qquad\qquad z > 0$$

$$V^i(z) + V^r(z) = V_0 e^{-j\beta_1 z} + \Gamma V_0 e^{j\beta_1 z} \qquad\qquad V^t(z) = T\, V_0 e^{-j\beta_2 z} \qquad\qquad (2.156)$$

$$I^i(z) + I^r(z) = \frac{V_0}{Z_{01}} e^{-j\beta_1 z} - \Gamma \frac{V_0}{Z_{01}} e^{j\beta_1 z} \qquad\qquad I^t(z) = T \frac{V_0}{Z_{02}} e^{-j\beta_2 z}. \qquad\qquad (2.157)$$

At $z = 0$, the voltage and current must be continuous. In other words, $V^i(0) + V^r(0) = V^t(0)$ and $I^i(0) + I^r(0) = I^t(0)$ which means that

$$1 + \Gamma = T \qquad\qquad (2.158)$$

and

$$\frac{1 - \Gamma}{Z_{01}} = \frac{T}{Z_{02}}. \qquad\qquad (2.159)$$

These can be solved for Γ and T with the result that

$$\Gamma = \frac{Z_{02} - Z_{01}}{Z_{02} + Z_{01}} \qquad\qquad (2.160)$$

and

$$T = \frac{2Z_{02}}{Z_{02} + Z_{01}}. \qquad\qquad (2.161)$$

Next, we will consider a transmission line with a load, as shown in Figure 2.14(b). The line has a characteristic impedance Z_0 and a phase constant β; the load is Z_L. Since z is always negative, it is convenient to define a new variable $\ell = -z$. The voltage reflection coefficient is defined by

$$\Gamma_0 = \frac{V^r(\ell = 0)}{V^i(\ell = 0)}. \qquad\qquad (2.162)$$

$V/I = Z_0$ for the incident wave, and $V/I = -Z_0$ for the reflected wave. Therefore, the voltage and current on the line can be written as

$$V(\ell) = V^i(\ell) + V^r(\ell) = V_0 e^{j\beta\ell} + \Gamma_0 V_0 e^{-j\beta\ell} \qquad\qquad (2.163)$$

$$I(\ell) = I^i(\ell) + I^r(\ell) = \frac{V_0}{Z_0} e^{j\beta\ell} - \Gamma_0 \frac{V_0}{Z_0} e^{-j\beta\ell}. \qquad\qquad (2.164)$$

At $\ell = 0$, it has to be true that $V(0)/I(0) = Z_L$ so that

$$\frac{V(0)}{I(0)} = \frac{1 + \Gamma_0}{\frac{1}{Z_0}(1 - \Gamma_0)} = Z_L$$

which can be solved for Γ_0

$$\Gamma_0 = \frac{Z_L - Z_0}{Z_L + Z_0}. \qquad\qquad (2.165)$$

For the transmission line and load in Figure 2.14(b), it is very useful to know the input impedance and reflection coefficient at any point on the line. Taking a more general viewpoint, the ratio of reflected and incident voltages at an arbitrary position ℓ can be used to define the reflection coefficient

$$\Gamma(\ell) = \frac{V^r(\ell)}{V^i(\ell)} = \frac{V_0 \Gamma_0 e^{-j\beta\ell}}{V_0 e^{j\beta\ell}}$$

so that

$$\Gamma(\ell) = \Gamma_0 e^{-j2\beta\ell}. \tag{2.166}$$

The input impedance is

$$Z(\ell) = \frac{V(\ell)}{I(\ell)} = \frac{V_0 e^{j\beta\ell} + \Gamma_0 V_0 e^{-j\beta\ell}}{\frac{V_0}{Z_0} e^{j\beta\ell} - \Gamma_0 \frac{V_0}{Z_0} e^{-j\beta\ell}} = Z_0 \frac{1 + \Gamma_0 e^{-j2\beta\ell}}{1 - \Gamma_0 e^{-j2\beta\ell}}$$

and using (2.166) it becomes

$$Z(\ell) = Z_0 \frac{1 + \Gamma(\ell)}{1 - \Gamma(\ell)}. \tag{2.167}$$

Solving for $\Gamma(\ell)$ provides the inverse relationship

$$\Gamma(\ell) = \frac{Z(\ell) - Z_0}{Z(\ell) + Z_0}. \tag{2.168}$$

In both (2.166) and (2.168), it is noted that $\Gamma(0) \equiv \Gamma_0 = (Z_L - Z_0)/(Z_L + Z_0)$.

Another way to express the input impedance of the line is to use (2.166) in (2.167), which becomes

$$Z(\ell) = Z_0 \frac{Z_L + jZ_0 \tan\beta\ell}{Z_0 + jZ_L \tan\beta\ell}. \tag{2.169}$$

Two useful special cases are a short-circuit and an open-circuit load. These 'stubs' with a length $0 \leq \beta\ell < \pi/2$ are usually used to realize an inductive or capacitive reactance

$$Z(\ell) = jZ_0 \tan\beta\ell; \quad \text{short-circuit stub} \tag{2.170}$$

$$Z(\ell) = -jZ_0 \cot\beta\ell; \quad \text{open-circuit stub}. \tag{2.171}$$

2.7.7   Physical Meaning of Z_0

A typical coaxial line such as RG58 has $Z_0 = 50\ \Omega$. A beginner might want to measure it with an ohmmeter and would not have much luck. It is, however, possible to devise an experiment whereby the 50 Ω could be measured. If we had a very long line and a 'time-domain ohmmeter', we would be able to see the 50 Ω impedance in the early time, prior to the arrival of reflected waves from the distant endpoint.

Let us consider a transmission line with a length ℓ and a short-circuit load, as in Figure 2.15. The generator sees an impedance $Z = jZ_0 \tan\beta\ell$ where $\beta = \omega/u$ and u is the wave velocity. From the voltage divider rule, the voltage at the line's input is

$$V(\omega) = V_g(\omega) \frac{jZ_0 \tan\omega\ell/u}{Z_0 + jZ_0 \tan\omega\ell/u}.$$

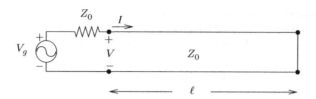

Figure 2.15   Transmission line with matched generator and short-circuit load.

Let us suppose that a unit step of voltage $v_g(t) = U(t)$ is the excitation. Using the Fourier transform pair

$$U(t) \leftrightarrow \pi\delta(\omega) + 1/j\omega$$

the input voltage becomes

$$V(\omega) = (\pi\delta(\omega) + 1/j\omega)\frac{jZ_0 \tan \omega\ell/u}{Z_0 + jZ_0 \tan \omega\ell/u} = \frac{1 - e^{-j2\omega\ell/u}}{j2\omega}.$$

Applying the integration property

$$\frac{F(\omega)}{j\omega} \leftrightarrow \int_{-\infty}^{t} f(\lambda)\, d\lambda$$

and shifting property

$$F(\omega)e^{-j\omega a} \leftrightarrow f(t - a)$$

gives

$$v(t) = \frac{1}{2}(U(t) - U(t - 2\ell/u)).$$

The current is $I = V/Z$, and a similar procedure gives

$$I = \frac{V}{Z} = \frac{1 + e^{-j2\omega\ell/u}}{j2\omega Z_0}.$$

Inversion yields

$$i(t) = \frac{1}{2Z_0}(U(t) + U(t - 2\ell/u)).$$

The functions $v(t)$, $i(t)$ and $v(t)/i(t)$ are shown in Figure 2.16. The results reveal that in the early time, the generator will think that it is connected to a 50 Ω resistor, and it is only after a delay time of $2\ell/u$

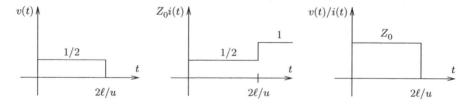

Figure 2.16   Voltage, current and v/i on the transmission line.

(the round-trip time) that it will know there is a short circuit. Therefore, one can measure $v(t)$ with a step generator and an oscilloscope, as long as the equipment has a rise time that is appreciably faster than $2\ell/u$. Other loads could be used; the key point is that the load should be far away so that its effect can be ignored.

2.8   Transverse Equivalent Network

The reflection and transmission coefficients for oblique-incident plane waves in Section 2.4 closely resemble the transmission-line results (2.160), (2.161). It is possible to recast the plane-wave results as a *transverse equivalent network* (TEN). Then, transmission-line design tools such as the Smith chart can be used for the electromagnetic design of structures involving oblique-incident plane waves.

For the case of perpendicular polarization, comparing (2.160) and (2.37) suggests a TEN having

$$Z_{01} = \eta_1/\cos\theta_i \tag{2.172}$$

$$Z_{02} = \eta_2/\cos\theta_t. \tag{2.173}$$

Similarly, for parallel polarization, comparing (2.160) and (2.49) suggests a TEN having

$$Z_{01} = \eta_1\cos\theta_i \tag{2.174}$$

$$Z_{02} = \eta_2\cos\theta_t. \tag{2.175}$$

The incident and transmission angles θ_i, θ_t are measured with respect to the z axis as in Figure 2.3.

The TEN was used here to replace the planar interface problem in Figure 2.3 by two transmission lines. The following example shows that the procedure is applicable to other cases as well.

Example 2.9 (TEN for a Grounded Slab) Find the reflection coefficient $\Gamma_\perp$ for a grounded dielectric slab of thickness d, as in Figure 2.17. The plane wave has an incidence angle θ_i and is perpendicularly polarized.

Solution: The transmission lines have impedances $Z_{01} = \eta_1/\cos\theta_i$ and $Z_{02} = \eta_2/\cos\theta_t$. The z-part of the propagation constant inside the dielectric is $\beta = k_2\cos\theta_t$. From transmission-line theory, the

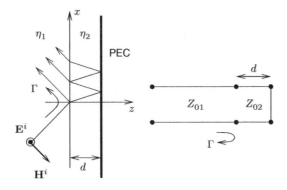

Figure 2.17   Grounded dielectric slab and TEN equivalent.

short-circuited transmission line of length d has an input impedance at $z = 0$ that is $Z_{in} = jZ_{02}\tan\beta d$. Therefore,

$$\Gamma_\perp = \frac{Z_{in} - Z_{01}}{Z_{in} + Z_{01}} = \frac{j(\eta_2/\cos\theta_t)\tan\beta d - (\eta_1/\cos\theta_i)}{j(\eta_2/\cos\theta_t)\tan\beta d + (\eta_1/\cos\theta_i)}.$$

This can also be written as

$$\Gamma_\perp = \frac{j(\mu_2/\mu_1)k_1 d\cos\theta_i - \beta d\cot\beta d}{j(\mu_2/\mu_1)k_1 d\cos\theta_i + \beta d\cot\beta d}.$$

■

The ideas in the previous example can be adapted to a dielectric slab without a ground plane. In this case, the transmitted ray exits the slab at an angle θ_i, so the transmission line has a load impedance $Z_L = \eta_1/\cos\theta_i$. From (2.169), the input impedance at $z = 0$ is

$$Z_{in} = Z_{02}\frac{Z_L + jZ_{02}\tan\beta d}{Z_{02} + jZ_L\tan\beta d} \tag{2.176}$$

from which Γ can be found. The details are left for Problem 2.21.

2.9 Absorbers

Absorbers are used for reducing reflection from conductors. Thinking in terms of transmission lines is a good way to explain how some absorbers work. Normal incidence is assumed here, but oblique incidence can also be understood by using a more general TEN equivalent.

Figure 2.18 shows two absorber types on a perfect electric conductor. The first one, the *Salisbury screen*, is an impedance sheet having $R_s = \eta_0 = 377\,\Omega$. It is spaced by $\lambda/4$ from a metal plate. The metal plate is equivalent to a short-circuit load. A $\lambda/4$ line turns the plate into an open circuit. The open circuit is in parallel with 377 Ω, so the final input impedance is 377 Ω and therefore $\Gamma = 0$. Impedance sheets that are 377 Ω are commercially available and are known as 'space cloth'. The $\lambda/4$ spacer can

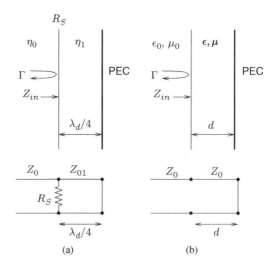

Figure 2.18 Two types of absorbers. (a) Salisbury screen, and (b) Dallenbach layer.

be any low-loss dielectric material, but the thickness should be $\lambda_d/4$ where the dielectric wavelength is $\lambda_d = \lambda_0/\sqrt{\epsilon_r}$. The Salisbury screen is necessarily narrowband, because it can only be $\lambda_d/4$ thick at one frequency. A variation of this design, the *Jaumann absorber*, uses a multi-layer stack of resistive sheets and spacers to improve the bandwidth.

It is possible to use sheets of dipole arrays, slots or their variants, instead of space cloth. These structures generally provide a complex shunt impedance $Z_s = R_s + jX_s$. Since the sheets are usually thought of as shunt circuit elements on a transmission-line equivalent, they are called *circuit-analogue absorbers*. They provide additional parameters, which increase the design flexibility. This can lead to improved absorber performance over wider frequency ranges and incident angles.

Figure 2.18(b) shows another approach, the *Dallenbach layer* (also called a $\mu = \epsilon$ absorber). It uses a material that has both electric and magnetic losses, so that $\epsilon_r = \epsilon_r' - j\epsilon_r''$ and $\mu_r = \mu_r' - j\mu_r''$. The material is designed so that it has $\eta = Z_0 = 377\ \Omega$. For the equivalent transmission line,

$$Z_0 = \sqrt{\frac{\mu_0}{\epsilon_0}}\sqrt{\frac{\mu_r' - j\mu_r''}{\epsilon_r' - j\epsilon_r''}}.$$

If $\mu_r = \epsilon_r$, then there will be no reflection at the air–material interface.

Although $\Gamma = 0$, there is still a transmission $T = 1 + \Gamma = 1$ into the material. The wave will enter the material, reflect from the metal and re-emerge into the air. However, the losses give a complex propagation constant

$$\gamma = j\omega\sqrt{(\mu' - j\mu'')(\epsilon' - j\epsilon'')}.$$

Therefore, the wave decays, and the re-emerging wave can be made small by choosing an adequate material thickness d and losses. Some materials in use are iron powder (carbonyl iron) and ferrites. They can be made as tiles, or as a magnetic dust in epoxy, a spray-on 'iron paint'.

The physical origin of losses is a point of interest. In a good conductor, the current is due to free-flowing electrons which encounter friction-like forces as they collide with each other, ions, and impurities. However, in a poor conductor the electrons are somewhat bound to their nuclei. Under an applied AC field, most of the electrons will oscillate about, and not flow. The oscillations are damped, so this constitutes a loss. The loss gives rise to a phase shift between $\mathbf{D}$ and $\mathbf{E}$ which implies that ϵ is complex. When we specify the loss as ϵ'', we do not know if the loss originates from collisions or from damping effects – nor does it matter.

For a magnetic material, μ can be complex. However, μ'' is associated with damping losses. Since there is no such thing as a magnetic current, μ'' cannot be associated with a current flow.

2.10   Phase and Group Velocity

Until now, we have been using the term 'wave velocity' or simply the velocity $u = \omega/\beta$. This has been adequate, as it gives the velocity of a phasefront and also the velocity of any signal that may be modulated onto the wave. It turns out that things are this simple if and only if β is linearly proportional to ω. We now consider what happens if this is not the case.

The *phase velocity* is

$$u_p = \frac{\omega}{\beta}, \tag{2.177}$$

and it gives the speed of a point of constant phase on a wave. The *group velocity* is

$$u_g = \frac{d\omega}{d\beta} = \frac{1}{d\beta(\omega)/d\omega} \tag{2.178}$$

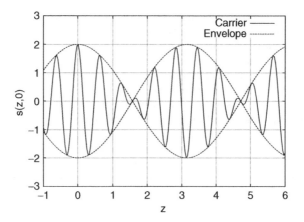

Figure 2.19   Plot of Equation (2.179) at $t = 0$, with $\beta_0 = 1$ and $\delta\beta = 0.1$.

and it is the speed associated with a group of frequencies in a narrowband modulated signal. The group velocity gives the speed at which information travels, for a modulated wave.

A derivation of (2.177) and (2.178) now follows. We begin with a signal that is a wave with a frequency ω_0 and phase constant β_0

$$S_0(z, \omega_0) = e^{j(\omega_0 t - \beta_0 z)}.$$

To understand what u_g and u_p are, it is necessary to consider a modulated narrowband signal. The simplest way to obtain one is by adding up two waves S_1 and S_2 that are at slightly different frequencies $\omega_1 = \omega_0 - \delta\omega$ and $\omega_2 = \omega_0 + \delta\omega$. We cannot find the corresponding β_1, β_2 from these frequencies because $\beta = \beta(\omega)$ is assumed to be an arbitrary function of ω and is not specified in any detail. However, we can say that each wave will have its own phase constant $\beta_1 = \beta_0 - \delta\beta$ and $\beta_2 = \beta_0 + \delta\beta$. The sum of these waves is

$$S(z, \omega) = S_1(z, \omega) + S_2(z, \omega) = e^{j(\omega_1 t - \beta_1 z)} + e^{j(\omega_2 t - \beta_2 z)}$$

$$= e^{j(\omega_0 - \delta\omega)t} e^{-j(\beta_0 - \delta\beta)z} + e^{j(\omega_0 + \delta\omega)t} e^{-j(\beta_0 + \delta\beta)z} = 2\cos(t\delta\omega - z\delta\beta)\, e^{j(\omega_0 t - \beta_0 z)}$$

or in the time domain,

$$s(z, t) = 2\cos(t\delta\omega - z\delta\beta)\cos(\omega_0 t - \beta_0 z). \tag{2.179}$$

Equation (2.179) amounts to an amplitude-modulated wave having a carrier $\cos(\omega_0 t - \beta_0 z)$ and an amplitude (or envelope) $2\cos(t\delta\omega - z\delta\beta)$. These features are illustrated in Figure 2.19 which shows $s(z, t)$ at $t = 0$ with $\beta_0 = 1$ and $\delta\beta = 0.1$.

The carrier velocity is found by letting $\omega_0 t - \beta_0 z = \text{const.}$ and evaluating $u_p = dz/dt$, which leads to (2.177). Similarly, the envelope's velocity is found by letting $t\delta\omega - z\delta\beta = \text{const.}$ and evaluating $u_g = dz/dt$. For small $\delta\omega$ and $\delta\beta$, it becomes (2.178).

Since u_g is associated with the modulation, it is the speed at which the information travels. It is required that $u_g \leq c$. On the other hand, there is no such restriction on the phase velocity, and $u_p > c$ is physically permissible; this will be further discussed in the context of waveguides, in Chapter 3.

Let us examine the phase and group velocities for a plane wave

$$\mathcal{E}_x = K_1 \cos(\omega t - \beta z).$$

The phase constant is

$$\beta(\omega) = \omega\sqrt{\mu\epsilon}.$$

For this β, it follows that the phase velocity is

$$u_p = \frac{\omega}{\beta} = \frac{1}{\sqrt{\mu\epsilon}}$$

and the group velocity is

$$u_g = \frac{d\omega}{d\beta} = \frac{1}{d\beta/d\omega} = \frac{1}{\sqrt{\mu\epsilon}}.$$

In this case, the phase and group velocities have turned out to be the same. This always happens when $\beta \propto \omega$. However, in some other situations, β might depend on ω in a more complicated way. Then, the phase and group velocities will be different. For instance, a plane wave in a lossy material has a complex propagation constant

$$\gamma = \alpha + j\beta = \sqrt{j\omega\mu(\sigma + j\omega\epsilon)}.$$

Squaring both sides,

$$\alpha^2 - \beta^2 + j2\alpha\beta = j\omega\mu(\sigma + j\omega\epsilon).$$

Equating the real and imaginary parts gives two equations which can be solved for α and β

$$\alpha = \omega\sqrt{\frac{\mu\epsilon}{2}}[-1 + \sqrt{1 + (\sigma/\omega\epsilon)^2}]^{1/2}$$

$$\beta = \omega\sqrt{\frac{\mu\epsilon}{2}}[1 + \sqrt{1 + (\sigma/\omega\epsilon)^2}]^{1/2}.$$

Clearly, β is not linearly proportional to ω, so consequently, $u_g \neq u_p$.

An important consequence of having a medium with $\beta \propto \omega$ is that u_g is frequency independent. The implication is that if we send a waveform such as a rectangular pulse, all of its frequency components will arrive at the same time, and the pulse will remain sharp. This is called a dispersionless medium. Conversely, when $\beta \not\propto \omega$, it is called a dispersive medium. In wave theory, the function $\beta(\omega)$ is called the *dispersion relation*.

The time delay for a signal travelling at the group velocity is called the *group delay*. If a wave travels a distance ℓ at a velocity u_g, the group delay is given by

$$\tau_g = \frac{\ell}{u_g} = \ell\frac{d\beta}{d\omega}. \tag{2.180}$$

The term 'group delay' is also used in filter theory. Suppose that the filter has a transfer function with an amplitude A and phase ψ as in $H(j\omega) = A(\omega)e^{j\psi(\omega)}$. If we interpret the phase shift as being on a fictitious transmission line of length ℓ, then $\psi = -\beta\ell$. Therefore, $d\beta = -d\psi/\ell$, and the filter's group delay is

$$\tau_g = -\frac{d\psi}{d\omega}. \tag{2.181}$$

Sometimes, dispersion characteristics are encountered that have $u_g > c$. In such cases, u_g can no longer be associated with the information velocity. This is a shortcoming of the narrowband explanation given here. It then becomes necessary to define a third velocity type, the *signal velocity* u_s. This was investigated by Brillouin. Having all these different types of velocities, it is good to summarize their definitions, which are in Table 2.2.

Table 2.2   Types of velocities.

Name	Definition
Speed of light	$c = \dfrac{1}{\sqrt{\mu_0 \epsilon_0}} = 2.99792458 \times 10^8$ m/s
Wave velocity	$u_0 = \dfrac{1}{\sqrt{\mu \epsilon}}$
Phase velocity	$u_p = \dfrac{\omega}{\beta}$
Group velocity	$u_g = \dfrac{d\omega}{d\beta} = \dfrac{1}{d\beta(\omega)/d\omega}$

In dispersionless media $u = u_0 = u_p = u_g$.

2.11   Further Reading

Approximate boundary conditions and their usage are discussed in Volakis and Senior (1995).

A good review of microstrip, stripline and related coupled-line configurations can be found in the book by Collin (2001). His book includes useful formulas for attenuation due to conductor and dielectric losses that were not covered here.

A description of electromagnetic absorbers and materials is in Knott et al. (2004).

More detailed discussions about phase, group and signal velocities are available in Johnson (1965), Papoulis (1962, sec. 7.6) and Stratton (1941).

References

Collin RE (1991) *Field Theory of Guided Waves*. Oxford University Press.
Collin RE (2001) *Foundations for Microwave Engineering*. John Wiley & Sons, Inc.
Demarest KR (1997) *Engineering Electromagnetics*. Prentice-Hall.
Johnson CC (1965) *Field and Wave Electrodynamics*. McGraw-Hill.
Knott EF, Shaeffer JF and Tuley MT (2004) *Radar Cross Section*. SciTech Publishing, Raleigh, NC.
Papoulis A (1962) *The Fourier Integral and its Applications*. McGraw-Hill.
Pozar D (2012) *Microwave Engineering*. John Wiley & Sons, Inc.
Richmond JH (1965) Efficient recursive solutions for plane and cylindrical multilayers. Technical Report 1968-1, ElectroScience Laboratory, Ohio State University, Columbus, OH.
Silvester P (1968) *Modern Electromagnetic Fields*. Prentice-Hall.
Stratton JA (1941) *Electromagnetic Theory*. McGraw-Hill. (Reprinted, IEEE Press 2007).
Volakis J and Senior TBA (1995) *Approximate Boundary Conditions in Electromagnetics*. IET.
Wait JR (1993) Use and misuse of impedance boundary conditions in electromagnetics. *IEEE Antennas Propag. Mag.* **35**(6), 78.

Problems

2.1   A plane wave in air has an amplitude of 6 V/m at $(x, y, z) = (0, 0, 0)$, is z-polarized and travels in the $+y$ direction.

(a) Find **E** and **H**.
(b) Calculate the electric field phasor if the frequency is 3 GHz and $y = 1.25$ cm.
(c) Find **S** and calculate the power density.

2.2   A perfectly conducting metal sheet is at $z = 0$. The region $z < 0$ is air and has a normal incidence plane wave $\mathbf{E}^i = \hat{\mathbf{x}} E_0 e^{-jkz}$.

   (a) Write out the expressions for $\mathbf{E}^r$, $\mathbf{H}^i$ and $\mathbf{H}^r$.
   (b) Find the surface current by using the boundary condition $\mathbf{J}_s = \hat{\mathbf{n}} \times \mathbf{H}$.
   (c) Sketch the total $|\mathbf{E}|$ and $|\mathbf{H}|$ for $-3\pi \le kz \le 0$. Indicate the positions of the maxima and minima. In terms of λ, what is the spacing of the minima?
   (d) Using the total $\mathbf{E}$ and $\mathbf{H}$ for $z < 0$, show that the Poynting vector $\mathbf{S}$ is zero. Because there is no power flow, this is called a *standing wave*.

2.3   The region $z < 0$ is air with $k = k_1$, and $z > 0$ is a dielectric with relative permittivity ϵ_r and $k = k_2$. A normal incidence plane wave for $z < 0$ is $\mathbf{E}^i = \hat{\mathbf{x}} E_0 e^{-jk_1 z}$.

   (a) Write out the $z < 0$ expressions for $\mathbf{E}^r$, $\mathbf{H}^i$ and $\mathbf{H}^r$.
   (b) Write out the $z > 0$ expressions for $\mathbf{E}^t$ and $\mathbf{H}^t$.
   (c) Show that $(1 - |\Gamma|^2)/\eta_1 = |T|^2/\eta_2$ which proves that power is conserved – that is, incident – reflected = transmitted power density.

2.4   A homogeneous plane wave has a strength of $2 + j0$ V/m at $(x, y, z) = (0, 0, 0)$. It is polarized in the z direction and travels in the direction $4\hat{\mathbf{x}} + 3\hat{\mathbf{y}}$. The wavelength is 0.2 m.

   (a) Find the propagation constant $\mathbf{k}$, and write down the mathematical expression for $\mathbf{E}$.
   (b) Evaluate the magnitude and phase of $\mathbf{E}$ at $(2, 5, 0)$ m.
   (c) Track the phasefront out from $(2, 5, 0)$ m to a new point $(3, y, 0)$ m, and find y.
   (d) A perfectly conducting plane is now placed at $x = 0$. Find an expression for the reflected field.
   (e) Find the surface current $\mathbf{J}_s = \hat{\mathbf{n}} \times (\mathbf{H}^i + \mathbf{H}^r)$ induced on the metal plane.
   (f) Find the Poynting vector $\mathbf{S}$ for the incident wave.

2.5   For a $\hat{\mathbf{p}}$-polarized plane wave $\mathbf{E} = \hat{\mathbf{p}} E_0 e^{-j\mathbf{k}\cdot\mathbf{r}}$, show that $\mathbf{H} = (\hat{\mathbf{k}} \times \mathbf{E})/\eta$. Do this by taking the curl of the given $\mathbf{E}$.

2.6   Find the polarization for a wave having

   (a) $\mathbf{E} = \hat{\mathbf{x}} 2 e^{-j\beta z} + \hat{\mathbf{y}} 2 e^{-j\beta z}$
   (b) $\mathbf{E} = \hat{\mathbf{x}} j 2 e^{-j\beta z} + \hat{\mathbf{y}} 2 e^{-j\beta z}$
   (c) $\mathbf{E} = \hat{\mathbf{x}} 2 e^{-j\beta z} + \hat{\mathbf{y}} j 5 e^{-j\beta z}$
   (d) $\mathbf{E} = \hat{\mathbf{x}} 2 e^{-j\beta z} - \hat{\mathbf{y}} j 5 e^{-j\beta z}$.

2.7   An electric field is $\mathbf{E} = \hat{\mathbf{y}} \, 2 \,\underline{/120°}\, e^{-j6x} + \hat{\mathbf{z}} \, 2 \,\underline{/30°}\, e^{-j6x}$. The medium is air. Find the

   (a) propagation direction
   (b) frequency-domain magnetic field
   (c) frequency
   (d) time-domain electric field
   (e) polarization.

2.8   A plane wave travels in Lucite and is incident on a dielectric–air interface. The incident angle θ_i is measured from the surface normal. The index of refraction is $m = 1.6$ and the frequency is 3 GHz. Calculate the critical angle. For the perpendicular polarization, calculate the propagation and attenuation constants β, α of the transmitted field. Do this every $10°$ for $\theta_c \le \theta_i \le 90°$. Tabulate the results.

2.9   A planar conductor is made of copper.

   (a) Find the skin depth at 100 Hz, 1 kHz and 10 kHz.
   (b) For normal incidence and a 1 mm thickness, find the transmission coefficient. Use the multilayer formulas with $N = 1$.

2.10   A dielectric slab at $0 \leq z \leq d$ has a relative permittivity ϵ_r. A normally incident plane wave is given by $E_x^i = E_0 e^{jkz}$.

    (a) Use the multilayer formulas with $N = 1$ to find the relation between Γ and T. Here, the reflection coefficient is defined as $\Gamma = E_x^r / E_x^i$ at $z = d$.

    (b) Show that for a thin slab, $T \approx 1 + \Gamma$.

2.11   Calculate the reflection coefficient for a dielectric slab. Assume the perpendicular polarization. Use the multilayer formulation (2.83) which is coded in PROGRAM mlslab; only minor modifications are needed. Assume a frequency of 1 GHz, a thickness $d = 1$cm and relative permittivity $\epsilon_r = 4$.

    (a) Over the range $0 \leq \theta_i \leq 90°$ plot $|\Gamma|$.

    (b) Calculate Γ, magnitude and phase, for normal incidence. The phase reference should be at $(0, 0, d)$ not $(0, 0, 0)$; see Figure 2.5.

2.12   Calculate the reflection coefficient for normal incidence on a dielectric slab. The thickness is $d = 1$cm, and the relative permittivity is $\epsilon_r = 4$. Use the exact formulation (2.83) which is coded in PROGRAM mlslab; only minor modifications are needed. For $1 - 10$ GHz, calculate and plot $|\Gamma|$.

2.13   For the parallel-plate transmission line in Example 2.4, find the surface current density on the bottom plate. The normal on the bottom plate is $\hat{x}$, and the fields inside the plate are zero. Then, find the current from the surface current density by integrating

$$I = \int_0^w J_{sz} \, dy.$$

Repeat the procedure for the top plate.

2.14   For the coaxial transmission line in Example 2.5, find the surface current density on the inner cylinder. The normal on the inner cylinder is $\hat{\rho}$, and the fields inside the metal are zero. Then, find the current from the surface current density by integrating

$$I = \int_0^{2\pi} J_{sz} \, a \, d\phi.$$

Repeat the procedure for the outer cylinder.

2.15   For a coaxial transmission line with inner radius a and outer radius b, the characteristic impedance Z_0 and shunt conductance G involve very similar integrals. The field is a plane wave, so $\eta \mathbf{H} = \hat{z} \times \mathbf{E}$. Use this in the expression

$$Z_0 = \frac{V}{I} = \frac{\int_a^b \mathbf{E} \cdot \hat{\rho} \, d\rho}{\oint_0^{2\pi} \mathbf{H} \cdot \hat{\phi} \, \rho d\phi},$$

and by comparing it with

$$G = \frac{I_t/\ell}{V} = \frac{\oint_0^{2\pi} \sigma_d \mathbf{E} \cdot \hat{\rho} \, \rho d\phi}{\int_a^b \mathbf{E} \cdot \hat{\rho} \, d\rho},$$

show that

$$G = \frac{\eta \sigma_d}{Z_0}.$$

This relationship holds in general for TEM transmission lines.

2.16 A coaxial line has an inner radius a and outer radius b. If the conductors have a finite conductivity σ_c, find the AC resistance per unit length R by using Wheeler's incremental inductance rule. The result is given in Table 2.1.

2.17 A round wire of radius $a = d/2$ is at a distance $D/2$ from a perfectly conducting ground plane. The capacitance between the wire and ground will be twice as high as for a pair of wires with a spacing D and no ground plane. (A proof of this by using image theory can be found in many books, e.g. Demarest (1997).)

    (a) Knowing Z_0 for the wire pair in Table 2.1, what is Z_0 for the wire above ground?

    (b) If the wire has a finite conductivity σ_c, find the AC resistance per unit length R by using Wheeler's incremental inductance rule.

    (c) Use this result to find R for two wires with a spacing D and no ground plane. The result is given in Table 2.1.

2.18 A microstrip line has a line width of 200 mils, a substrate relative permittivity of 2.20 and a substrate thickness of 31 mils. Find the characteristic impedance and the effective permittivity.

2.19 A stripline has a line width of 100 mils, a substrate relative permittivity of 2.20 and the total thickness is 62 mils, a sandwich of two 31 mils substrates. Find the characteristic impedance.

2.20 Find the characteristic impedance of a coplanar line. The substrate relative permittivity is 10.0, the central conductor is 3 mils wide and the gaps are 5 mils each. The substrate is assumed to be thick.

2.21 Find the reflection coefficient for an ungrounded dielectric slab by using the TEN method. Assume the perpendicular polarization. This is similar to Example 2.9, except that the transmission line's load impedance depends on the air impedance η_1 and the angle of the transmitted ray.

    (a) With a thickness of $d = 1$cm, $\epsilon_r = 4$ and a frequency of 1 GHz, calculate and plot $|\Gamma|$ for incidence angles $0 \leq \theta_i \leq 90°$.

    (b) Confirm your result numerically by comparing it with the multilayer formula (2.83); it should be identical.

2.22 A planar conductor is made of copper and is 1 mm thick. Find the transmission coefficient at 100 Hz, 1 kHz and 10 kHz for normal incidence. Use an impedance sheet approximation with the TEN equivalent. Compare with the exact result from Problem 2.9.

2.23 Calculate the reflection coefficient for the dielectric slab described in Problem 2.11. This time an impenetrable IBC will be used; the fields are zero when $z < d$.

    Suppose that as an approximation, $\bar{Z}_s$ does not depend on the incidence angle. Using Γ for normal incidence from Problem 2.11, find

$$\bar{Z}_s = \frac{1 + \Gamma}{1 - \Gamma}.$$

Then, using this $\bar{Z}_s$, the reflection coefficient is

$$\Gamma = \frac{\bar{Z}_s \cos\theta - 1}{\bar{Z}_s \cos\theta + 1}.$$

    (a) Calculate and plot $|\Gamma|$ for $0 \leq \theta \leq 90°$.

    (b) Compare this with the exact result.

2.24 For the dielectric slab described in Problem 2.12, recalculate Γ, using the impedance sheet approximation (2.97). Use a TEN model and a hand calculation to obtain Γ when $d = \lambda_d/4$; here, $\lambda_d = \lambda_0/\sqrt{\epsilon_r}$ is the wavelength in the dielectric.

2.25 Design a Salisbury screen for 10 GHz using Plexiglas with $\epsilon_r = 2.76$ as a spacer. Compute and plot $|\Gamma|$ for the perpendicular polarization as a function of incidence angle, from $0°$ (broadside) to $90°$.

2.26 A Salisbury screen is designed to work at 3 GHz. The surface gets splashed with some seawater, having $\epsilon_r = 81$ and $\sigma = 4\,\text{S/m}$. The coating of water is 1 mm thick. Calculate $|\Gamma|$ for normal incidence, in dB. Treat the 377 Ω sheet and seawater as impedance sheets that are in parallel.

2.27 Design a Dallenbach absorber for 10 GHz. Use $\mu_r = \epsilon_r = 4 - j1$. Find the material thickness d required so that the round-trip amplitude loss in the material $e^{-2\alpha d}$ is 30 dB; here, $\alpha = \text{Re}(\gamma)$. Assuming normal incidence, compute and plot $|\Gamma|$ as a function of frequency, from 5 GHz to 30 GHz.

2.28 A material has ohmic loss σ, as well as electric and magnetic damping losses, so that $\epsilon = \epsilon' - j\epsilon''$ and $\mu = \mu' - j\mu''$ are complex. Develop a more general form of Poynting's theorem (1.68) that allows for $\epsilon'' \neq 0$ and $\mu'' \neq 0$.

2.29 For the wire shown in Figure 2.8, we can apply Ampere's law to conclude $I = 2\pi a H_\phi$. The electric field at the surface is $E_z = Z_s H_\phi$. Knowing the fields at $\rho = a$, find the Poynting vector and find the power dissipated in the wire. Compare this with the circuit formula, $P = \frac{1}{2}\,\text{Re}[VI^*]$.

2.30 Find expressions for the phase and group velocities, u_p, u_g in a lossy medium. Calculate u_g and u_p for seawater at 60 Hz and 1 MHz. The conductivity is $\sigma = 4$ S/m, and the relative permittivity is $\epsilon_r = 72$.

2.31 A plane wave in a lossless medium is given by $E_x = E_0 e^{-jkz}$, where $k = \omega\sqrt{\mu\epsilon}$.
  (a) Find the phase velocity, magnetic field and the Poynting vector.
  (b) Repeat, assuming that $\mu < 0$ and $\epsilon < 0$. This is called a *double-negative* medium. Show that $\hat{\mathbf{k}}$ and the Poynting vector are in opposite directions. The directions of $\mathbf{E}$, $\mathbf{H}$ and $\hat{\mathbf{k}}$ in a cross product obey a 'left-hand' rule.

3

Waveguides and Resonators

This chapter presents field solutions for closed and open waveguides and the related topic of cavity resonators. It begins with a description of the method of separation of variables, which is needed for solving the wave equation. The method is then used to find the modes in rectangular and circular waveguides. Excitation methods for waveguides are also discussed. Next, 2D parallel-plate and dielectric-slab waveguides are presented, along with a discussion of surface-wave fields. The ridge waveguide and finline are important structures that have a single-mode bandwidth that is larger than that of a rectangular or circular waveguide. Approximate solutions for the propagation constant β are obtained from the transverse resonance method.

Waveguide filters and impedance-matching networks often have discontinuities such as a step or an iris, and an understanding of their effects is important. Approximate but highly accurate variational solutions for irises are summarized. The mode matching technique is a useful tool for analysing waveguide discontinuities, and its application is demonstrated for a rectangular waveguide step and an iris.

Cavity resonators are merely waveguides with the ends closed off, so it is a small step to construct standing-wave solutions from the travelling-wave ones. The resonant frequencies and the effect of conductor and dielectric losses on the quality factor Q are described. Often, a small dielectric or metallic 'perturbation' is used in a cavity to slightly change or 'tune' the resonant frequency. The perturbation method for predicting resonance shifts is described.

3.1 Separation of Variables

The separation of variables is a general-purpose method for solving partial differential equations. It is prerequisite knowledge for solving the wave equation in waveguides. The basic premise is that a solution $\psi(x, y, z)$ can be written as a product of single-variable functions. If the partial differential equation can be separated out into three differential equations, one involving only x, another with y and a third one with z, it is said to be 'separable'. The scalar Helmholtz equation is separable in rectangular, cylindrical and spherical coordinates, as well as several other more esoteric coordinate systems.[1]

Let us proceed with solving the wave equation in rectangular coordinates

$$\left(\frac{\partial^2}{\partial x^2} + \frac{\partial^2}{\partial y^2} + \frac{\partial^2}{\partial z^2} + k^2 \right) \psi(x, y, z) = 0, \tag{3.1}$$

---

[1] The Helmholtz equation is separable in 11 coordinate systems: rectangular, circular cylinder, elliptic cylinder, parabolic cylinder, spherical, conical, parabolic, prolate spheroidal, oblate spheroidal, ellipsoidal and paraboloidal (Morse and Feshbach 1953, Chapter 5).

Applied Frequency-Domain Electromagnetics, First Edition. Robert Paknys.
© 2016 John Wiley & Sons, Ltd. Published 2016 by John Wiley & Sons, Ltd.
Companion Website: www.wiley.com/go/paknys9981

where $k = \omega/u_0$ is the wavenumber and $u_0 = 1/\sqrt{\mu\epsilon}$ is the wave velocity. The method of separation of variables presumes that the 3D solution can be expressed as a product of 1D solutions

$$\psi(x, y, z) = X(x)Y(y)Z(z). \tag{3.2}$$

Substitution of (3.2) in (3.1) leads to

$$X''YZ + XY''Z + XYZ'' + k^2 XYZ = 0$$

and dividing by XYZ gives

$$\underbrace{\frac{X''(x)}{X(x)}}_{-k_x^2} + \underbrace{\frac{Y''(y)}{Y(y)}}_{-k_y^2} + \underbrace{\frac{Z''(z)}{Z(z)}}_{-k_z^2} + k^2 = 0,$$

where k_x, k_y and k_z are known as the *separation constants*. The equation has to be true for an arbitrary (x, y, z). Consequently, each term has to be a constant.[2] It can be rewritten as

$$\frac{d^2 X}{dx^2} + k_x^2 X = 0 \tag{3.3}$$

$$\frac{d^2 Y}{dy^2} + k_y^2 Y = 0 \tag{3.4}$$

$$\frac{d^2 Z}{dz^2} + k_z^2 Z = 0, \tag{3.5}$$

where

$$k_x^2 + k_y^2 + k_z^2 = k^2. \tag{3.6}$$

We see that $X(x)$, $Y(y)$ and $Z(z)$ are solutions to three ordinary differential equations (3.3)–(3.5), and the *separation equation* (3.6) must be satisfied as well. In waveguides, (3.6) plays a key role in determining the dispersive characteristics of the wave propagation, so it is also known as the *dispersion relation*.

The general solution $\psi(x, y, z)$ is of the form

$$\psi(x, y, z) = \left\{ \begin{array}{c} \cos k_x x \\ \sin k_x x \\ e^{\pm j k_x x} \end{array} \right\} \left\{ \begin{array}{c} \cos k_y y \\ \sin k_y y \\ e^{\pm j k_y y} \end{array} \right\} \left\{ \begin{array}{c} \cos k_z z \\ \sin k_z z \\ e^{\pm j k_z z} \end{array} \right\}. \tag{3.7}$$

Any two functions in each bracket are independent solutions. The constants k_x, k_y and k_z are chosen so that the boundary conditions are met.[3]

Example 3.1 (Free Space) Solve the wave equation in free space. Assume that the wave travels in the positive x, y and z directions.

---

[2] For instance, if $y = $ const. and $z = $ const., then k_y and k_z are constants. Therefore, k_x also has to be a constant – even though x is not. This leads to (3.3). The same idea applies to (3.4) and (3.5).

[3] It was assumed that all of the separation constants are positive. This is usually the most useful choice. However, other types of solutions are possible. For instance, if $k_x = 0$, then $X(x) = Ax + B$. If $k_x^2 < 0$, we can let $\alpha_x^2 = -k_x^2 > 0$, and then, the possible solutions for $X(x)$ are $\cosh \alpha_x x$, $\sinh \alpha_x x$, or $e^{\pm \alpha_x x}$.

Solution: From (3.7), we select the terms having positive propagation directions $e^{-jk_x x}$, $e^{-jk_y y}$ and $e^{-jk_z z}$. The product solution is then

$$\psi(x,y,z) = Ke^{-jk_x x}\, e^{-jk_y y}\, e^{-jk_z z} = Ke^{-j(k_x x + k_y y + k_z z)}$$

which is a plane wave. The constant K is arbitrary.   ∎

Example 3.2 (Rectangular Region) Solve the wave equation in the rectangular region $0 \leq x \leq a, 0 \leq y \leq b$ subject to the boundary conditions

$$\psi(0,y,z) = 0; \quad \psi(a,y,z) = 0$$
$$\psi(x,0,z) = 0; \quad \psi(x,b,z) = 0.$$

In addition, the z dependence is required to be a $+z$ travelling wave.

Solution: Since $\psi(x,y,z) = X(x)Y(y)Z(z)$, the boundary conditions can be thought of as

$$X(0) = 0; \quad X(a) = 0$$
$$Y(0) = 0; \quad Y(b) = 0.$$

The condition $X(0) = 0$ implies that we should select $X(x) = \sin k_x x$. From $X(a) = 0$, we conclude that $k_x = m\pi/a$; $m = 1, 2, 3, \ldots$. Similarly, $Y(0) = Y(b) = 0$ leads to $Y(y) = \sin k_y y$ in which $k_y = n\pi/b$; $n = 1, 2, 3, \ldots$. From the travelling-wave requirement, we select $Z(z) = e^{-jk_z z}$. The solution meeting all these conditions is

$$\psi(x,y,z) = \sin\frac{m\pi x}{a} \sin\frac{n\pi y}{b}\, e^{-jk_z z}.$$

The integers $m = 0$ and $n = 0$ lead to trivial solutions and are excluded. Since

$$k_x^2 + k_y^2 + k_z^2 = k^2,$$

we can obtain k_z from

$$k_z = \sqrt{k^2 - \left(\frac{m\pi}{a}\right)^2 - \left(\frac{n\pi}{b}\right)^2}.$$

∎

3.2   Rectangular Waveguide

A rectangular waveguide is illustrated in Figure 3.1(a). It is a hollow pipe with perfectly conducting walls at $x = 0, a$ and $y = 0, b$. The fields will be found by using the separation of variables method to solve the wave equation.

   Before jumping into the mathematical solution, it is helpful to have a physical picture of how a waveguide works. The way a wave propagates down a waveguide is as a plane wave that bounces back and forth between the conducting walls in a zigzag fashion, along the z direction. This is shown in Figure 3.1(b) for a wave that is y-polarized. More generally, the zigzag could be between the side walls, the top and bottom walls or even from all four walls.

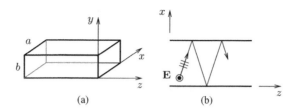

Figure 3.1 (a) Rectangular metallic waveguide and (b) illustration of a plane wave propagating down the guide.

In general, an arbitrary waveguide field can be expressed as a sum of 'modes', and these fall into one of two categories. Transverse electric (TE) modes have a transverse electric field with $E_x \neq 0$, $E_y \neq 0$ and $E_z = 0$. Transverse magnetic (TM) modes have a transverse magnetic field with $H_x \neq 0$, $H_y \neq 0$ and $H_z = 0$. A general field that is neither TE nor TM can be expressed as a linear combination of these two mode types.[4] The mode illustrated in Figure 3.1(b) is TE, as the electric field is transverse to the z direction.

It turns out that all of the TE field components can be found from the axial magnetic field component H_z. For this reason, TE modes are also called H-modes. Similarly, TM modes can be found from E_z and are called E-modes. To see how this works, we have to write out the two Maxwell curl equations as six scalar equations. Since travelling-wave solutions are sought, it will be assumed that all of the field components behave as $e^{-j\beta z}$. Therefore, in any z derivatives, $\partial/\partial z \rightarrow -j\beta$. (In the present notation, $\beta \equiv k_z$.) The six equations can be used to obtain (see Problem 3.7)

$$E_x = \frac{-j}{k_c^2}\left(\beta\frac{\partial E_z}{\partial x} + \omega\mu\frac{\partial H_z}{\partial y}\right) \tag{3.8}$$

$$E_y = \frac{j}{k_c^2}\left(-\beta\frac{\partial E_z}{\partial y} + \omega\mu\frac{\partial H_z}{\partial x}\right) \tag{3.9}$$

$$H_x = \frac{-j}{k_c^2}\left(\beta\frac{\partial H_z}{\partial x} - \omega\epsilon\frac{\partial E_z}{\partial y}\right) \tag{3.10}$$

$$H_y = \frac{j}{k_c^2}\left(-\beta\frac{\partial H_z}{\partial y} - \omega\epsilon\frac{\partial E_z}{\partial x}\right), \tag{3.11}$$

where $k_c^2 = k_x^2 + k_y^2$ or equivalently, $k_c^2 = k^2 - \beta^2$. k_c is called the cutoff wavenumber (and is also known as the transverse wavenumber).

Let us begin by finding the possible modes. Since we only want the modes, a source is not necessary. The fields should satisfy the homogeneous vector Helmholtz equations

$$\nabla^2\mathbf{E} + k^2\mathbf{E} = 0$$

and

$$\nabla^2\mathbf{H} + k^2\mathbf{H} = 0.$$

These are equivalent to six scalar equations

$$\nabla^2 E_x + k^2 E_x = 0 \tag{3.12a}$$

---

[4] Sometimes, these are called TE_z and TM_z to emphasize that they are transverse electric or transverse magnetic with respect to z. If the subscript is omitted, it is understood to be z.

$$\nabla^2 E_y + k^2 E_y = 0 \tag{3.12b}$$

$$\nabla^2 E_z + k^2 E_z = 0 \tag{3.12c}$$

$$\nabla^2 H_x + k^2 H_x = 0 \tag{3.12d}$$

$$\nabla^2 H_y + k^2 H_y = 0 \tag{3.12e}$$

$$\nabla^2 H_z + k^2 H_z = 0. \tag{3.12f}$$

For the TM modes, $H_z = 0$, and from (3.8)–(3.11), it follows that all of the field components can be obtained from E_z alone. Therefore, we only need to solve the scalar wave equation (3.12c), and the other five can be ignored.

The boundary conditions require that $\mathbf{E}_{tan} = 0$ on the four metal walls at $x = 0, x = a, y = 0, y = b$. Using (3.8)–(3.11), the boundary conditions in terms of E_z become

$$E_x \propto \frac{\partial E_z}{\partial x} = 0; \quad y = 0, b \tag{3.13}$$

$$E_y \propto \frac{\partial E_z}{\partial y} = 0; \quad x = 0, a. \tag{3.14}$$

Solving the wave equation (3.12c) for E_z and applying the boundary conditions lead to

$$E_z = A_{mn} \sin\frac{m\pi x}{a} \sin\frac{n\pi y}{b} e^{\mp j\beta_{mn} z}, \tag{3.15}$$

where m, n are any integers, and

$$\beta_{mn} = \sqrt{k^2 - \left(\frac{m\pi}{a}\right)^2 - \left(\frac{n\pi}{b}\right)^2}. \tag{3.16}$$

A_{mn} is an arbitrary constant, $k = \omega\sqrt{\mu\epsilon}$ and (3.16) follows from (3.6) with $k_z = \beta_{mn}$. Equation (3.16) is the dispersion relation, which allows us to calculate the propagation constant β_{mn} as a function of frequency (or k).

The lower frequency limit for propagation (the waveguide cutoff) is when $\beta \to 0$. It is convenient to define a cutoff wavenumber k_c such that

$$k_{c,mn}^2 = \left(\frac{m\pi}{a}\right)^2 + \left(\frac{n\pi}{b}\right)^2. \tag{3.17}$$

and express the dispersion relation (3.16) as

$$\beta_{mn} = \sqrt{k^2 - k_{c,mn}^2}. \tag{3.18}$$

At cutoff, $k = k_c$, and from (3.17), the corresponding cutoff frequency is

$$f_{c,mn} = \frac{u_0}{2}\sqrt{\left(\frac{m}{a}\right)^2 + \left(\frac{n}{b}\right)^2} \tag{3.19}$$

where $u_0 = 1/\sqrt{\mu\epsilon}$ is the velocity of a plane wave in a medium with μ, ϵ.

At frequencies lower than f_c, the argument of the square root in (3.18) becomes negative, so that

$$j\beta = \alpha = \pm\sqrt{\left(\frac{m\pi}{a}\right)^2 + \left(\frac{n\pi}{b}\right)^2 - k^2}. \tag{3.20}$$

In this case, the field does not propagate. Rather, it decays as $e^{-\alpha z}$. Implicit in the square root is a $\pm$ sign, but only a decaying solution $e^{-\alpha z}$ with $\alpha \geq 0$ is physically admissible.

From E_z, the $+z$ propagating modes are readily obtained as

$$E_{xmn} = A_{mn} \frac{-j\beta_{mn}m\pi}{ak_{c,mn}^2} \cos\frac{m\pi x}{a} \sin\frac{n\pi y}{b} e^{-j\beta_{mn}z} \tag{3.21}$$

$$E_{ymn} = A_{mn} \frac{-j\beta_{mn}n\pi}{bk_{c,mn}^2} \sin\frac{m\pi x}{a} \cos\frac{n\pi y}{b} e^{-j\beta_{mn}z} \tag{3.22}$$

$$H_{ymn} = \frac{E_{xmn}}{Z_{mn}} \tag{3.23}$$

$$H_{zmn} = 0 \tag{3.24}$$

$$Z_{mn} = \beta_{mn}\eta/k, \tag{3.25}$$

where $\eta = \sqrt{\mu/\epsilon}$.

In a similar manner, the TE modes can be found from a solution of (3.12f) for H_z which is constructed to give $\mathbf{E}_{tan} = 0$ on the metal walls. The TE modes are denoted with a prime. The result is found to be

$$H'_{zmn} = A'_{mn} \cos\frac{m\pi x}{a} \cos\frac{n\pi y}{b} e^{-j\beta_{mn}z} \tag{3.26}$$

from which

$$H'_{xmn} = A'_{mn} \frac{j\beta_{mn}m\pi}{ak_{c,mn}^2} \sin\frac{m\pi x}{a} \cos\frac{n\pi y}{b} e^{-j\beta_{mn}z} \tag{3.27}$$

$$H'_{ymn} = A'_{mn} \frac{j\beta_{mn}n\pi}{bk_{c,mn}^2} \cos\frac{m\pi x}{a} \sin\frac{n\pi y}{b} e^{-j\beta_{mn}z} \tag{3.28}$$

$$E'_{xmn} = H'_{ymn}Z'_{mn} \tag{3.29}$$

$$E'_{ymn} = -H'_{xmn}Z'_{mn} \tag{3.30}$$

$$E'_{zmn} = 0 \tag{3.31}$$

$$Z'_{mn} = k\eta/\beta_{mn}. \tag{3.32}$$

It turns out that β_{mn} for the TE and TM modes of rectangular waveguides are the same. The solutions for $-z$ travelling waves are readily obtained by replacing $\beta \rightarrow -\beta$ in the field expressions. The lowest non-zero modes are TE_{10}, TE_{01} and TM_{11}. In practice, it is usually assumed that $a > b$ so that the lowest cutoff frequency is associated with the TE_{10} (and not the TE_{01}) mode.

A waveguide can have two different modes with the same β, and these are called 'degenerate modes'. These can occur, for instance, in a square waveguide. When $a = b$, the TE_{mn}, TE_{nm}, TM_{mn} and TM_{nm} modes all have the same β.

3.2.1 Dominant TE_{10} Mode

To obtain maximum bandwidth with single-mode operation, using $b = a/2$ with the TE_{10} mode is the best choice. It maximizes the bandwidth of the 10 mode while suppressing the next higher 01 mode (which is then at the same cutoff frequency as the inevitable 20 mode). We could allow $b < a/2$, but the smaller cross-sectional area would mean that the power is reduced, for a given field strength. In practice,

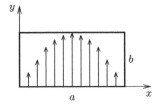

Figure 3.2   Electric field for the TE_{10} mode.

$b \approx a/2$ is normally used. Waveguide sizes are specified in several ways; one is the 'WR' (waveguide rectangular) designation.[5]

A good way to quickly visualize some of the cutoff frequencies is to notice that for the $n = 0$ modes, there is no field variation along y and that cutoff occurs when the waveguide width is $a = m\lambda/2$. Similarly, the $m = 0$ modes do not vary with x, and their cutoff is when the height is $b = n\lambda/2$.

With $m, n = 1, 0$ we obtain $E_x = 0$ and

$$E_y = A_{10} \frac{-j\pi k \eta}{a k_c^2} \sin \frac{\pi x}{a} e^{-j\beta z}. \tag{3.33}$$

Figure 3.2 illustrates the electric field in a $z = $ const. plane of the waveguide. By writing the sine function as complex exponentials, it becomes

$$E_y = A_{10} \frac{-j\pi k \eta}{a k_c^2} \frac{e^{j\pi x/a} - e^{-j\pi x/a}}{j2} e^{-j\beta z}$$

or

$$E_y = A_{10} \frac{-\pi k \eta}{2a k_c^2} (e^{j(\pi x/a - \beta z)} - e^{-j(\pi x/a - \beta z)}). \tag{3.34}$$

Equation (3.34) can be interpreted as two plane waves; one propagating in the $-x$ and $+z$ directions, and the other one propagating in $+x$ and $+z$ directions. This can also be interpreted as a wave that propagates along a zigzag path, as it bounces off the metal walls at $x = 0$ and $x = a$. If θ is the zigzag angle with respect to the z axis, then $\tan \theta = k_x/k_z = (\pi/a)/\beta$. At cutoff, $\theta \to \pi/2$, so the wave bounces back and forth between the metal walls at $x = 0$, $x = a$ without moving along in the z direction. At frequencies well above cutoff, $\theta \to 0$ and the wave travels almost straight along the z direction with very little zigzagging.

The phase velocity (2.177) and group velocity (2.178) can be found from the dispersion relation $\beta(\omega)$ in (3.16). The result is

$$u_p = \frac{u_0}{\cos \theta}$$

$$u_g = u_0 \cos \theta,$$

where $u_0 = 1/\sqrt{\mu \epsilon}$ is the velocity of the wave along the zigzag.

The interpretation of u_p is shown in Figure 3.3(a). Suppose that in a given time t_0, a phasefront travels a distance AB along the zigzag. At that same time, it will travel a larger distance $A'B'$ along the guiding

---

[5] For example, a WR90 waveguide has internal dimensions of $0.900'' \times 0.400''$. A WR15 waveguide is $0.148'' \times 0.074''$. As a general rule, $2b \approx a$ and the WR digits are closely related to the larger of a and b. To know the exact numbers, it is necessary to consult the literature, for example, Brady (1968); Lowman and Simons (2007).

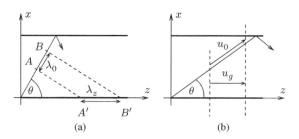

Figure 3.3 Interpretation of (a) phase velocity and (b) group velocity for the TE_{10} mode.

direction z. Therefore, the velocity along $A'B'$ is faster, and the *guide wavelength* λ_z is always longer than λ_0. The phase velocity is thus

$$u_p = \frac{\omega}{\beta} = \frac{2\pi u_0/\lambda_0}{2\pi/\lambda_z} = \frac{\lambda_z}{\lambda_0} = u_0 \sec\theta.$$

Figure 3.3(b) interprets the group velocity. The plane wave travels at a speed u_0; however, the effective speed along the z direction is only $u_g = u_0\cos\theta$. In free space, $u_0 = c$ and as expected, $u_g \leq c$.

3.2.2   Fourier Series of Modes

Rectangular waveguide modes can be used to construct a 2D Fourier series representation of an arbitrary function $f(x,y)$ on a $z = $ const. plane. This is accomplished by using the orthogonality of trigonometric functions on the period $p = 2L$

$$\int_{-L}^{L} \cos\frac{m\pi x}{L}\cos\frac{n\pi x}{L}\,dx = \begin{cases} (1+\delta_{m0})L; & m = n \\ 0; & m \neq n \end{cases} \tag{3.35}$$

$$\int_{-L}^{L} \sin\frac{m\pi x}{L}\sin\frac{n\pi x}{L}\,dx = \begin{cases} L; & m = n \neq 0 \\ 0; & m \neq n \end{cases} \tag{3.36}$$

$$\int_{-L}^{L} \cos\frac{m\pi x}{L}\sin\frac{n\pi x}{L}\,dx = 0 \tag{3.37}$$

where the 'Kronecker delta' $\delta_{m0} = 1$ when $m = 0$; otherwise, $\delta_{m0} = 0$. If the lower limit of integration is zero instead of $-L$, then the value of the integrals is halved.

For instance, suppose that a y-directed current sheet in a rectangular waveguide is placed at $z = 0$. It is given by

$$\mathbf{J}_s = \hat{\mathbf{y}}f(x,y).$$

The source will excite modes that travel away from the current sheet. The jump discontinuity in $\mathbf{H}_{tan}$ at $z = 0$ should equal $\mathbf{J}_s$. Using (3.27) with $z = 0^+$,

$$H_x(z = 0^+) = \sum_m \sum_n A_{mn}\frac{j\beta_{mn}m\pi}{ak_{c,mn}^2}\sin\frac{m\pi x}{a}\cos\frac{n\pi y}{b}e^{-j\beta_{mn}0^+}.$$

At $z = 0^-$ with $\beta \to -\beta$ for the $-z$-travelling wave,

$$H_x(z = 0^-) = \sum_m \sum_n A_{mn}\frac{-j\beta_{mn}m\pi}{ak_{c,mn}^2}\sin\frac{m\pi x}{a}\cos\frac{n\pi y}{b}e^{j\beta_{mn}0^-}.$$

The jump condition is

$$\hat{\mathbf{z}} \times (\mathbf{H}_1(z = 0^+) - \mathbf{H}_2(z = 0^-)) = \mathbf{J}_s,$$

or

$$H_x(z = 0^+) - H_x(z = 0^-) = J_y$$

and since H_x is odd symmetric about the current sheet,

$$2H_x(z = 0^+) = J_y.$$

Therefore, at $z = 0^+$,

$$\sum_m \sum_n 2A_{mn} \frac{j\beta_{mn}m\pi}{ak_{c,mn}^2} \sin \frac{m\pi x}{a} \cos \frac{n\pi y}{b} = f(x, y).$$

Next, we multiply both sides by $\sin(p\pi x/a)\cos(q\pi y/b)$ and integrate over the cross section of the waveguide. Because of orthogonality (3.35)–(3.37), all of the left-hand side terms will be zero, except when $m = p$ and $n = q$. This gives

$$2A_{mn} \frac{j\beta_{mn}m\pi}{ak_{c,mn}^2} \int_0^b \int_0^a \sin^2 \frac{m\pi x}{a} \cos^2 \frac{n\pi y}{b} dx\, dy$$

$$= \int_0^b \int_0^a f(x, y) \sin \frac{m\pi x}{a} \cos \frac{n\pi y}{b} dx\, dy.$$

Evaluating the left-hand side with (3.35), (3.36) and rearranging gives

$$A_{mn} = \frac{ak_{c,mn}^2}{j2\beta_{mn}m\pi} \frac{\int_0^b \int_0^a f(x, y) \sin(m\pi x/a) \cos(n\pi y/b) dx\, dy}{(a/2)(b/2)(1 + \delta_{n0})}. \tag{3.38}$$

Example 3.3 (Current Sheet in Waveguide) Find the TE waveguide fields produced by a current sheet that is given by $J_{sy} = \sin(3\pi x/a)$.

Solution: In this case, $f(x, y) = \sin(3\pi x/a)$. Evaluating (3.38), only $A_{30} \neq 0$, and it is

$$A_{30} = \frac{ak_{c,30}^2}{j2\beta_{30}3\pi}.$$

Using the amplitude A_{30} and (3.26)–(3.32), the fields are

$$H_x = \frac{1}{2} \sin \frac{3\pi x}{a} e^{-j\beta_{30}z}$$

$$H_z = \frac{ak_{c,30}^2}{j2\beta_{30}3\pi} \cos \frac{3\pi x}{a} e^{-j\beta_{30}z}$$

$$E_y = -\frac{k\eta}{2\beta_{30}} \sin \frac{3\pi x}{a} e^{-j\beta_{30}z}.$$

■

3.3   Cylindrical Waves

In this section, solutions for the wave equation in cylindrical coordinates are developed. All the field components can be found from E_z and H_z. Analogous to the rectangular case in (3.8)–(3.11), it can be shown that

$$E_\rho = \frac{-j}{k_c^2}\left(\beta\frac{\partial E_z}{\partial\rho} + \frac{\omega\mu}{\rho}\frac{\partial H_z}{\partial\phi}\right) \tag{3.39}$$

$$E_\phi = \frac{j}{k_c^2}\left(-\frac{\beta}{\rho}\frac{\partial E_z}{\partial\phi} + \omega\mu\frac{\partial H_z}{\partial\rho}\right) \tag{3.40}$$

$$H_\rho = \frac{-j}{k_c^2}\left(\beta\frac{\partial H_z}{\partial\rho} - \frac{\omega\epsilon}{\rho}\frac{\partial E_z}{\partial\phi}\right) \tag{3.41}$$

$$H_\phi = \frac{j}{k_c^2}\left(-\frac{\beta}{\rho}\frac{\partial H_z}{\partial\phi} - \omega\epsilon\frac{\partial E_z}{\partial\rho}\right). \tag{3.42}$$

The wave equation is

$$\left[\frac{1}{\rho}\frac{\partial}{\partial\rho}\left(\rho\frac{\partial}{\partial\rho}\right) + \frac{1}{\rho^2}\frac{\partial^2}{\partial\phi^2} + \frac{\partial^2}{\partial z^2} + k^2\right]\psi(\rho,\phi,z) = 0. \tag{3.43}$$

Substituting in the product solution $\psi(\rho,\phi,z) = R(\rho)\Phi(\phi)Z(z)$ and dividing by $R\Phi Z$ leads to

$$\frac{1}{\rho R}\frac{d}{d\rho}\left(\rho\frac{dR}{d\rho}\right) + \underbrace{\frac{1}{\rho^2}\frac{1}{\Phi}\frac{d^2\Phi}{d\phi^2}}_{-\nu^2} + \underbrace{\frac{1}{Z}\frac{d^2Z}{dz^2}}_{-k_z^2} + k^2 = 0, \tag{3.44}$$

where ν and k_z are separation constants. For the $R(\rho)$ equation, it is convenient to define

$$k_\rho^2 = k^2 - k_z^2. \tag{3.45}$$

The equations can then be written as

$$\frac{1}{\rho R}\frac{d}{d\rho}\left(\rho\frac{dR}{d\rho}\right) - \frac{\nu^2}{\rho^2} + k_\rho^2 = 0 \tag{3.46a}$$

$$\frac{1}{\Phi}\frac{d^2\Phi}{d\phi^2} = -\nu^2 \tag{3.46b}$$

$$\frac{1}{Z}\frac{d^2Z}{dz^2} = -k_z^2. \tag{3.46c}$$

Equation (3.46a) is Bessel's differential equation. Possible solutions include Bessel functions of the first kind $J_\nu(k_\rho\rho)$ and second kind $Y_\nu(k_\rho\rho)$. These are linearly independent solutions. They can be thought of as being analogous to the rectangular coordinate wave solutions $\cos k_x x$ and $\sin k_x x$. They have many important properties, described in Appendix C. The J_ν functions are usually used to describe standing waves. The Y_ν also resemble standing waves, except that they are singular when $\rho \to 0$. Inward- and outward-travelling waves are expressed in terms of Hankel functions of the first kind and second kind

$$H_\nu^{(1)}(k_\rho\rho) = J_\nu(k_\rho\rho) + jY_\nu(k_\rho\rho) \tag{3.47}$$

$$H_\nu^{(2)}(k_\rho\rho) = J_\nu(k_\rho\rho) - jY_\nu(k_\rho\rho). \tag{3.48}$$

The behaviour of these functions is difficult to visualize; however, some insight can be gained by looking at their graphs and their asymptotic approximations (see Appendix C). If the argument $x = k_\rho\rho$ is large, then the Hankel functions behave as

$$H_\nu^{(1)}(x) \sim \sqrt{\frac{2}{\pi x}}\, e^{j(x - \nu\pi/2 - \pi/4)} \tag{3.49}$$

$$H_\nu^{(2)}(x) \sim \sqrt{\frac{2}{\pi x}}\, e^{-j(x - \nu\pi/2 - \pi/4)}. \tag{3.50}$$

This shows their characteristics, as inward- and outward-travelling waves. Taking the real or imaginary part of $H^{(1)}$ gives expressions for J_ν and Y_ν

$$J_\nu(x) \sim \sqrt{\frac{2}{\pi x}}\, \cos(x - \nu\pi/2 - \pi/4) \tag{3.51}$$

$$Y_\nu(x) \sim \sqrt{\frac{2}{\pi x}}\, \sin(x - \nu\pi/2 - \pi/4) \tag{3.52}$$

which resemble standing waves.

The general form of the cylindrical wave solution is

$$\psi(\rho, \phi, z) = \begin{Bmatrix} J_\nu(k_\rho\rho) \\ Y_\nu(k_\rho\rho) \\ H_\nu^{(1)}(k_\rho\rho) \\ H_\nu^{(2)}(k_\rho\rho) \end{Bmatrix} \begin{Bmatrix} \cos\nu\phi \\ \sin\nu\phi \\ e^{\pm j\nu\phi} \end{Bmatrix} \begin{Bmatrix} \cos k_z z \\ \sin k_z z \\ e^{\pm jk_z z} \end{Bmatrix}. \tag{3.53}$$

Any two functions in each bracket are independent solutions.

3.4 Circular Waveguide

In this section, we find the modes for a z-directed circular metal waveguide. This is a hollow pipe with a perfectly conducting wall at $\rho = a$, as illustrated in Figure 3.4(a).

Among all the possible solutions (3.53), the J Bessel function is chosen because it is the only one that is finite at the origin. The solutions for $\Phi(\phi)$ have to be 2π periodic in ϕ, so $\nu = m$ is an integer. $Z(z)$ shall represent a $+z$ travelling wave, so $e^{-j\beta z}$ is appropriate; here, $\beta \equiv k_z$. The product solution of the wave equation is therefore

$$\psi(\rho, \phi, z) = J_m(k_\rho\rho) \begin{Bmatrix} \cos m\phi \\ \sin m\phi \end{Bmatrix} e^{-j\beta z}. \tag{3.54}$$

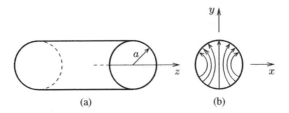

(a)	(b)

Figure 3.4   (a) Circular metallic waveguide and (b) electric field lines for the dominant TE_{11} mode.

For TM modes, E_z and E_ϕ have to be zero at $\rho = a$. All the field components are found from $\psi = E_z$, and using (3.39)–(3.42),

$$E_{\rho mn} = \frac{\beta_{mn}}{jk_{c,mn}} J'_m(k_{c,mn}\rho) \begin{Bmatrix} \cos m\phi \\ \sin m\phi \end{Bmatrix} e^{-j\beta_{mn}z} \tag{3.55}$$

$$E_{\phi mn} = \frac{m\beta_{mn}}{jk^2_{c,mn}\rho} J_m(k_{c,mn}\rho) \begin{Bmatrix} -\sin m\phi \\ \cos m\phi \end{Bmatrix} e^{-j\beta_{mn}z} \tag{3.56}$$

$$E_{zmn} = J_m(k_{c,mn}\rho) \begin{Bmatrix} \cos m\phi \\ \sin m\phi \end{Bmatrix} e^{-j\beta_{mn}z} \tag{3.57}$$

$$H_{\rho mn} = -\frac{E_{\phi mn}}{Z_{mn}} \tag{3.58}$$

$$H_{\phi mn} = \frac{E_{\rho mn}}{Z_{mn}} \tag{3.59}$$

$$H_{zmn} = 0 \tag{3.60}$$

$$Z_{mn} = \beta_{mn}\eta/k. \tag{3.61}$$

The requirement $\mathbf{E}_{tan} = 0$ at $\rho = a$ implies that

$$J_m(k_{c,mn}a) = 0. \tag{3.62}$$

The first few roots of (3.62) are given in Table C.1. The values $x_n = k_{c,mn}a$ can be used to find the propagation constant

$$\beta_{mn} = \sqrt{k^2 - k^2_{c,mn}}. \tag{3.63}$$

The lowest frequency cutoff is for the TM_{01} mode, with $J_0(2.405) = 0$ so that $\lambda_c = 2.613a$.

For TE modes, E_ϕ is zero at $\rho = a$. All the field components are found from $\psi = H_z$. A prime is used to distinguish the modes from the TM case. It follows from (3.39)–(3.42) that

$$H'_{\rho mn} = \frac{\beta'_{mn}}{jk'_{c,mn}} J'_m(k'_{c,mn}\rho) \begin{Bmatrix} \cos m\phi \\ \sin m\phi \end{Bmatrix} e^{-j\beta'_{mn}z} \tag{3.64}$$

$$H'_{\phi mn} = \frac{m\beta'_{mn}}{k'^2_{c,mn}\rho} J_m(k'_{c,mn}\rho) \begin{Bmatrix} -\sin m\phi \\ \cos m\phi \end{Bmatrix} e^{-j\beta'_{mn}z} \tag{3.65}$$

$$H'_{zmn} = J_m(k'_{c,mn}\rho) \begin{Bmatrix} \cos m\phi \\ \sin m\phi \end{Bmatrix} e^{-j\beta'_{mn}z} \tag{3.66}$$

$$E'_{\rho mn} = H'_{\phi mn} Z'_{mn} \tag{3.67}$$

$$E'_{\phi mn} = -H'_{\rho mn} Z'_{mn} \tag{3.68}$$

$$E'_{zmn} = 0 \tag{3.69}$$

$$Z'_{mn} = k\eta/\beta'_{mn}. \tag{3.70}$$

The requirement $\mathbf{E}_{tan} = 0$ at $\rho = a$ implies that

$$J'_m(k'_{c,mn}a) = 0. \tag{3.71}$$

The first few roots of (3.71) are given in Table C.1. The values $x_n = k'_{c,mn}a$ can be used to find the propagation constant

$$\beta'_{mn} = \sqrt{k^2 - k'^2_{c,mn}}. \tag{3.72}$$

It is noted that unlike the rectangular waveguide, $\beta_{mn} \neq \beta'_{mn}$.

The lowest frequency cutoff occurs for the TE_{11} mode, with $J'_1(1.841) = 0$ so that $\lambda_c = 3.413a$. This is lower than any TM mode, so the TE_{11} mode is the dominant mode. Figure 3.4(b) shows how the electric field lines look.

A Fourier series can be constructed to represent a circular waveguide field component ψ. For example,

$$\psi(\rho, \phi, z) = \sum_m \sum_n A_{mn} J_m(\alpha_n \rho) \begin{Bmatrix} \cos m\phi \\ \sin m\phi \end{Bmatrix} e^{\mp j\beta_{mn} z}. \tag{3.73}$$

For the $\cos m\phi$ and $\sin m\phi$ expansions, we can use the orthogonality integrals (3.35)–(3.37) from the rectangular case with the replacement $x \to \phi$; the period is $2L = 2\pi$. For the ρ expansion, we need to use the orthogonality of Bessel functions

$$\int_0^a J_m(\alpha_p \rho) J_m(\alpha_q \rho) \rho \, d\rho = \begin{cases} (a^2/2) J'^2_m(\alpha_p a); & p = q \\ 0; & p \neq q \end{cases}, \tag{3.74}$$

where $p, q = 1, 2, 3, \ldots$. The orthogonality relation holds, provided that the numbers α_p satisfy

$$J_m(\alpha_p a) = 0. \tag{3.75}$$

A Fourier–Bessel series resembles the more familiar trigonometric series. For example, if $m = 1$, $a = 1$ and the α_p satisfy $J_1(\alpha_p) = 0$, the functions $J_1(\alpha_p \rho)$ resemble $\sin(p\pi x/a)$.

3.4.1   Coaxial Line

In Chapter 2, we saw that a coaxial line supports a TEM wave. It is less obvious is that at sufficiently high frequencies, it can turn into a waveguide, supporting TE_{11} and higher modes. This 'overmoding' is something that should be avoided in a TEM line. We should therefore find out what the TE_{11} cutoff frequency is and stay below it.

The TE modes in the region $a \leq \rho \leq b$ inside the line can be obtained from the axial magnetic field. We start with the general solution (3.53) which for our purposes can be written as

$$H_z = (A J_m(k_c \rho) + B Y_m(k_c \rho)) \begin{Bmatrix} \cos m\phi \\ \sin m\phi \end{Bmatrix} e^{-j\beta z}, \tag{3.76}$$

where A and B are constants, and the separation equation is $k_c^2 + \beta^2 = k^2$. The transverse wavenumber k_c is unknown. This is similar to (3.66), except that we have kept $Y_m(k_c \rho)$ as a possible solution. This is necessary because $\rho = 0$ does not occur.

From H_z, we can find E_ϕ

$$E_\phi = -\frac{k\eta}{jk_c}(A J'_m(k_c \rho) + B Y'_m(k_c \rho)) \begin{Bmatrix} \cos m\phi \\ \sin m\phi \end{Bmatrix} e^{-j\beta z}. \tag{3.77}$$

Requiring $E_\phi = 0$ at $\rho = a$ and $\rho = b$ gives two equations

$$A J'_m(k_c a) + B Y'_m(k_c a) = 0$$

$$AJ'_m(k_c b) + BY'_m(k_c b) = 0$$

from which

$$\frac{J'_m(k_c a)}{J'_m(k_c b)} = \frac{Y'_m(k_c a)}{Y'_m(k_c b)}$$

or

$$J'_m(k_c a)Y'_m(k_c b) - J'_m(k_c b)Y'_m(k_c a) = 0. \qquad (3.78)$$

Equation (3.78) can be solved numerically for the roots k_c. An approximate solution for the smallest k_c and $m = 1$ is

$$k_c \approx \frac{2}{a+b} \qquad (3.79)$$

which gives the $m, n = 1, 1$ cutoff wavenumber. At cutoff, $\beta = 0$, so the cutoff frequency follows from (3.72).

3.5   Waveguide Excitation

The inputs and outputs of waveguides ultimately need to be connected to coaxial cables. This is done with a coax-to-waveguide 'transition'. A transition can be thought of as an antenna inside the waveguide that couples to the waveguide's desired mode.

A few rectangular waveguide transitions for coupling to the dominant TE_{10} mode are shown in Figure 3.5. In case (a), a loop couples to the axial magnetic field H_z. The loop is on the side wall where H_z of the TE_{10} mode is maximum. In case (b), the probe is similar to a monopole antenna that couples to E_y, which is maximum at $x = a/2$. For both (a) and (b), the probe can be positioned at $d = \lambda_z/4$ so that the $-z$-travelling wave that gets reflected from the conducting back wall reinforces the $+z$-travelling wave.

Figure 3.5(c) shows a probe with modifications that provide an improved bandwidth. The probe is shorter and thicker than the one in case (b), and it has a dielectric sleeve, made of a low-loss material such as polyethylene. The spacing $d \sim \lambda_z/8$ is smaller than before, and further matching can be designed within the coaxial connector section. With this type of transition, it is possible to obtain a good match over a waveguide's operating bandwidth, for example, 8–12 GHz for an X-band WR90 guide.

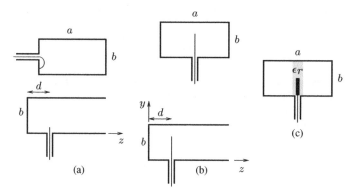

Figure 3.5   Coax to waveguide transitions. (a) Loop coupling to H_z. (b) Probe coupling to E_y. (c) Probe coupling with dielectric sleeve.

If the probe is thin, the input impedance can be found analytically (Collin 1991, Chapter 7). In practice, probes are usually not so thin, so that a better bandwidth can be achieved. In these cases, a numerical solution using the method of moments gives good results for design purposes.

The fundamental idea presented here is that a probe couples to $\mathbf{E}$ whereas a loop couples to $\mathbf{H}$. This can be readily applied to other types of waveguides. For instance, in a circular waveguide, a radial probe or a transverse loop can be used to couple to the fields of the TE_{11} mode. Since the concept is the same as for the rectangular case, the details will not be covered here.

3.6   2D Waveguides

We now consider 2D waveguides that are invariant in the y direction with the guiding in the z direction. Since $\partial/\partial y = 0$, Equations (3.8)–(3.11) become

$$E_x = -\frac{j\beta}{k_x^2}\frac{\partial E_z}{\partial x} \tag{3.80}$$

$$E_y = \frac{j\omega\mu}{k_x^2}\frac{\partial H_z}{\partial x} \tag{3.81}$$

$$H_x = -\frac{j\beta}{k_x^2}\frac{\partial H_z}{\partial x} \tag{3.82}$$

$$H_y = -\frac{j\omega\epsilon}{k_x^2}\frac{\partial E_z}{\partial x} \tag{3.83}$$

where $k_x^2 = k^2 - \beta^2$.

3.6.1   Parallel-Plate Waveguide

A parallel-plate waveguide is shown in Figure 3.6. Perfectly conducting plates are at $x = 0$ and $x = a$; there is no variation in the y direction. The medium between the plates is characterized by μ and ϵ. Three types of modes exist: TEM, TM and TE.

For the TEM mode in Figure 3.6(a), an x-polarized plane wave will meet the required boundary condition $\mathbf{E}_{tan} = 0$. It is

$$E_x = E_0\, e^{-jkz} \tag{3.84}$$

$$H_y = \frac{E_0}{\eta}\, e^{-jkz}, \tag{3.85}$$

where $\eta = \sqrt{\mu/\epsilon}$ and $k = \omega\sqrt{\mu\epsilon}$. It is not possible to have a y-polarized TEM wave propagating along the z direction because the condition $\mathbf{E}_{tan} = 0$ cannot be met on the conductors.

In Figure 3.6(b), the TM modes are found by solving the wave equation (3.12c) for E_z, which in 2D becomes

$$\left(\frac{\partial^2}{\partial x^2} + \frac{\partial^2}{\partial z^2}\right)E_z + k^2 E_z = 0.$$

Using a product solution $E_z(x, z) = X(x)Z(z)$ leads to

$$X''Z + XZ'' + k^2 XZ = 0$$

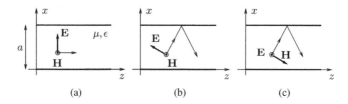

Figure 3.6 Parallel-plate waveguide with (a) TEM mode, (b) TM mode and (c) TE mode.

from which

$$\underbrace{\frac{X''(x)}{X(x)}}_{-k_x^2} + \underbrace{\frac{Z''(z)}{Z(z)}}_{-\beta^2} + k^2 = 0.$$

The boundary conditions require that $E_z = 0$ at the metal plates. The solutions are the TM_n modes

$$E_z(x, z) = E_0 \sin\frac{n\pi x}{a} e^{-j\beta z}; \quad n = 1, 2, 3, \ldots, \tag{3.86}$$

where $k_x = n\pi/a$ and $\beta = \sqrt{k^2 - (n\pi/a)^2}$. From (3.80), (3.83) and (3.86), the other field components can be found and are

$$E_x(x, z) = -\frac{j\beta E_0}{n\pi/a}\cos\frac{n\pi x}{a} e^{-j\beta z} \tag{3.87}$$

$$H_y(x, z) = -\frac{j\omega\epsilon E_0}{n\pi/a}\cos\frac{n\pi x}{a} e^{-j\beta z}. \tag{3.88}$$

In Figure 3.6(c), the TE modes are obtained by a similar procedure. The wave equation (3.12f) is solved for H_z; then (3.81) and (3.82) give E_y and H_x. We require $E_y = 0$ at $x = 0$ and $x = a$. Using (3.81), this translates into the boundary condition

$$\frac{\partial H_z}{\partial x} = 0; \ x = 0, a.$$

The solutions are the TE_n modes, having

$$H_z(x, z) = H_0 \cos\frac{n\pi x}{a} e^{-j\beta z}; \ n = 1, 2, 3, \ldots. \tag{3.89}$$

From (3.81) (3.82) and (3.89), the other field components can be found and are

$$E_y(x, z) = -\frac{j\omega\mu H_0}{n\pi/a}\sin\frac{n\pi x}{a} e^{-j\beta z} \tag{3.90}$$

$$H_x(x, z) = \frac{j\beta H_0}{n\pi/a}\sin\frac{n\pi x}{a} e^{-j\beta z}. \tag{3.91}$$

For both the TM and TE cases, the field's x dependence can be written in terms of complex exponentials. This leads to an interpretation of the modes as plane waves propagating along zigzag paths, as in Figure 3.6(b) and (c). The development is left as an exercise in Problem 3.16.

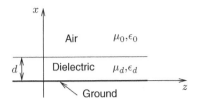

Figure 3.7 Dielectric-slab waveguide on a ground plane, PEC or PMC.

3.6.2 Dielectric Slab on PEC Ground

Figure 3.7 shows a dielectric slab of thickness d on top of a ground plane at $x = 0$. It is assumed that there is no variation in the y direction and that the ground is a perfect electric conductor (PEC). The slab is characterized by ϵ_d and μ_d.[6]

We want to find the surface-wave modes. Surface waves have the property that they propagate in the $+z$ direction inside the slab, and decay exponentially, outside in the $+x$ direction. The decay is often rapid, so the wave appears to be tightly 'bound' to the surface; hence, the name.

Let us first consider the TM case. All the TM field components come from E_z and (3.80)–(3.82). The boundary conditions require that $E_z = 0$ on the metal ground plane, and E_z, H_y are continuous at the dielectric–air interface. In the dielectric region, we need to solve

$$\nabla^2 E_z + k_d^2 E_z = 0, \tag{3.92}$$

where $k_d^2 = \omega^2 \mu_d \epsilon_d$. With $\partial/\partial y = 0$, this becomes

$$\left(\frac{\partial^2}{\partial x^2} + \frac{\partial^2}{\partial z^2} \right) E_z + k_d^2 E_z = 0.$$

Using a product solution $E_z(x, z) = X(x)Z(z)$ leads to

$$X''Z + XZ'' + k_d^2 XZ = 0,$$

and the separation of variables method gives

$$\underbrace{\frac{X''(x)}{X(x)}}_{-k_x^2} + \underbrace{\frac{Z''(z)}{Z(z)}}_{-\beta_d^2} + k_d^2 = 0$$

with the separation constants obeying

$$k_x^2 + \beta_d^2 = k_d^2 = \omega^2 \mu_d \epsilon_d. \tag{3.93}$$

Similarly, in the air region, we need to solve

$$\nabla^2 E_z + k_0^2 E_z = 0, \tag{3.94}$$

---

[6] If $\mu_d \neq \mu_0$, then it also has magnetic properties. Strictly speaking, it can no longer be called a dielectric, but we will do so anyway for convenience, to accommodate the use of duality later on.

where $k_0^2 = \omega^2 \mu_0 \epsilon_0$. Applying the separation of variables method, this becomes

$$\underbrace{\frac{X''(x)}{X(x)}}_{+\alpha^2} + \underbrace{\frac{Z''(z)}{Z(z)}}_{-\beta_0^2} + k_0^2 = 0,$$

where

$$- \alpha^2 + \beta_0^2 = k_0^2 = \omega^2 \mu_0 \epsilon_0. \tag{3.95}$$

Choosing the x separation constant as $\alpha^2 > 0$ gives x solutions that are exponential, having $X(x) = e^{\pm \alpha x}$. The solution $X(x) = e^{-\alpha x}$ is retained, as it provides the exponential decay in the air region that is required for a surface wave.

The tangential fields have to be continuous for all y, z on the interface. This is only possible if the waves in both regions travel with the same β; this is called the phase-matching condition. Therefore, we must have $\beta_d = \beta_0 = \beta$. With phase matching enforced, the separation equations (3.93) and (3.95) become

$$k_x^2 + \beta^2 = k_d^2 = \omega^2 \mu_d \epsilon_d \tag{3.96}$$

$$-\alpha^2 + \beta^2 = k_0^2 = \omega^2 \mu_0 \epsilon_0. \tag{3.97}$$

The constants α, β and k_x are unknown. To find them, it will be convenient to eliminate β. Taking (3.96) minus (3.97) and multiplying by d^2 gives

$$(k_x d)^2 + (\alpha d)^2 = (k_0 d)^2 (\mu_r \epsilon_r - 1) = \ell^2 \tag{3.98}$$

where $\epsilon_r = \epsilon_d / \epsilon_0$ and $\mu_r = \mu_d / \mu_0$. Equation (3.98) represents a circle of radius ℓ in the $k_x d, \alpha d$ plane. A solution in the dielectric that gives $E_z = 0$ on the metal is

$$E_{zd} = E_0 \sin k_x x \, e^{-j\beta z} \tag{3.99}$$

and (3.80), (3.83) give the other field components

$$E_{xd} = \frac{\beta E_0}{jk_x} \cos k_x x \, e^{-j\beta z} \tag{3.100}$$

$$H_{yd} = \frac{\omega \epsilon_d E_0}{jk_x} \cos k_x x \, e^{-j\beta z}. \tag{3.101}$$

In the air region,

$$E_{z0} = K E_0 e^{-\alpha x} e^{-j\beta z} \tag{3.102}$$

and from (3.80), (3.83) with $k_x^2 = -\alpha^2$,

$$E_{x0} = \frac{\beta K E_0}{j\alpha} e^{-\alpha x} e^{-j\beta z} \tag{3.103}$$

and

$$H_{y0} = \frac{\omega \epsilon_0 K E_0}{j\alpha} e^{-\alpha x} e^{-j\beta z}. \tag{3.104}$$

The boundary conditions can be used to find K and k_x. Imposing the continuity of H_y at $x = d$ gives

$$\frac{\epsilon_d}{k_x} \cos k_x d = \frac{K \epsilon_0}{\alpha} e^{-\alpha d} \tag{3.105}$$

and continuity of E_z at $x = d$ gives

$$e^{\alpha d} \sin k_x d = K. \tag{3.106}$$

Using (3.106), the fields in the air region (3.102)–(3.104) become

$$E_{z0} = E_0 \sin k_x d \ e^{-\alpha(x-d)} e^{-j\beta z} \tag{3.107}$$

$$E_{x0} = \frac{\beta E_0}{j\alpha} \sin k_x d \ e^{-\alpha(x-d)} e^{-j\beta z} \tag{3.108}$$

$$H_{y0} = \frac{\omega \epsilon_0 E_0}{j\alpha} \sin k_x d \ e^{-\alpha(x-d)} e^{-j\beta z}. \tag{3.109}$$

Dividing (3.106) by (3.105) gives

$$(k_x d) \tan(k_x d) = \epsilon_r (\alpha d). \tag{3.110}$$

By drawing the circle (3.98) and the tangent function (3.110) as in Figure 3.8, a graphical solution for k_x and α can be found; here, $x = k_x d$, $y = \alpha d$, $\ell = 1.2$ and $\epsilon_r = 4$. Once k_x and α are known, β can be obtained from (3.96) or (3.97). An admissible surface-wave solution should be restricted to the upper half region $y = \alpha d \geq 0$ to ensure that $\alpha \geq 0$ for a decaying wave. The graphical solution gives an estimate which can then be refined with a numerical root-finding algorithm such as the secant method or Newton-Raphson method (described in Appendix G).

The mode (3.99)–(3.101) inside the slab can be interpreted as a zigzag plane wave that propagates at an angle θ with respect to the z axis; then, $\tan \theta = k_x / \beta$. The surface wave undergoes total internal reflection, so θ is smaller than the critical angle $\cos \theta_c = 1/\sqrt{\epsilon_r}$ (if $\mu_r = 1$).

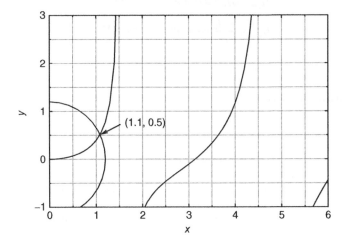

Figure 3.8   Plot of $y = (x/4) \tan x$ and $x^2 + y^2 = 1.2^2$, showing the approximate solution $(1.1, 0.5)$.

Let us now consider more carefully the frequency dependence of this mode. In Figure 3.8, as the frequency is lowered, the circle's radius ℓ decreases. When $0 < \ell < \pi/2$, there is always a solution, so this mode has no lower cutoff. Going the other way, as the frequency is increased, ℓ is increases and eventually $\ell \geq \pi$ – so, a second surface-wave mode appears. This defines the cutoff frequency of the second mode. As the frequency is further increased, more modes will appear. In general, the TM surface-wave cutoff frequencies are defined by $\ell = 0, \pi, 2\pi, \ldots$. We call these the TM_n surface waves, where $n = 0, 1, 2, \ldots$.

So far, we have only allowed for decaying solutions which occur in the upper half region $y = \alpha d \geq 0$ in Figure 3.8. However, we can see that if ℓ is only slightly less than π, there will be a solution for the second mode with $\alpha < 0$. This mode grows exponentially in x and is called an *improper surface wave*. This is in contrast to the decaying modes which are *proper surface waves*. As ℓ is further decreased, we reach a point where the circle just touches the tangent function. At frequencies lower than this, there are no solutions for real values of α. However, solutions with complex α and β become possible. These are called *leaky waves*. Therefore, as the frequency is lowered towards the second mode's cutoff, the surface wave does not simply disappear. It evolves into an improper surface wave and, beyond that, a leaky wave. The leaky wave is not confined by total internal reflection and leaks out of the slab.

Both improper surface waves and leaky waves grow exponentially in the x direction. These wave types cannot exist in isolation. However, they do occur in a physically meaningful way if there is a source of finite extent, such as a dipole. These wave types are further discussed in Section 12.6.

A similar procedure can be applied to find the propagation constants and fields for TE modes. In this case, H_z gives the other field components, E_y and H_x. A solution in the dielectric that gives $E_y = 0$ on the metal is

$$H_{zd} = H_0 \cos k_x x \, e^{-j\beta z} \tag{3.111}$$

$$H_{xd} = \frac{j\beta H_0}{k_x} \sin k_x x \, e^{-j\beta z} \tag{3.112}$$

$$E_{yd} = \frac{\omega \mu_d H_0}{jk_x} \sin k_x x \, e^{-j\beta z}. \tag{3.113}$$

In the air region,

$$H_{z0} = H_0 \cos k_x d \, e^{-\alpha(x-d)} e^{-j\beta z} \tag{3.114}$$

$$H_{x0} = \frac{\beta H_0}{j\alpha} \cos k_x d \, e^{-\alpha(x-d)} e^{-j\beta z} \tag{3.115}$$

$$E_{y0} = \frac{j\omega \mu_0 H_0}{\alpha} \cos k_x d \, e^{-\alpha(x-d)} e^{-j\beta z}. \tag{3.116}$$

In addition, k_x and α must satisfy

$$-(k_x d) \cot(k_x d) = \mu_r(\alpha d). \tag{3.117}$$

Equations (3.96), (3.97) and (3.98) also apply for the TE case, and the simultaneous solution of (3.98) and (3.117) gives k_x, α and β.

In the TE case, the graphical solution involves circles of radius ℓ and the cotangent function. By drawing the circles (3.98) and the cotangent function (3.117) as in Figure 3.9, a graphical solution can be found; here, $x = k_x d$, $y = \alpha d$, $\ell = 2.2$ and $\mu_r = 1$. Once k_x and α are known, β follows from (3.96) or (3.97).

The TE surface-wave cutoff frequencies are defined by $\ell = \pi/2, 3\pi/2, 5\pi/2, \ldots$. We call these the TE_n surface waves, where $n = 1, 2, 3, \ldots$. The lowest TE mode is TE_1 and has its cutoff when $\ell = \pi/2$.

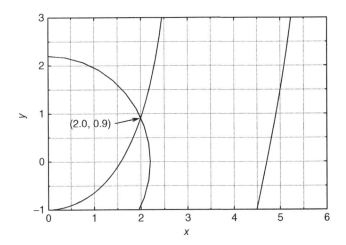

Figure 3.9   Plot of $y = -x \cot x$ and $x^2 + y^2 = 2.2^2$, showing the approximate solution $(2.0, 0.9)$.

This is in contrast to the lowest TM mode which is TM_0 and has no lower cutoff. The modes arranged by increasing cutoff frequency are TM_0, TE_1, TM_1, TE_2, $\cdots$.

Example 3.4 (TE$_1$ Mode Propagation)   A grounded dielectric slab has a thickness of 62 mils and $\epsilon_r = 4$. Find α, β and the phase velocity for the TE_1 mode at 38.5 GHz.

Solution: The thickness is $d = 2.54 \times 62/1000 = 0.1575$ cm. Using (3.98) (and $c = 1/\sqrt{\mu_0 \epsilon_0}$), we find that

$$\ell = k_0 d \sqrt{\mu_r \epsilon_r - 1} = \frac{2\pi f}{c} d \sqrt{\mu_r \epsilon_r - 1} \approx 2.2$$

The graphical solution of (3.98) and (3.117) for this value of ℓ can be found in Figure 3.9, and within graphical accuracy, $k_x d = 2.0$, $\alpha d = 0.9$, from which

$$\alpha = 0.9/d = 570 \text{ np/m}.$$

Using (3.97)

$$\beta = \sqrt{k_0^2 + \alpha^2} = \sqrt{(2\pi f/c)^2 + \alpha^2} = 990 \text{ rad/m}.$$

The phase velocity is $u_p = \omega/\beta = 2.45 \times 10^8$ m/s.   ∎

Example 3.5 (TE$_1$ Mode Cutoff Frequency)   A grounded dielectric slab has a thickness of 62 mils and $\epsilon_r = 4$. Find the cutoff frequency for the TE_1 mode. Also find the mode's zigzag angle with respect to the z axis, at cutoff.

Solution: The thickness is $d = 2.54 \times 62/1000 = 0.1575$ cm. According to (3.98) and (3.117), the lowest TE mode cutoff occurs when $\ell = \pi/2$. From (3.98) (and $c = 1/\sqrt{\mu_0 \epsilon_0}$)

$$\ell = k_0 d \sqrt{\mu_r \epsilon_r - 1} = \frac{\pi}{2}$$

and since $k_0 = 2\pi/\lambda_0 = 2\pi f/c$, cutoff is at

$$f = \frac{c}{4d\sqrt{\mu_r \epsilon_r - 1}} = 27.5 \text{ GHz}.$$

Inside the dielectric, the $\sin k_x x$ term of (3.113) implies that the wave travels along a zigzag path with propagation constants β along z and k_x along x. Therefore, the angle θ with respect to the z axis can be found from the relation $\tan\theta = k_x/\beta$. From (3.96) $k_x = \sqrt{k_d^2 - \beta^2}$ and since $\alpha = 0$ at cutoff, (3.97) gives $\beta = k_0$. Therefore,

$$\tan\theta = \frac{k_x}{k_0} = \sqrt{\mu_r \epsilon_r - 1} = \sqrt{4-1}$$

or $\theta = 60°$.

Note that this is the critical angle for obtaining total internal reflection from the dielectric–air boundary. The same result could have been obtained from $\theta = \cos^{-1}(1/\sqrt{\epsilon_r})$ as discussed in Section 2.4.4.

If the frequency is increased above 27.5 GHz, then $\theta > 60°$ and total internal reflection will be maintained. At frequencies below 27.5 GHz, the mode will no longer be trapped inside the dielectric slab.   ∎

3.6.3   Dielectric Slab on PMC Ground

Figure 3.7 shows a dielectric slab of thickness d on top of a ground plane at $x = 0$. Suppose that the ground plane is a perfect magnetic conductor (PMC). In this case, $\mathbf{H}_{tan} = 0$ on the ground. The fields immediately follow by applying duality to the equations for a dielectric slab on a PEC. The TM PEC equations become the TE PMC case, having $H_z = 0$ on the magnetic conductor. Inside the dielectric, (3.99)–(3.101) become

$$H_{zd} = H_0 \sin k_x x \, e^{-j\beta z} \tag{3.118}$$

$$H_{xd} = \frac{\beta H_0}{jk_x} \cos k_x x \, e^{-j\beta z} \tag{3.119}$$

$$E_{yd} = \frac{j\omega\mu_d H_0}{k_x} \cos k_x x \, e^{-j\beta z}. \tag{3.120}$$

In the air region, (3.102)–(3.104) become

$$H_{z0} = H_0 \sin k_x d \, e^{-\alpha(x-d)} e^{-j\beta z} \tag{3.121}$$

$$H_{x0} = \frac{\beta H_0}{j\alpha} \sin k_x d \, e^{-\alpha(x-d)} e^{-j\beta z} \tag{3.122}$$

$$E_{y0} = \frac{j\omega\mu_0 H_0}{\alpha} \sin k_x d \, e^{-\alpha(x-d)} e^{-j\beta z}. \tag{3.123}$$

The dual of (3.110) is

$$(k_x d)\tan(k_x d) = \mu_r(\alpha d). \tag{3.124}$$

In a similar manner, the TE PEC equations (3.111)–(3.116) and duality give the TM PMC case, having $H_y = 0$ on the magnetic conductor. Inside the dielectric, (3.111)–(3.113) become

$$E_{zd} = E_0 \cos k_x x \, e^{-j\beta z} \tag{3.125}$$

$$E_{xd} = \frac{j\beta E_0}{k_x} \sin k_x x \, e^{-j\beta z} \tag{3.126}$$

$$H_{yd} = \frac{j\omega \epsilon_d E_0}{k_x} \sin k_x x \, e^{-j\beta z}. \tag{3.127}$$

In the air region, (3.114)–(3.116) become

$$E_{z0} = E_0 \cos k_x d \, e^{-\alpha(x-d)} e^{-j\beta z} \tag{3.128}$$

$$E_{x0} = \frac{\beta E_0}{j\alpha} \cos k_x d \, e^{-\alpha(x-d)} e^{-j\beta z} \tag{3.129}$$

$$H_{y0} = \frac{\omega \epsilon_0 E_0}{j\alpha} \cos k_x d \, e^{-\alpha(x-d)} e^{-j\beta z}. \tag{3.130}$$

The dual of (3.117) is

$$-(k_x d) \cot(k_x d) = \epsilon_r(\alpha d). \tag{3.131}$$

3.6.4   Ungrounded Dielectric Slab

A dielectric slab with no ground plane and a thickness $2d$ is shown in Figure 3.10. It turns out that the previously developed modes for the PEC and PMC ground can be used. It is convenient at this point to introduce an even/odd terminology, which refers to the behaviour of the transverse field components. So far, we have discussed four types of modes, in this order: TM even and TE odd for the PEC ground; TE odd and TM even for the PMC ground. This is summarized in Table 3.1.

The ungrounded slab requires continuity of $\mathbf{E}_{tan}$ and $\mathbf{H}_{tan}$ at $x = \pm d$, and exponential decay, as $|x| \to \infty$. Everything we need can be found from the modes in Table 3.1. These satisfy all the necessary conditions in the upper region $0 \le x < \infty$. No changes are needed for $-d \le x \le 0$ For $x < -d$, the fields have to decay as $x \to -\infty$. These can be obtained from the ones for $x > d$ by replacing $\alpha \to -\alpha$ and $d \to -d$.

For an ungrounded slab, the even and odd modes are linearly independent solutions of the wave equation. Since there is no boundary at $x = 0$, both types of solutions are admissible, and any linear combination of even and odd modes in Table 3.1 is a valid solution. For instance, if the two lowest TM modes are present, a general solution inside the dielectric could be of the form

$$E_{zd} = K_1 \sin k_{x1} x \, e^{-j\beta_1 z} + K_2 \cos k_{x2} x \, e^{-j\beta_2 z},$$

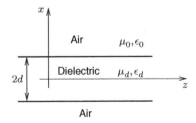

Figure 3.10   Dielectric-slab waveguide with thickness $2d$.

Table 3.1   Summary of TE and TM grounded slab fields for $0 \leq x < \infty$.

PEC ground	PMC ground
TM even	**TE even**
$E_{zd} = E_0 \sin k_x x \, e^{-j\beta z}$	$H_{zd} = H_0 \sin k_x x \, e^{-j\beta z}$
$E_{xd} = \dfrac{\beta E_0}{jk_x} \cos k_x x \, e^{-j\beta z}$	$H_{xd} = \dfrac{\beta H_0}{jk_x} \cos k_x x \, e^{-j\beta z}$
$H_{yd} = \dfrac{\omega \epsilon_d E_0}{jk_x} \cos k_x x \, e^{-j\beta z}$	$E_{yd} = \dfrac{j\omega \mu_d H_0}{k_x} \cos k_x x \, e^{-j\beta z}$
$E_{z0} = E_0 \sin k_x d \, e^{-\alpha(x-d)} e^{-j\beta z}$	$H_{z0} = H_0 \sin k_x d \, e^{-\alpha(x-d)} e^{-j\beta z}$
$E_{x0} = \dfrac{\beta E_0}{j\alpha} \sin k_x d \, e^{-\alpha(x-d)} e^{-j\beta z}$	$H_{x0} = \dfrac{\beta H_0}{j\alpha} \sin k_x d \, e^{-\alpha(x-d)} e^{-j\beta z}$
$H_{y0} = \dfrac{\omega \epsilon_0 E_0}{j\alpha} \sin k_x d \, e^{-\alpha(x-d)} e^{-j\beta z}$	$E_{y0} = \dfrac{j\omega \mu_0 H_0}{\alpha} \sin k_x d \, e^{-\alpha(x-d)} e^{-j\beta z}$
$(k_x d) \tan(k_x d) = \epsilon_r (\alpha d)$	$(k_x d) \tan(k_x d) = \mu_r (\alpha d)$
TE odd	**TM odd**
$H_{zd} = H_0 \cos k_x x \, e^{-j\beta z}$	$E_{zd} = E_0 \cos k_x x \, e^{-j\beta z}$
$H_{xd} = \dfrac{j\beta H_0}{k_x} \sin k_x x \, e^{-j\beta z}$	$E_{xd} = \dfrac{j\beta E_0}{k_x} \sin k_x x \, e^{-j\beta z}$
$E_{yd} = \dfrac{\omega \mu_d H_0}{jk_x} \sin k_x x \, e^{-j\beta z}$	$H_{yd} = \dfrac{j\omega \epsilon_d E_0}{k_x} \sin k_x x \, e^{-j\beta z}$
$H_{z0} = H_0 \cos k_x d \, e^{-\alpha(x-d)} e^{-j\beta z}$	$E_{z0} = E_0 \cos k_x d \, e^{-\alpha(x-d)} e^{-j\beta z}$
$H_{x0} = \dfrac{\beta H_0}{j\alpha} \cos k_x d \, e^{-\alpha(x-d)} e^{-j\beta z}$	$E_{x0} = \dfrac{\beta E_0}{j\alpha} \cos k_x d \, e^{-\alpha(x-d)} e^{-j\beta z}$
$E_{y0} = \dfrac{j\omega \mu_0 H_0}{\alpha} \cos k_x d \, e^{-\alpha(x-d)} e^{-j\beta z}$	$H_{y0} = \dfrac{\omega \epsilon_0 E_0}{j\alpha} \cos k_x d \, e^{-\alpha(x-d)} e^{-j\beta z}$
$-(k_x d) \cot(k_x d) = \mu_r (\alpha d)$	$-(k_x d) \cot(k_x d) = \epsilon_r (\alpha d)$

The even/odd terminology is associated with the transverse field components.

where K_1 and K_2 are arbitrary constants. The propagation constants k_{x1} and β_1 satisfy the TM even equation (3.110); k_{x2} and β_2 satisfy the TM odd equation (3.131).

How much of each mode is excited is another matter. The constants K_1 and K_2 are arbitrary; however, if a specific source is given, then K_1 and K_2 will depend on the type of source being used.

3.7 Transverse Resonance Method

A primary quantity of interest for waveguides is the propagation constant β. This can always be found from the relation

$$\beta = \sqrt{k^2 - k_c^2}; \quad k = \omega\sqrt{\mu\epsilon}.$$

If we know k_c, then β can be found. For rectangular and circular waveguides, k_c can be readily found from the analytical solutions. For other types of waveguides, an analytical solution might be more difficult to find. In these cases, an alternate way of finding k_c would be desirable.

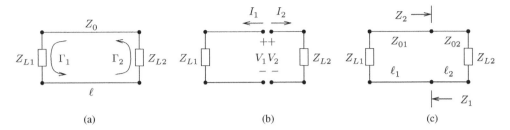

Figure 3.11 Resonance on a transmission line. (a) One line, with characteristic impedance Z_0. (b) Two lines, impedance viewpoint. (c) Two lines, joined to form a resonator.

As a waveguide mode bounces back and forth between the walls, it forms a standing wave. This is called 'transverse resonance'. If the standing wave is along the x direction, it can be described in terms of an equivalent x-directed transmission line, and k_c can be found. Resonances can be in the x and/or y directions or in the ρ direction for a circular waveguide.

The main idea behind resonance is that after a round trip, a wave should end up with the same phase that it originally had. For the transmission-line circuit in Figure 3.11(a), if the loads Z_{L1} and Z_{L2} are lossless, then over a round-trip length of 2ℓ, the resonance condition is $\Gamma_1\Gamma_2 e^{-j2\beta\ell} = 1$. If either load has some loss, then $\Gamma_1\Gamma_2 e^{-j2\beta\ell} < 1$ but the round-trip phase should still be zero (or any multiple of $360°$).

In a more general case, there could be two transmission lines having differing characteristic impedances. For this situation, it is easier to develop the resonance idea using impedance concepts. In Figure 3.11(b), resonance will occur if $V_1 = V_2$ and $I_1 = -I_2$. This is accomplished by connecting the lines, as in Figure 3.11(c). We can then say that

$$Z_1 = \frac{V_1}{I_1} = \frac{V_2}{-I_2} = -Z_2$$

or

$$Z_1 + Z_2 = 0.$$

Sometimes, the 'left-looking impedance' Z_1 and 'right-looking impedance' Z_2 are denoted as $\overleftarrow{Z}$ and $\overrightarrow{Z}$, so that the transverse resonance condition becomes

$$\overleftarrow{Z} + \overrightarrow{Z} = 0. \tag{3.132}$$

The nodes where $\overleftarrow{Z}$ and $\overrightarrow{Z}$ are defined can be positioned anywhere along the transmission line. Some choices are better than others, in terms of keeping the algebra simple. The transverse resonance method requires that we identify characteristic impedances for the equivalent transmission lines. This can be done with a transverse equivalent network (TEN) that was discussed in Section 2.8.

Example 3.6 (Transverse Resonance, Parallel-Plate Waveguide) Find the TM cutoff wavenumbers for the parallel-plate waveguide in Figure 3.12.

Solution: The separation equation is $k_x^2 + \beta^2 = k^2 = \omega^2\mu\epsilon$ and modes have an angle θ such that $\sin\theta = k_x/k$. The transmission-line equivalent is along the x direction, so the characteristic impedance Z_0 of the transmission line should be in terms of the field components that are transverse to x; these are E_z and H_y. Similar to (2.174),

$$Z_0 = -\frac{E_z}{H_y} = \eta\sin\theta = \eta\frac{k_x}{k} = \sqrt{\frac{\mu}{\epsilon}}\frac{k_x}{\omega\sqrt{\mu\epsilon}} = \frac{k_x}{\omega\epsilon}.$$

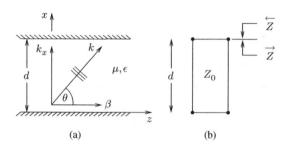

Figure 3.12   (a) Parallel-plate waveguide and (b) transmission-line equivalent.

The input impedance of a short-circuited transmission line stub, of length ℓ, from (2.170) is $Z(\ell) = jZ_0 \tan \beta\ell$. To find $\overleftarrow{Z}$, we note that there is a short-circuit stub of length d. Since the transmission-line equivalent is along x, the associated wavenumber is k_x, so we should replace $\beta\ell$ by $k_x d$. This leads to

$$\overleftarrow{Z} = j\frac{k_x}{\omega\epsilon} \tan k_x d.$$

The other impedance is $\overrightarrow{Z} = 0$ because it is looking into a short circuit. The transverse resonance condition (3.132) becomes

$$\overleftarrow{Z} + \overrightarrow{Z} = j\frac{k_x}{\omega\epsilon} \tan k_x d + 0 = 0.$$

This has roots at $k_x d = n\pi$, so the cutoff wavenumbers are

$$k_x = \frac{n\pi}{d}; \quad n = 0, 1, 2, \dots.$$

This k_x is identical to the result in (3.86), obtained from the wave equation solution.   ∎

Example 3.7 (Transverse Resonance, Grounded Slab) Find an equation for the TM cutoff wavenumbers for a dielectric slab on a PEC ground, in Figure 3.13.

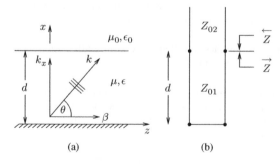

Figure 3.13   (a) Dielectric slab on PEC ground and (b) transmission-line equivalent.

Solution: Looking down into the dielectric at $x = d$, the structure looks the same as in the previous example. The dispersion relation, characteristic impedance and $\overleftarrow{Z}$ are unchanged: $k_x^2 + \beta^2 = k^2 = \omega^2 \mu \epsilon$, $Z_{01} = k_x/(\omega\epsilon)$ and $\overleftarrow{Z} = j(k_x/\omega\epsilon)\tan k_x d$.

In the air region $x \geq d$, the TM wave should decay in x and propagate in z, so we can say that the magnetic field has to be of the form

$$\mathbf{H} = \hat{y}H_0 e^{-\alpha x} e^{-j\beta z},$$

and if substituted into the wave equation in (3.12e), it yields the condition $-\alpha^2 + \beta^2 = k_0^2 = \omega^2 \mu_0 \epsilon_0$.

We can take the curl of $\mathbf{H}$ to find $\mathbf{E}$, in particular, $\partial H_y/\partial x = j\omega\epsilon_0 E_z$ which can be used to obtain the transmission-line characteristic impedance in the air region

$$Z_{02} = -\frac{E_z}{H_y} = -\frac{1}{j\omega\epsilon_0}\frac{\partial H_y/\partial x}{H_y} = \frac{\alpha}{j\omega\epsilon_0}.$$

The transmission line at $x \geq d$ is infinitely long, and its characteristic impedance is Z_{02}, so

$$\overrightarrow{Z} = Z_{02} = \frac{\alpha}{j\omega\epsilon_0}.$$

Applying the transverse resonance condition (3.132),

$$\overleftarrow{Z} + \overrightarrow{Z} = j\frac{k_x}{\omega\epsilon}\tan k_x d + \frac{\alpha}{j\omega\epsilon_0} = 0.$$

This can be rewritten as

$$k_x d \tan k_x d = \epsilon_r \alpha d.$$

This k_x is identical to the result (3.105) from the wave equation solution.   ∎

Both of these examples showed that if we know the transmission-line characteristic impedances, we can find k_x and hence β. This is easier than solving the wave equation boundary value problem. Although the transverse resonance method is good for finding β, it does not help us find the fields for a particular excitation. If we want to know that, then the wave equation has to be solved and there is no shortcut.

3.8   Other Waveguide Types

A rectangular waveguide with $a = 2b$ can have at best a 2:1 bandwidth for single-mode operation. One way to improve the bandwidth is to use a *ridge waveguide*. Another configuration is the *finline*. Both of these are now discussed.

For any waveguide, the propagation constant can be obtained from

$$\beta = \sqrt{k^2 - k_c^2}.$$

Therefore, if we can find the cutoff wavenumber k_c, we will know β. The formulations presented here follow Collin (2001, Chapter 3), which uses the transverse resonance method to find k_c.

3.8.1   Ridge Waveguide

One way to improve the bandwidth of a rectangular waveguide is to put a metallic ridge, as shown in Figure 3.14(a). The ridge has a width w, and the vertical gap size is s. The field in the gap resembles a parallel-plate transmission line which has no lower cutoff. Within a waveguide, a mode cannot be TEM;

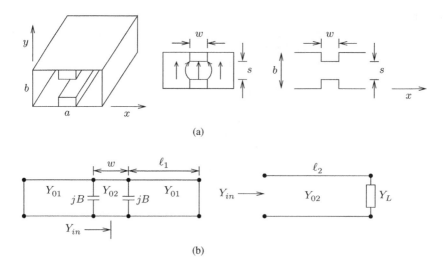

Figure 3.14   (a) Ridge waveguide, electric field lines and parallel-plate model. (b) Transverse network.

nevertheless, the ridge lowers the cutoff frequency of the dominant mode and raises the cutoff frequency of the next higher mode. In practice, bandwidths of 5:1 or more can be possible.

When applying the transverse resonance method, operation at cutoff is assumed so that $\beta = 0$ and the fields have no z dependence. It is a very good approximation to represent the ridge waveguide as three parallel-plate waveguides along the x direction, with plate spacings b, s and b. This in turn can be analysed as three transmission lines, as in Figure 3.14(b). The shunt capacitors are associated with step discontinuities. The characteristic impedances are proportional to the plate spacings, so $Y_{02}/Y_{01} = b/s$.

At cutoff the mode behaves as a standing wave in the x direction with a wavenumber k_c, which can be found from the transverse resonance method. The transverse resonance condition (3.132) can just as well be written as $\overleftarrow{Y} + \overrightarrow{Y} = 0$. Choosing the terminal plane at $x = a/2$ means that $\overleftarrow{Y} = \overrightarrow{Y}$ and in turn $Y_{in} = 0$. Referring to Figure 3.14(b), the load Y_L is the parallel combination of a short-circuit stub and one capacitor. The stub's length is $\ell_1 = (a - w)/2$. The stub's admittance follows from (2.170) with $\beta \to k_c$ and is $Y = -jY_{01} \cot k_c \ell_1$. In parallel with the capacitor, this becomes

$$Y_L = jB - jY_{01} \cot k_c \ell_1.$$

The transmission line of length $\ell_2 = w/2$ represents the ridge. From (2.169) with $\beta \to k_c$, the line Y_{02} terminated by Y_L has an admittance

$$Y_{in} = Y_{02} \frac{Y_L + jY_{02} \tan k_c \ell_2}{Y_{02} + jY_L \tan k_c \ell_2}.$$

Enforcing $Y_{in} = 0$ gives

$$B - Y_{01} \cot k_c (a - w)/2 + Y_{02} \tan k_c w/2 = 0. \tag{3.133}$$

The capacitance associated with the step can be approximately solved using a quasistatic assumption and conformal mapping (Marcuvitz 1951) with the result that

$$B = Y_{01} \frac{2b}{\lambda_c} \left(1 - \ln 4u + \frac{1}{3} u^2 + \frac{1}{2}(1 - u^2) \frac{b^2}{\lambda_c^2} \right); \quad u = \frac{s}{b} \le 0.5. \tag{3.134}$$

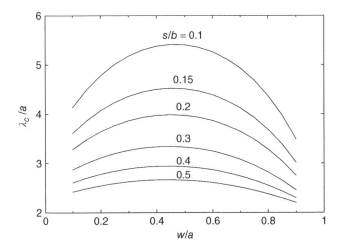

Figure 3.15 Ridge waveguide with $a = 2b$. Normalized cutoff wavelength λ_c/a versus ridge width w for several gap sizes s.

The transcendental equation (3.133) together with (3.134) for B can be solved numerically for k_c, using the secant method; see Appendix G. Some numerical results from PROGRAM ridgewg are shown in Figure 3.15. With no ridge, cutoff occurs when $\lambda_c/a = 2$. The results show that λ_c is increased significantly (the cutoff frequency $f_c = u_0/\lambda_c$ is lowered) by the ridge.

3.8.2   Finline

If w in a ridge waveguide is made very small, it becomes a *finline* as shown in Figure 3.16(a). Similar to a ridge, the conducting fins increase the dominant mode's bandwidth. If the gap s is small, it can provide a place to connect two-terminal surface-mount devices such as diodes. The fins can be thin and fragile, so in practice they can be backed by a thin dielectric card to provide mechanical support. The finline gap s can be tapered along the z direction, to make impedance-matching networks.

Like the ridge guide, finding k_c can be done with the transverse resonance method. Figure 3.16(b) shows the transverse network. The fin is modelled as a shunt capacitance, and the propagation along x has a wavenumber k_c. It is convenient to gain access to terminals at the exact centre $x = a/2$ by splitting the shunt capacitance into two. The short-circuit stub in Figure 3.16(b) has an admittance of $Y = -jY_0 \cot k_c a/2$, which adds to the parallel capacitor. The transverse resonance condition $Y_{in} = 0$ becomes

$$Y_{in} = jB/2 - jY_0 \cot k_c a/2 = 0. \tag{3.135}$$

A good approximation for the shunt susceptance of the fin is

$$B = Y_0 \frac{k_c b}{\pi}\left(-\ln \alpha_2 + \sum_{n=1}^{4} Q_n P_n^2 + \frac{Q_1 P_1^2}{1 + Q_1 \alpha_2^2}\right), \tag{3.136}$$

where

$$Q_n = \frac{\pi}{\Gamma_n b} - \frac{1}{n}$$

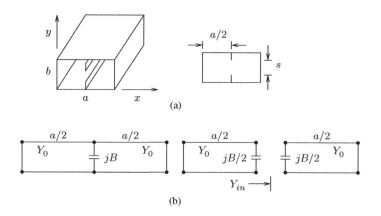

Figure 3.16 (a) Finline geometry. (b) Transverse network.

$$\Gamma_n = \sqrt{\left(\frac{n\pi}{b}\right)^2 - k_c^2}$$

$$P_1 = \alpha_1; \quad P_2 = 2\alpha_1^2 + \alpha_2^2 - 1; \quad P_3 = 4\alpha_1^3 + 6\alpha_1\alpha_2^2 - 3\alpha_1$$

$$P_4 = 8\alpha_1^4 + 3\alpha_2^4 + 24\alpha_1^2\alpha_2^2 - 8\alpha_1^2 - 4\alpha_2^2 + 1$$

$$\alpha_1 = \cos^2\pi s/2b; \quad \alpha_2 = \sin^2\pi s/2b.$$

Numerically solving the transcendental equation (3.135) with B from (3.136) gives k_c. The normalized cutoff wavelength $\lambda_c/a = 2\pi/k_c a$ is shown in Figure 3.17. In the limiting case as $s \to b$, the fins disappear and the dominant mode cutoff $\lambda_c = 2a$ is recovered.

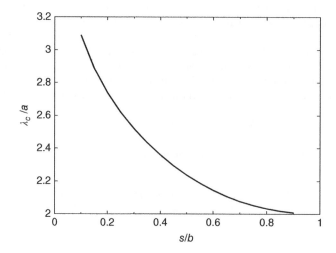

Figure 3.17 Finline with $a = 2b$. Normalized cutoff wavelength λ_c/a versus gap size s.

3.9   Waveguide Discontinuities

This section describes the effect of waveguide discontinuities on mode propagation. The discussion
will be restricted to rectangular waveguides, with the understanding that the analysis techniques can
be extended to other cross-section shapes if needed.

The first type of discontinuity occurs in a waveguide having transverse dimensions $a \times b$ and a transverse obstacle such as a thin metal sheet with an aperture or a metal post. The second type of discontinuity
is a step, whereby two waveguides having different cross-section sizes are joined.

3.9.1   Irises and Posts

Suppose that the cross section of a waveguide is partially obstructed at $z = 0$ by an infinitely thin metal
plate with an opening (an 'iris'). This is shown in Figure 3.18(a). The iris causes a reflection with properties that resemble a shunt capacitor or inductor on a transmission line, as in Figure 3.18(b). This is useful
for making waveguide impedance-matching circuits and filters. Besides irises, it is also possible to use
thin metal posts. We now discuss the properties of these elements.

It is assumed that the waveguide dimensions a and b are chosen so that only the dominant TE_{10} mode
propagates. Under this condition, the incident and reflected waves in the $z \leq 0$ region can be written as

$$E_y = E_0 \sin \frac{\pi x}{a} e^{-j\beta z} + \Gamma E_0 \sin \frac{\pi x}{a} e^{j\beta z},$$

where Γ is the ratio of the reflected and incident electric fields at $z = 0^-$.

It is common to represent a waveguide with an equivalent transmission line circuit, as in Figure 3.18(b).
Cases (a) and (b) are equivalent in the sense that they should have the same reflection coefficient. On a
transmission line, there is a 1:1 relation between Γ and the normalized admittance $\bar{Y}$ so that

$$\bar{Y} = \bar{G} + j\bar{B} = \frac{1 - \Gamma}{1 + \Gamma}.$$

Conversely, we also have that

$$\Gamma = \frac{1 - \bar{Y}}{1 + \bar{Y}}.$$

It does not really matter if we specify Γ or $\bar{Y}$; we should think of these as freely interchangeable quantities, as one can always be found from the other. In a waveguide, Γ is always physically meaningful and
hence so is $\bar{Y}$. For a waveguide, there is no unique way to define Y_0 and the unnormalized admittance
$Y = \bar{Y}Y_0$. Fortunately, it turns out that finding Y is unnecessary and can simply be avoided.

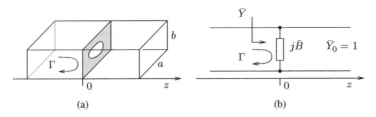

(a)   (b)

Figure 3.18   (a) Reflection by a thin metal iris at $z = 0$. (b) Equivalent transmission line and shunt
susceptance $j\bar{B}$.

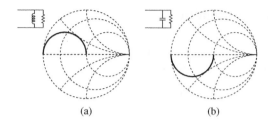

                   (a)   (b)

Figure 3.19   Possible reflection coefficients for thin metal irises; (a) shunt inductance and (b) shunt capacitance.

Table 3.2   Inductive shunt susceptances.

Type	Geometry	Normalized susceptance
Asymmetric iris		$\bar{B} = -\dfrac{2\pi}{\beta a}\cot^2\dfrac{\pi c}{2a}\left(1 + \csc^2\dfrac{\pi c}{2a}\right)$
Symmetric iris		$\bar{B} = -\dfrac{2\pi}{\beta a}\cot^2\dfrac{\pi c}{2a}\left(1 + \dfrac{a\gamma_3 - 3\pi}{4\pi}\sin^2\dfrac{\pi c}{a}\right)$
Thin circular post		$\bar{B} = -\dfrac{4\pi}{\beta a}\left(\ln\dfrac{a}{\pi t} - 1 + \dfrac{6\pi a - 2a^2\gamma_3}{3\pi^2 t^2 \gamma_3}\sin^2\dfrac{3\pi t}{a}\right)^{-1}$
Small hole		$\bar{B} = -\dfrac{3ab}{8\beta r^3}$

where $\beta = \sqrt{k_0^2 - (\pi/a)^2}$ and $\gamma_3 = \sqrt{(3\pi/a)^2 - k_0^2}$.

In Figure 3.18(b), the shunt $j\bar{B}$ is in parallel with a transmission line. The line has a characteristic admittance Y_0 which could be anything, but by definition, the normalized admittance has to be $\bar{Y}_0 = 1$. The parallel combination of the $z \geq 0$ line and the shunt has an admittance of $\bar{Y} = 1 + j\bar{B}$, so the admittance looking to the right at $z = 0^-$ is

$$\bar{Y} = 1 + j\bar{B} = \frac{1 - \Gamma}{1 + \Gamma}.$$

Shunt elements of the type in Figure 3.18(a) always produce a reflection having $\text{Re }\bar{Y} = 1$. This has to be so, because if the obstacle disappears, then $j\bar{B} \to 0$ and we must have $\Gamma \to 0$. When $\bar{B} > 0$, the shunt element is 'capacitive', whereas when $\bar{B} < 0$, it is 'inductive'. Clearly, these are not actual capacitors or inductors, but the terminology conveniently describes the actual reflection by the waveguide shunt elements. The possible values of Γ for waveguide shunt susceptances are illustrated in Figure 3.19, using a Smith chart.

Some shunt inductive elements are shown in Table 3.2. The waveguide cross section is $0 \leq x \leq a$ and $0 \leq y \leq b$, with $b < a$. The irises are infinitely thin metal sheets. The asymmetric iris is metal for $c \leq x \leq a$, and the region $0 \leq x \leq c$ is open. The symmetric iris has an opening of width c. The thin

Table 3.3 Capacitive shunt susceptances.

Type	Geometry	Normalized susceptance
Asymmetric iris		$\bar{B} = \dfrac{4\beta b}{\pi}\left[\ln\csc\dfrac{\pi c}{2b} + \left(\dfrac{\pi}{b\gamma_1} - 1\right)\cos^4\dfrac{\pi c}{2b}\right]$
Symmetric iris		$\bar{B} = \dfrac{2\beta b}{\pi}\left[\ln\csc\dfrac{\pi c}{2b} + \left(\dfrac{2\pi}{b\gamma_2} - 1\right)\cos^4\dfrac{\pi c}{2b}\right]$

where $\beta = \sqrt{k_0^2 - (\pi/a)^2}$, $\gamma_1 = \sqrt{(\pi/b)^2 - \beta^2}$, and $\gamma_2 = \sqrt{(2\pi/b)^2 - \beta^2}$.

circular post is metal, is at the centre $x = a/2$ and has a diameter of $2t$. The circular iris is a small hole with a radius r such that $r \ll \lambda_0$. In the first three cases, the irises have no variation in y (the E plane) and the associated fields are a combination of TE_{m0} modes. The non-propagating $m \geq 2$ modes are associated with stored magnetic energy which leads to an inductive behaviour.

Some shunt capacitive elements are shown in Table 3.3. The waveguide cross section is $0 \leq x \leq a$ and $0 \leq y \leq b$, with $b < a$. The irises are infinitely thin metal sheets. The asymmetric iris is metal for $c \leq y \leq b$, and the region $0 \leq y \leq c$ is open. The symmetric iris has an opening of width c. The irises have no variation in x (the H plane), and the associated fields are a combination of TE_{1n} modes. The non-propagating $n \geq 1$ modes are associated with stored electric energy which leads to a capacitive behaviour.

One particularly useful shunt element is a short vertical post. It can be made with a screw at $x = a/2$ in the broad wall. It is very easy to adjust and provides a variable shunt capacitance. The expression for $j\bar{B}$ is complicated and is not given here, but can be found in the references at the end of the chapter. The capacitive behaviour of such a small 'bump' can be appreciated by solving Problem 3.41.

The formulas in Tables 3.2 and 3.3 are from Collin (2001). These are approximate formulas but are highly satisfactory for most engineering purposes. They were obtained by using a variational method, introduced by J. Schwinger in the 1940s. The variational method is discussed in Chapter 7.

Example 3.8 (Asymmetric Iris, Reflection) Find the reflection coefficient for a WR90 inductive asymmetric waveguide iris, at 9.37 GHz. The iris opening is $c = 0.5a$. The end of the waveguide is terminated in a matched load.

Solution: The inductive iris and matched load form the equivalent circuit shown in Figure 3.20.

The WR90 waveguide has $a = 0.9''$ and $b = 0.4''$. At 9.37 GHz, $k_0 = 2\pi/\lambda_0 = 196$ rad/m and $\beta = \sqrt{k_0^2 - (\pi/a)^2} = 140$ rad/m. With $c = 0.5a$, Table 3.2 gives

$$\bar{B} = -\frac{2\pi}{\beta a}\cot^2\frac{\pi c}{2a}\left(1 + \csc^2\frac{\pi c}{2a}\right) = -5.88$$

so that the admittance is $\bar{Y} = 1 - j5.88$ and

$$\Gamma = \frac{1 - \bar{Y}}{1 + \bar{Y}} = 0.947\underline{/161.21°}.$$

The reflection coefficient is in the second quadrant of the complex Γ plane, as expected in Figure 3.19(a). ∎

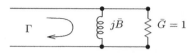

Figure 3.20 Equivalent circuit.

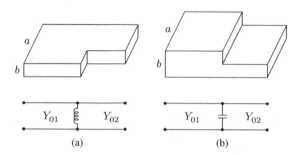

Figure 3.21 Waveguide step discontinuities in the (a) H plane and (b) E plane.

3.9.2 Waveguide Step

Some junctions of dissimilar waveguides and their equivalent circuits are shown in Figure 3.21. It will be assumed that a and b are chosen so that the dominant TE_{10} mode is propagating in the larger guide. For the H-plane step in Figure 3.21(a), the TE_{10} mode might or might not propagate in the narrower guide, depending on its width. For the E-plane step Figure 3.21(b), the TE_{10} mode will propagate in the narrower guide, because the TE_{10} cutoff frequency depends on a but not b. Similar to an iris, an H-plane discontinuity is associated with stored magnetic energy and a shunt inductance; an E-plane discontinuity is associated with stored electric energy and a shunt capacitance. In each waveguide, the transverse wave impedance $Z = -E_y/H_x$ will be different, so the transmission line model needs to account for this by having different characteristic admittances Y_{01} and Y_{02}. Accurate approximate formulas for the equivalent circuit parameters can be obtained by using the variational method. The formulas are not given here, but can be found in the references at the end of the chapter.

3.10 Mode Matching

Mode matching (MM) is a powerful and versatile technique for finding the fields in waveguides having discontinuities. Some examples are a step discontinuity, a waveguide iris, or a junction between a waveguide and the flare of a horn antenna.

The basic idea is to use mode expansions with unknown coefficients on either side of the discontinuity. By enforcing the continuity of $\mathbf{E}_{tan}$ and $\mathbf{H}_{tan}$ across the opening at the junction and $\mathbf{E}_{tan} = 0$ on any metal portions, a solution for the coefficients is obtained. Complete coverage of the topic is not possible here, but some of the main ideas will be demonstrated by applications to an H-plane waveguide step and an inductive asymmetric iris.

3.10.1 H-Plane Step

The mode-matching technique is illustrated here by applying it to the H-plane step in Figure 3.21(a). Because the structure is invariant in the vertical direction, we can find a solution in terms of TE_{m0} modes. The $x - z$ plane (horizontal) is shown in Figure 3.22. The left waveguide at $z \leq 0$ has $0 \leq x \leq a$, and

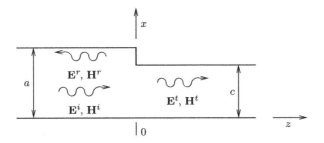

Figure 3.22   Rectangular waveguide with an H-plane step at $z = 0$.

the right waveguide at $z \geq 0$ has $0 \leq x \leq c$. It is assumed that the right waveguide is narrower so that $c \leq a$. Both waveguides have $0 \leq y \leq b$.

$\mathbf{E}_{tan}$ and $\mathbf{H}_{tan}$ have to be continuous across the opening at $z = 0$. Denoting $z = 0^-$ as $(0-)$ and $z = 0^+$ as $(0+)$,

$$E_y^i(0-) + E_y^r(0-) = E_y^t(0+); \quad 0 \leq x \leq c \tag{3.137}$$

$$H_x^i(0-) + H_x^r(0-) = H_x^t(0+); \quad 0 \leq x \leq c. \tag{3.138}$$

At the metal step, $\mathbf{E}_{tan} = 0$ so that

$$E_y^i(0-) + E_y^r(0-) = 0; \quad c \leq x \leq a. \tag{3.139}$$

The fields are expanded as a Fourier series of modes

$$E_y^i = \sin \frac{\pi x}{a} e^{-j\beta_{a1} z} \tag{3.140}$$

$$E_y^r = \sum_{n=1}^{\infty} A_n \sin \frac{n\pi x}{a} e^{j\beta_{an} z} \tag{3.141}$$

$$E_y^t = \sum_{m=1}^{\infty} B_m \sin \frac{m\pi x}{c} e^{-j\beta_{cm} z} \tag{3.142}$$

and

$$H_x^i = \frac{-1}{Z_{a1}} \sin \frac{\pi x}{a} e^{-j\beta_{a1} z} \tag{3.143}$$

$$H_x^r = \sum_{n=1}^{\infty} \frac{A_n}{Z_{an}} \sin \frac{n\pi x}{a} e^{j\beta_{an} z} \tag{3.144}$$

$$H_x^t = \sum_{m=1}^{\infty} \frac{-B_m}{Z_{cm}} \sin \frac{m\pi x}{c} e^{-j\beta_{cm} z}. \tag{3.145}$$

Here, $\beta_{an} = \sqrt{k_0^2 - (n\pi/a)^2}$ and $\beta_{cm} = \sqrt{k_0^2 - (m\pi/c)^2}$. In the left waveguide, the transverse magnetic field for a $\pm z$-travelling mode n is $H_x = \mp E_y/Z_{an}$ where

$$Z_{an} = \frac{k_0 \eta}{\beta_{an}}.$$

In the right waveguide,

$$Z_{cm} = \frac{k_0 \eta}{\beta_{cm}}.$$

The boundary conditions (3.137) and (3.139) are applied to the tangential electric fields (3.140)–(3.142) at $z = 0$ so that

$$\sin \frac{\pi x}{a} + \sum_{n=1}^{\infty} A_n \sin \frac{n\pi x}{a} = \begin{cases} \displaystyle\sum_{m=1}^{\infty} B_m \sin \frac{m\pi x}{c}; \ 0 \le x \le c \\ 0; \ c \le x \le a. \end{cases} \tag{3.146}$$

Multiplying both sides by $\sin p\pi x/a$ and integrating over $(0, a)$ gives

$$\frac{a}{2}\delta_{p1} + \frac{a}{2}A_p = \sum_{m=1}^{\infty} B_m I_{pm}, \tag{3.147}$$

where

$$I_{pm} = \int_0^c \sin \frac{p\pi x}{a} \sin \frac{m\pi x}{c} \, dx. \tag{3.148}$$

Note that $E_y = 0$ on $c \le x \le a$, so the upper limit in (3.148) is equal to c. The integral I_{pm} is easily done in closed form.

The boundary condition (3.138) is applied to the tangential magnetic fields (3.143)–(3.145) at $z = 0$ so that

$$\frac{-1}{Z_{a1}} \sin \frac{\pi x}{a} + \sum_{n=1}^{\infty} \frac{A_n}{Z_{an}} \sin \frac{n\pi x}{a} = \sum_{m=1}^{\infty} \frac{-B_m}{Z_{cm}} \sin \frac{m\pi x}{c}; \quad 0 \le x \le c. \tag{3.149}$$

Multiplying both sides by $\sin q\pi x/c$ and integrating over $(0, c)$ gives

$$\frac{-I_{1q}}{Z_{a1}} + \sum_{n=1}^{\infty} \frac{A_n}{Z_{an}} I_{nq} = -\frac{B_q c}{2Z_{cq}}. \tag{3.150}$$

B_m in (3.147) can be eliminated by using (3.150) (with $q \to m$). The result, after replacing the summation index $m \to q$ and changing $p \to m$, is

$$\frac{a}{2}A_m + \sum_{n=1}^{\infty}\sum_{q=1}^{\infty} \frac{2Z_{cq}}{cZ_{an}} I_{mq} I_{nq} A_n = -\frac{a}{2}\delta_{m1} + \sum_{q=1}^{\infty} I_{mq} I_{1q} \frac{2Z_{cq}}{cZ_{a1}}. \tag{3.151}$$

An actual solution would use a finite number of modes N. In this case, (3.151) can be written as an $N \times N$ matrix equation

$$S_m = \sum_{n=1}^{N} R_{mn} A_n, \tag{3.152}$$

where $m = 1, 2, \ldots, N$ and

$$S_m = \sum_{q=1}^{N} \frac{2Z_{cq}}{cZ_{a1}} I_{mq} I_{1q} - \frac{a}{2}\delta_{m1} \tag{3.153}$$

$$R_{mn} = \sum_{q=1}^{N} \frac{2Z_{cq}}{cZ_{an}} I_{mq} I_{nq} + \frac{a}{2}\delta_{mn}. \tag{3.154}$$

It is straightforward to solve (3.152)–(3.154) and obtain the A_n. The incident wave (3.140) is a unit-amplitude TE_{10} mode, so the reflection coefficient for the dominant mode is simply $\Gamma = A_1$. The normalized impedance looking right at $z = 0^-$ follows from the reflection coefficient and is

$$\bar{Z} = \bar{R} + j\bar{X} = \frac{1 + \Gamma}{1 - \Gamma}.$$

It is important to choose the correct sign of the square root when calculating β, so that $\operatorname{Im} \beta < 0$. This is necessary to ensure that the evanescent modes decay, away from the waveguide junction. The requirement is easily coded in the following way (for β_{an} in this case)

```
cj=(0.,1.)
tmp=k0**2-(n*pi/a)**2
IF(tmp .GE. 0.)THEN
beta_a=SQRT(tmp)
ELSE
beta_a=-cj*SQRT(-tmp)
ENDIF
...
```

Figure 3.23 shows the normalized impedance of the waveguide step, as a function of c/a. The frequency is 9.37 GHz, and the waveguide type is WR90. The H-plane discontinuity gives rise to stored magnetic energy and an inductive behaviour. The numerical results show that $\bar{X} > 0$, which is inductive. When $c/a < 0.7$, the right waveguide is below cutoff, so no power can go into it; as a result, $\bar{R} = 0$. When $c/a = 1$, the step has disappeared and no reflection takes place, so $\bar{R} + j\bar{X} = 1 + j0$ which implies that $\Gamma = 0$. The solution used $N = 10$ modes. Computations were done with PROGRAM wgstep.

It is also possible to define a transmission coefficient T. If only the 10 mode is propagating in each waveguide, (3.140)–(3.142) imply that E_y for the dominant mode should obey $1 + A_1 = B_1$, which can also be written as $1 + \Gamma = T$.

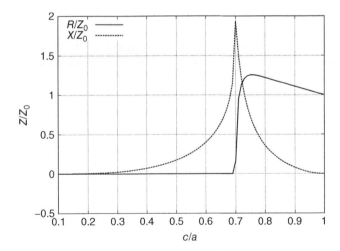

Figure 3.23   H-plane step, mode-matching solution for the normalized impedance $\bar{Z} = Z/Z_0$ looking right at $z = 0$. The left waveguide is WR90, and $f = 9.37$ GHz.

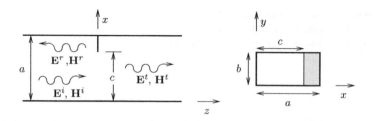

Figure 3.24   Rectangular waveguide with inductive iris at $c \leq x \leq a$; $z = 0$.

3.10.2   Inductive Iris

The MM technique is illustrated here by applying it to an H-plane (inductive) iris in a rectangular waveguide. The structure is shown in Figure 3.24. The waveguide cross section is $0 \leq x \leq a$, $0 \leq y \leq b$. At $z = 0$, an infinitely thin metal iris forms a partial obstruction. The region $0 \leq x \leq c$ is open, and $c \leq x \leq a$ is blocked. Because the structure is invariant in y, we can find a solution in terms of TE_{m0} modes.

At the opening, $0 \leq x \leq c$, $\mathbf{E}_{tan}$ and $\mathbf{H}_{tan}$ are continuous. Denoting $z = 0^-$ as $(0-)$ and $z = 0^+$ as $(0+)$,

$$E_y^i(0-) + E_y^r(0-) = E_y^t(0+) \tag{3.155}$$

$$H_x^i(0-) + H_x^r(0-) = H_x^t(0+). \tag{3.156}$$

At the iris, $c \leq x \leq a$, $\mathbf{E}_{tan} = 0$ so that

$$E_y^i(0-) + E_y^r(0-) = E_y^t(0+) = 0. \tag{3.157}$$

Surface current flows on the thin iris and causes a jump discontinuity in $\mathbf{H}_{tan}$ in accordance with (1.43) so that

$$H_x^i(0-) + H_x^r(0-) = H_x^t(0+) - J_{sy}. \tag{3.158}$$

The surface current J_{sy} is responsible for producing the iris scattered field. We will call this H_x^s.

The next step is to find a way to eliminate J_{sy} in (3.158) so that we can have a boundary condition involving only fields. From symmetry considerations, the scattered field produced by J_{sy} has the property

$$H_x^s(0-) = -H_x^s(0+). \tag{3.159}$$

Furthermore, the scattered field is related to the reflected and transmitted field definitions in the following way:

$$H_x^r = H_x^s; \quad z \leq 0 \tag{3.160}$$

$$H_x^i + H_x^s = H_x^t; \quad z \geq 0. \tag{3.161}$$

Using (3.160) and (3.161) in (3.158) gives

$$H_x^i(0-) + H_x^s(0-) = H_x^i(0+) + H_x^s(0+) - J_{sy}. \tag{3.162}$$

The jump in the total H_x at $z = 0$ is related to J_{sy}. Since H_x^i is continuous at $z = 0$, the jump condition (3.162) and symmetry condition (3.159) lead to

$$J_{sy} = H_x^s(0+) - H_x^s(0-) = 2H_x^s(0+).$$ (3.163)

Using (3.163) in (3.162) to eliminate J_{sy} gives

$$H_x^i(0-) + H_x^s(0-) = H_x^i(0+) - H_x^s(0+).$$ (3.164)

Following the definitions in (3.160) and (3.161) and noting that H_x^i is continuous at $z = 0$, we can rewrite (3.164) as

$$H_x^i(0-) + H_x^r(0-) = 2H_x^i(0) - H_x^t(0+); \quad c \le x \le a.$$ (3.165)

Equation (3.165) replaces (3.158). It is preferable because it eliminates the problematic J_{sy} term and gives the boundary condition purely in terms of the field.

The boundary conditions are now applied to the fields, which are expanded in terms of modes

$$E_y^i = \sin\frac{\pi x}{a}e^{-j\beta_1 z}$$ (3.166)

$$E_y^r = \sum_{n=1}^{\infty} A_n \sin\frac{n\pi x}{a}e^{j\beta_n z}$$ (3.167)

$$E_y^t = \sum_{m=1}^{\infty} B_m \sin\frac{m\pi x}{a}e^{-j\beta_m z}$$ (3.168)

and

$$H_x^i = \frac{-1}{Z_1}\sin\frac{\pi x}{a}e^{-j\beta_1 z}$$ (3.169)

$$H_x^r = \sum_{n=1}^{\infty}\frac{A_n}{Z_n}\sin\frac{n\pi x}{a}e^{j\beta_n z}$$ (3.170)

$$H_x^t = \sum_{m=1}^{\infty}\frac{-B_m}{Z_m}\sin\frac{m\pi x}{a}e^{-j\beta_m z}.$$ (3.171)

Here, $\beta_n = \sqrt{k_0^2 - (n\pi/a)^2}$. The transverse magnetic field for a $\pm z$-travelling mode n is $H_x = \mp E_y/Z_n$ where

$$Z_n = \frac{k_0\eta}{\beta_n}.$$

The boundary conditions (3.155) and (3.157) are applied to the tangential electric fields (3.168)–(3.170) at $z = 0$ so that

$$\sin\frac{\pi x}{a} + \sum_{n=1}^{\infty} A_n \sin\frac{n\pi x}{a} = \begin{cases} \sum_{m=1}^{\infty} B_m \sin\frac{m\pi x}{a}; & 0 \le x \le c \\ 0; & c \le x \le a. \end{cases}$$ (3.172)

Multiplying both sides by $\sin p\pi x/a$ and integrating over $(0, a)$ gives

$$\frac{a}{2}\delta_{p1} + \frac{a}{2}A_p = \sum_{m=1}^{\infty} B_m I_{pm},$$ (3.173)

where

$$I_{pm} = \int_0^c \sin \frac{p\pi x}{a} \sin \frac{m\pi x}{a} \, dx. \tag{3.174}$$

The integral I_{pm} is easily done in closed form. The upper limit is c because $E_y = 0$ for $c \le x \le a$.

The boundary conditions (3.156) and (3.165) are applied to the tangential magnetic fields (3.169)–(3.171) at $z = 0$ so that

$$\frac{-1}{Z_1} \sin \frac{\pi x}{a} + \sum_{n=1}^\infty \frac{A_n}{Z_n} \sin \frac{n\pi x}{a} = \begin{cases} \sum\limits_{m=1}^\infty \frac{-B_m}{Z_m} \sin \frac{m\pi x}{a} ; 0 \le x \le c \\ \sum\limits_{m=1}^\infty \frac{B_m}{Z_m} \sin \frac{m\pi x}{a} - \frac{2}{Z_1} \sin \frac{\pi x}{a} ; c \le x \le a. \end{cases} \tag{3.175}$$

Multiplying both sides by $\sin q\pi x/a$ and integrating over $(0, a)$, gives

$$\frac{-2I_{1n}}{Z_1} + \frac{a\delta_{1q}}{2Z_1} + \frac{a}{2Z_q} A_q = \sum_{m=1}^\infty \frac{B_m}{Z_m} \left(-2I_{mq} + \frac{a}{2}\delta_{mq} \right). \tag{3.176}$$

A_q in (3.176) can be eliminated by using (3.173) (with $p \to q$). The result, after renaming $q \to n$ and interchanging $m \leftrightarrow n$, is

$$\frac{2I_{1m}}{Z_1} - \frac{a\delta_{1m}}{2Z_1} + \frac{a\delta_{m1}}{2Z_m} = \sum_{n=1}^\infty B_n \left(\frac{I_{mn}}{Z_m} + \frac{2I_{nm}}{Z_n} - \frac{a}{2Z_n}\delta_{nm} \right). \tag{3.177}$$

An actual solution would use a finite number of modes N. In this case, (3.177) can be written as an $M \times N$ matrix equation

$$P_m = \sum_{n=1}^N B_n Q_{mn} \tag{3.178}$$

where $m = 1, 2, \ldots, M$ and

$$P_m = \frac{2I_{1m}}{Z_1} - \frac{a\delta_{1m}}{2Z_1} + \frac{a\delta_{m1}}{2Z_m} \tag{3.179}$$

$$Q_{mn} = \frac{I_{mn}}{Z_m} + \frac{2I_{nm}}{Z_n} - \frac{a}{2Z_n}\delta_{nm}. \tag{3.180}$$

A square matrix and a unique solution result when $M = N$. It is also possible to choose $M > N$ and obtain a least squares solution for the overdetermined system. After solving for the B_n, we can use (3.173) to find the A_n. Since the incident wave (3.166) is a unit-amplitude TE_{10} mode, the reflection coefficient for that mode is $\Gamma = A_1$. The normalized admittance $\bar{Y} = Y/Y_0$ follows from the reflection coefficient

$$\bar{Y} = \bar{G} + j\bar{B} = \frac{1 - \Gamma}{1 + \Gamma}.$$

Just as we saw for the H-plane step, it is important to choose the correct sign of the square root when calculating β, so that $\text{Im } \beta < 0$. This is necessary to ensure that the evanescent modes decay, away from the iris.

Figure 3.25 shows the normalized iris susceptance $\bar{B} = B/Y_0$ as a function of c/a. The frequency is 9.37 GHz, and the waveguide type is WR90. Also shown is the approximate but highly accurate variational solution from Table 3.2

$$\bar{B} = -\frac{2\pi}{\beta a} \cot^2 \frac{\pi c}{2a} \left(1 + \csc^2 \frac{\pi c}{2a} \right).$$

The values of $\bar{B}$ from the mode-matching solution are in good agreement with the variational technique.

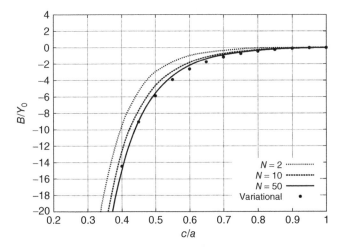

Figure 3.25 Inductive iris normalized susceptance $\bar{B} = B/Y_0$. Mode-matching and variational solutions, for a WR90 waveguide and $f = 9.37$ GHz. The number of modes is N.

If the matrix equation (3.178) is solved as a square $N \times N$ system, the numerical convergence is poor and unacceptable. This is because H_x is singular near the sharp edge of the thin iris, and the Fourier series needs many terms. Masterman and Clarricoats (1971) found that convergence can be greatly improved by solving an overdetermined system of equations with the least squares method. That is, with M equations and N modes, we allow $M > N$. An excessively large M is inefficient, but on the other hand, $M = N$ gives poor results. They showed that $M/N \geq a/c$ is sufficient. Since M and N are integers, we can take a conservative roundoff and program it as M=INT(N*a/c)+1. The results in Figure 3.25 are for several values of N, with M following this criterion. Computations were done with PROGRAM iris.

An iris of finite thickness can be solved as two waveguide step discontinuities. The convergence is better but does not help solve the infinitely thin iris problem.

3.11 Waveguide Cavity

A waveguide cavity resonator can be made by closing off the ends of a waveguide. Let us assume that we have a rectangular waveguide with a TE_{mn} mode. The ends at $z = 0$ and $z = \ell$ are closed off with metal endcaps, as in Figure 3.26. It is necessary that $E_x = 0$ and $E_y = 0$ at the endcaps. This condition can be met with two oppositely travelling waves, which form a standing wave.

The standing-wave field can be developed from the travelling-wave solution in (3.26)–(3.32). By taking right- and left-travelling waves

$$H_z = A^+ \cos\frac{m\pi x}{a} \cos\frac{n\pi y}{b} e^{-j\beta z} + A^- \cos\frac{m\pi x}{a} \cos\frac{n\pi y}{b} e^{j\beta z}$$

and choosing the constants as $A^+ = -A/j2$ and $A^- = A/j2$, we obtain

$$H_z = A \cos\frac{m\pi x}{a} \cos\frac{n\pi y}{b} \sin\beta z. \tag{3.181}$$

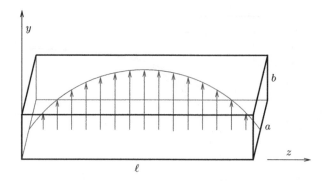

Figure 3.26 Rectangular cavity resonator, E_y at $x = a/2$ for the TE_{101} mode.

Knowing H_z, the electric field can be found and is

$$E_x = \frac{jk\eta n\pi A}{bk_c^2} \cos\frac{m\pi x}{a} \sin\frac{n\pi y}{b} \sin\beta z \qquad (3.182)$$

and

$$E_y = -\frac{jk\eta m\pi A}{ak_c^2} \sin\frac{m\pi x}{a} \cos\frac{n\pi y}{b} \sin\beta z. \qquad (3.183)$$

E_x and E_y vanish as needed at $z = 0$. If $\beta = p\pi/\ell$ with $p = 1, 2, 3, \ldots$, then they vanish at and $z = \ell$ as well. The TE_{mnp} mode can exist if and only if the separation equation (3.5) is satisfied. With $k_x = m\pi/a$, $k_y = n\pi/b$ and $\beta = k_z$, it becomes

$$\left(\frac{m\pi}{a}\right)^2 + \left(\frac{n\pi}{b}\right)^2 + \left(\frac{p\pi}{\ell}\right)^2 = k^2 = \omega^2\mu\epsilon = \left(\frac{\omega}{u_0}\right)^2.$$

This defines the mode's resonant frequency $f_0 = \omega/2\pi$, which can be written as

$$f_0 = \frac{u_0}{2}\sqrt{\left(\frac{m}{a}\right)^2 + \left(\frac{n}{b}\right)^2 + \left(\frac{p}{\ell}\right)^2}. \qquad (3.184)$$

Equation (3.184) shows that there are an infinite number of resonant frequencies f_0. These occur when cavity dimensions are a multiple of a half wavelength so that the standing wave 'fits', that is, $a = m\pi/k_x$, $b = n\pi/k_y$ and $\ell = p\pi/\beta = p\lambda_z/2$ must all hold.

A similar derivation for TM_{mnp} modes shows that (3.184) also applies to the TM case. It is found that the lowest possible resonances occur for the TE_{101} and TM_{111} modes.

3.11.1 Rectangular Cavity Q

The TE_{101} mode in Figure 3.26 can be excited by putting a coaxial probe at the middle of the $y = 0$ face: $(x, y, z) = (a/2, 0, \ell/2)$. However, unless the source is precisely tuned to the resonant frequency, the field will be zero. Because we have assumed an ideal cavity with no losses, the resonance will be infinitely narrow. Equivalently, the quality factor Q is infinite. Any real cavity will have conductor and/or dielectric losses which broaden the resonance and make the Q finite. Q is an important figure of merit for a cavity resonator and is defined as

$$Q = \frac{\omega(W_e + W_m)}{P_d} = \frac{\omega \times \text{average energy stored}}{\text{energy dissipated/second}}. \tag{3.185}$$

The waveguide walls have a finite conductivity, and the cavity might be filled with a lossy dielectric. Both of these contribute to the dissipated power P_d. The electric and magnetic fields in the cavity are associated with stored electric and magnetic energy W_e and W_m. These are similar to the stored AC energy $W_e = \frac{1}{4}C|V|^2$ in a capacitor, and $W_m = \frac{1}{4}L|I|^2$ in an inductor.

We now derive an expression for the Q of the TE_{10p} modes. The fields are

$$E_y = E_0 \sin\frac{\pi x}{a} \sin\frac{p\pi z}{\ell}. \tag{3.186}$$

$$H_x = \frac{-j\beta E_0}{k\eta} \sin\frac{\pi x}{a} \cos\frac{p\pi z}{\ell} \tag{3.187}$$

$$H_z = \frac{j\pi E_0}{ka\eta} \cos\frac{\pi x}{a} \sin\frac{p\pi z}{\ell} \tag{3.188}$$

and E_0 is an arbitrary complex constant.

We begin by finding P_d due to conductor loss. We are assuming a good conductor with $\sigma \gg \omega\epsilon_0$, so the loss can be found from a surface impedance approximation. If a conductor's resistance is R and it carries a current I, then the dissipated power is $P_d = |I|^2 R/2$. From (2.110), we saw that the AC impedance of a conductor of width w and length ℓ is $Z = Z_s \ell/w$, from which $R = R_s \ell/w$. The relation between the current and surface current density is $I = J_s w$. Therefore,

$$P_d = \frac{|I|^2}{2} R = \frac{|J_s w|^2}{2} R_s \frac{\ell}{w} = \frac{|J_s|^2}{2} R_s \,\ell w.$$

If we allow for a surface current that is not constant on the conductor surface S_0, then ℓw should be replaced by a differential surface element dS. Aside from the vector direction, the surface current is the same as the tangential magnetic field, so that

$$P_d = \frac{R_s}{2} \int_{S_0} |\mathbf{J}_s|^2 \, dS = \frac{R_s}{2} \int_{S_0} |\mathbf{H}_{tan}|^2 \, dS. \tag{3.189}$$

By carrying out this integral on the six sides of the cavity, P_d can be found. It is noted that the faces at $y = 0, b$ have both H_y and H_z components; the other faces only have one component. The power dissipated by the TE_{10p} mode is found to be

$$P_d = \frac{R_s |E_0|^2 \pi^2}{2k^2\eta^2} \left(\frac{p^2 ab}{\ell^2} + \frac{b\ell}{a^2} + \frac{p^2 a}{2\ell} + \frac{\ell}{2a} \right). \tag{3.190}$$

The stored electric energy in the cavity resonator's volume V_0 is

$$W_e = \frac{1}{4} \text{Re} \int_{V_0} \mathbf{D}^* \cdot \mathbf{E} \, dV = \frac{\epsilon}{4} \int_{V_0} E_y^* E_y \, dV = \frac{\epsilon}{16} ab\ell |E_0|^2. \tag{3.191}$$

The stored magnetic energy within V_0 is

$$W_m = \frac{1}{4} \text{Re} \int_{V_0} \mathbf{H}^* \cdot \mathbf{B} \, dV = \frac{\mu}{4} \int_{V_0} |\mathbf{H}|^2 \, dV. \tag{3.192}$$

At resonance, $W_e = W_m$ so there is no need to evaluate (3.192). From (3.185), (3.190) and (3.191), the Q associated with conductor loss is

$$Q_c = \frac{2\omega_0 W_e}{P_d} = \frac{k^3 ab\ell\eta}{4\pi^2 R_s} \left(\frac{p^2 ab}{\ell^2} + \frac{b\ell}{a^2} + \frac{p^2 a}{2\ell} + \frac{\ell}{2a} \right)^{-1}. \tag{3.193}$$

The cavity might contain a lossy dielectric, so the next step is to account for this. The dielectric's conductivity is $\sigma = \omega\epsilon_0\epsilon_r \tan\delta$. The power dissipated in the lossy volume V_0 is

$$P_d = \frac{1}{2} \int_{V_0} \mathbf{J}^* \cdot \mathbf{E} \, dV = \frac{\sigma}{2} \int_{V_0} E_y^* E_y \, dV = \frac{\sigma}{8} |E_0|^2 ab\ell. \tag{3.194}$$

Using (3.185) with $W_e = W_m$ and (3.190), (3.194) gives the Q due to dielectric loss

$$Q_d = \frac{2\omega_0 W_e}{P_d} = \frac{2\omega_0 \frac{\epsilon}{16} ab\ell |E_0|^2}{\frac{\sigma}{8} |E_0|^2 ab\ell} = \frac{\omega_0 \epsilon}{\sigma} = \frac{1}{\tan\delta}. \tag{3.195}$$

It turns out that Q_d in (3.195) is independent of the cavity's shape. This is quite nice, as the formula can be used for other nonrectangular shapes.

We should combine the dissipative effects of both conductor and dielectric loss. Adding their respective P_ds gives the total Q

$$\frac{1}{Q} = \frac{1}{Q_c} + \frac{1}{Q_d}. \tag{3.196}$$

Example 3.9 (Cavity Q) Find the length ℓ and the Q for two rectangular cavities. Both resonate at 11.0 GHz, are made of copper and use the TE_{101} mode. One uses an empty X-band WR90 waveguide. The other one uses a V-band WR15 waveguide that is filled with a ceramic having $\epsilon_r = 30$ and $\tan\delta = 6.7 \times 10^{-5}$.

Solution: The WR90 copper waveguide has $a = 0.900''$, $b = 0.400''$ and $\sigma = 5.8 \times 10^7$ S/m. At 11 GHz, $k_0 = \omega\sqrt{\mu_0\epsilon_0} = 230.5$ rad/m, $\eta = \sqrt{\mu_0/\epsilon_0} = 376.7$ Ω and $R_s = \mathrm{Re}\sqrt{j\omega\mu_0/\sigma} = 0.02736$ Ω. From (3.184) with m, n, p = 1, 0, 1, the cavity length is $\ell = 0.668''$. From (3.193), the Q is

$$Q = Q_c = 7720.$$

The WR15 copper waveguide has $a = 0.148''$ and $b = 0.074''$. With $\epsilon_r = 30$, we recalculate $k = k_0\sqrt{\epsilon_r} = 1263$ rad/m, $\eta = \eta_0/\sqrt{\epsilon_r} = 68.78$ Ω and R_s is unchanged. The cavity length from (3.184) is $\ell = 0.131''$.

From (3.193), the Q associated with conductor loss is

$$Q_c = 1440.$$

From (3.195), the Q associated with dielectric loss is

$$Q_d = \frac{1}{\tan\delta} = 14,900.$$

From (3.196), the total Q is

$$Q = \frac{1}{1/Q_c + 1/Q_d} = 1310.$$

The empty cavity with only conduction loss has the better Q. For the dielectric-loaded cavity, conduction loss is more serious than dielectric loss.

   The dielectric-loaded cavity has the advantage of being more compact. The dielectric shrinks the wavelength by a factor of $\sqrt{\epsilon_r}$ which allows it to be about five times smaller. Unfortunately, the stored energy W_e is also reduced, which causes Q_c to drop by a factor of about $\sqrt{\epsilon_r}$ (see Problem 3.37). The dielectric loss $\tan \delta$ contributes to a further reduction via the Q_d term.   ■

3.11.2   Cylindrical Cavity Resonator

A cylindrical cavity can also be used for a resonator. For TE modes, we can use (3.66) to construct a standing-wave representation in z

$$H_z = A J_m(k_c \rho) \left\{ \begin{matrix} \cos m\phi \\ \sin m\phi \end{matrix} \right\} \sin \beta z \tag{3.197}$$

and for brevity, the TE cutoff wavenumber $k'_{c,mn}$ is called k_c. From H_z, the other field components can be found and are

$$E_\rho = \frac{mk\eta A}{jk_c^2 \rho} J_m(k_c \rho) \left\{ \begin{matrix} -\sin m\phi \\ \cos m\phi \end{matrix} \right\} \sin \beta z \tag{3.198}$$

$$E_\phi = -\frac{k\eta A}{jk_c} J'_m(k_c \rho) \left\{ \begin{matrix} \cos m\phi \\ \sin m\phi \end{matrix} \right\} \sin \beta z. \tag{3.199}$$

   Both E_ρ and E_ϕ must vanish at $z = 0$ and $z = \ell$, so that $\beta = p\pi/\ell$ with $\ell = 1, 2, 3, \ldots$. The boundary condition $E_\phi = 0$ at $\rho = a$ is met by enforcing (3.71), repeated here

$$J'_m(k_c a) = 0. \tag{3.200}$$

The first few roots x_n of $J'_m(x_n) = 0$ for $n = 1, 2, 3, \ldots$ are given in Table C.1, from which $k_c = x_n/a$.
   The resonant frequencies can be found from (3.72), which is the separation equation. It can be written as

$$\beta^2 + k_c^2 = k^2 = \omega^2 \mu \epsilon = \left(\frac{\omega}{u_0} \right)^2$$

or

$$f_0 = \frac{\omega}{2\pi} = \frac{u_0}{2\pi} \sqrt{k_c^2 + \left(\frac{p\pi}{\ell} \right)^2}, \tag{3.201}$$

where $u_0 = 1/\sqrt{\mu_0/\epsilon_0}$.

3.11.3   Cylindrical Cavity Q

A cylindrical cavity can have TE_{mnp} and TM_{mnp} modes. Since a circular waveguide's dominant mode is TE_{11}, one might expect to use the TE_{111} mode for a cavity. However, it turns out that this is not the best choice. A much better Q can be obtained with the higher TE_{011} mode. With $m = 0$, this mode has no ϕ variation, and $E_\rho = 0$. The electric field for the TE_{01p} mode is of the form

$$E_\phi(\rho, z) = E_0 J'_0(k_c \rho) \sin \frac{p\pi z}{\ell},$$

where $J'_0(k_c a) = 0$ gives k_c and $p = 1, 2, 3, \ldots$.

To find Q, the integrals for P_d and W_e can be readily carried out using the standing-wave fields in cylindrical coordinates. The details are omitted for brevity, but can be found in Collin (2001, Section 7.4). In summary, if one chooses the radius as

$$a \approx \frac{\ell}{2} \tag{3.202}$$

then, from Collin's Figure 7.18, the TE_{011} mode has

$$Q_c \approx \frac{2}{3} \frac{\pi \eta}{R_s}. \tag{3.203}$$

At 10 GHz and using copper, $Q \approx 30,000$ which is much higher than for the rectangular cavity of Example 3.9.

3.11.4   Dielectric Resonator

The dielectric-loaded cavity in Example 3.9 showed a significant conductor loss. A possible solution is to simply remove the metal, leaving us with a dielectric resonator. This is an attractive possibility that will now be developed. In the following, it will be assumed that the resonator is a cylinder that is made of a high-ϵ_r material, with radius a and endcaps at $z = 0$ and $z = \ell$.

With a high ϵ_r, the fields are strongly confined to the dielectric. To get a qualitative idea as to why this is so, consider the Fresnel reflection coefficient, either (2.37) or (2.49). Inside the dielectric and looking out, if $\epsilon_r \gg 1$, then $\Gamma \approx 1$. This makes $\mathbf{H}_{tan} \approx 0$ at the boundary, so it is similar to having a magnetic conductor. Therefore, a dielectric resonator can be approximately modelled by a high-ϵ_r dielectric surrounded by a PMC boundary.

Because the fields are not fully confined to the dielectric, a closed-form solution is not possible. Nevertheless, we can still develop an approximate model that provides insights into how the resonator works and suggest how the dimensions and ϵ_r should be chosen. An approximate model can be very useful for preliminary design work; a numerical solution can then be used for refinement towards a final design.

To develop the field solution, we start with the TM modes in (3.55)–(3.63). These satisfy $\mathbf{E}_{tan} = 0$ on the PEC waveguide walls. By taking the dual, we obtain TE modes having $\mathbf{H}_{tan} = 0$ on PMC walls. Like the metal cavity, the preferred mode is $m, n = 0, 1$. Using (3.55)–(3.63), we can construct a standing-wave field from $\pm z$-travelling waves. The result is

$$H_z = A J_0(k_c \rho) \cos \beta z \tag{3.204}$$

$$H_\rho = -\frac{\beta A}{k_c} J_0'(k_c \rho) \sin \beta z \tag{3.205}$$

$$E_\phi = \frac{jk\eta A}{k_c} J_0'(k_c \rho) \cos \beta z \tag{3.206}$$

where $k_c a$ are roots from Table C.1 that satisfy

$$J_0(k_c a) = 0 \tag{3.207}$$

and (3.63) tells us that

$$k_c^2 + \beta^2 = k^2 = \omega^2 \mu_0 \epsilon_0 \epsilon_r. \tag{3.208}$$

This solution has been constructed in such a way that it satisfies the PMC boundary condition $\mathbf{H}_{tan} = 0$ on the sidewall at $\rho = a$ and on the endcap at $z = 0$. We also need $H_\rho = 0$ at $z = \ell$, which implies that $\beta = p\pi/\ell$; $p = 1, 2, 3, \ldots$. The TE_{01p} resonant frequencies then follow from (3.208) and are

$$f_0 = \frac{\omega}{2\pi} = \frac{u_0}{2\pi\sqrt{\epsilon_r}}\sqrt{k_c^2 + \left(\frac{p\pi}{\ell}\right)^2}, \tag{3.209}$$

where $u_0 = 1/\sqrt{\mu_0\epsilon_0}$. Equation (3.209) is almost identical to (3.201) obtained earlier, except that k_c is now obtained from the roots of (3.207), and f_0 is lowered by the $\sqrt{\epsilon_r}$ factor.

In any real dielectric resonator, there will be some leakage from the sidewalls and endcaps that cause a resonance shift. For the TE_{011} mode, the resonance is no longer at exactly $\ell = \lambda_z/2$, but at some slightly shorter length. In common notation, this is acknowledged by calling the mode $\mathrm{TE}_{01\delta}$ where $\ell = \delta\lambda_z/2$ and δ is a non-integer number that is slightly less than 1.

Example 3.10 (Dielectric Resonator) Find the $\mathrm{TE}_{01\delta}$ resonant frequency of a dielectric resonator that is made of titanium dioxide, having $\epsilon_r = 95$ and $\tan\delta = 0.0001$. The cylindrical resonator's length and diameter are $0.325''$. Also, find the Q.

Solution: The length is $\ell = 8.255$ mm, and the radius is $a = 4.128$ mm. From Table C.1, $J_0(2.405) = 0$, so $k_c a = 2.405$ gives $k_c = 582.7$ rad/m. The $p = 1$ cutoff frequency from (3.209) is

$$f_0 = \frac{u_0}{2\pi\sqrt{\epsilon_r}}\sqrt{k_c^2 + \left(\frac{\pi}{\ell}\right)^2} = 3.41 \text{ GHz}.$$

There are no conductors present, so the only loss is from the dielectric. The Q can be estimated from (3.195) and is

$$Q = \frac{1}{\tan\delta} = 10,000.$$

Experimental results (Cohn 1968, Figure 2) show a resonant frequency of about 3.40 GHz which agrees with the simple estimate obtained here. ∎

When ϵ_r is large, the resonator is very small in terms of the free-space wavelength λ_0, which means the field outside the resonator is quasistatic and the radiation is weak. In microwave circuits, the resonator usually ends up in a shielding box that is somewhat larger than the resonator and does not touch it. This completely eliminates any radiation loss, and because the field is weak, it does not add much conduction loss.

A dielectric resonator's $\mathrm{TE}_{01\delta}$ mode can be excited by placing its flat face on the substrate adjacent to a microstrip line; see Figure 3.27. The coupling occurs through the microstrip's fringing magnetic field.

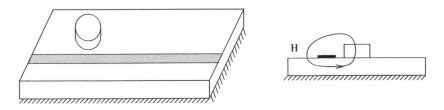

Figure 3.27   Dielectric resonator with microstrip excitation.

The coupling is controlled by the gap size between the resonator and the line. This type of structure is often used in the input circuit of a *dielectric resonator oscillator* (or DRO).

If a resonator's ϵ_r is reduced to some smaller value as 30 or less, the resonator becomes larger (in terms of λ_0) and starts to radiate much better. This is known as a *dielectric resonator antenna*. Depending on the radiation pattern requirements, various modes can be used, and the feed type and position are chosen to excite the desired mode. The $TM_{01\delta}$ mode produces a monopole pattern, whereas the $TE_{01\delta}$ and hybrid 11δ modes radiate like a slot on a ground plane (Stutzman and Thiele 2012).

A dielectric resonator antenna is usually on a ground plane, excited by a monopole probe that penetrates the resonator or by a ground-plane slot positioned under the resonator. Obtaining a tight fit between the probe and dielectric to avoid air gaps can be difficult, so the slot feed is easier to make. Further information about dielectric resonators can be found in the references at the end of the chapter.

3.12 Perturbation Method

For regular-shape cavities such as rectangular and cylindrical, the exact field solution and resonant frequencies can be readily found. In practice, these cavities are often fine-tuned by introducing a small 'perturbation' such as a dielectric or metal post. With a perturbation present, a closed-form field solution is no longer possible. However, the perturbation method allows us to find the change in the resonant frequency, knowing only the original *unperturbed* cavity field. In addition, knowing the resonance shift provides us with a new way to infer the equivalent inductance or capacitance of the perturbing structure. The following sections describe two types of perturbations, involving small changes in the cavity's interior materials or the cavity's shape.

3.12.1 Material Perturbation

Figure 3.28(a) shows an empty metal cavity, and it is presumed that we know the fields $\mathbf{E}_0, \mathbf{H}_0$ at ω_0. In Figure 3.28(b), a perturbing material having μ, ϵ is present. This perturbation slightly disturbs the fields and resonant frequency, which are now $\mathbf{E}, \mathbf{H}$ and ω. The objective is to find ω, having limited or no knowledge of $\mathbf{E}, \mathbf{H}$.

We begin by writing Maxwell's equations for the two situations

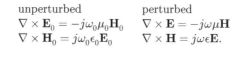

$$
\begin{array}{ll}
\text{unperturbed} & \text{perturbed} \\
\nabla \times \mathbf{E}_0 = -j\omega_0\mu_0\mathbf{H}_0 & \nabla \times \mathbf{E} = -j\omega\mu\mathbf{H} \\
\nabla \times \mathbf{H}_0 = j\omega_0\epsilon_0\mathbf{E}_0 & \nabla \times \mathbf{H} = j\omega\epsilon\mathbf{E}.
\end{array}
$$

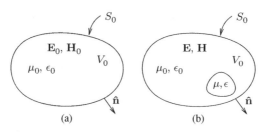

Figure 3.28 (a) Empty metal cavity resonant at ω_0 with fields $\mathbf{E}_0, \mathbf{H}_0$. (b) Metal cavity resonant at ω with material $\mu(x, y, z), \epsilon(x, y, z)$ and fields $\mathbf{E}, \mathbf{H}$.

Taking conjugates and dot products as follows (note $j^* = -j$)

$$\mathbf{H} \cdot (\nabla \times \mathbf{E}_0 = -j\omega_0\mu_0\mathbf{H}_0)^* \quad \mathbf{H}_0^* \cdot (\nabla \times \mathbf{E} = -j\omega\mu\mathbf{H})$$
$$\mathbf{E} \cdot (\nabla \times \mathbf{H}_0 = j\omega_0\epsilon_0\mathbf{E}_0)^* \quad \mathbf{E}_0^* \cdot (\nabla \times \mathbf{H} = j\omega\epsilon\mathbf{E})$$

the first minus fourth equation is

$$\mathbf{H} \cdot \nabla \times \mathbf{E}_0^* - \mathbf{E}_0^* \cdot \nabla \times \mathbf{H} = j\omega_0\mu_0\mathbf{H}_0^* \cdot \mathbf{H} - j\omega\epsilon\mathbf{E}_0^* \cdot \mathbf{E}$$

and using $\nabla \cdot (\mathbf{A} \times \mathbf{B}) = \mathbf{B} \cdot \nabla \times \mathbf{A} - \mathbf{A} \cdot \nabla \times \mathbf{B}$, it becomes

$$\nabla \cdot (\mathbf{E}_0^* \times \mathbf{H}) = j\omega_0\mu_0\mathbf{H}_0^* \cdot \mathbf{H} - j\omega\epsilon\mathbf{E}_0^* \cdot \mathbf{E}. \tag{3.210}$$

The same procedure applied to the second and third equations gives

$$\nabla \cdot (\mathbf{H}_0^* \times \mathbf{E}) = -j\omega_0\epsilon_0\mathbf{E}_0^* \cdot \mathbf{E} + j\omega\mu\mathbf{H}_0^* \cdot \mathbf{H}. \tag{3.211}$$

Taking (3.210) minus (3.211) and integrating over the cavity volume V_0 gives

$$\int_{V_0} \nabla \cdot (\mathbf{E}_0^* \times \mathbf{H}) - \nabla \cdot (\mathbf{H}_0^* \times \mathbf{E}) \, dV$$
$$= \int_{V_0} j(\omega_0\mu_0 - \omega\mu)\mathbf{H}_0^* \cdot \mathbf{H} - j(\omega\epsilon - \omega_0\epsilon_0)\mathbf{E}_0^* \cdot \mathbf{E} \, dV. \tag{3.212}$$

The left-hand side of (3.212) can be evaluated with the divergence theorem. The surface element for the outward normal is $\mathbf{dS} = \hat{\mathbf{n}} \, dS$. Using the vector identity $(\mathbf{A} \times \mathbf{B}) \cdot \mathbf{C} = (\mathbf{C} \times \mathbf{A}) \cdot \mathbf{B}$ with $\mathbf{C} = \hat{\mathbf{n}}$,

$$\int_{V_0} \nabla \cdot (\mathbf{E}_0^* \times \mathbf{H}) - \nabla \cdot (\mathbf{H}_0^* \times \mathbf{E}) \, dV = \int_{S_0} (\mathbf{E}_0^* \times \mathbf{H}) \cdot \hat{\mathbf{n}} \, dS - \int_{S_0} (\mathbf{H}_0^* \times \mathbf{E}) \cdot \hat{\mathbf{n}} \, dS$$
$$= \int_{S_0} (\hat{\mathbf{n}} \times \mathbf{E}_0^*) \cdot \mathbf{H} \, dS - \int_{S_0} (\mathbf{E} \times \hat{\mathbf{n}}) \cdot \mathbf{H}_0^* \, dS.$$

On the perfectly conducting cavity walls, $\mathbf{E}_0 \times \hat{\mathbf{n}} = 0$ and $\mathbf{E} \times \hat{\mathbf{n}} = 0$, so both surface integrals are zero. Consequently, the left-hand side of (3.212) is zero. Denoting the material perturbations as $\Delta\epsilon = \epsilon - \epsilon_0$ and $\Delta\mu = \mu - \mu_0$, Equation (3.212) can be rearranged as

$$\frac{\omega - \omega_0}{\omega} = -\frac{\int_{V_0} \Delta\epsilon\mathbf{E}_0^* \cdot \mathbf{E} + \Delta\mu\mathbf{H}_0^* \cdot \mathbf{H} \, dV}{\int_{V_0} \epsilon_0\mathbf{E}_0^* \cdot \mathbf{E} + \mu_0\mathbf{H}_0^* \cdot \mathbf{H} \, dV}. \tag{3.213}$$

We have allowed for the possibility that $\Delta\epsilon$ and $\Delta\mu$ are functions of position (x, y, z). Usually, the perturbations are zero for all but a small portion of V_0. Therefore, the integral in the numerator of (3.213) only needs to be evaluated over portions of V_0 where $\Delta\epsilon$ and $\Delta\mu$ are non-zero.

Equation (3.213) is exact, and if the perturbed field $\mathbf{E}, \mathbf{H}$ is known, then the resonance shift can be known exactly. In practice, the perturbed fields are not known but can often be estimated. If we assume

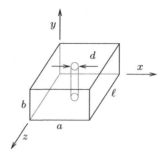

Figure 3.29   Dielectric post in a cavity.

that the perturbation does not change the fields, then $\mathbf{E} \approx \mathbf{E}_0$ and $\mathbf{H} \approx \mathbf{H}_0$ and (3.213) becomes

$$\frac{\omega - \omega_0}{\omega} \approx -\frac{\int_{V_0} \Delta\epsilon |\mathbf{E}_0|^2 + \Delta\mu |\mathbf{H}_0|^2 \, dV}{\int_{V_0} \epsilon_0 |\mathbf{E}_0|^2 + \mu_0 |\mathbf{H}_0|^2 \, dV}. \tag{3.214}$$

The right-hand side is always negative, so it shows that introducing a material perturbation always causes a *drop* in the resonant frequency.

In some cases, the approximate result (3.214) can be unsatisfactory. One remedy is to use a better 'quasistatic' approximation to estimate $\mathbf{E}$ and $\mathbf{H}$ in (3.213). This is further considered in Problem 3.39.

Example 3.11   (Dielectric Post Perturbation)   An air-filled rectangular cavity of dimensions $a \times b \times \ell$ supports the TE_{101} mode; the geometry is in Figure 3.29. A slender dielectric post of height b, diameter d and relative permittivity ϵ_r is placed vertically at $x = a/2$, $z = \ell/2$. Find the effect of the post on the resonant frequency.

Solution: The unperturbed electric field is

$$\mathbf{E}_0 = \hat{\mathbf{y}} A \sin\left(\frac{\pi x}{a}\right) \sin\left(\frac{\pi z}{\ell}\right).$$

Equation (3.213) can be simplified for this application. There is no magnetic material, so $\Delta\mu = 0$. The integral in the denominator is related to the stored electric and magnetic energies, which are equal at resonance, so we can double the electric term and drop the magnetic term. On the left-hand side, we can let $1/\omega \approx 1/\omega_0$, with the result that

$$\frac{\omega - \omega_0}{\omega_0} = -\frac{\int_{V_d} (\epsilon - \epsilon_0) \mathbf{E}_0^* \cdot \mathbf{E} \, dV}{2 \int_{V_0} \epsilon_0 \mathbf{E}_0^* \cdot \mathbf{E} \, dV}$$

where V_d is the volume occupied by the perturbing dielectric and V_0 is the total volume.

The perturbed field $\mathbf{E}$ is not known. However, if the post is thin in terms of wavelength (the quasistatic approximation), then $\mathbf{E} \approx \text{const.}$ inside the post. $\mathbf{E}$ is also tangential to the post's surface and is continuous there, so it is justifiable and in fact a good approximation to let $\mathbf{E} \approx \mathbf{E}_0$ inside the post. With this approximation,

$$\frac{\omega - \omega_0}{\omega_0} = -\frac{\int_{V_d} (\epsilon - \epsilon_0) \mathbf{E}_0^* \cdot \mathbf{E}_0 \, dV}{2 \int_{V_0} \epsilon_0 \mathbf{E}_0^* \cdot \mathbf{E}_0 \, dV}.$$

In the denominator, we make use of $\int_0^a \sin^2(\pi x/a)dx = a/2$ and $\int_0^\ell \sin^2(\pi z/\ell)dz = \ell/2$; the perturbed portion of the total volume is small and can be ignored in the integration. In the numerator, $\mathbf{E}_0^* \cdot \mathbf{E}_0 \approx |A|^2$ at all points within the rod so that

$$\frac{\omega - \omega_0}{\omega_0} = -\frac{\epsilon - \epsilon_0}{2\epsilon_0} \frac{\pi(d/2)^2 b}{(a/2)(\ell/2)b} = -\frac{\epsilon_r - 1}{2a\ell}\pi d^2.$$

The result is negative, indicating that the dielectric post causes a drop in the resonant frequency. The behaviour can be interpreted as a shunt capacitor at the midpoint of a transmission-line resonator of length ℓ.

∎

3.12.2   Geometry Perturbation

Figure 3.30(a) shows an empty metal cavity, and it is presumed we know the fields $\mathbf{E}_0, \mathbf{H}_0$ at ω_0. In case (b), the cavity has a small 'dent'. The shape perturbation slightly disturbs the fields and resonant frequency, which are now $\mathbf{E}, \mathbf{H}$ and ω. The objective is to find ω, having limited or no knowledge of $\mathbf{E}, \mathbf{H}$.

As in the previous section, we begin by writing Maxwell's equations for the two situations. In both cases, the ambient medium is μ_0, ϵ_0 so that

$$
\begin{array}{ll}
\text{unperturbed} & \text{perturbed} \\
\nabla \times \mathbf{E}_0 = -j\omega_0\mu_0\mathbf{H}_0 & \nabla \times \mathbf{E} = -j\omega\mu_0\mathbf{H} \\
\nabla \times \mathbf{H}_0 = j\omega_0\epsilon_0\mathbf{E}_0 & \nabla \times \mathbf{H} = j\omega\epsilon_0\mathbf{E}.
\end{array}
$$

Taking conjugates and dot products as follows (note $j^* = -j$)

$$
\begin{array}{ll}
\mathbf{H} \cdot (\nabla \times \mathbf{E}_0 = -j\omega_0\mu_0\mathbf{H}_0)^* & \mathbf{H}_0^* \cdot (\nabla \times \mathbf{E} = -j\omega\mu_0\mathbf{H}) \\
\mathbf{E} \cdot (\nabla \times \mathbf{H}_0 = j\omega_0\epsilon_0\mathbf{E}_0)^* & \mathbf{E}_0^* \cdot (\nabla \times \mathbf{H} = j\omega\epsilon_0\mathbf{E})
\end{array}
$$

the first minus fourth equation is

$$\mathbf{H} \cdot \nabla \times \mathbf{E}_0^* - \mathbf{E}_0^* \cdot \nabla \times \mathbf{H} = j\omega_0\mu_0\mathbf{H}_0^* \cdot \mathbf{H} - j\omega\epsilon_0\mathbf{E}_0^* \cdot \mathbf{E}$$

and using $\nabla \cdot (\mathbf{A} \times \mathbf{B}) = \mathbf{B} \cdot \nabla \times \mathbf{A} - \mathbf{A} \cdot \nabla \times \mathbf{B}$, it becomes

$$\nabla \cdot (\mathbf{E}_0^* \times \mathbf{H}) = j\omega_0\mu_0\mathbf{H}_0^* \cdot \mathbf{H} - j\omega\epsilon_0\mathbf{E}_0^* \cdot \mathbf{E}. \tag{3.215}$$

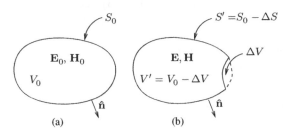

(a)   (b)

Figure 3.30   (a) Empty metal cavity resonant at ω_0 with fields $\mathbf{E}_0, \mathbf{H}_0$. (b) Metal cavity resonant at ω with a small dent, and fields $\mathbf{E}, \mathbf{H}$.

The same procedure applied to the second and third equations gives

$$\nabla \cdot (\mathbf{H}_0^* \times \mathbf{E}) = -j\omega_0\epsilon_0\mathbf{E}_0^* \cdot \mathbf{E} + j\omega\mu_0\mathbf{H}_0^* \cdot \mathbf{H}. \tag{3.216}$$

Taking (3.215) minus (3.216) and integrating over the *perturbed* cavity volume $V' = V_0 - \Delta V$ gives

$$\int_{V'} \nabla \cdot (\mathbf{E}_0^* \times \mathbf{H}) - \nabla \cdot (\mathbf{H}_0^* \times \mathbf{E}) \, dV$$

$$= \int_{V'} j(\omega_0\mu_0 - \omega\mu_0)\mathbf{H}_0^* \cdot \mathbf{H} - j(\omega\epsilon_0 - \omega_0\epsilon_0)\mathbf{E}_0^* \cdot \mathbf{E} \, dV. \tag{3.217}$$

The left-hand side of (3.217) can be evaluated with the divergence theorem. Using the outward normal,

$$\int_{V'} \nabla \cdot (\mathbf{E}_0^* \times \mathbf{H}) - \nabla \cdot (\mathbf{H}_0^* \times \mathbf{E}) \, dV = \int_{S'} (\mathbf{E}_0^* \times \mathbf{H}) \cdot \hat{\mathbf{n}} \, dS - \int_{S'} (\mathbf{H}_0^* \times \mathbf{E}) \cdot \hat{\mathbf{n}} \, dS$$

$$= \int_{S'} (\hat{\mathbf{n}} \times \mathbf{E}_0^*) \cdot \mathbf{H} \, dS - \int_{S'} (\mathbf{E} \times \hat{\mathbf{n}}) \cdot \mathbf{H}_0^* \, dS.$$

Since $\mathbf{E} \times \hat{\mathbf{n}} = 0$ on the perturbed surface S', the last integral on the second line is zero. Since $S' = S_0 - \Delta S$, the second-last integral can be written as

$$\int_{S'} (\hat{\mathbf{n}} \times \mathbf{E}_0^*) \cdot \mathbf{H} \, dS = \int_{S_0} (\hat{\mathbf{n}} \times \mathbf{E}_0^*) \cdot \mathbf{H} \, dS - \int_{\Delta S} (\hat{\mathbf{n}} \times \mathbf{E}_0^*) \cdot \mathbf{H} \, dS.$$

Since $\hat{\mathbf{n}} \times \mathbf{E}_0 = 0$ on S_0, the integral over S_0 is zero. Consequently, the integral over S' is simply the negative of the integral over ΔS. We are now able to write the left-hand side of (3.217) as

$$-\int_{\Delta S} (\mathbf{E}_0^* \times \mathbf{H}) \cdot \hat{\mathbf{n}} \, dS$$

so that (3.217) becomes

$$-\int_{\Delta S} (\mathbf{E}_0^* \times \mathbf{H}) \cdot \hat{\mathbf{n}} \, dS = \int_{V'} j(\omega_0\mu_0 - \omega\mu_0)\mathbf{H}_0^* \cdot \mathbf{H} - j(\omega\epsilon_0 - \omega_0\epsilon_0)\mathbf{E}_0^* \cdot \mathbf{E} \, dV$$

which can be rearranged as

$$\omega - \omega_0 = \frac{-j\int_{\Delta S} (\mathbf{E}_0^* \times \mathbf{H}) \cdot \hat{\mathbf{n}} \, dS}{\int_{V'} \epsilon_0 \mathbf{E}_0^* \cdot \mathbf{E} + \mu_0 \mathbf{H}_0^* \cdot \mathbf{H} \, dV}. \tag{3.218}$$

Equation (3.218) is exact. To obtain a more usable form, we can assume $\mathbf{E} \approx \mathbf{E}_0$ and $\mathbf{H} \approx \mathbf{H}_0$. Then, the surface integral can be written as

$$-\int_{\Delta S} (\mathbf{E}_0^* \times \mathbf{H}) \cdot \hat{\mathbf{n}} \, dS \approx -\int_{\Delta S} (\mathbf{E}_0^* \times \mathbf{H}_0) \cdot \hat{\mathbf{n}} \, dS = j\omega_0 \int_{\Delta V} \epsilon_0 |\mathbf{E}_0|^2 - \mu_0 |\mathbf{H}_0|^2 \, dV$$

where the conversion to a volume integral is via the complex Poynting theorem. Equation (3.218) becomes

$$\frac{\omega - \omega_0}{\omega_0} = -\frac{\int_{\Delta V} \epsilon_0 |\mathbf{E}_0|^2 - \mu_0 |\mathbf{H}_0|^2 \, dV}{\int_{V'} \epsilon_0 |\mathbf{E}_0|^2 + \mu_0 |\mathbf{H}_0|^2 \, dV}. \tag{3.219}$$

For convenience, V' can be changed to V_0 in the denominator. The small change in the volume has a minor effect, and the integral is then related to the total stored energy.

If we know the unperturbed fields, then from the shape and position of the dent we can find out how the resonance is affected. If an inward dent ΔV causes an increase in stored electric energy, then the integral in the numerator will become larger and the resonant frequency will drop. Similarly, an inward dent that causes an increase in stored magnetic energy will make the resonant frequency rise. For an outward dent, the effects are the opposite.

3.13 Further Reading

Variational solutions for many types of waveguide discontinuities can be found in the iconic *Waveguide Handbook* by Marcuvitz (1951). This book has been reprinted several times since its initial release. This topic is also well described in (Collin 1991, Chapter 8). Irises and posts are used for making matching networks and filters, and design procedures are available in Matthaei et al. (1980). A general coverage of waveguides and microwave circuits can be found in Collin (2001), Elliott (1993) and Pozar (2012).

The mode-matching method is reviewed by Arndt (2005). Readers wanting to know more about dielectric resonators can consult Kajfez and Guillon (1986). The design of transitions for waveguides, microstrip and other types of lines is given by Izadian and Izadian (1988). The perturbation method and applications are covered in Harrington (2001, Chapter 7).

References

Arndt F (2005) Mode matching methods. In *Encyclopedia of RF and Microwave Engineering* (ed. Chang K) John Wiley & Sons, Inc.

Brady MM (1968) Rectangular waveguide flange nomenclature (correspondence). *IEEE Trans. Microwave Theory Tech.* **13**(4), 469–471.

Cohn SB (1968) Microwave bandpass filters containing high-Q dielectric resonators. *IEEE Trans. Microwave Theory Tech.* **MTT-16**(4), 218–227.

Collin RE (1991) *Field Theory of Guided Waves*. Oxford University Press.

Collin RE (2001) *Foundations for Microwave Engineering*. John Wiley & Sons, Inc.

Elliott RS (1993) *An Introduction to Guided Waves and Microwave Circuits*. Prentice-Hall.

Harrington RF (2001) *Time-Harmonic Electromagnetic Fields*. IEEE Press.

Izadian JS and Izadian SM (1988) *Microwave Transition Design*. Artech House.

Kajfez D and Guillon P (1986) *Dielectric Resonators*. Artech House.

Lowman RV and Simons RN (2007) Transmission lines and waveguides. In *Antenna Engineering Handbook* (ed. Volakis J) McGraw-Hill pp. 51.1–51.54.

Marcuvitz N (1951) *Waveguide Handbook*. McGraw-Hill.

Masterman PH and Clarricoats PJB (1971) Computer field-matching solution of waveguide transverse discontinuities. *Proc. IEE* **118**(1), 51–63.

Matthaei GL, Young L and Jones EMT (1980) *Microwave Filters, Impedance-Matching Networks, and Coupling Structures*. Artech House.

Morse PM and Feshbach H (1953) *Methods of Theoretical Physics*. McGraw-Hill.
Pozar D (2012) *Microwave Engineering*. John Wiley & Sons, Inc.
Stutzman WL and Thiele GA (2012) *Antenna Theory and Design*. John Wiley & Sons, Inc.

Problems

3.1 In a rectangular waveguide, the electric field is $\boldsymbol{\mathcal{E}} = \hat{y} E_0 \sin(\pi x/a) \cos(\omega t - \beta z)$. Find $\boldsymbol{\mathcal{H}}$, and use the wave equation to find β and the guide wavelength.

3.2 A rectangular waveguide has $H_z = H_0 \cos(\pi x/a)e^{-j\beta z}$. Find **E**, and use the wave equation to find β and the guide wavelength.

3.3 A rectangular waveguide has an electric field $\mathbf{E} = \hat{y} E_0 \sin(\pi x/a)e^{-j\beta z}$.
   (a) Find the magnetic field.
   (b) Find the total power travelling down the waveguide, in terms of the incident electric field strength E_0. Hint: $\int \sin^2 ax\,dx = \frac{1}{2a}(ax - \frac{1}{2}\sin 2ax)$.

3.4 For a TE rectangular waveguide, convert the boundary condition $\mathbf{E}_{tan} = 0$ into boundary conditions on H_z. Show that Equation (3.26) meets the necessary requirements. Derive all of the electric and magnetic field components from this H_z.

3.5 A WR90 waveguide has $a = 0.900''$ and $b = 0.400''$. Note that $a \approx 2b$ which maximizes the bandwidth of the 10 mode while suppressing the other higher modes.
   (a) Calculate and compare the cutoff frequencies for the modes $mn = 10, 01, 11, 12, 21$.
   (b) Now, assume that the frequency is halfway between cutoff for the 10 mode and 20 mode. For the 10 mode, find the angle θ between the z axis and the zigzag plane-wave direction.

3.6 A rectangular waveguide has $a = 2b$. Describe all the possibilities for degenerate modes.

3.7 Derive the equations for a rectangular waveguide in Section 3.2 that give the transverse field components E_x, E_y, H_x, H_y in terms of E_z and H_z.
   (a) First assume that H_z is known and that $E_z = 0$. Expand out the two Maxwell curl equations as six scalar equations. Assuming an $e^{-j\beta z}$ propagation, replace $\partial/\partial z \to -j\beta$. Then, express the transverse components E_x, E_y, H_x, H_y in terms of H_z.
   (b) Now assume that E_z is known and that $H_z = 0$. Apply duality to your previous results, and obtain the transverse components in terms of E_z.

3.8 It is not obligatory to derive all the field components from E_z and H_z. For instance, a field solution could have $E_x = 0$. This would be transverse electric with respect to x or TE_x. These are called *longitudinal section electric* (LSE) modes. Similarly, the other type of solution could have $H_x = 0$, the transverse magnetic with respect to x or TM_x. These are called *longitudinal section magnetic* (LSM) modes.
   (a) Construct expressions for the LSE modes by combining the known TE_z and TM_z solutions to get $E_x = 0$. Express all field components in terms of H_x.

(b) Construct expressions for the LSM modes by combining the known TE_z and TM_z solutions to get $H_x = 0$. Express all field components in terms of E_x.

3.9 A rectangular waveguide has a thin filament of current of strength I_0 that is constant for $0 \le y \le b$. It is given by $\mathbf{J} = \hat{\mathbf{y}} I_0 \delta(x - a/2)\delta(z)$. The current will excite TE modes having (for $z > 0$)

$$H'_{zmn} = A_{mn} \cos \frac{m\pi x}{a} \cos \frac{n\pi y}{b} e^{-j\beta_{mn} z}.$$

All the other field components can be found from H'_{zmn}. Since the current does not vary with y, it will only excite TE_{m0} modes. Using the Fourier series approach, find the coefficients A_{mn} and show that

$$E_y = \frac{-\omega\mu_0 I_0}{a} \sum_{n=1}^{\infty} \frac{1}{\beta_n} \sin \frac{n\pi}{2} \sin \frac{n\pi x}{a} e^{-j\beta_n |z|},$$

where $\beta_n = \sqrt{k_0^2 - (n\pi/a)^2}$.

3.10 Using the dispersion relation for $\beta(\omega)$ in a rectangular waveguide, differentiate it to find the group velocity u_g. Put the result in terms of the zigzag angle θ.

3.11 The rectangular waveguide in Figure 3.31(a) is air-filled for the region $z < 0$ and is dielectric-filled with a material $\epsilon = \epsilon_0 \epsilon_r$, for $z > 0$. Assuming dominant mode propagation, and a wave incident from the air side, find the reflection and transmission coefficients for the mode. Assume a WR90 waveguide, a frequency of 10 GHz and a dielectric with $\epsilon_r = 4$.

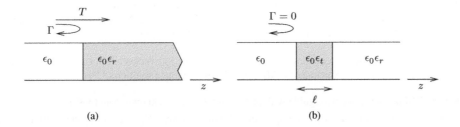

Figure 3.31   Waveguide, (a) air-dielectric interface, (b) quarter-wave transformer.

3.12 Design a quarter wave transformer for a TE_{10} rectangular waveguide, as in Figure 3.31(b). The transformer is to match an air section to a dielectric with a relative permittivity ϵ_r. The transformer is a block of dielectric with a relative permittivity ϵ_t, and its length is $\ell = \lambda_z/4$ inside the dielectric. Zero reflection is obtained when the transformer dielectric is chosen so that the TE wave impedances satisfy

$$Z_t = \sqrt{Z_a Z_d},$$

where Z_t, Z_a and Z_d are the transverse wave impedances for the transformer, air and dielectric load, respectively. Assume a WR90 waveguide, a frequency of 10 GHz and a load with $\epsilon_r = 4$.

3.13 A rectangular waveguide has two sources. Source A is a y-directed current sheet $\mathbf{J}_s^A = \hat{\mathbf{y}}\sin(\pi x/a)$ at $z = c$ that produces a field

$$\mathbf{E}^A = -\hat{\mathbf{y}}\frac{k\eta}{2\beta_{10}}\sin\frac{\pi x}{a}e^{-j\beta_{10}|z-c|}.$$

A second source B is a thin filament of current of strength I_0 that is constant for $0 \le y \le b$. It is given by $\mathbf{J}^B = \hat{\mathbf{y}}I_0\delta(x - a/2)\delta(z)$ and produces a field

$$\mathbf{E}^B = \hat{\mathbf{y}}\sum_{n=1}^{\infty}A_n\sin\frac{n\pi x}{a}e^{-j\beta_{n0}|z|},$$

where the A_n are unknown.

Find the field produced by the filament by using the reciprocity theorem. The theorem requires that the sources and fields obey

$$\int_{V_0}\mathbf{E}^A\cdot\mathbf{J}^B\,dV = \int_{V_0}\mathbf{E}^B\cdot\mathbf{J}^A\,dV,$$

where V_0 encloses all sources. Show that

$$A_n = \frac{-\omega\mu_0 I_0}{a\beta_n}\sin\frac{n\pi}{2}.$$

3.14 An RG58 coaxial transmission line has inner and outer conductor diameters of $2a = 33.2$ mils and $2b = 116$ mils, respectively. The polyethylene dielectric has $\epsilon_r = 2.25$. Find the TE_{11} mode cutoff frequency. Operation below this frequency ensures TEM operation.

3.15 For concentric cylinders, compute the roots of (3.78). Plot $k_c a$ versus b/a for $1 \le b/a \le 10$. Use the secant method; see Appendix G. Also, plot the approximate solution (3.79).

3.16 For a TM parallel-plate waveguide with plate spacing a, H_y is known from (3.88).
  (a) By writing $\cos(m\pi x/a)$ as complex exponentials, express H_y as a pair of plane waves. Find the angle θ between the z axis and the plane-wave direction.
  (b) Repeat for E_y in the TE case.

3.17 A TM parallel-plate waveguide has a plate spacing of 2.5 mm, and it is filled with a material having a relative permittivity $\epsilon_r = 3.5$. Find the cutoff frequencies for the TM_1 and TM_2 modes.

3.18 Find the real $y \ge 0$ roots of the simultaneous equations

$$x^2 + y^2 = 4 \tag{1}$$

$$y = 3\tan x. \tag{2}$$

You can write these in the form $f(x) = 0$:

$$3\tan x - \sqrt{4 - x^2} = 0.$$

Make a careful graph of (1) and (2). Their intersection will give you a good guess value for the root. Then, refine the guess, using the secant method; see Appendix G.

3.19 A dielectric slab is on an $x = 0$ ground plane. The slab has a thickness of d and surface-wave modes are guided in the $+z$ direction. For the TM case, the propagation constant can be found from the solution of

$$(k_x d)^2 + (\alpha d)^2 = (k_0 d)^2 (\mu_r \epsilon_r - 1) = \ell^2$$

$$(k_x d) \tan(k_x d) = \epsilon_r (\alpha d).$$

Assume that the frequency is 10 GHz, $\epsilon_r = 4$ and $d = 4.50$ mm.

(a) Use the secant method (see Appendix G) to find the propagation constant β.
(b) Find the zigzag angle of the wave inside the slab, $\theta = \tan^{-1}(k_x/\beta)$.
(c) Find the critical angle needed to support total internal reflection inside the slab.
(d) At what frequency would the next higher mode appear?

3.20 Modify your dielectric slab program from Problem 3.19 to calculate $\beta(\omega)$. Use this to plot β, the phase velocity and group velocity for 9.5–10.5 GHz.

3.21 The lowest TM mode in a grounded dielectric slab has no lower cutoff frequency. The plane-wave zigzag angle approaches a limiting value as $\omega \to 0$. Find this angle by approximating the transcendental equations when $k_x d$ and αd are small.

3.22 For a dielectric slab on a PEC ground, show that the surface impedance at the air-dielectric interface is $Z_s = j\alpha/\omega\epsilon_0$ for the TM case and $Z_s = -j\omega\mu_0/\alpha$ for the TE case. Therefore, an inductive/capacitive boundary supports a surface wave in the TM/TE case.

3.23 A capacitor has metal plates at $x = 0$ and $x = d$, with edges at $y = \pm w/2$, $z = \pm \ell/2$. Terminal A is at $(0, 0, 0)$, and terminal B is at $(d, 0, 0)$. Assuming that ℓ/λ is not small, the structure will act as a transmission line in the z direction, and the electric field between the plates will be a standing wave. Neglecting fringing effects, this is

$$E_x = E_0 \left(e^{-jkz} + e^{jkz}\right) = 2E_0 \cos kz$$

where $k = \omega\sqrt{\mu\epsilon}$.

(a) Find V_{AB}.
(b) Find the surface charge $\rho_s(0, y, z)$ on the bottom plate and the total charge Q.
(c) Find the capacitance.
(d) Show that at low frequencies $C \approx \epsilon w\ell/d$.
(e) At what frequency does the total charge on one plate become zero?

3.24 A parallel-plate waveguide has metal plates at $x = 0$ and $x = a$. The magnetic field between the plates is

$$H_y = e^{-j(k_x x + \beta z)} + Be^{j(k_x x - \beta z)}.$$

(a) Find the relation between k_x and β.
(b) Enforce boundary conditions at $x = 0$, $x = a$ to obtain the constants B and k_x.
(c) Find the surface current and the surface charge on the plate at $x = 0$.

3.25 A parallel-plate waveguide has metal plates at $x = 0$ and $x = a$. The electric field between the plates is $\mathbf{E} = \hat{x}E_0 e^{-j\beta z}$.

(a) Find $\mathbf{H}$ and the surface current on the bottom plate.
(b) From $\mathbf{D}$, find the surface charge on the bottom plate.
(c) Find the surface charge on the bottom plate by using the continuity equation for surface current,
$\nabla \cdot \mathbf{J}_s = -j\omega\rho_s$.

3.26 A WR90 waveguide has a ridge with $s/b = 0.2$ and $w/a = 0.3$. Using the approximation $a \approx 2b$ and the graphical data in Figure 3.15, find the cutoff frequency in GHz.

3.27 A WR90 waveguide has a ridge with $s/b = 0.2$ and $w/a = 0.3$. Using PROGRAM ridgewg and the correct ratio $a/b = 0.9/0.4$, calculate the cutoff frequency in GHz.

3.28 A WR90 waveguide has a finline with $s/b = 0.3$. Using the approximation $a \approx 2b$ and the graphical data in Figure 3.17, find the cutoff frequency in GHz.

3.29 A WR90 waveguide with a finline has $s/b = 0.3$. Modify PROGRAM ridgewg to solve for k_c of a finline. Using the correct ratio $a/b = 0.9/0.4$, calculate the cutoff frequency in GHz.

3.30 Two rectangular WR90 waveguides are connected through H-plane steps to a smaller waveguide of reduced width c, as in Figure 3.32. This device is intended to work as an attenuator. The frequency is 9.37 GHz and $c/a = 0.6$. If the reduced-width section has a length $\ell = a$, find the dB attenuation as the wave propagates from $z = 0$ to $z = \ell$. This is an approximate solution, neglecting any reflections at the junctions.

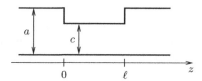

Figure 3.32   Waveguide attenuator.

3.31 For the waveguide attenuator in Problem 3.30, draw a transmission-line equivalent for the portion $-\infty < z \le \ell/2$. To model the discontinuity at $z = 0$, use numbers from the mode-matching data in Figure 3.23. The frequency is 9.37 GHz, $c/a = 0.6$ and the length is $\ell = a$.

  (a) Evaluate Γ and $T = 1 + \Gamma$ at the $z = 0$ junction.
  (b) With an incident TE_{10} mode $E_y^i = E_0 \sin(\pi x/a)e^{-j\beta_{10}z}$, find the transmitted wave $E_y^t = E_0 T \sin(\pi x/c)e^{-\gamma_{10}z}$. Find the attenuation in dB, at the midpoint $z = \ell/2$.
  (c) The structure is symmetric, so we can double the attenuation found earlier, to find the total. Compare this to the result obtained in Problem 3.30.
  (d) This solution neglects multiple reflections between the steps. For this particular problem, is it a significant source of error?

3.32 For the waveguide attenuator considered in Problem 3.31, calculate and plot the dB attenuation for a range of $0.2 \le c/a \le 0.7$, at 9.37 GHz. Use PROGRAM wgstep.

3.33 Use the transverse resonance method to find the TE cutoff wavenumbers k_x for a parallel-plate waveguide.

3.34 Use the transverse resonance method to obtain the expression (3.117) for the TE cutoff wavenumbers k_x for a dielectric slab on a PEC ground. The slab thickness is d, and the relative dielectric constant is ϵ_r.

3.35 For the rectangular cavity resonator, use the known magnetic fields (3.187) and (3.188) to evaluate the stored magnetic energy W_m in (3.192). From your result, confirm that $W_e = W_m$ at resonance.

3.36 For the two resonators in Example 3.9, calculate the dissipated power P_d and stored energy W_e. From these quantities, calculate Q_c.

3.37 Using (3.193) and assuming that the frequency is fixed, show that $Q_c \propto 1/\sqrt{\epsilon_r}$. Therefore, a resonant cavity that is made smaller by filling it with a dielectric will have a smaller Q_c. Note that the cavity dimensions shrink by $\sqrt{\epsilon_r}$ when a dielectric is introduced.

3.38 Using (3.193) and assuming an air-filled cavity, show that $Q_c \propto 1/\sqrt{f}$. Therefore, a proportionally smaller resonator to be used at a higher frequency will have a smaller Q_c.

3.39 An air-filled rectangular cavity resonator of dimensions $a \times b \times \ell$ supports the TE_{101} mode; the geometry is in Figure 3.33(a). A thin dielectric slab of thickness d and relative permittivity ϵ_r covers the bottom, at $y = 0$. Find the resonant frequency shift caused by the dielectric. The unperturbed cavity field is

$$\mathbf{E}_0 = \hat{\mathbf{y}} A \sin\left(\frac{\pi x}{a}\right) \sin\left(\frac{\pi z}{\ell}\right).$$

The normal component of the electric flux density $\mathbf{D}$ has to be continuous at the dielectric-air interface at $y = d$, similar to the electrostatic problem of two series capacitors. Therefore, the field E_y^{in} inside and E_y^{out} outside the dielectric are related at the boundary by $\epsilon_r E_y^{in} = E_y^{out}$. The perturbed field inside the dielectric is approximately

$$\mathbf{E} \approx \hat{\mathbf{y}} \frac{A}{\epsilon_r} \sin\left(\frac{\pi x}{a}\right) \sin\left(\frac{\pi z}{\ell}\right).$$

Show that the resonance shift is

$$\frac{\omega - \omega_0}{\omega_0} = -\frac{1}{2}\frac{\epsilon_r - 1}{\epsilon_r}\frac{d}{b}.$$

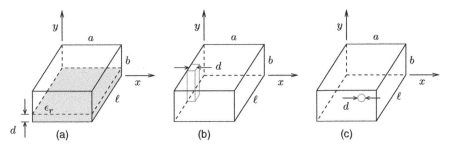

Figure 3.33 Cavity with (a) dielectric slab on the bottom, (b) square metallic post at $z = \ell/2$ on the side wall and (c) hemispherical metallic bump at bottom centre.

3.40 An air-filled rectangular cavity of dimensions $a \times b \times \ell$ supports the TE_{101} mode; the geometry is in Figure 3.33(b). A slender metallic post of height b is placed vertically at $x = 0$, $z = \ell/2$. The post has a square cross section with area d^2.

(a) Find the effect of the post on the resonant frequency.

(b) On a z-directed transmission-line equivalent, does it act as a shunt capacitance or inductance?

3.41 An air-filled rectangular cavity of dimensions $a \times b \times \ell$ supports the TE_{101} mode; the geometry is in Figure 3.33(c). A small metallic hemispherical bump of diameter d is at $(x, y, z) = (a/2, 0, \ell/2)$.

(a) Find the effect of the bump on the resonant frequency.

(b) On a z-directed transmission-line equivalent, does it act as a shunt capacitance or inductance?

3.42 A corrugated conductor is made from an array of parallel-plate waveguides, as shown in Figure 3.34. Assuming that $a < \lambda/2$, only a TEM mode with E_z and H_y will propagate in the guides. Neglecting evanescent modes, show that this structure acts like an inductive boundary at $x = 0$, having an impedance that is approximately $Z_s = j\eta_0 \tan k_0 d$.

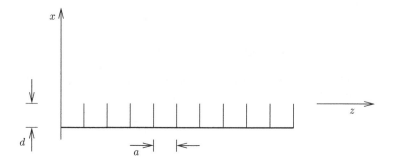

Figure 3.34   Corrugated conductor.

With $d = \lambda_0/4$ and the TM polarization, the corrugated surface can be used to approximate a perfect magnetic conductor.

4

Potentials, Concepts and Theorems

In electrostatics, the electric field can be found by solving Poisson's equation

$$\nabla^2 \Phi = \frac{-\rho_v}{\epsilon}$$

for the scalar potential Φ and then finding the electric field from $\mathbf{E} = -\nabla\Phi$. The advantage of this two-step process is that Poisson's equation is easier to solve than the differential equation for $\mathbf{E}$:

$$\nabla^2 \mathbf{E} - \nabla(\nabla \cdot \mathbf{E}) = 0.$$

This chapter will develop these ideas for the time-harmonic case, where $\omega \neq 0$. It will be seen that it is still advantageous to use potentials rather than trying to directly solve the wave equation for $\mathbf{E}$. However, the electric field should now be expressed as

$$\mathbf{E} = -\nabla\Phi - j\omega\mathbf{A}$$

which uses both a scalar potential Φ and a *vector potential* $\mathbf{A}$.

When developing field formulations, potentials are often used in conjunction with several field theorems and principles. We present these in detail, in particular: image theory, uniqueness, surface, volume and induction equivalents and the physical optics approximation.

Potentials are also used for constructing field solutions in source-free regions, in particular, waveguides. Several cases are presented to demonstrate these techniques.

4.1 Vector Potentials A and F

We now proceed to solve the Maxwell curl equations

$$\nabla \times \mathbf{E} = -j\omega\mu\mathbf{H} \tag{4.1}$$

$$\nabla \times \mathbf{H} = \mathbf{J} + j\omega\epsilon\mathbf{E} \tag{4.2}$$

when a source is present. In previous chapters we considered the possibility of $\mathbf{J} = \sigma\mathbf{E}$ being a conduction current in a lossy medium. Now we will consider a more complicated situation involving an

Applied Frequency-Domain Electromagnetics, First Edition. Robert Paknys.
© 2016 John Wiley & Sons, Ltd. Published 2016 by John Wiley & Sons, Ltd.
Companion Website: www.wiley.com/go/paknys9981

impressed source such as an antenna. This is best accomplished by using the vector potential $\mathbf{A}$. First, a wave equation for $\mathbf{A}$ is obtained. The wave equation can be solved by standard techniques. Then, the fields can be obtained from $\mathbf{A}$. This two-step process is desirable because the differential equation for $\mathbf{A}$ is easier to solve than the ones for $\mathbf{E}$, $\mathbf{H}$.

Let us assume that the ambient medium is air and that a z-directed ideal dipole is at the origin. The dipole strength is p_e and it can be represented by a volume electric current that is an impulse function

$$\mathbf{J} = \hat{z} \, p_e \, \delta(x)\delta(y)\delta(z).$$

In the air $\mathbf{J} = \sigma \mathbf{E} = 0$, nevertheless $\mathbf{J} \neq 0$ at the origin has to somehow be accounted for. A dipole is used here to present the concept, but what follows below can apply to any impressed source.

The curl of (4.1) in (4.2) along with the double-curl identity (B.32) gives the wave equation for $\mathbf{E}$

$$\nabla^2 \mathbf{E} + k^2 \mathbf{E} = j\omega\mu\mathbf{J} + \nabla(\nabla \cdot \mathbf{E}) \tag{4.3}$$

and the curl of (4.2) in (4.1) gives the wave equation for $\mathbf{H}$

$$\nabla^2 \mathbf{H} + k^2 \mathbf{H} = -\nabla \times \mathbf{J} + \nabla(\nabla \cdot \mathbf{H}), \tag{4.4}$$

where $k^2 = \omega^2 \mu\epsilon$.

When a source is present, the solution of the Maxwell curl equations (4.3) and (4.4) is more difficult. In this case, it is expedient to define the vector potential $\mathbf{A}$. From the Helmholtz theorem (see Appendix B) a vector function can be uniquely determined by its divergence and curl. To this end, we define

$$\mathbf{H} = \frac{1}{\mu}\nabla \times \mathbf{A}. \tag{4.5}$$

Equation (4.5) in (4.1) gives

$$\nabla \times (\mathbf{E} + j\omega\mathbf{A}) = 0 \tag{4.6}$$

and since $\nabla \times \nabla f \equiv 0$ for any function f, (4.1) will be automatically satisfied if we define

$$\mathbf{E} = -\nabla\Phi - j\omega\mathbf{A}. \tag{4.7}$$

This is a generalization of the electrostatic formula $\mathbf{E} = -\nabla\Phi$ to the case where $\omega \neq 0$. The curl of (4.5) in (4.2) gives

$$\frac{1}{\mu}\nabla \times \nabla \times \mathbf{A} = \mathbf{J} + j\omega\epsilon\mathbf{E} \tag{4.8}$$

and using the double-curl identity (B.32) leads to

$$-\nabla^2 \mathbf{A} + \nabla\nabla \cdot \mathbf{A} = \mu\mathbf{J} + j\omega\mu\epsilon\mathbf{E}. \tag{4.9}$$

Using (4.7) for $\mathbf{E}$, (4.9) becomes

$$-\nabla^2 \mathbf{A} + \nabla\nabla \cdot \mathbf{A} = \mu\mathbf{J} + j\omega\mu\epsilon(-\nabla\Phi - j\omega\mathbf{A}). \tag{4.10}$$

We are still free to define the divergence of $\mathbf{A}$ and can choose

$$\nabla \cdot \mathbf{A} = -j\omega\mu\epsilon\Phi. \tag{4.11}$$

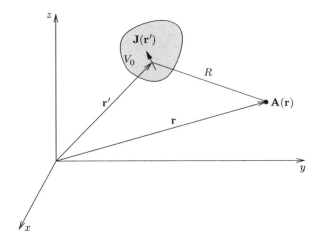

Figure 4.1 Coordinates for the vector potential integral in Equation (4.13).

This is known as the *Lorenz gauge.*[1] It reduces (4.10) to

$$\nabla^2 \mathbf{A} + k^2 \mathbf{A} = -\mu \mathbf{J} \tag{4.12}$$

which is the inhomogeneous Helmholtz equation for $\mathbf{A}$.

A particular solution of (4.12) can be found (it takes a few steps; see Section 4.14). The result is

$$\mathbf{A}(\mathbf{r}) = \frac{\mu}{4\pi} \int_{V_0} \mathbf{J}(\mathbf{r}') \frac{e^{-jkR}}{R} dV' \tag{4.13}$$

where $R = |\mathbf{r} - \mathbf{r}'|$.

The coordinates for the vector potential integral are shown in Figure 4.1. Here, $\mathbf{r} = \hat{\mathbf{x}}x + \hat{\mathbf{y}}y + \hat{\mathbf{z}}z$ gives the position of the field point and is regarded as a constant. The vector $\mathbf{r}' = \hat{\mathbf{x}}x' + \hat{\mathbf{y}}y' + \hat{\mathbf{z}}z'$ is the variable of integration within the domain V_0 containing all the points that are occupied by the source distribution $\mathbf{J}(\mathbf{r}')$. The integration is written for a volume current $\mathbf{J}$; however, it is understood that one can include surface and line currents, if $\mathbf{J}(\mathbf{r}')dV'$ is replaced by $\mathbf{J}_s(\mathbf{r}')dS'$ or $\hat{\boldsymbol{\ell}}I(\ell')d\ell'$. Dimensionally, all three versions of this quantity are in A-m.

A differential equation for the potential Φ can be found from Gauss' law

$$\nabla \cdot \mathbf{E} = \frac{\rho_v}{\epsilon}. \tag{4.14}$$

Using (4.7) for $\mathbf{E}$ and (4.11) for $\nabla \cdot \mathbf{A}$ gives us

$$\nabla^2 \Phi + k^2 \Phi = \frac{-\rho_v}{\epsilon}. \tag{4.15}$$

A particular solution is

$$\Phi(\mathbf{r}) = \frac{1}{4\pi\epsilon} \int_{V_0} \rho_v(\mathbf{r}') \frac{e^{-jkR}}{R} dV'. \tag{4.16}$$

---

[1] In much of the literature, this is erroneously called the Lorentz gauge. Nevels and Shin (2001) point out that the correct credit is to Lorenz.

Once **A** is known, **H** can be found from (4.5). **E** can be found by using (4.7), along with (4.11) for Φ, with the result that

$$E = -j\omega A + \frac{1}{j\omega\mu\epsilon}\nabla(\nabla \cdot A).$$ (4.17)

Sometimes, (4.7) is preferable over (4.17) because the integral for Φ can be easier to evaluate than the derivatives in $\nabla(\nabla \cdot A)$.

Analogous to the above developments, duality can be applied to find the fields due to a magnetic current. In this case, $A \to F$ and it follows that

$$E = \frac{-1}{\epsilon}\nabla \times F$$ (4.18)

with a corresponding wave equation

$$\nabla^2 F + k^2 F = -\epsilon M.$$ (4.19)

The magnetic field is the dual of (4.17)

$$H = -j\omega F + \frac{1}{j\omega\mu\epsilon}\nabla(\nabla \cdot F).$$ (4.20)

A particular solution of (4.19) is

$$F(r) = \frac{\epsilon}{4\pi}\int_{V_0} M(r')\frac{e^{-jkR}}{R}dV'.$$ (4.21)

In general, the total **E** is obtained from (4.17) plus (4.18) and the total **H** follows from (4.5) plus (4.20) so that

$$E = -j\omega A + \frac{1}{j\omega\mu\epsilon}\nabla\nabla \cdot A - \frac{1}{\epsilon}\nabla \times F$$ (4.22)

$$H = \frac{1}{\mu}\nabla \times A - j\omega F + \frac{1}{j\omega\mu\epsilon}\nabla\nabla \cdot F.$$ (4.23)

We have seen that the wave equations (4.12) and (4.19) have particular solutions (4.13), (4.21). Often, we are interested in finding the radiation field at large distances r. If $r \gg r'$ for all $r' \in V_0$, then $R = r - r'$ is approximately parallel to r so that the integrals can be simplified by making the *far-field approximation*. This is illustrated in Figure 4.2, from which

$$R = |r - r'| \approx r - \hat{r} \cdot r'.$$ (4.24)

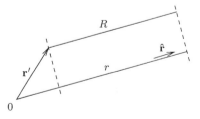

Figure 4.2   Far-field approximation for the vector potential integral.

Then, (4.13) and (4.21) become the *radiation integrals*[2]

$$\mathbf{A}(\mathbf{r}) \approx \frac{\mu}{4\pi} \frac{e^{-jkr}}{r} \int_{V_0} \mathbf{J}(\mathbf{r}') e^{jk\hat{\mathbf{r}}\cdot\mathbf{r}'} \, dV' \tag{4.25}$$

$$\mathbf{F}(\mathbf{r}) \approx \frac{\epsilon}{4\pi} \frac{e^{-jkr}}{r} \int_{V_0} \mathbf{M}(\mathbf{r}') e^{jk\hat{\mathbf{r}}\cdot\mathbf{r}'} \, dV'. \tag{4.26}$$

The fields are generally obtainable from (4.22) and (4.23), but for large r they can be greatly simplified by retaining only the $1/r$ terms and dropping the higher-order contributions. By examining the field expressions and vector potential of a dipole, one finds that

$$\mathbf{E} \approx -j\omega\mathbf{A}_T + O\left(1/r^2\right) \tag{4.27}$$

$$\mathbf{H} \approx -j\omega\mathbf{F}_T + O\left(1/r^2\right), \tag{4.28}$$

where the subscript 'T' indicates that only the components transverse to r are taken. Alternatively,

$$E_\theta \approx -j\omega A_\theta; \quad E_\phi \approx -j\omega A_\phi. \tag{4.29}$$

$$H_\theta \approx -j\omega F_\theta; \quad H_\phi \approx -j\omega F_\phi. \tag{4.30}$$

$\mathbf{E}$ due to $\mathbf{F}$ and $\mathbf{H}$ due to $\mathbf{A}$ can be readily found by applying the plane wave cross-product relation $\mathbf{E} = \eta\mathbf{H} \times \hat{\mathbf{k}}$.

The transverse part of $\mathbf{A}$ and $\mathbf{F}$ can usually be identified by inspection. Alternatively, a more mathematical way can be used. One approach is to remove the radial part from the total

$$\mathbf{A}_T = \mathbf{A} - \hat{\mathbf{r}}(\hat{\mathbf{r}} \cdot \mathbf{A}).$$

Another way is to use

$$\mathbf{A}_T = -\hat{\mathbf{r}} \times \hat{\mathbf{r}} \times \mathbf{A}.$$

It is useful to know that both $\hat{\mathbf{r}} \times \mathbf{A}$ and $\hat{\mathbf{r}} \times \hat{\mathbf{r}} \times \mathbf{A}$ are perpendicular to $\hat{\mathbf{r}}$. These relations are useful whenever transverse components need to be extracted. This can be used to obtain a convenient form of the radiation integrals (Silver 1949, p. 88)

$$\mathbf{E} = \frac{jk}{4\pi} \int_{V_0} (\eta\hat{\mathbf{R}} \times \hat{\mathbf{R}} \times \mathbf{J}(\mathbf{r}') + \hat{\mathbf{R}} \times \mathbf{M}(\mathbf{r}')) \frac{e^{-jkR}}{R} dV' \tag{4.31}$$

and

$$\mathbf{H} = \frac{-jk}{4\pi} \int_{V_0} \left(\hat{\mathbf{R}} \times \mathbf{J}(\mathbf{r}') - \frac{1}{\eta}\hat{\mathbf{R}} \times \hat{\mathbf{R}} \times \mathbf{M}(\mathbf{r}')\right) \frac{e^{-jkR}}{R} dV', \tag{4.32}$$

where $\mathbf{R} = \mathbf{r} - \mathbf{r}'$. Since the integrand has a $1/R$ dependence, the above formulas are valid as long as $R > 1.6\lambda$, the far field of an ideal dipole. In the far field of V_0, we may also use $\hat{\mathbf{R}} \approx \hat{\mathbf{r}}$ and $R \approx r - \hat{\mathbf{r}} \cdot \mathbf{r}'$. Note that the plane wave relationship $\mathbf{E} = \eta\mathbf{H} \times \hat{\mathbf{R}}$ also applies.

---

[2] When far-field radiation patterns are presented, the factor e^{-jkr}/r is never included. The factor can be reintroduced if the field values at a particular distance r are required. In 2D the factor is $e^{-jk\rho}/\sqrt{\rho}$.

The reader is cautioned to remember that the radiation integrals only apply to currents that are in free space. If a physical obstacle is present, it must first be converted into an equivalent free-space representation. This can be done with the aid of the surface equivalence or volume equivalence principles; these will be described in Sections 4.8 and 4.11. Then, (4.13) and (4.21) (or (4.25) and (4.26)) may be used.

The integrals (4.13) and (4.21) provide the particular (or forced) response to the wave equations (4.12) and (4.19). There can also be a homogeneous solution that satisfies the required boundary conditions and

$$\nabla^2 \mathbf{A} + k^2 \mathbf{A} = 0 \tag{4.33}$$

$$\nabla^2 \mathbf{F} + k^2 \mathbf{F} = 0. \tag{4.34}$$

These differential equations can be solved by the method of separation of variables and other techniques that are appropriate for boundary value problems. Any multiple of the homogeneous solution can be added to the particular solution.

Example 4.1 (Dipole Point Source) Find the fields produced by a z-directed infinitesimal current at $(0, 0, 0)$ in free space. The strength (the dipole moment) is p_e, and the current is specified by

$$\mathbf{J} = \hat{z} p_e \delta(\mathbf{r}).$$

Solution: $\mathbf{A}$ is found from (4.13). V_0 is a small volume that encloses the source. Since the unit impulse satisfies $\int_{V_0} \delta(\mathbf{r}) \, dV = 1$, (4.13) becomes

$$\mathbf{A} = \hat{z} p_e \frac{\mu_0}{4\pi} \frac{e^{-jkr}}{r}.$$

The fields follow from $\nabla \times \mathbf{A} = \mu_0 \mathbf{H}$ and $\nabla \times \mathbf{H} = j\omega\epsilon_0 \mathbf{E}$. Then,

$$H_\phi = \frac{p_e k^2}{4\pi} \sin\theta \, e^{-jkr} \left(\frac{j}{kr} + \frac{1}{(kr)^2} \right)$$

$$E_r = \frac{p_e k^2 \eta}{2\pi} \cos\theta \, e^{-jkr} \left(\frac{1}{(kr)^2} - \frac{j}{(kr)^3} \right)$$

$$E_\theta = \frac{p_e k^2 \eta}{4\pi} \sin\theta \, e^{-jkr} \left(\frac{j}{kr} + \frac{1}{(kr)^2} - \frac{j}{(kr)^3} \right).$$

It is noted that when $kr > 10$ (or $r > 1.6\lambda$) the $1/(kr)^2$ and $1/(kr)^3$ terms can be ignored. ∎

4.2 Hertz Potentials

The Hertz potential $\mathbf{\Pi}^e$ is sometimes used for finding fields. It is an alternative to the potentials $\mathbf{A}$, Φ. It is defined as

$$\mathbf{A} = \mu\epsilon \frac{\partial \mathbf{\Pi}^e}{\partial t}; \quad \Phi = -\nabla \cdot \mathbf{\Pi}^e. \tag{4.35}$$

The Hertz potential satisfies the Lorenz gauge. Knowing the equations that describe $\mathbf{A}$ and Φ, the corresponding ones involving $\mathbf{\Pi}^e$ can be found. In the time-harmonic case it leads to

$$\nabla^2 \mathbf{\Pi}^e + k^2 \mathbf{\Pi}^e = -\frac{\mathbf{J}}{j\omega\epsilon} \tag{4.36}$$

$$\mathbf{H} = j\omega\epsilon \nabla \times \mathbf{\Pi}^e \tag{4.37}$$

$$\mathbf{E} = k^2 \mathbf{\Pi}^e + \nabla\nabla \cdot \mathbf{\Pi}^e. \tag{4.38}$$

Applying duality to the above equations gives the Hertz potential $\mathbf{\Pi}^m$ for magnetic currents. It follows that $\mathbf{A} = j\omega\mu\epsilon\mathbf{\Pi}^e$ and $\mathbf{F} = j\omega\mu\epsilon\mathbf{\Pi}^m$, so it can be seen that $\mathbf{\Pi}^e$ and $\mathbf{\Pi}^m$ are almost the same thing as $\mathbf{A}$ and $\mathbf{F}$.

4.3   Vector Potentials and Boundary Conditions

To solve a given boundary value problem, there are many ways to choose the vector potentials. The choice is not unique, and it is also problem dependent. The potentials should be chosen in such a way that the resulting fields have enough degrees of freedom to allow that the boundary conditions be met. Another consideration is that the formulation should be as simple as possible.

As examples, the fields of a multilayer dielectric slab that is stratified along z could be found from two potential functions, A_x and A_z. It could also be found from A_z and F_z. A formulation in terms of A_z and F_z is also useful for z-directed waveguides, for cylinders along z or wedges with the edge along z. These points will now be clarified by describing some implications of choosing z-directed and other vector potentials.

4.3.1 A_z and F_z

The fields in general follow from the vector potentials by using (4.22) and (4.23), repeated here

$$\mathbf{E} = -j\omega\mathbf{A} + \frac{1}{j\omega\mu\epsilon}\nabla\nabla\cdot\mathbf{A} - \frac{1}{\epsilon}\nabla\times\mathbf{F} \tag{4.39}$$

$$\mathbf{H} = \frac{1}{\mu}\nabla\times\mathbf{A} - j\omega\mathbf{F} + \frac{1}{j\omega\mu\epsilon}\nabla\nabla\cdot\mathbf{F}. \tag{4.40}$$

To gain a better understanding of how we might construct field solutions from potentials, let us assume, for the moment, that we have z-directed vector potentials $\mathbf{A} = \hat{\mathbf{z}}A_z$ and $\mathbf{F} = \hat{\mathbf{z}}F_z$. Expanding (4.39) and (4.40) in rectangular coordinates for this case leads to

$$\mathbf{E} = \hat{\mathbf{x}}\left(\frac{1}{j\omega\mu\epsilon}\frac{\partial^2 A_z}{\partial x\partial z} - \frac{1}{\epsilon}\frac{\partial F_z}{\partial y}\right) + \hat{\mathbf{y}}\left(\frac{1}{j\omega\mu\epsilon}\frac{\partial^2 A_z}{\partial y\partial z} + \frac{1}{\epsilon}\frac{\partial F_z}{\partial x}\right)$$

$$+ \hat{\mathbf{z}}\frac{1}{j\omega\mu\epsilon}\left(\frac{\partial^2}{\partial z^2} + k^2\right)A_z \tag{4.41}$$

$$\mathbf{H} = \hat{\mathbf{x}}\left(\frac{1}{\mu}\frac{\partial A_z}{\partial y} + \frac{1}{j\omega\mu\epsilon}\frac{\partial^2 F_z}{\partial x\partial z}\right) + \hat{\mathbf{y}}\left(\frac{-1}{\mu}\frac{\partial A_z}{\partial x} + \frac{1}{j\omega\mu\epsilon}\frac{\partial F_z}{\partial y\partial z}\right)$$

$$+ \hat{\mathbf{z}}\frac{1}{j\omega\mu\epsilon}\left(\frac{\partial^2}{\partial z^2} + k^2\right)F_z. \tag{4.42}$$

In cylindrical coordinates,

$$\mathbf{E} = \hat{\rho}\left(\frac{1}{j\omega\mu\epsilon}\frac{\partial^2 A_z}{\partial\rho\partial z} - \frac{1}{\epsilon\rho}\frac{\partial F_z}{\partial\phi}\right) + \hat{\phi}\left(\frac{1}{j\omega\mu\epsilon\rho}\frac{\partial^2 A_z}{\partial\phi\partial z} + \frac{1}{\epsilon}\frac{\partial F_z}{\partial\rho}\right)$$

$$+ \hat{\mathbf{z}}\frac{1}{j\omega\mu\epsilon}\left(\frac{\partial^2}{\partial z^2} + k^2\right)A_z \tag{4.43}$$

$$\mathbf{H} = \hat{\rho}\left(\frac{1}{\mu\rho}\frac{\partial A_z}{\partial\phi} + \frac{1}{j\omega\mu\epsilon}\frac{\partial^2 F_z}{\partial\rho\partial z}\right) + \hat{\phi}\left(\frac{-1}{\mu}\frac{\partial A_z}{\partial\rho} + \frac{1}{j\omega\mu\epsilon\rho}\frac{\partial^2 F_z}{\partial\phi\partial z}\right)$$

$$+ \hat{\mathbf{z}}\frac{1}{j\omega\mu\epsilon}\left(\frac{\partial^2}{\partial z^2} + k^2\right)F_z. \tag{4.44}$$

From (4.41) and (4.42) or (4.43) and (4.44) it is seen that if $A_z \neq 0$ and $F_z = 0$, then $H_z = 0$, so TM fields are obtained. Likewise, $A_z = 0$ and $F_z \neq 0$ gives $E_z = 0$ and TE fields. Therefore, a general solution that is neither TE nor TM can be constructed from a superposition of these two cases.

Some remarks are in order, on the relationship between A_z, F_z and sources. From Section 4.1 we know that the wave equations

$$\nabla^2 A_z + k^2 A_z = -\mu J_z \tag{4.45}$$

$$\nabla^2 F_z + k^2 F_z = -\epsilon M_z \tag{4.46}$$

have the particular solutions

$$A_z(\mathbf{r}) = \frac{\mu}{4\pi}\int_{V_0} J_z(\mathbf{r}')\frac{e^{-jkR}}{R}dV' \tag{4.47}$$

$$F_z(\mathbf{r}) = \frac{\epsilon}{4\pi}\int_{V_0} M_z(\mathbf{r}')\frac{e^{-jkR}}{R}dV'. \tag{4.48}$$

It is important to remember that (4.47) and (4.48) only apply to sources in *free space*; otherwise they cannot be used. So, in free space it is correct to associate the source J_z with A_z and M_z with F_z. However, in general this association does not hold. For example, consider that the TE_{10} mode in a rectangular waveguide can be found from F_z. However, this field could be excited by a current sheet $J_{sy} = \sin(\pi x/a)$ or $M_{sx} = \sin(\pi x/a)$.

The vector potential was presented as a way to find the fields due to sources. However, it is also possible to use vector potentials to find the fields in source-free regions. By making the right-hand side of (4.45) and (4.46) equal to zero, the homogeneous differential equations can be solved for A_z and/or F_z, from which all the field components can then be found. It is common to use vector potentials in this way to construct solutions for source-free waveguide problems. For instance, A_z and F_z could be used to find the TM_z and TE_z fields in a waveguide. This is an alternative to the earlier approach in Chapter 3 that was based on E_z and H_z.

Example 4.2 (Rectangular Waveguide) Use a vector potential to find the possible TE_{mn} modes in a source-free waveguide, having metal walls at $x = 0$, $x = a$ and $y = 0$, $y = b$.

Solution: From (4.41), we see that choosing $A_z = 0$ and $F_z \neq 0$ will make TE fields. F_z should satisfy

$$\nabla^2 F_z + k^2 F_z = 0$$

and possible solutions for $+z$ travelling waves are

$$F_z(x, y, z) = \begin{Bmatrix} \cos k_x x \\ \sin k_x x \end{Bmatrix}\begin{Bmatrix} \cos k_y y \\ \sin k_y y \end{Bmatrix}e^{-j\beta z}$$

where $k_x^2 + k_y^2 + \beta^2 = k^2$. We must use (4.41) to convert the boundary condition $\mathbf{E}_{tan} = 0$ on the metal walls into conditions on F_z. This leads to the requirements

$$E_y \propto \frac{\partial F_z}{\partial x} = 0; \quad x = 0, a$$

$$E_x \propto \frac{\partial F_z}{\partial y} = 0; \quad y = 0, b.$$

A solution meeting these conditions is

$$F_z(x, y, z) = K_0 \cos k_x x \cos k_y y \, e^{-j\beta z}$$

in which $k_x = m\pi/a$, $k_y = n\pi/b$ and K_0 is an arbitrary constant. The field is then

$$E_x(x, y, z) = K_0 \frac{n\pi}{\epsilon b} \cos \frac{m\pi x}{a} \sin \frac{n\pi y}{b} \, e^{-j\beta z}$$

$$E_y(x, y, z) = -K_0 \frac{m\pi}{\epsilon a} \sin \frac{m\pi x}{a} \cos \frac{n\pi y}{b} \, e^{-j\beta z}. \qquad \blacksquare$$

Example 4.3 (Coaxial Line) Use a vector potential to find the transverse electromagnetic (TEM) field in a coaxial transmission line. The line consists of concentric metal cylinders with the walls at $\rho = a$ and $\rho = b$.

Solution: From (4.43) we can see that if A_z is of the form

$$A_z = \psi(\rho, \phi) e^{-jkz}$$

then e^{-jkz} propagation will be obtained and E_z will be zero. If $F_z = 0$ then from (4.44), H_z will also be zero. As a result, the field from this choice of potentials will be TEM.

A_z is a solution of

$$\nabla^2 A_z + k^2 A_z = 0$$

or

$$\frac{1}{\rho} \frac{\partial}{\partial \rho} \left(\rho \frac{\partial A_z}{\partial \rho} \right) + \frac{1}{\rho^2} \frac{\partial^2 A_z}{\partial \phi^2} + \frac{\partial^2 A_z}{\partial z^2} + k^2 A_z = 0.$$

At the metal surfaces we require $E_\phi = 0$ at $\rho = a$ and $\rho = b$. From (4.43), this becomes

$$E_\phi \propto \frac{\partial A_z}{\partial \phi} = 0; \quad \rho = a, b.$$

If we restrict our interest to ϕ-independent solutions of A_z, then the boundary conditions will be met. The solution for a $+z$ travelling wave becomes

$$A_z = (K_0 + K_1 \ln \rho) \, e^{-jkz}.$$

From (4.43),

$$\mathbf{E} = \hat{\rho}\frac{1}{j\omega\mu\epsilon}\frac{\partial^2 A_z}{\partial\rho\partial z} = \frac{-K_1}{\sqrt{\mu\epsilon}\rho}e^{-jkz}$$

and from (4.44),

$$\mathbf{H} = -\hat{\phi}\frac{1}{\mu}\frac{\partial A_z}{\partial\rho} = \frac{-K_1}{\mu\rho}e^{-jkz}.$$

It follows that $E_\rho/H_\phi = \sqrt{\mu/\epsilon} = \eta$. ∎

The above example also shows why there are no TEM modes in a waveguide. If the centre conductor vanishes, then $a \to 0$. Consequently, the potential $A_z = (K_0 + K_1 \ln \rho)e^{-jkz}$ requires $K_1 = 0$. This potential leads to fields that are zero.

Another point of interest here is that coaxial cables are normally operated so that the field is TEM and independent of ϕ. However, other TEM ϕ-dependent modes are possible. These modes can be found by choosing the potentials as $A_z = 0$ and $F_z = \psi(\phi)e^{-jkz}$; see Problem 4.13. At sufficiently high frequencies, the cable can also support TE and TM waveguide modes; this is called 'overmoding' and was discussed in Section 3.4.1. Because each mode has its own β and corresponding phase velocity $u_p = \omega/\beta$, multimode operation causes undesirable dispersion. Higher modes can be avoided by staying below their cutoff frequency.

4.3.2   Hybrid Modes, A_y and F_y

Vector potentials provide a convenient way to construct other types of solutions. For example, we could use F_y to obtain TE_y fields or A_y to obtain TM_y fields. It will soon be shown that this is useful for finding the modes in dielectric-loaded waveguides.

Examples using y-directed potentials will now be developed. With $\mathbf{A} = \hat{y}A_y$ and $\mathbf{F} = \hat{y}F_y$, we can expand (4.39) and (4.40). Alternatively, we can take (4.41) and (4.42) and perform a cyclical interchange of $x \to y \to z \to x$ twice to obtain

$$\mathbf{E} = \hat{z}\left(\frac{1}{j\omega\mu\epsilon}\frac{\partial^2 A_y}{\partial z\partial y} - \frac{1}{\epsilon}\frac{\partial F_y}{\partial x}\right) + \hat{x}\left(\frac{1}{j\omega\mu\epsilon}\frac{\partial^2 A_y}{\partial x\partial y} + \frac{1}{\epsilon}\frac{\partial F_y}{\partial z}\right)$$

$$+ \hat{y}\frac{1}{j\omega\mu\epsilon}\left(\frac{\partial^2}{\partial y^2} + k^2\right)A_y \tag{4.49}$$

$$\mathbf{H} = \hat{z}\left(\frac{1}{\mu}\frac{\partial A_y}{\partial x} + \frac{1}{j\omega\mu\epsilon}\frac{\partial^2 F_y}{\partial z\partial y}\right) + \hat{x}\left(\frac{-1}{\mu}\frac{\partial A_y}{\partial z} + \frac{1}{j\omega\mu\epsilon}\frac{\partial F_y}{\partial x\partial y}\right)$$

$$+ \hat{y}\frac{1}{j\omega\mu\epsilon}\left(\frac{\partial^2}{\partial y^2} + k^2\right)F_y. \tag{4.50}$$

If $F_y = 0$ then $H_y = 0$ and A_y gives TM_y modes. Similarly, if $A_y = 0$ then $E_y = 0$ and F_y gives TE_y modes. A general solution can then be constructed from TM_y and TE_y modes. These are called *hybrid modes* and this term is used whenever modes have both $E_z \neq 0$ and $H_z \neq 0$.

Example 4.4 (TM$_y$ Modes) Find the TM_y modes for a rectangular metal waveguide with cross- sectional dimensions $a \times b$ that is filled with a medium μ, ϵ. The walls are at $x = 0, a$ and $y = 0, b$.

Solution: For TM_y modes we let $F_y = 0$. We need a solution of

$$\nabla^2 A_y + k^2 A_y = 0.$$

To get $\mathbf{E}_{tan} = 0$ on the metal walls, we choose

$$A_y = A_0 \sin \frac{m\pi x}{a} \cos \frac{n\pi y}{b} e^{-j\beta z}.$$

A_0 is an arbitrary constant, and the separation equation is

$$\left(\frac{m\pi}{a}\right)^2 + \left(\frac{n\pi}{b}\right)^2 + \beta^2 = k^2 = \omega^2 \mu\epsilon.$$

From (4.49) and (4.50) the fields are

$$E_z = A_0 \frac{\beta n\pi/b}{\omega\mu\epsilon} \sin \frac{m\pi x}{a} \sin \frac{n\pi y}{b} e^{-j\beta z}$$

$$E_x = -A_0 \frac{(m\pi/a)(n\pi/b)}{j\omega\mu\epsilon} \cos \frac{m\pi x}{a} \sin \frac{n\pi y}{b} e^{-j\beta z}$$

$$E_y = A_0 \frac{k^2 - (n\pi/b)^2}{j\omega\mu\epsilon} \sin \frac{m\pi x}{a} \cos \frac{n\pi y}{b} e^{-j\beta z}$$

$$H_z = A_0 \frac{m\pi/a}{\mu} \cos \frac{m\pi x}{a} \cos \frac{n\pi y}{b} e^{-j\beta z}$$

$$H_x = A_0 \frac{j\beta}{\mu} \sin \frac{m\pi x}{a} \cos \frac{n\pi y}{b} e^{-j\beta z}$$

$$H_y = 0. \qquad\qquad \blacksquare$$

In the above example, suppose we let $m, n = 1, 0$. The $\mathrm{TM}_{x,10}$ mode is then

$$E_y = -A_0 j\omega \sin \frac{\pi x}{a} e^{-j\beta z}$$

$$H_z = A_0 \frac{\pi/a}{\mu} \cos \frac{\pi x}{a} e^{-j\beta z}$$

$$H_x = A_0 \frac{j\beta}{\mu} \sin \frac{\pi x}{a} e^{-j\beta z}.$$

This is identical to the dominant $\mathrm{TE}_{z,10}$ mode which could have been obtained from the F_z in Example 4.2. Therefore, it can be seen that there may be more than one choice of potential to obtain a field.

A more challenging problem that benefits from the $A_y - F_y$ formulation is the partially filled rectangular waveguide in Figure 4.3. The region $0 \le y \le d$ is a dielectric having a permittivity ϵ_1, and the medium in the rest of the waveguide is air. Since a and b are arbitrary, the geometry can appear either way, with the air-dielectric boundary along the long or short dimension. Consequently, if we find the mode equations for Figure 4.3(a), it is unnecessary to solve case (b).

Besides enforcing $\mathbf{E}_{tan} = 0$ on the conducting walls, the continuity of $\mathbf{E}_{tan}$ and $\mathbf{H}_{tan}$ is required at $y = d$. It turns out that it is not possible to meet these boundary conditions with a combination of TE_z and TM_z modes, except for the TE_{0n} case. However, a solution in terms of TE_y and TM_y modes is possible, as will now be shown.

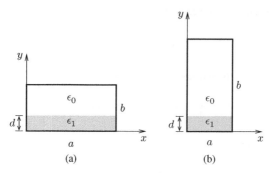

Figure 4.3 Rectangular metallic waveguide with cross section $a \times b$, partially filled with a dielectric, assuming (a) $a > b$ and (b) $a < b$.

For the TM_y modes, we need to find an A_y that makes $\mathbf{E}_{tan} = 0$ on the metal walls and $\mathbf{E}_{tan}$, $\mathbf{H}_{tan}$ continuous at $y = d$. Expecting phase matching at $y = d$, we assume that $k_x = m\pi/a$ and $k_z = \beta$ are the same in the air and dielectric. The separation equations are

$$\left(\frac{m\pi}{a}\right)^2 + k_{y1}^2 + \beta^2 = k_1^2 = \omega^2 \mu_0 \epsilon_1; \quad 0 \le y \le d \tag{4.51}$$

$$\left(\frac{m\pi}{a}\right)^2 + k_{y0}^2 + \beta^2 = k_0^2 = \omega^2 \mu_0 \epsilon_0; \quad d \le y \le b. \tag{4.52}$$

To obtain $\mathbf{E}_{tan} = 0$ on the metal walls, the potentials should be of the form

$$A_y = A_1 \sin \frac{m\pi x}{a} \cos k_{y1} y \, e^{-j\beta z}; \quad 0 \le y \le d \tag{4.53}$$

$$A_y = A_0 \sin \frac{m\pi x}{a} \cos k_{y0}(b - y) \, e^{-j\beta z}; \quad d \le y \le b. \tag{4.54}$$

A_1 and A_0 are arbitrary constants. From (4.49) and (4.50) the fields in the dielectric are

$$E_{z1} = A_1 \frac{\beta k_{y1}}{\omega \mu_0 \epsilon_1} \sin \frac{m\pi x}{a} \sin k_{y1} y \, e^{-j\beta z} \tag{4.55}$$

$$E_{x1} = -A_1 \frac{(m\pi/a)k_{y1}}{j\omega \mu_0 \epsilon_1} \cos \frac{m\pi x}{a} \sin k_{y1} y \, e^{-j\beta z} \tag{4.56}$$

$$E_{y1} = A_1 \frac{k^2 - k_{y1}^2}{j\omega \mu_0 \epsilon_1} \sin \frac{m\pi x}{a} \cos k_{y1} y \, e^{-j\beta z} \tag{4.57}$$

$$H_{z1} = A_1 \frac{m\pi/a}{\mu_0} \cos \frac{m\pi x}{a} \cos k_{y1} y \, e^{-j\beta z} \tag{4.58}$$

$$H_{x1} = A_1 \frac{j\beta}{\mu_0} \sin \frac{m\pi x}{a} \cos k_{y1} y \, e^{-j\beta z} \tag{4.59}$$

and in the air region

$$E_{z0} = -A_0 \frac{\beta k_{y0}}{\omega \mu_0 \epsilon_0} \sin \frac{m\pi x}{a} \sin k_{y0}(b - y) e^{-j\beta z} \tag{4.60}$$

$$E_{x0} = A_0 \frac{(m\pi/a)k_{y0}}{j\omega \mu_0 \epsilon_0} \cos \frac{m\pi x}{a} \sin k_{y0}(b - y) e^{-j\beta z} \tag{4.61}$$

$$E_{y0} = A_0 \frac{k^2 - k_{y0}^2}{j\omega \mu_0 \epsilon_0} \sin \frac{m\pi x}{a} \cos k_{y0}(b - y) e^{-j\beta z} \tag{4.62}$$

$$H_{z0} = A_0 \frac{m\pi/a}{\mu_0} \cos \frac{m\pi x}{a} \cos k_{y0}(b - y) e^{-j\beta z} \tag{4.63}$$

$$H_{x0} = A_0 \frac{j\beta}{\mu_0} \sin \frac{m\pi x}{a} \cos k_{y0}(b - y) e^{-j\beta z}. \tag{4.64}$$

Imposing $E_{z1} = E_{z0}$ and $E_{x1} = E_{x0}$ at $y = d$ leads to identical requirements

$$A_1 \frac{k_{y1}}{\epsilon_1} \sin k_{y1} d = -A_0 \frac{k_{y0}}{\epsilon_0} \sin k_{y0}(b - d). \tag{4.65}$$

Requiring $H_{z1} = H_{z0}$ and $H_{x1} = H_{x0}$ at $y = d$ leads to

$$A_1 \cos k_{y1} d = A_0 \cos k_{y0}(b - d). \tag{4.66}$$

Dividing (4.65) by (4.66) eliminates A_0 and A_1 so that

$$\frac{k_{y1}}{\epsilon_1} \tan k_{y1} d = -\frac{k_{y0}}{\epsilon_0} \tan k_{y0}(b - d). \tag{4.67}$$

From the separation equations (4.51) and (4.52),

$$k_{y1}^2 - k_{y0}^2 = k_0^2(\epsilon_r - 1); \quad \epsilon_r = \epsilon_1/\epsilon_0. \tag{4.68}$$

Using (4.68), either k_{y1} or k_{y0} can be eliminated from (4.67). The resulting transcendental equation can then be solved numerically for k_{y1} and k_{y0}. Once these are known, β can be found from (4.51) or (4.52). Either A_0 or A_1 remains arbitrary, but the relation between them can be found from (4.65) or (4.66).

The β values will be somewhere between those of an empty waveguide and a waveguide completely filled with the dielectric. This can be explained by the perturbation theory in Section 3.12. Some numerical results for partially filled rectangular waveguides can be found in Harrington (2001, Section 4.6).

It is useful to note that (4.67) can be found by using the transverse resonance method. Figure 4.4 shows the partially filled waveguide and its transverse network. Looking downwards at $y = d$ the characteristic impedance (similar to Example 3.6) is $Z_{01} = k_{y1}/\omega \epsilon_1$. The transmission line's electrical length is $k_{y1} d$ and it is a short-circuit stub, so from (2.170) we have $\overleftarrow{Z} = j(k_{y1}/\omega \epsilon_1) \tan k_{y1} d$. Looking upwards at $y = d$ the characteristic impedance is $Z_{00} = k_{y0}/\omega \epsilon_0$. The transmission line's electrical length is $k_{y0}(b - d)$ so the short-circuit stub has $\overrightarrow{Z} = j(k_{y0}/\omega \epsilon_0) \tan k_{y1}(b - d)$. The transverse resonance condition (3.132) is $\overleftarrow{Z} + \overrightarrow{Z} = 0$ which gives (4.67).

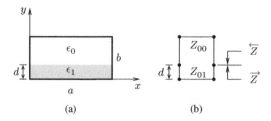

Figure 4.4   (a) Rectangular metallic waveguide with cross section $a \times b$, partially filled with a dielectric. (b) Transmission line equivalent.

We have thus far considered TM_y modes. The whole procedure can be repeated with a potential F_y to obtain TE_y modes. One finds that the separation equations (4.51) and (4.52) are unchanged. In place of (4.67), we now have

$$k_{y1} \cot k_{y1} d = -k_{y0} \cot k_{y0}(b-d). \tag{4.69}$$

The derivation is left as an exercise in Problem 4.15.

4.4   Uniqueness Theorem

The uniqueness theorem states that we will have a unique solution for the fields in a closed volume V bounded by a surface S if we specify (a) the value of $\mathbf{E}_{tan}$ on S, or (b) $\mathbf{H}_{tan}$ on S, or (c) $\mathbf{E}_{tan}$ on one part of S plus $\mathbf{H}_{tan}$ on the rest of S. These are minimum conditions, which means that specifying both $\mathbf{E}_{tan}$ *and* $\mathbf{H}_{tan}$ on S is also acceptable. By 'specify' we mean that a tangential field is given as zero, a constant or a function in terms of the surface coordinates. In addition, the theorem requires that we specify all the sources in V. If there are no sources in V, then only the boundary conditions on S are needed to ensure uniqueness.

The proof of the uniqueness theorem is as follows. Figure 4.5 shows two situations. The currents $\mathbf{J}_0$, $\mathbf{J}_1$ together produce the fields $\mathbf{E}_1, \mathbf{H}_1$ everywhere. Likewise, $\mathbf{J}_0, \mathbf{J}_2$ produce $\mathbf{E}_2, \mathbf{H}_2$. $\mathbf{J}_0$ is in V but $\mathbf{J}_1$ and $\mathbf{J}_2$ are not. The ambient medium in V is characterized by μ, ϵ and σ. The permittivity and permeability

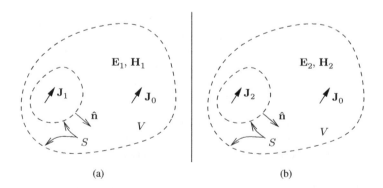

Figure 4.5   Uniqueness theorem. (a) $\mathbf{J}_0$ and $\mathbf{J}_1$ generate $\mathbf{E}_1, \mathbf{H}_1$ everywhere. (b) $\mathbf{J}_0$ and $\mathbf{J}_2$ generate $\mathbf{E}_2, \mathbf{H}_2$ everywhere.

can be assumed to be real. The field laws in V for each case are

$$\nabla \times \mathbf{E}_1 = -j\omega\mu\mathbf{H}_1 \qquad \nabla \times \mathbf{H}_1 = j\omega\epsilon\mathbf{E}_1 + \sigma\mathbf{E}_1 + \mathbf{J}_0$$
$$\nabla \times \mathbf{E}_2 = -j\omega\mu\mathbf{H}_2 \qquad \nabla \times \mathbf{H}_2 = j\omega\epsilon\mathbf{E}_2 + \sigma\mathbf{E}_2 + \mathbf{J}_0.$$

The difference field $\mathbf{E}_1 - \mathbf{E}_2$, $\mathbf{H}_1 - \mathbf{H}_2$ satisfies

$$\nabla \times (\mathbf{E}_1 - \mathbf{E}_2) = -j\omega\mu(\mathbf{H}_1 - \mathbf{H}_2)$$
$$\nabla \times (\mathbf{H}_1 - \mathbf{H}_2) = j\omega\epsilon(\mathbf{E}_1 - \mathbf{E}_2) + \sigma(\mathbf{E}_1 - \mathbf{E}_2)$$

which shows that the difference field does not depend on the source $\mathbf{J}_0$. If we can find conditions that make $\mathbf{E}_1 = \mathbf{E}_2$ and $\mathbf{H}_1 = \mathbf{H}_2$, then uniqueness will be obtained.

Applying Poynting's theorem to the difference field gives

$$\oint_S (\mathbf{E}_1 - \mathbf{E}_2) \times (\mathbf{H}_1 - \mathbf{H}_2)^* \cdot d\mathbf{S} = j\omega \int_V \mu|\mathbf{H}_1 - \mathbf{H}_2|^2 - \epsilon|\mathbf{E}_1 - \mathbf{E}_2|^2 \, dV + \int_V \sigma|\mathbf{E}_1 - \mathbf{E}_2|^2 \, dV.$$

$$(4.70)$$

Suppose we specify the value of $\mathbf{E}_{tan}$ on S. Then, it is the same for both cases; $\mathbf{E}_{tan} = \mathbf{E}_{tan1} = \mathbf{E}_{tan2}$. This will make the left-hand side of (4.70) equal to zero (use $\mathbf{A} \times \mathbf{B} \cdot \mathbf{C} = -\mathbf{B} \cdot \mathbf{A} \times \mathbf{C}$). Likewise, specifying $\mathbf{H}_{tan}$ will make it zero. We can even specify both $\mathbf{E}_{tan}$ and $\mathbf{H}_{tan}$.

On the right-hand side, $|\mathbf{E}_1 - \mathbf{E}_2| \geq 0$ and $|\mathbf{H}_1 - \mathbf{H}_2| \geq 0$ so the only way for the real part to be zero is if $\mathbf{E}_1 = \mathbf{E}_2$. The remaining imaginary part will only become zero when we add the condition $\mathbf{H}_1 = \mathbf{H}_2$. The conclusion is that if we specify $\mathbf{E}_{tan}$ and/or $\mathbf{H}_{tan}$ on S, it forces $\mathbf{E}_1 = \mathbf{E}_2$ and $\mathbf{H}_1 = \mathbf{H}_2$ in V. Therefore, the field in V will be unique. In other words, the fields in V are independent of any changes to the sources and/or materials outside V, *provided that we maintain the same* $\mathbf{E}_{tan}$ *or* $\mathbf{H}_{tan}$ *on S that we had before.*

The proof breaks down in a lossless resonator, which has $\sigma = 0$ and a field $\mathbf{E}, \mathbf{H}$ with equal amounts of stored average electric and magnetic energy

$$\int_V \mu|\mathbf{H}|^2 \, dV = \int_V \epsilon|\mathbf{E}|^2 \, dV.$$

As a result, it becomes possible to have a non-zero difference field $\mathbf{E} = \mathbf{E}_1 - \mathbf{E}_2$, $\mathbf{H} = \mathbf{H}_1 - \mathbf{H}_2$ that makes the right-hand side of (4.70) equal to zero. Physically, this means a resonator can have modes that came from sources that were turned off at $t = -\infty$. The resonant modes only exist at certain discrete frequencies ω, so the uniqueness theorem can be applied, with an exception for these cases.

To illustrate the usage of the uniqueness theorem, consider as an example a dipole source $\mathbf{J}_0$ that illuminates a perfectly conducting metal sphere with a dielectric coating. The bounding surface $S = S_1 + S_2$ has two parts. S_1 is a sphere just outside the metal, but inside the dielectric, and S_2 is a sphere at infinity. The region between S_1 and S_2 is the volume V, where a unique solution is sought.

On S_1, the conducting sphere sets the condition $\mathbf{E}_{tan} = 0$. On the surface S_2 at infinity, $\mathbf{E}_{tan} = 0$ and $\mathbf{H}_{tan} = 0$ because the fields behave as $1/r$ and have decayed to zero. Therefore, the solution in V will be unique. The remaining requirements are that $\mathbf{E}_{tan}$ and $\mathbf{H}_{tan}$ shall be continuous at the dielectric-air interface, as there are no surface currents on a dielectric-air interface. This completes the requirements for a unique solution of Maxwell's equations in V.

What if we had a dielectric sphere with no metal core? In this case, S_1 is gone. If the fields vanish on S_2, then uniqueness requirements are satisfied. As before, the field laws require that $\mathbf{E}_{tan}$ and $\mathbf{H}_{tan}$ shall be continuous at the dielectric-air boundary.

Another example is a dipole source $\mathbf{J}_0$ in the presence of a perfectly conducting sphere. Here, we specify $\mathbf{E}_{tan} = 0$ on the sphere and zero fields at infinity; this confirms uniqueness. We cannot specify $\mathbf{H}_{tan}$ on the sphere because we don't know it until we solve the problem.

It is worth noting that the expressions

$$(\mathbf{E}_1 - \mathbf{E}_2) \times \hat{\mathbf{n}} = \mathbf{M}_s$$

$$\hat{\mathbf{n}} \times (\mathbf{H}_1 - \mathbf{H}_2) = \mathbf{J}_s$$

specify the jump discontinuity of $\mathbf{E}_{tan}$ and $\mathbf{H}_{tan}$, but *not* the boundary values of $\mathbf{E}_{tan}$ and $\mathbf{H}_{tan}$. Therefore, specifying $\mathbf{M}_s$ and/or $\mathbf{J}_s$ does not establish uniqueness. On the other hand, if $\mathbf{E}_{tan}$ and $\mathbf{H}_{tan}$ are known to be zero on one side of S (typically from the null-field assumption $\mathbf{E} = \mathbf{H} = 0$), the null fields together with the jump information will determine $\mathbf{E}_{tan}$ and $\mathbf{H}_{tan}$ on the other side, and uniqueness is obtained.

The field of a dipole in the presence of a sphere is called the *Green's function* of a sphere. Likewise, the field of a dipole in free space is called the free-space Green's function. Presented this way, a Green's function is simply the function that gives the field of a point source, possibly in the presence of some body. We have been using Green's functions extensively in this chapter without saying the words; it is only being mentioned at this point to raise the reader's awareness of this terminology. Green's functions will be discussed in detail in Chapter 10.

4.5   Radiation Condition

It was mentioned that at infinity, the radiated fields should decay to zero. There is a more formal requirement governing the field behaviour at infinity, and it is known as the *Sommerfeld radiation condition*

$$\lim_{r \to \infty} r \left(\frac{\partial F}{\partial r} + jkF \right) = 0. \tag{4.71}$$

F represents any of E_θ, E_ϕ, H_θ or H_ϕ in the far field. It is a statement about how the fields must behave at a large distance from the sources. It means that the outward-travelling waves have to decay at least as fast as $1/r$.

There is a corresponding radiation condition in 2D

$$\lim_{\rho \to \infty} \sqrt{\rho} \left(\frac{\partial F}{\partial \rho} + jkF \right) = 0 \tag{4.72}$$

that says cylindrical waves should decay at least as fast as $1/\sqrt{\rho}$. In 1D, for waves travelling in the $\pm x$ directions, the radiation condition is

$$\lim_{x \to \infty} \left(\frac{dF}{dx} \pm jkF \right) = 0. \tag{4.73}$$

Radiation conditions are useful when applying the uniqueness theorem to open regions.

4.6   Image Theory

When a source is above a flat, perfectly conducting ground plane, it can be converted into the source plus its 'image' radiating in free space. This is illustrated in Figure 4.6 for an electric dipole above a perfect electric conductor. Let us first consider the vertical dipole in case (a). On the PEC ground, $\mathbf{E}_{tan}$ has to be zero. Upon consideration of the near-zone electric field of the dipole pair in case (b), one finds that

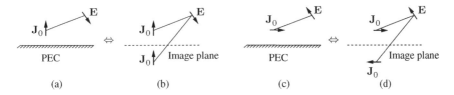

Figure 4.6 (a) Vertical electric dipole above a perfect electric conductor and (b) equivalent in free space. (c) Horizontal electric dipole above a perfect electric conductor and (d) equivalent in free space.

the horizontal component of the electric field is zero along the 'image plane' that is midway between the sources. In the upper half space, (a) and (b) have the same source $\mathbf{J}_0$ and the same boundary condition at the plane, so from the uniqueness theorem, they have the same field.

An alternative way to look at this is $\mathbf{E}_{tan} = 0$ along the image plane in (b). Therefore, inserting a PEC at the image plane will not perturb the field in any way. Surface currents will be induced on the top and bottom surfaces of the plane. In the upper half space, the total field is produced by the top dipole and the $\mathbf{J}_s$ that is induced on the PEC's top surface. In the lower half space, the total field is produced by the bottom dipole and the $\mathbf{J}_s$ on the bottom surface. Since this is a PEC plane, the bottom dipole can be removed without having any effect in the upper half space. This establishes the equivalence with (a).

Similar logic applies to the horizontal electric dipole above a PEC ground in Figure 4.6(c) and its equivalent (d). Consideration of the near-zone electric field of the dipole pair in case (d) indicates that the horizontal component of the electric field is zero along the image plane. In the upper half space, (c) and (d) have the same source $\mathbf{J}_0$ and the same boundary condition at the plane, so they have the same fields.

A subtle point to be aware of is that in Figure 4.6, an induced surface current $\mathbf{J}_s = \hat{n} \times \mathbf{H}$ on the ground plane radiates the reflected field, whereas in the image equivalent the lower dipole $\mathbf{J}_0$ radiates the reflected field. Therefore, the same reflected field can be produced by two very different currents: $\mathbf{J}_s$ on the PEC, or the image $\mathbf{J}_0$.

These concepts can be extended to electric and magnetic sources in the presence of perfect electric/magnetic conducting ground planes, as in Figure 4.7. A PEC ground requires $\mathbf{E}_{tan} = 0$ and a PMC

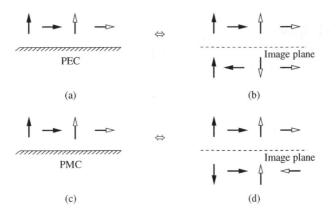

Figure 4.7 (a) Electric dipoles ⟶ and magnetic dipoles ⟹ above a perfect electric conductor and (b) equivalent in free space. (c) Electric and magnetic dipoles above a perfect magnetic conductor and (d) equivalent in free space.

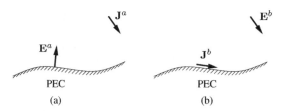

Figure 4.8 A curved perfect electric conductor. (a) Source $\mathbf{J}^a$ produces the field $\mathbf{E}^a$. (b) Source $\mathbf{J}^b$ produces the field $\mathbf{E}^b$.

ground requires $\mathbf{H}_{tan} = 0$. Considering the near fields of the dipole sources, the orientations of the image sources are then chosen to so that the boundary conditions are met at the image plane.

Important special cases occur when sources are placed directly *on* a ground plane. When a source and its image are oriented the same way, the field will be doubled. When they are oppositely directed, the resulting field is zero. With reference to Figure 4.7, a PEC ground will double the field of a vertical electric current or a horizontal magnetic current. No field can be produced by a horizontal electric current or a vertical magnetic current. Likewise, a PMC ground will double the field of a vertical magnetic current or a horizontal electric current. No field can be produced by a horizontal magnetic current or a vertical electric current on a PMC.

A tangential electric current on a flat PEC cannot radiate, and this was shown to be true by using image theory. It turns out that this is still true for a non-planar PEC; however, image theory can no longer be used for the proof. It can be proven by using reciprocity. Figure 4.8 shows a dipole $\mathbf{J}^a$ that produces a field $\mathbf{E}^a$ and a second dipole $\mathbf{J}^b$ that is tangential to the PEC surface and has a field $\mathbf{E}^b$. From reciprocity,

$$\mathbf{E}^a \cdot \mathbf{J}^b = \mathbf{E}^b \cdot \mathbf{J}^a.$$

Dotting $\mathbf{E}^a$ with $\mathbf{J}^b$ extracts the tangential part of $\mathbf{E}^a$, which is necessarily zero on the conductor. Since $\mathbf{J}^a \neq 0$ it implies that $\mathbf{E}^b$ must be zero. Therefore, the field produced by $\mathbf{J}^b$ is zero. Similarly, one can use reciprocity to show that no field is produced by a vertical magnetic current on a PEC and a horizontal magnetic or vertical electric current on a PMC.

4.7 Physical Optics

Referring to Figure 4.9, suppose we have a flat perfect electric conductor, illuminated by an incident field $\mathbf{E}^i$, $\mathbf{H}^i$. The reflected field is $\mathbf{E}^r$, $\mathbf{H}^r$ and the total field is $\mathbf{E} = \mathbf{E}^i + \mathbf{E}^r$, $\mathbf{H} = \mathbf{H}^i + \mathbf{H}^r$. In case (a), the conductor is flat and of infinite extent, so image theory applies. On the conductor's surface the tangential

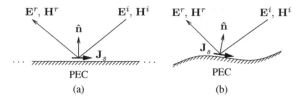

Figure 4.9 An incident field $\mathbf{E}^i$, $\mathbf{H}^i$ illuminates a perfect electric conductor. (a) Infinite, flat conductor. (b) Finite, smoothly curved conductor.

parts of $\mathbf{E}^i$ and $\mathbf{E}^r$ will cancel to produce $\mathbf{E}_{tan} = 0$. The tangential parts of $\mathbf{H}^i$ and $\mathbf{H}^r$ are equal and will add. The exact surface current is therefore

$$\mathbf{J}_s = \hat{\mathbf{n}} \times \mathbf{H} = \hat{\mathbf{n}} \times (\mathbf{H}^i + \mathbf{H}^r) = 2\hat{\mathbf{n}} \times \mathbf{H}^i. \tag{4.74}$$

The reflected field obeys the laws of geometrical optics, so this is called the *GO current*.

An electrically large and smooth conducting body of finite extent is shown in Figure 4.9(b). In this case, it is assumed that the notion of GO reflection approximately applies, at least locally, at the reflection point. Then, we can approximate $\mathbf{J}_s$ by (4.74) on the lit side, and in accordance with optics ideas, let $\mathbf{J}_s \approx 0$ on the shadow side. Therefore,

$$\mathbf{J}_s \approx \begin{cases} 2\hat{\mathbf{n}} \times \mathbf{H}^i; & \text{lit side} \\ 0; & \text{shadow side.} \end{cases} \tag{4.75}$$

In case (b), at any lit point on the body, the GO current appears as though it were on a local tangent plane of infinite extent.

The approximate current (4.75) can be used in a radiation integral to find the scattered field from a finite, curved body. The finiteness is accounted for by the limits of integration, which extend over the lit part of the body. The procedure of using the GO approximate currents in a radiation integral is called *physical optics* (PO).

With reference to Figure 4.9(b), $\mathbf{J}_s \approx 0$ on the shadow side of the conductor, which implies that $\mathbf{H}_{tan} \approx 0$ on the bottom surface. This does *not* imply that the fields below the conductor are zero. In fact the GO current (4.75) will radiate *everywhere* including the 'shadow' region below the finite-size body. So the bottom side is not really a shadow at all, as far as the fields are concerned.

The notion of PO is often extended to other situations. In a more general sense, PO can mean, 'take the current on an infinite structure and terminate it with the radiation integral'. For instance, one could find the surface current induced by an incident plane wave, on an infinitely long and thin wire. Using this current in a radiation integral but truncating the integration to a length ℓ, we would then know the scattered field for a wire of length ℓ. The limitation is that the true current on a terminated wire experiences reflection at the endpoints, and this is neglected in the PO approximation. Nevertheless, PO often gives very good results – in particular, near the direction of maximum (i.e. specular) scattering.

4.8   Surface Equivalent

The uniqueness theorem told us that we can find the fields in V from a knowledge of all sources in V plus $\mathbf{E}_{tan}$ and/or $\mathbf{H}_{tan}$ on S. But it tells us *nothing* about how to do it. To find the fields we need to use what is essentially the vector version of *Huygens principle*. Electromagnetic versions of Huygens principle have been attributed to Stratton and Chu, Schelkunoff, and Love. The details will now be given.

We have seen that the vector potential integrals

$$\mathbf{A}(\mathbf{r}) = \frac{\mu}{4\pi} \int_{V_0} \mathbf{J}(\mathbf{r}') \frac{e^{-jkR}}{R} dV' \tag{4.76}$$

$$\mathbf{F}(\mathbf{r}) = \frac{\epsilon}{4\pi} \int_{V_0} \mathbf{M}(\mathbf{r}') \frac{e^{-jkR}}{R} dV' \tag{4.77}$$

can be used to obtain the fields due to sources that are in a homogeneous medium characterized by μ, ϵ. The *surface equivalence principle* allows us to convert a physical scattering body into an equivalent that involves only currents that act in the homogeneous medium. Then, (4.76) and (4.77) apply and the fields can be found from the currents.

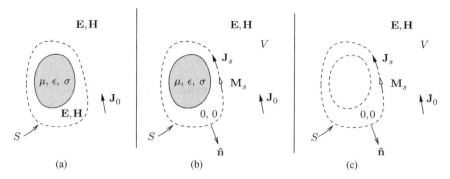

Figure 4.10 (a) Original problem with the scatterer, source $\mathbf{J}_0$ and fields $\mathbf{E}, \mathbf{H}$. The ambient medium is free space. (b) Equivalent problem, having the same $\mathbf{E}, \mathbf{H}$ within V (outside S) and zero fields outside V (inside S). (c) Equivalent problem in free space.

In the following discussion we will assume, without loss of generality, that the ambient medium is free space so that $\mu = \mu_0$ and $\epsilon = \epsilon_0$ in Equations (4.76) and (4.77). Figure 4.10 shows a scattering body in free space, the body having electrical properties μ, ϵ, σ. A source $\mathbf{J}_0$ illuminates the body. It is assumed that $\mathbf{E}, \mathbf{H}$ outside S in the original problem (a) is somehow known.

In case (b), a mathematical surface S surrounds the scatterer (but it does not have to conform to the scatterer's surface). It is postulated that inside V (outside S) we have the original field $\mathbf{E}, \mathbf{H}$, whereas outside V (inside S), both $\mathbf{E} = 0$ and $\mathbf{H} = 0$. Since the fields inside S are zero, the tangential fields on the inside surface of S are zero. In order for the discontinuous fields to exist, the surface currents

$$\mathbf{J}_s = \hat{\mathbf{n}} \times (\mathbf{H} - 0) \tag{4.78}$$

$$\mathbf{M}_s = (\mathbf{E} - 0) \times \hat{\mathbf{n}} \tag{4.79}$$

are required on S. These are the *surface equivalent currents*.

The currents $\mathbf{J}_s$, $\mathbf{M}_s$ and $\mathbf{J}_0$ together will produce $\mathbf{E}, \mathbf{H}$ outside S. However, the vector potential integrals (4.76) and (4.77) cannot be used to find $\mathbf{E}, \mathbf{H}$ because the currents are not in free space. So, the third step is to simply remove the scatterer, as shown in case (c). We are allowed to do this because (b) and (c) have the same source $\mathbf{J}_0$ and the same $\mathbf{E}_{tan}$ and $\mathbf{H}_{tan}$ on S. Therefore, according to the uniqueness theorem, they will have the same fields.[3] With the sources in free space, the vector potential integrals (4.76) and (4.77) can now be used to find the fields.

It may seem peculiar that removal of the scatterer does not change the fields. However, something does change. What happens is the individual contributions made by $\mathbf{J}_s$, $\mathbf{M}_s$ and $\mathbf{J}_0$ for cases (b) and (c) will be different. In other words, the field is unique, but the connection between each source and its corresponding field contribution is not unique.

When constructing the surface equivalent, we specified $\mathbf{J}_0$ as well as both $\mathbf{E}_{tan}$ and $\mathbf{H}_{tan}$ on S. Therefore, the field in V is unique. But this is not the only way to achieve uniqueness. According to the uniqueness theorem, specifying just $\mathbf{E}_{tan}$ or $\mathbf{H}_{tan}$ should be sufficient. Although the uniqueness theorem tells us that such a formulation exists, it doesn't tell us anything at all about how to find it. Finding a solution in terms of just $\mathbf{E}_{tan}$ or $\mathbf{H}_{tan}$ will now be described.

Suppose that in Figure 4.10(c) we filled up the space inside S with a perfect electric conductor. We are allowed do this because the uniqueness of $\mathbf{E}, \mathbf{H}$ in V is unaffected by the material properties inside S. In other words, the total $\mathbf{E}, \mathbf{H}$ in V will be the same, whether or not the conductor is there.

---

[3] The reader is reminded we have specified $\mathbf{E}_{tan}$ and $\mathbf{H}_{tan}$ on both sides of S. On the inside face of S, $\mathbf{E}_{tan}$ and $\mathbf{H}_{tan}$ are zero. It is the $\mathbf{E}_{tan}$ and $\mathbf{H}_{tan}$ on the outside face of S that is pertinent to uniqueness in V.

The conductor changes the individual field contributions from $\mathbf{J}_s, \mathbf{M}_s$ and $\mathbf{J}_0$ – but not the total. The source $\mathbf{J}_s$ cannot radiate anything because it is being 'shorted out' by the perfect conductor. So the remaining contributions from $\mathbf{M}_s$ and $\mathbf{J}_0$ have to be different from before, in just the right way, to make up for the missing contribution of $\mathbf{J}_s$. Since the currents now radiate in the *presence* of the conducting body, we *cannot* use (4.76) and (4.77) to find the fields. So, in a sense the problem has become more difficult.

Another possibility is to put a perfect magnetic conductor inside S, on which $\mathbf{H}_{tan} = 0$. This shorts out (by duality) the contribution of $\mathbf{M}_s$ and gives a field purely in terms of $\mathbf{J}_s$ and $\mathbf{J}_0$. The currents radiate in the presence of the magnetic conductor, so again we cannot use (4.76) and (4.77) to find the fields.

We could put other things inside. For instance, we could put a dielectric sphere of any size, anywhere inside the bounding surface S. There is no requirement that the scatterer's surface conform to S. We would now have to know how $\mathbf{J}_s, \mathbf{M}_s$ and $\mathbf{J}_0$ radiate in the presence of the sphere. Such a choice is theoretically acceptable but does not lead to any simplifications or advantages.

In an electromagnetic problem $\mathbf{E}, \mathbf{H}$ are generally unknown so $\mathbf{J}_s$ and $\mathbf{M}_s$ are also unknown. They can sometimes be found or estimated from the physics of the problem. For the special case of a perfectly conducting scatterer, it is useful to shrink S so that it conforms to the metallic surface. Since $\mathbf{E}_{tan} = 0$ on the metal, we immediately know that $\mathbf{M}_s = 0$ on S. In addition, if the scatterer is large, an approximate technique such as physical optics might be used to estimate the electric current $\mathbf{J}_s$ on the conducting surface. It is noted that for a perfect electric conductor, the true physical surface currents are equal to the equivalent currents.

To gain familiarity with the surface equivalence concept, it will be exercised in the following two examples.

Example 4.5 (Dielectric Half Space) A distant source $\mathbf{J}_0$ produces a normal-incident plane wave at an air-dielectric interface as in Figure 4.11(a). The fields are

$$z < 0 \qquad\qquad\qquad\qquad z > 0$$

$$\mathbf{E} = \hat{\mathbf{x}} E_0 e^{-jk_0 z} + \hat{\mathbf{x}} \Gamma E_0 e^{jk_0 z} \qquad\qquad \mathbf{E} = \hat{\mathbf{x}} T E_0 e^{-jk_1 z}$$

$$\mathbf{H} = \hat{\mathbf{y}} \frac{E_0}{\eta_0} e^{-jk_0 z} - \hat{\mathbf{y}} \Gamma \frac{E_0}{\eta_0} e^{jk_0 z} \qquad\qquad \mathbf{H} = \hat{\mathbf{y}} T \frac{E_0}{\eta_1} e^{-jk_1 z},$$

where $k_{0,1}$ are the wavenumbers, $\eta_{0,1}$ are the wave impedances and Γ, T are the Fresnel reflection and transmission coefficients. Find surface equivalents for $z < 0$ and $z > 0$.

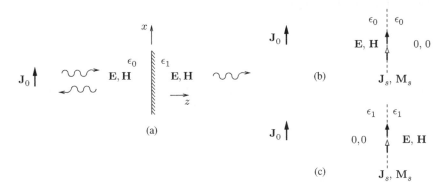

(a)

(b)

(c)

Figure 4.11 (a) Distant source $\mathbf{J}_0$ illuminates a dielectric half space with permittivity ϵ_1. (b) Free-space equivalent for $z < 0$. (c) Equivalent in medium with ϵ_1, for $z > 0$.

Solution: The surface equivalent for $z < 0$ is in free space and is shown in case (b). The surface currents follow from (4.78) and (4.79). With $\hat{\mathbf{n}} = -\hat{\mathbf{z}}$,

$$\mathbf{J}_s = \hat{\mathbf{x}} \frac{E_0(1 - \Gamma)}{\eta_0}; \quad \mathbf{M}_s = \hat{\mathbf{y}} E_0(1 + \Gamma).$$

$\mathbf{J}_0$, $\mathbf{J}_s$ and $\mathbf{M}_s$ generate $\mathbf{E}, \mathbf{H}$ for $z < 0$. Since $\mathbf{J}_0$ is now in free space, it no longer produces the total field. It produces the incident part of $\mathbf{E}, \mathbf{H}$. The surface currents produce the reflected part of $\mathbf{E}, \mathbf{H}$. Since the source $\mathbf{J}_0$ is within the equivalence region, it must be included as a contributor to the total $\mathbf{E}, \mathbf{H}$.

The $z > 0$ equivalent in case (c) has a homogeneous dielectric region with a permittivity ϵ_1. The surface currents are obtained from (4.78) and (4.79). With $\hat{\mathbf{n}} = \hat{\mathbf{z}}$,

$$\mathbf{J}_s = -\hat{\mathbf{x}} \frac{E_0 T}{\eta_1}; \quad \mathbf{M}_s = -\hat{\mathbf{y}} E_0 T.$$

$\mathbf{J}_s$ and $\mathbf{M}_s$ generate $\mathbf{E}, \mathbf{H}$ for $z > 0$. The source $\mathbf{J}_0$ is outside the equivalence region so it must be excluded.

It is noted that in case (b) $\mathbf{J}_0$ $\mathbf{J}_s$ and $\mathbf{M}_s$ add up to produce zero fields when $z > 0$; in case (c) $\mathbf{J}_s$ plus $\mathbf{M}_s$ produce no field when $z < 0$.   ∎

Example 4.6 (Surface Equivalent, Conductor) Use a surface equivalent and physical optics to calculate the scattered field and RCS in the $\phi = 0$ plane for a perfectly conducting flat plate in free space that is in the $z = 0$ plane, along $-a/2 \le x \le a/2, -b/2 \le y \le b/2$. The incident field is an x-polarized plane wave travelling in the $-z$ direction

$$\mathbf{E}^i = \hat{\mathbf{x}} E_0 e^{jk_0 z}; \quad \mathbf{H}^i = -\hat{\mathbf{y}} \frac{E_0}{\eta_0} e^{jk_0 z},$$

where $k_0 = \omega\sqrt{\mu_0 \epsilon_0}$ and $\eta_0 = \sqrt{\mu_0/\epsilon_0}$.

Solution: First, the plate is surrounded by a closed surface S. Inside S the fields are zero and outside S the fields are $\mathbf{E}, \mathbf{H}$. Because of the discontinuous fields, there will be $\mathbf{J}_s$ and $\mathbf{M}_s$ on S. Now we shrink S down to the metal surface. Since $\mathbf{E}_{tan} = 0$ on the metal, we immediately conclude that $\mathbf{M}_s = 0$. Also, specifying $\mathbf{E}_{tan} = 0$ on S ensures uniqueness outside S.

The electric surface current $\mathbf{J}_s$ is harder to find. We can approximate it, using geometrical optics. According to GO, the lit-side current is

$$\mathbf{J}_s \approx 2\hat{\mathbf{n}} \times \mathbf{H}^i = \hat{\mathbf{x}} \frac{2E_0}{\eta_0}$$

and the shadow side current is $\mathbf{J}_s \approx 0$.

To evaluate the radiation integral for $\mathbf{A}$, the current is at $z' = 0$ so $\mathbf{r}' = \hat{\mathbf{x}} x' + \hat{\mathbf{y}} y'$. The observer is in the $\phi = 0$ plane so $\hat{\mathbf{r}} = \hat{\mathbf{x}} \sin\theta + \hat{\mathbf{z}} \cos\theta$. Using the GO current in the radiation integral leads to the PO approximation

$$\mathbf{A} = \hat{\mathbf{x}} \frac{e^{-jk_0 r}}{r} \frac{\mu_0 E_0}{2\pi\eta_0} \int_{x'=-a/2}^{a/2} \int_{y'=-b/2}^{b/2} e^{jk_0 x' \sin\theta} dx' dy'$$

which can be evaluated, with the result that

$$\mathbf{A} = \hat{\mathbf{x}} \frac{\mu_0 E_0 ab}{2\pi\eta_0} \frac{\sin X}{X} \frac{e^{-jk_0 r}}{r}$$

where $X = (k_0 a/2) \sin \theta$. In the far field, $E_\theta \approx -j\omega A_\theta$ and $A_\theta = A_x \cos \theta$, so that

$$E_\theta = -j\omega \cos \theta \, \frac{\mu_0 E_0 ab}{2\pi\eta_0} \, \frac{\sin X}{X} \, \frac{e^{-jk_0 r}}{r}.$$

From the scattered field, the RCS is readily found to be

$$\sigma = \lim_{r\to\infty} 4\pi r^2 \frac{|\mathbf{E}^s|^2}{|\mathbf{E}^i|^2} = \frac{4\pi(ab)^2}{\lambda_0^2} \left(\cos\theta \frac{\sin X}{X} \right)^2.$$

∎

4.9 Love's Equivalent

If the source $\mathbf{J}_0$ is inside S, an equivalent outside S can be found, as in Figure 4.12. This is known as *Love's equivalence principle*. This formulation is useful for finding the radiation patterns of antennas, especially aperture antennas such as horns, reflectors and open-ended waveguides. Case (a) shows an impressed source $\mathbf{J}_0$ and a scattering body with μ, ϵ, σ. The source is inside S but can be either inside or outside the scatterer. The source and scatterer together could be used to represent an arbitrary antenna. In case (b) an equivalent in free space is postulated, having the same $\mathbf{E}, \mathbf{H}$ as the original case (a) when outside S and having zero fields inside S. Both problems have no sources in V and the same $\mathbf{E}_{tan}$ and $\mathbf{H}_{tan}$ on S. It follows from the uniqueness theorem that both configurations will have the same field in V.

For case (b) and just outside S, $\mathbf{E}_{tan}$ and $\mathbf{H}_{tan}$ are known because $\mathbf{E}$ and $\mathbf{H}$ are known. Just inside, the fields are zero, so $\mathbf{E}_{tan} = 0$ and $\mathbf{H}_{tan} = 0$. Because $\mathbf{E}_{tan}, \mathbf{H}_{tan}$ have jump discontinuities at S, there must be surface currents in accordance with

$$\mathbf{J}_s = \hat{n} \times \mathbf{H}; \quad \mathbf{M}_s = \mathbf{E} \times \hat{n}.$$

Outside S, the fields come from $\mathbf{J}_s$ and $\mathbf{M}_s$. Therefore, the surface sources alone radiate the complete $\mathbf{E}, \mathbf{H}$ in V and the radiation from $\mathbf{J}_0$ must not be included. Comparing Love's equivalent to the scattering equivalent, the only difference is that in the scattering equivalent, the fields in V come from $\mathbf{J}_0, \mathbf{J}_s$ and $\mathbf{M}_s$, whereas in Love's equivalent there are no sources in V so the complete $\mathbf{E}, \mathbf{H}$ comes from just $\mathbf{J}_s$ and $\mathbf{M}_s$.

Example 4.7 (Surface Equivalent, Antenna) Use a surface equivalent to calculate the field radiated by the aperture antenna in Figure 4.13. The antenna has a conducting body with an aperture along

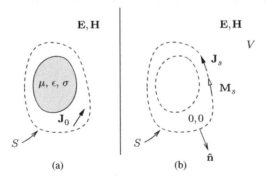

Figure 4.12 (a) Original problem with the scatterer, source $\mathbf{J}_0$ and fields $\mathbf{E}, \mathbf{H}$. (b) Equivalent problem in free space, having the same $\mathbf{E}, \mathbf{H}$ outside S and zero fields inside S.

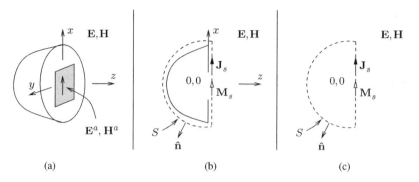

Figure 4.13 (a) An aperture in a metallic body. (b) External equivalent with surface sources $\mathbf{J}_s$, $\mathbf{M}_s$. The fields are zero inside S. (c) External equivalent in free space.

$-a/2 \le x \le a/2$, $-b/2 \le y \le b/2$. The aperture field is assumed to be approximately known and is given as a plane wave

$$\mathbf{E}^a = \hat{\mathbf{x}} E_0 e^{-jkz}; \quad \mathbf{H}^a = \hat{\mathbf{y}} \frac{E_0}{\eta} e^{-jkz},$$

where in free space, $k = \omega \sqrt{\mu_0 \epsilon_0}$ and $\eta = \sqrt{\mu_0/\epsilon_0}$. The field point is in the $\phi = 0$ plane.

Solution: We enclose the antenna with a surface S. In Figure 4.13(b) the true $\mathbf{E}_{tan}$ and $\mathbf{H}_{tan}$ are non-zero everywhere on S. If we shrink S down to the antenna surface, then $\mathbf{E}_{tan} = 0$ everywhere except at the aperture. On the metal and in the aperture, the true $\mathbf{H}_{tan} \ne 0$, but we will assume $\mathbf{H}_{tan} \approx 0$ on the metal away from the aperture. Using the assumed aperture fields $\mathbf{E}^a$, $\mathbf{H}^a$ at $z = 0$ gives

$$\mathbf{M}_s = \mathbf{E}^a \times \hat{\mathbf{n}} = \hat{\mathbf{x}} E_0 \times \hat{\mathbf{z}} = -\hat{\mathbf{y}} E_0$$

$$\mathbf{J}_s = \hat{\mathbf{n}} \times \mathbf{H}^a = \hat{\mathbf{z}} \times \hat{\mathbf{y}} E_0/\eta = -\hat{\mathbf{x}} E_0/\eta.$$

To evaluate the radiation integrals, the currents are at $z' = 0$ so $\mathbf{r}' = \hat{\mathbf{x}} x' + \hat{\mathbf{y}} y'$. The observer is in the $\phi = 0$ plane so $\hat{\mathbf{r}} = \hat{\mathbf{x}} \sin\theta + \hat{\mathbf{z}} \cos\theta$. The radiation integrals become

$$\mathbf{A} = -\hat{\mathbf{x}} \frac{e^{-jkr}}{r} \frac{\mu E_0}{4\pi\eta} \int_{x'=-a/2}^{a/2} \int_{y'=-b/2}^{b/2} e^{jkx' \sin\theta} dx' dy'$$

$$= -\hat{\mathbf{x}} \frac{e^{-jkr}}{r} \frac{\mu E_0 ab}{4\pi\eta} \frac{\sin X}{X},$$

where $X = (ka/2) \sin\theta$. Using $E_\theta \approx -j\omega A_\theta$ and $A_\theta = A_x \cos\theta$, we obtain

$$E_\theta = j\omega \cos\theta \frac{\mu E_0 ab}{4\pi\eta} \frac{e^{-jkr}}{r} \frac{\sin X}{X}.$$

Similarly, the contribution from $\mathbf{F}$ is

$$\mathbf{F} = -\hat{\mathbf{y}} \frac{e^{-jkr}}{r} \frac{\epsilon E_0}{4\pi} \int_{x'=-a/2}^{a/2} \int_{y'=-b/2}^{b/2} e^{jkx' \sin\theta} dx' dy'$$

$$= -\hat{\mathbf{y}} \frac{e^{-jkr}}{r} \frac{\epsilon E_0 ab}{4\pi} \frac{\sin X}{X}.$$

Using $H_\phi \approx -j\omega F_\phi$ with $F_\phi = F_y$ and $E_\theta \approx \eta H_\phi$, we obtain

$$E_\theta = j\omega\eta \frac{\epsilon E_0 ab}{4\pi} \frac{e^{-jkr}}{r} \frac{\sin X}{X}.$$

Combining the $\mathbf{J}_s$ and $\mathbf{M}_s$ parts gives

$$E_\theta = jk\frac{E_0 ab}{4\pi} \frac{e^{-jkr}}{r} \frac{\sin X}{X}(\cos\theta + 1).$$

Discussion: In the $(\cos\theta + 1)$ factor, the first term comes from $\mathbf{J}_s$, whereas the second term is from $\mathbf{M}_s$. This shows how the two source types radiate differently. These terms are the obliquity factors for the x-directed $\mathbf{J}_s$ and the y-directed $\mathbf{M}_s$.

As an alternate solution, we could put a PEC just inside S. The aperture sources would now be acting in the presence of a PEC that is the same size and shape as the original conducting body as shown in Figure 4.13 but with the aperture filled with metal. If the flat face is large, it would be a good approximation (by image theory) to double the $\mathbf{M}_s$ contribution; the $\mathbf{J}_s$ would not radiate. The radiated field would then be

$$E_\theta = jk\frac{E_0 ab}{2\pi} \frac{e^{-jkr}}{r} \frac{\sin X}{X}.$$

Likewise, we could put a PMC sheet inside S. This would eliminate the $\mathbf{M}_s$ contribution and double the $\mathbf{J}_s$ part, with the result that

$$E_\theta = jk\frac{E_0 ab}{2\pi} \frac{e^{-jkr}}{r} \frac{\sin X}{X}\cos\theta.$$

In the main-beam region near $\theta = 0$, all three formulations are in close agreement. In the back region near $\theta = 180°$, we expect a weak field, but only the formulation using both $\mathbf{J}_s$ and $\mathbf{M}_s$ predicts this. We used image theory with an infinite PEC or PMC sheet to eliminate the contributions of $\mathbf{J}_s$ or $\mathbf{M}_s$. Therefore, those results are only valid for $z > 0$.

Our first result is approximate because $\mathbf{J}_s$ and $\mathbf{M}_s$ are approximate. If the exact currents were known everywhere on S, then the radiated field would also be exact. For the second and third results, the currents and the infinite image plane are approximate. Strictly speaking, the doubling effect is only true for an infinite PEC or PMC plane. Since the first method has the fewest approximations, it is expected to be the most accurate. If the aperture were in an infinite PEC ground, then the second method would be the most accurate. ∎

4.10 Induction Equivalent

Suppose that a scattering body is bounded by surface S, as in Figure 4.14(a). The fields can be decomposed into incident and scattered parts

$$\mathbf{E} = \mathbf{E}^i + \mathbf{E}^s; \quad \mathbf{H} = \mathbf{H}^i + \mathbf{H}^s.$$

The incident field is produced by $\mathbf{J}_0$, which produces $\mathbf{E}^i, \mathbf{H}^i$ in the absence of the scatterer. The induction equivalent in Figure 4.14(b) is constructed so that it radiates the scattered field $\mathbf{E}^s, \mathbf{H}^s$ outside S and the total $\mathbf{E}, \mathbf{H}$ inside S.

Applying the boundary conditions at S, we obtain

$$\mathbf{J}_s = \hat{\mathbf{n}} \times (\mathbf{H}^s - \mathbf{H}); \quad \mathbf{M}_s = (\mathbf{E}^s - \mathbf{E}) \times \hat{\mathbf{n}}.$$

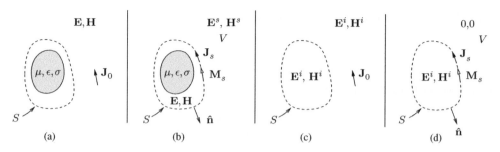

Figure 4.14 (a) Original problem with the scatterer, source $\mathbf{J}_0$ and fields $\mathbf{E}, \mathbf{H}$. (b) Induction equivalent, with $\mathbf{E}^s, \mathbf{H}^s$ outside S and $\mathbf{E}, \mathbf{H}$ inside S. (c) Original source and fields $\mathbf{E}^i, \mathbf{H}^i$. (d) Induction equivalent.

Given the definitions of the total, incident and scattered fields, these become

$$\mathbf{J}_s = -\hat{\mathbf{n}} \times \mathbf{H}^i; \quad \mathbf{M}_s = -\mathbf{E}^i \times \hat{\mathbf{n}}.$$

Since the incident field is known, the induction equivalent currents $\mathbf{J}_s, \mathbf{M}_s$ are known exactly. However, these currents radiate in the *presence* of the body so (4.76) and (4.77) can *not* be used to find the fields. Finding the field due to $\mathbf{J}_s, \mathbf{M}_s$ in the presence of the body is a difficult problem that usually doesn't have an exact solution. It is interesting to compare this with the surface equivalent – where the currents are difficult to find because they are related to the total field, which is unknown. However, they radiate in free space so it is easy to find their field from the radiation integrals (4.76) and (4.77).

Another way to think of the induction equivalent is that it replaces the incident field due to $\mathbf{J}_0$ by equivalent currents on S. This is shown in Figure 4.14 for the original incident field in case (c) and its equivalent in (d). Viewed this way, the induction equivalent is merely a surface equivalent for the incident field (Altintas and Paknys 2004). The equivalence can be proven by applying the boundary conditions on S to the fields shown in (d).

4.11 Volume Equivalent

Figure 4.15(a) shows a body in the presence of impressed sources $\mathbf{J}_i, \mathbf{M}_i$ which produce the total field $\mathbf{E}, \mathbf{H}$ inside and outside the body. Case (b) shows the same impressed sources $\mathbf{J}_i, \mathbf{M}_i$; they produce the

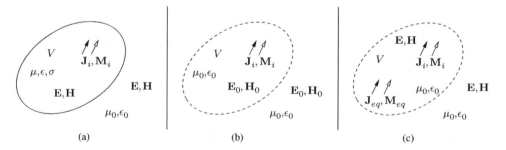

Figure 4.15 (a) Body in the presence of impressed sources $\mathbf{J}_i, \mathbf{M}_i$. (b) Impressed sources $\mathbf{J}_i, \mathbf{M}_i$ produce the incident field $\mathbf{E}_0, \mathbf{H}_0$ in free space. (c) Free-space equivalent: volume equivalent currents $\mathbf{J}_{eq}, \mathbf{M}_{eq}$ inside V plus the impressed sources $\mathbf{J}_i, \mathbf{M}_i$ produce the original field $\mathbf{E}, \mathbf{H}$.

field $\mathbf{E}_0, \mathbf{H}_0$ in free space. In case (a), at any point inside V we have

$$\nabla \times \mathbf{E} = -j\omega\mu\mathbf{H} - \mathbf{M}_i \tag{4.80}$$

$$\nabla \times \mathbf{H} = j\omega\epsilon\mathbf{E} + \mathbf{J}_i. \tag{4.81}$$

For case (b) we have

$$\nabla \times \mathbf{E}_0 = -j\omega\mu_0\mathbf{H}_0 - \mathbf{M}_i \tag{4.82}$$

$$\nabla \times \mathbf{H}_0 = j\omega\epsilon_0\mathbf{E}_0 + \mathbf{J}_i. \tag{4.83}$$

The sources can be inside or outside V; it does not matter.

The *scattered* field is the difference between the total and incident field so that $\mathbf{E} - \mathbf{E}_0 = \mathbf{E}^s$ and $\mathbf{H} - \mathbf{H}_0 = \mathbf{H}^s$. Taking (4.80) minus (4.82) gives

$$\nabla \times (\mathbf{E} - \mathbf{E}_0) = -j\omega(\mu\mathbf{H} - \mu_0\mathbf{H}_0)$$

or

$$\nabla \times \mathbf{E}^s = -j\omega\mu_0\mathbf{H}^s - \mathbf{M}_{eq}, \tag{4.84}$$

where we have defined a volume magnetic current

$$\mathbf{M}_{eq} = j\omega(\mu - \mu_0)\mathbf{H}. \tag{4.85}$$

Similarly, (4.81) minus (4.83) leads us to

$$\nabla \times \mathbf{H}^s = j\omega\epsilon_0\mathbf{E}^s + \mathbf{J}_{eq} \tag{4.86}$$

where we have defined a volume electric current

$$\mathbf{J}_{eq} = j\omega(\epsilon - \epsilon_0)\mathbf{E}. \tag{4.87}$$

These *volume equivalent currents* $\mathbf{J}_{eq}, \mathbf{M}_{eq}$ act in free space. They exist only within V where $\epsilon \neq \epsilon_0$ and $\mu \neq \mu_0$. If we know the true $\mathbf{E}, \mathbf{H}$ inside the body, then we can replace it with a free-space equivalent, as in Figure 4.15(c). $\mathbf{J}_{eq}$ and $\mathbf{M}_{eq}$ generate the true scattered field $\mathbf{E}^s, \mathbf{H}^s$ *everywhere* inside and outside V.

In Chapter 1 we discussed the 'polarizations' $\mathbf{P}$ and $\mathbf{M}$ which appear in

$$\mathbf{D} = \epsilon\mathbf{E} = \epsilon_0\mathbf{E} + \mathbf{P}$$

$$\mathbf{B} = \mu\mathbf{H} = \mu_0(\mathbf{H} + \mathbf{M})$$

and account for the presence of materials. It follows that $\mathbf{P} = (\epsilon - \epsilon_0)\mathbf{E}$ and $\mu_0\mathbf{M} = (\mu - \mu_0)\mathbf{H}$, permitting us to rewrite (4.85) and (4.87) as

$$\mathbf{M}_{eq} = j\omega\mu_0\mathbf{M}$$

$$\mathbf{J}_{eq} = j\omega\mathbf{P}.$$

Since $\mathbf{J}_{eq}$ and $\mathbf{M}_{eq}$ are related to the polarizations, they are also known as *volume polarization currents*.

Example 4.8 (Conductor, Volume Equivalence) A special case of interest is a nonmagnetic conductor. The permeability is $\mu = \mu_0$. There are no dielectric polarization effects in a conductor, so its permittivity is $\epsilon \approx \epsilon_0$. The conductivity is σ. It follows that the complex permittivity is $\epsilon_c = \epsilon_0 - j\sigma/\omega$. The volume polarization currents become

$$\mathbf{M}_{eq} = j\omega(\mu - \mu_0)\mathbf{H} = 0$$

and

$$\mathbf{J}_{eq} = j\omega(\epsilon_c - \epsilon_0)\mathbf{E} = \sigma\mathbf{E}.$$

So, from the volume polarization currents, the well-known result for conduction current $\mathbf{J} = \sigma\mathbf{E}$ is recovered. ∎

Example 4.9 (Dielectric, Volume Equivalence) Find the scattering cross section σ of a small dielectric sphere of radius $a \ll \lambda$. The sphere is at the origin and it is assumed that the permittivity is near unity. The incident field is a plane wave

$$E_z^i = E_0 e^{-jkx}$$

and $k = 2\pi/\lambda$ is the free-space wavenumber.

Solution: Since $\epsilon_r \approx 1$ but $\epsilon_r \neq 1$, the contrast is low and the so-called *Born approximation* for weak scattering can be applied. This permits us to neglect the scattered field within the dielectric and approximate the total field by the incident field. Since $a \ll \lambda$ we can let $e^{-jkx} \approx 1$ within the sphere, and then

$$\mathbf{E} \approx \mathbf{E}^i = \hat{z}E_0.$$

The volume equivalent current becomes

$$\mathbf{J} = \hat{z}j\omega\epsilon_0(\epsilon_r - 1)E_0.$$

Using $\mathbf{J}$ in the radiation integral with $e^{-jkR}/R \approx 1/r$, we obtain

$$E_z^s = k^2 E_0(\epsilon_r - 1)\,\frac{4}{3}\pi a^3\,\frac{e^{-jkr}}{r}$$

from which the scattering cross section can be found

$$\sigma = \lim_{r \to \infty} 4\pi r^2\,\frac{|\mathbf{E}^s|^2}{|\mathbf{E}^i|^2} = \frac{16\pi^2}{9}k^4 a^6(\epsilon_r - 1)^2.$$

The result shows that short wavelengths scatter most strongly. ∎

4.12   Radiation by Planar Sources

Radiation integrals are often needed for sources that are in the $x - y$ plane. These are useful for aperture antennas such as the horn, reflector and open-ended waveguide. It is assumed that surface currents $\mathbf{J}_s$, $\mathbf{M}_s$ produce the original field for $z > 0$ and the null field for $z < 0$. It can be shown that

$$E_\theta = jk\frac{e^{-jkr}}{4\pi r}\{P_x \cos\phi + P_y \sin\phi + \eta\cos\theta(Q_y \cos\phi - Q_x \sin\phi)\} \qquad (4.88)$$

$$E_\phi = jk\frac{e^{-jkr}}{4\pi r}\{\cos\theta(P_y\cos\phi - P_x\sin\phi) - \eta(Q_y\sin\phi + Q_x\cos\phi)\} \qquad (4.89)$$

$$P_{x,y} = \iint_{S_0} E^a_{x,y}(x',y')e^{jk(ux'+vy')}dx'dy' \qquad (4.90)$$

$$Q_{x,y} = \iint_{S_0} H^a_{x,y}(x',y')e^{jk(ux'+vy')}dx'dy' \qquad (4.91)$$

$$u = \sin\theta\cos\phi; \quad v = \sin\theta\sin\phi, \qquad (4.92)$$

where $E^a_{x,y}$, $H^a_{x,y}$ are the fields in the aperture S_0 in the $z = 0$ plane. The far-zone magnetic fields follow from the plane wave relationships $E_\theta = \eta H_\phi$ and $E_\phi = -\eta H_\theta$. Similar results for sources in other planes can be obtained by using a coordinate rotation.

4.13   2D Sources and Fields

In this section we will examine the relation between 2D sources and fields. Since this is 2D we set $\partial/\partial z = 0$. The source point is at $\boldsymbol{\rho}' = \hat{\mathbf{x}}x' + \hat{\mathbf{y}}y'$ and the field point is at $\boldsymbol{\rho} = \hat{\mathbf{x}}x + \hat{\mathbf{y}}y$. The distance R is given by $R = |\boldsymbol{\rho} - \boldsymbol{\rho}'|$. We will consider both z-directed and transverse sources as shown in Figure 4.16. The total field will be found by integration. To do this we will first derive field expressions for infinitesimal sources $J_z d\ell' = I_0$ (in Amps) and $J_\ell d\ell' = I_0$ (in Amps).

In the above, $d\ell'$ is an incremental length along the integration path C_0. Both $J_z d\ell'$ and $J_\ell d\ell'$ are current ribbons of width $d\ell'$. For currents distributed over the $x - y$ plane (the 2D version of volume currents), the appropriate elements are $J_{zs}dS'$ Amps and $J_{\ell s}dS'$ Amps.

4.13.1   z-Directed Source

A z-directed electric line source of strength I_0 Amps produces a field that is TM, having $E_z, H_x, H_y \neq 0$ and all other components equal to zero. The field is obtainable from A_z which must satisfy

$$\nabla^2 A_z + k^2 A_z = -\mu I_0 \delta(|\boldsymbol{\rho} - \boldsymbol{\rho}'|). \qquad (4.93)$$

A solution for A_z that corresponds to outward-travelling waves is

$$A_z = \frac{\mu I_0}{j4} H_0^{(2)}(kR). \qquad (4.94)$$

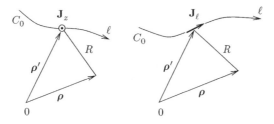

Figure 4.16   Coordinates showing the z-directed and ℓ-directed electric current ribbons J_z and J_ℓ.

The electric field can be found from

$$E_z = \frac{k^2}{j\omega\mu\epsilon} A_z \tag{4.95}$$

with the result that

$$E_z = \frac{-k\eta}{4} I_0 H_0^{(2)}(kR). \tag{4.96}$$

By duality, a z-directed magnetic line source of strength M_0 Volts produces a field that is TE, having $H_z, E_x, E_y \neq 0$ and all other components equal to zero. The magnetic field is

$$H_z = \frac{-k}{4\eta} M_0 H_0^{(2)}(kR). \tag{4.97}$$

4.13.2　Transverse Source

For a transverse current it is easier to first examine the special case of an x-directed dipole and then generalize the result to an arbitrary orientation. A source $I_0 = J_x dx$ produces a TE field with $H_z, E_x, E_y \neq 0$ and all other components equal to zero. The field can be found in terms of A_x which must satisfy

$$\nabla^2 A_x + k^2 A_x = -\mu I_0 \delta(|\boldsymbol{\rho} - \boldsymbol{\rho}'|). \tag{4.98}$$

A solution for A_x that corresponds to outward-travelling waves is

$$A_x = \frac{\mu I_0}{j4} H_0^{(2)}(kR). \tag{4.99}$$

The magnetic field can be found from

$$H_z = \frac{-1}{\mu} \frac{\partial A_x}{\partial y}. \tag{4.100}$$

Using $H_0'(x) = -H_1(x)$ we obtain

$$H_z = \frac{-jkI_0}{4} H_1^{(2)}(kR) \left(\frac{y-y'}{R}\right). \tag{4.101}$$

The result (4.101) can be generalized by replacing the dipole's direction $\hat{\mathbf{x}}$ with the arbitrary transverse direction $\hat{\boldsymbol{\ell}}$. Let $\hat{\mathbf{R}}$ be the unit vector from the source to the field point. Defining α as the angle between $\hat{\boldsymbol{\ell}}$ and $\hat{\mathbf{R}}$, we obtain $\sin\alpha = (y-y')/R$. This can also be written as $\sin\alpha = \hat{\mathbf{z}} \cdot (\hat{\boldsymbol{\ell}} \times \hat{\mathbf{R}})$. The field for an $\hat{\boldsymbol{\ell}}$-directed dipole then becomes

$$H_z = \frac{-jkI_0}{4} H_1^{(2)}(kR) \, \hat{\mathbf{z}} \cdot (\hat{\boldsymbol{\ell}} \times \hat{\mathbf{R}}). \tag{4.102}$$

4.13.3　Radiation Integrals

These results can be used to find the fields due to currents along a path C_0 in the $x-y$ plane, as shown in Figure 4.16. Using (4.96) with $I_0 \to J_z d\ell'$ and (4.102) with $I_0 \to J_\ell d\ell'$ leads to

$$E_z = \frac{-k\eta}{4} \int_{C_0} H_0^{(2)}(kR) J_z(\boldsymbol{\rho}') d\ell' \tag{4.103}$$

and

$$H_z = \frac{-jk}{4} \int_{C_0} \hat{\mathbf{z}} \cdot (\hat{\boldsymbol{\ell}}' \times \hat{\mathbf{R}}) \; H_1^{(2)}(kR) \; J_{\ell}(\boldsymbol{\rho}') d\ell'. \tag{4.104}$$

By duality, the fields due to magnetic current densities are found to be

$$H_z = \frac{-k}{4\eta} \int_{C_0} H_0^{(2)}(kR) M_z(\boldsymbol{\rho}') d\ell' \tag{4.105}$$

and

$$E_z = \frac{jk}{4} \int_{C_0} \hat{\mathbf{z}} \cdot (\hat{\boldsymbol{\ell}}' \times \hat{\mathbf{R}}) \; H_1^{(2)}(kR) \; M_{\ell}(\boldsymbol{\rho}') d\ell'. \tag{4.106}$$

When kR is large we can employ the large-argument form of the Hankel functions. If far-field conditions are satisfied, we can use the approximations $\hat{\mathbf{R}} \parallel \hat{\boldsymbol{\rho}}$ and $R \approx \rho - \hat{\boldsymbol{\rho}} \cdot \boldsymbol{\rho}'$. Then, (4.103)–(4.106) become

$$E_z = -e^{j\pi/4}\eta\sqrt{\frac{k}{8\pi}} \frac{e^{-jk\rho}}{\sqrt{\rho}} \int_{C_0} e^{jk\hat{\boldsymbol{\rho}}\cdot\boldsymbol{\rho}'} \; J_z(\boldsymbol{\rho}') d\ell' \tag{4.107}$$

$$H_z = \frac{-e^{j\pi/4}}{\eta}\sqrt{\frac{k}{8\pi}} \frac{e^{-jk\rho}}{\sqrt{\rho}} \int_{C_0} e^{jk\hat{\boldsymbol{\rho}}\cdot\boldsymbol{\rho}'} \; M_z(\boldsymbol{\rho}') d\ell' \tag{4.108}$$

$$H_z = e^{j\pi/4}\sqrt{\frac{k}{8\pi}} \frac{e^{-jk\rho}}{\sqrt{\rho}} \int_{C_0} \hat{\mathbf{z}} \cdot (\hat{\boldsymbol{\ell}}' \times \hat{\boldsymbol{\rho}}) \; e^{jk\hat{\boldsymbol{\rho}}\cdot\boldsymbol{\rho}'} \; J_{\ell}(\boldsymbol{\rho}') d\ell' \tag{4.109}$$

$$E_z = -e^{j\pi/4}\sqrt{\frac{k}{8\pi}} \frac{e^{-jk\rho}}{\sqrt{\rho}} \int_{C_0} \hat{\mathbf{z}} \cdot (\hat{\boldsymbol{\ell}}' \times \hat{\boldsymbol{\rho}}) \; e^{jk\hat{\boldsymbol{\rho}}\cdot\boldsymbol{\rho}'} \; M_{\ell}(\boldsymbol{\rho}') d\ell'. \tag{4.110}$$

4.13.4   2D and 3D Potentials

The 2D potential given in (4.94) can be obtained as a special case of 3D, by integrating over an infinitely long current filament on the z axis. Replacing $\mathbf{J}dV' \rightarrow \hat{\mathbf{z}}I_0 dz'$ in the radiation integral and letting $z = 0$, we obtain

$$A_z(\mathbf{r}) = \frac{\mu}{4\pi} \int_{z'=-\infty}^{\infty} I_0 \frac{e^{-jk\sqrt{x^2+y^2+z'^2}}}{\sqrt{x^2+y^2+z'^2}} dz'. \tag{4.111}$$

By employing the identity (Gradshteyn and Ryzhik 1980, Eq. 8.421.11)

$$\frac{1}{4\pi} \int_{z'=-\infty}^{\infty} \frac{e^{-jk\sqrt{D^2+z'^2}}}{\sqrt{D^2+z'^2}} dz' = \frac{1}{j4} H_0^{(2)}(kD) \tag{4.112}$$

in (4.111), we obtain (4.94).

Example 4.10 (Metal Strip) In this example we will find the echo width of a 2D metal strip for the TE and TM cases. The strip is a perfect conductor at $y = 0$ and along $-a/2 \le x \le a/2$.

Solution, TM case: The incident field is

$$E_z^i = E_0 e^{jk(x\cos\phi_i + y\sin\phi_i)}.$$

From this, we can find the magnetic field and the GO surface current

$$J_z(x') = \frac{2E_0}{\eta} \sin\phi_i \; e^{jkx'\cos\phi_i}.$$

The field point is at (ρ, ϕ), so $\hat{\rho} = \hat{x} \cos \phi + \hat{y} \sin \phi$. The integration point is at $\rho' = \hat{x}x'$. The radiation integral becomes

$$E_z^s = -e^{j\pi/4} \eta \sqrt{\frac{k}{8\pi}} \frac{e^{-jk\rho}}{\sqrt{\rho}} \int_{x'=-a/2}^{a/2} e^{jkx' \cos \phi} \frac{2E_0}{\eta} \sin \phi_i \; e^{jkx' \cos \phi_i} dx'$$

or

$$E_z^s = -2E_0 a \; e^{j\pi/4} \sqrt{\frac{k}{8\pi}} \frac{e^{-jk\rho}}{\sqrt{\rho}} \sin \phi_i \frac{\sin X}{X},$$

where $X = (ka/2)(\cos \phi_i + \cos \phi)$. The maximum occurs in the specular direction where $\cos \phi = -\cos \phi_i$. The echo width is

$$\sigma = \lim_{\rho \to \infty} 2\pi\rho \frac{|\mathbf{E}^s|^2}{|\mathbf{E}^i|^2} = \frac{2\pi a^2}{\lambda} \left(\sin \phi_i \frac{\sin X}{X} \right)^2.$$

Solution, TE case: The incident field is

$$H_z^i = H_0 e^{jk(x \cos \phi_i + y \sin \phi_i)}.$$

From this, we can find the magnetic field and the GO surface current

$$J_x(x') = 2H_0 \; e^{jkx' \cos \phi_i}.$$

The field point is at (ρ, ϕ), so $\hat{\rho} = \hat{x} \cos \phi + \hat{y} \sin \phi$. The integration point is at $\hat{\rho}' = \hat{x}x'$. The current element is x-directed, so $\hat{\ell} = \hat{x}$. The radiation integral becomes

$$H_z^s = e^{j\pi/4} \sqrt{\frac{k}{8\pi}} \frac{e^{-jk\rho}}{\sqrt{\rho}} \int_{x'=-a/2}^{a/2} \sin \phi \; e^{jkx' \cos \phi} \; 2H_0 \; e^{jkx' \cos \phi_i} dx'$$

or

$$H_z^s = 2H_0 a \; e^{j\pi/4} \sqrt{\frac{k}{8\pi}} \frac{e^{-jk\rho}}{\sqrt{\rho}} \sin \phi \frac{\sin X}{X},$$

where $X = (ka/2)(\cos \phi_i + \cos \phi)$. The echo width is

$$\sigma = \lim_{\rho \to \infty} 2\pi\rho \frac{|\mathbf{H}^s|^2}{|\mathbf{H}^i|^2} = \frac{2\pi a^2}{\lambda} \left(\sin \phi \frac{\sin X}{X} \right)^2.$$

■

4.14 Derivation of Vector Potential Integral

This section shows how to solve the inhomogeneous wave equation

$$\nabla^2 \mathbf{A}(\mathbf{r}) + k^2 \mathbf{A}(\mathbf{r}) = -\mu \mathbf{J}(\mathbf{r}). \tag{4.113}$$

It can be considered as three scalar equations of the form

$$\nabla^2 A(\mathbf{r}) + k^2 A(\mathbf{r}) = -\mu J(\mathbf{r}),$$

where $A = A_x, A_y$ or A_z and $J = J_x, J_y$ or J_z. Let us assume for the moment that the source is a z-directed point current of unit strength, at the origin $(x, y, z) = (0, 0, 0)$ so that $J_z = \delta(r)$. Since this is a point source, it is expected that the solution will not depend on θ or ϕ. Then, $\mathbf{r} \to r$ and A_z satisfies

$$\frac{1}{r^2}\frac{d}{dr}\left(r^2\frac{dA_z}{dr}\right) + k^2 A_z(r) = -\mu\delta(r). \tag{4.114}$$

When $r \neq 0$ this can be written as

$$\frac{1}{r^2}\frac{d}{dr}\left(r^2\frac{dA_z}{dr}\right) + k^2 A_z(r) = 0. \tag{4.115}$$

The solution of (4.115) that corresponds to an outward-travelling wave is

$$A_z = C\frac{e^{-jkr}}{r} \tag{4.116}$$

where C is a constant. Since (4.115) is homogeneous, C could be arbitrary; however, (4.114) is inhomogeneous, so C has to be found.

To obtain C, we integrate both sides of (4.114) over a volume that is a small sphere of radius r_0, enclosing the source $\delta(r)$:

$$\int_0^{2\pi}\int_0^{\pi}\int_0^{r_0} \nabla^2 A_z\, dV + k^2 \int_0^{2\pi}\int_0^{\pi}\int_0^{r_0} A_z\, dV = -\mu \int_0^{2\pi}\int_0^{\pi}\int_0^{r_0} \delta(r)\, dV. \tag{4.117}$$

We then let $r_0 \to 0$. Since $A_z \propto 1/r$ and $dV = r^2 \sin\theta\, dr d\theta d\phi$, the second integral vanishes. From the properties of the delta function[4] the third integral becomes unity. The first integral can be converted to a surface integral over the sphere of radius r_0 by using $\nabla^2 A_z = \nabla \cdot \nabla A_z$ and the divergence theorem, with the result that (4.117) becomes

$$\int_0^{2\pi}\int_0^{\pi} \nabla A_z \cdot \hat{\mathbf{r}} dS|_{r=r_0} = -\mu \tag{4.118}$$

or

$$\int_0^{2\pi}\int_0^{\pi} \hat{\mathbf{r}}\frac{dA_z}{dr} \cdot \hat{\mathbf{r}}\ r\ \sin\theta d\theta d\phi|_{r=r_0} = -\mu. \tag{4.119}$$

Using (4.116) for A_z, we obtain

$$C\int_0^{2\pi}\int_0^{\pi}\frac{d}{dr}\left(\frac{e^{-jkr}}{r}\right) r\ \sin\theta d\theta d\phi|_{r=r_0} = -\mu. \tag{4.120}$$

---

[4] In the integration of the delta function over a volume, the θ and ϕ dependence was omitted. More precisely, the delta function should be written as the product of three one-dimensional impulses so that

$$\int \delta(\mathbf{r})dV = \int \delta(x)\delta(y)\delta(z)\, dx\, dy\, dz = 1.$$

In spherical coordinates, $\delta(\mathbf{r}) = \delta(r)\delta(\theta)\delta(\phi)/r^2\sin\theta$ because

$$\int \delta(\mathbf{r})dV = \int \frac{\delta(r)\delta(\theta)\delta(\phi)}{r^2\sin\theta}\, r^2\sin\theta\, dr\, d\theta\, d\phi = 1.$$

This integral is easily evaluated, and letting $r_0 \to 0$ leads to the result $C = \mu/4\pi$ so that (4.116) becomes

$$A_z = \frac{\mu}{4\pi} \frac{e^{-jkr}}{r}. \tag{4.121}$$

If the z-directed current element is located at some position $\mathbf{r}'$ instead of at the origin, then this result can be generalized to

$$A_z(\mathbf{r}) = \frac{\mu}{4\pi} \frac{e^{-jk|\mathbf{r}-\mathbf{r}'|}}{|\mathbf{r}-\mathbf{r}'|}. \tag{4.122}$$

For an arbitrary (but z-directed) current distribution, A_z can be found by superposition

$$A_z(\mathbf{r}) = \frac{\mu}{4\pi} \int_V J_z(\mathbf{r}') \frac{e^{-jk|\mathbf{r}-\mathbf{r}'|}}{|\mathbf{r}-\mathbf{r}'|} dV'. \tag{4.123}$$

Since A_x is related to J_x and A_y is related to J_y, an arbitrarily oriented current has

$$\mathbf{A}(\mathbf{r}) = \frac{\mu}{4\pi} \int_V \mathbf{J}(\mathbf{r}') \frac{e^{-jk|\mathbf{r}-\mathbf{r}'|}}{|\mathbf{r}-\mathbf{r}'|} dV'. \tag{4.124}$$

In 2D with a line source at the origin and $\partial/\partial z = 0$, the wave equation solution is instead

$$A_z = C H_0^{(2)}(k\rho). \tag{4.125}$$

By enclosing the line source with a cylinder and shrinking the radius to zero, it can be shown that $C = \mu/j4$ which leads to the solution

$$A_z(\boldsymbol{\rho}) = \frac{\mu}{j4} \int_S J_z(\boldsymbol{\rho}') H_0^{(2)}(k|\boldsymbol{\rho} - \boldsymbol{\rho}'|) dS'. \tag{4.126}$$

4.15   Solution Without Using Potentials

It was mentioned that the advantage of using $\mathbf{A}$ is that the differential equation for $\mathbf{A}$ is easier to solve than the ones for $\mathbf{E}, \mathbf{H}$. It is, nevertheless, possible to solve for the fields $\mathbf{E}, \mathbf{H}$ directly without recourse to potential theory.[5] This is now described. In (4.3) we obtained

$$\nabla^2 \mathbf{E} + k^2 \mathbf{E} = j\omega\mu\mathbf{J} + \nabla(\nabla \cdot \mathbf{E}). \tag{4.127}$$

Although it avoids using $\mathbf{A}$, it is harder to solve than

$$\nabla^2 \mathbf{A} + k^2 \mathbf{A} = -\mu\mathbf{J}. \tag{4.128}$$

How do we solve (4.127)? We do know that a solution to the scalar problem

$$\nabla^2 F(\mathbf{r}) + k^2 F(\mathbf{r}) = -P(\mathbf{r}) \tag{4.129}$$

is

$$F(\mathbf{r}) = \int_{V_0} P(\mathbf{r}') \frac{e^{-jkR}}{4\pi R} dV'. \tag{4.130}$$

---

[5] Thanks are due to Prof. D. R. Wilton who provided this derivation in a private communication.

Treating each rectangular component of (4.128) as a scalar equation

$$\nabla^2 A_x + k^2 A_x = -\mu J_x$$

$$\nabla^2 A_y + k^2 A_y = -\mu J_y$$

$$\nabla^2 A_z + k^2 A_z = -\mu J_z$$

suggests an identity

$$\mathbf{E}(\mathbf{r}) = -\int_{V_0} (j\omega\mu\mathbf{J}(\mathbf{r}') + \nabla'\nabla' \cdot \mathbf{E}(\mathbf{r}'))\frac{e^{-jkR}}{4\pi R}dV'. \tag{4.131}$$

This expression is not useful for finding $\mathbf{E}$ from $\mathbf{J}$ because $\mathbf{E}$ is in the integrand. However, this troublesome term can be replaced by using Gauss's law

$$\nabla' \cdot \mathbf{E}(\mathbf{r}') = \frac{\rho_v(\mathbf{r}')}{\epsilon}$$

and the charge term can be integrated by parts. Using

$$f\nabla g + g\nabla f = \nabla(fg)$$

and

$$\int_{V_0} \nabla f \, dV = \oint_S f \, \mathbf{dS}$$

gives us

$$\int_{V_0} f\nabla g \, dV + \int_{V_0} g\nabla f \, dV = \oint_S fg \, \mathbf{dS}.$$

Let $f = \rho_v(\mathbf{r}')$ and $g = e^{-jkR}/4\pi R$. With primed variables,

$$\int_{V_0} \rho_v(\mathbf{r}')\nabla'g(R)dV' + \int_{V_0} g(R)\nabla'\rho_v(\mathbf{r}')dV' = \oint_S \rho_v(\mathbf{r}')g(R) \, \mathbf{dS}'. \tag{4.132}$$

The charge distribution ρ_v is of finite extent and completely contained within V_0. For a time-harmonic field the charges are moving and if ρ_v terminates abruptly at the surface S (as it would when using volume equivalent currents) there will be surface charges on S. To avoid having to account for the surface charges separately, we should assume that V_0 is slightly larger than the volume occupied by ρ_v so that the surface integral in (4.132) will be zero. Noting that $\nabla g = -\nabla' g$ we obtain

$$\int_{V_0} \nabla'(\rho_v(\mathbf{r}'))g(R)dV' = -\int_{V_0} \rho_v(\mathbf{r}')\nabla'g(R)dV' = \nabla\int_{V_0} \rho_v(\mathbf{r}')g(R)dV'. \tag{4.133}$$

Using (4.133) in (4.131) leads to

$$\mathbf{E}(\mathbf{r}) = -j\omega\mu\int_{V_0} \mathbf{J}(\mathbf{r}') \frac{e^{-jkR}}{4\pi R}dV' - \frac{1}{\epsilon}\nabla\int_{V_0} \rho_v(\mathbf{r}') \frac{e^{-jkR}}{4\pi R}dV' \tag{4.134}$$

which is the same thing as $\mathbf{E} = -j\omega\mathbf{A} - \nabla\Phi$. The previous derivation using potential theory is probably preferable because it is easier to follow and it avoids the integration by parts.

4.16   Further Reading

For supplementary reading on the various equivalence principles and theorems, the book by Harrington (2001) is recommended. Further discussions of potential theory can also be found in Rothwell and Cloud (2008).

References

Altintas A and Paknys R (2004) The relationship between the induction and equivalence theorems. *URSI National General Assembly and Scientific Symposium, Ankara Turkey.* (In Turkish).

Gradshteyn IS and Ryzhik IM (1980) *Table of Integrals, Series, and Products.* Academic Press.

Harrington RF (2001) *Time-Harmonic Electromagnetic Fields.* IEEE Press.

Nevels R and Shin C (2001) Lorenz, Lorentz, and the gauge. *IEEE Antennas Propag. Mag.* **43**(3), 70–73.

Rothwell EJ and Cloud MJ (2008) *Electromagnetics.* CRC Press.

Silver S (1949) *Microwave Antenna Theory and Design.* McGraw-Hill.

Problems

4.1   A z-directed dipole of strength p_e at (0,0,0) has a vector potential

$$\mathbf{A} = \hat{\mathbf{z}}\, p_e \frac{\mu_0}{4\pi} \frac{e^{-jkr}}{r}.$$

Find H_ϕ, E_θ and E_r. Hint: first convert the unit vector $\hat{\mathbf{z}}$ to spherical coordinates and then evaluate $\nabla \times \mathbf{A}$.

4.2   For a z-directed dipole of strength $p_e = I\ell$, use the electric field expression to show that when the angular frequency $\omega \to 0$, the electrostatic field of a dipole is recovered

$$\mathbf{E} = \frac{Q\ell}{4\pi\epsilon_0 r^3}(\hat{\mathbf{r}}\, 2\cos\theta + \hat{\boldsymbol{\theta}}\sin\theta).$$

Use the relation $I = j\omega Q$ where I is the current which causes the charges $\pm Q$ to accumulate at the ends of the dipole.

4.3   For a z-directed dipole of strength $p_e = I\ell$, use the magnetic field expression to show that when the angular frequency $\omega \to 0$, the magnetostatic field of a current element (the Biot–Savart law) is recovered

$$\mathbf{H} = \hat{\boldsymbol{\phi}} \frac{I\ell}{4\pi r^2}\sin\theta.$$

4.4   Show that the magnetic field is related to electric current by

$$\mathbf{H}(\mathbf{r}) = \frac{1}{4\pi}\int_{V'} \nabla\left(\frac{e^{-jkR}}{R}\right) \times \mathbf{J}(\mathbf{r}')dV'.$$

4.5   A z-directed electric dipole of strength p_e is at $(0,0,0+)$ and is slightly above a PEC ground plane at $z = 0$. Find the surface current that is induced on the ground plane.

4.6   A loop of radius $a \ll \lambda$ is in the $x - y$ plane and centred about the origin. It carries a current I_0 in the $+\phi$ direction.

(a) Show that in the $\phi = 0$ plane and the far field,

$$A_\phi = A_y = \mu I_0 a \frac{e^{-jkr}}{4\pi r} \int_0^{2\pi} \cos \phi' \, e^{jka \sin \theta \cos \phi'} \, d\phi'.$$

(b) From A_ϕ obtain the radiation field

$$E_\phi = k^2 \eta I_0 S \frac{e^{-jkr}}{4\pi r} \sin \theta,$$

where $S = \pi a^2$ is the loop area.

Hint: The integral in terms of $J_1(x)$ is given in Appendix C and for small x, $J_1(x) \approx x/2$.

4.7   A loop of radius $a \ll \lambda$ is in the $x - y$ plane and centred about the origin. It carries a current I_0 in the $+\phi$ direction.

(a) Show that in the $\phi = 0$ plane

$$A_\phi = A_y = \mu \frac{I_0 a}{4\pi} \int_0^{2\pi} f \cos \phi' d\phi',$$

where $f = e^{-jkR}/R$ and $R = \sqrt{r^2 + a^2 - 2ra \sin \theta \cos \phi'}$.

(b) Find an approximate expression for $f(a)$ for small a by using a Maclaurin series about $a = 0$; that is, $f(a) = f(0) + f'(0)a + O(a^2)$. Use this to evaluate the integral and show that

$$A_\phi \approx \mu \frac{I_0 \pi a^2}{4\pi} e^{-jkr} \left(\frac{jk}{r} + \frac{1}{r^2} \right) \sin \theta.$$

Here, $S = \pi a^2$ is the loop area and $I_0 \pi a^2 = I_0 S$ is called the *magnetic moment* of the loop.

4.8   For the small loop in the previous problem, use A_ϕ to show that H_r, H_θ and E_ϕ in the $\phi = 0$ plane are

$$H_r = \frac{I_0 S}{2\pi} \cos \theta \, e^{-jkr} \left(\frac{jk}{r^2} + \frac{1}{r^3} \right)$$

$$H_\theta = \frac{I_0 S}{4\pi} \sin \theta \, e^{-jkr} \left(\frac{-k^2}{r} + \frac{jk}{r^2} + \frac{1}{r^3} \right)$$

$$E_\phi = \frac{\eta I_0 S}{4\pi} \sin \theta \, e^{-jkr} \left(\frac{k^2}{r} - \frac{jk}{r^2} \right).$$

4.9   Using the time-harmonic form of the Hertz potential $\mathbf{\Pi}^e$ for electric currents, apply duality to find the equations governing $\mathbf{\Pi}^m$ for magnetic currents.

4.10   It is not always possible to express a field in terms of A_z and F_z. In an attempt to find plane wave
solutions, solve the wave equation

$$\nabla^2 A_z + k^2 A_z = 0.$$

Consider two cases, where A_z satisfies the plane wave assumptions (i) $\partial/\partial x = \partial/\partial y = 0$ or (ii)
$\partial/\partial y = \partial/\partial z = 0$. Show that the potential function A_z leads to $\mathbf{E} = 0$.

4.11   By considering the properties of z-directed potentials, examine the expressions in Section 4.3.1
and find potentials that give a uniform plane wave with (i) only an E_x and (ii) only an E_y. In each
case, you will have to decide on the propagation direction, and whether to use A_z or F_z.

4.12   Find a vector potential A_z for the TM_{mn} modes in a rectangular waveguide. That is, solve
$\nabla^2 A_z + k^2 A_z = 0$, and choose from the possibilities

$$A_z = \begin{Bmatrix} \cos(m\pi x/a) \\ \sin(m\pi x/a) \end{Bmatrix} \begin{Bmatrix} \cos(n\pi y/b) \\ \sin(n\pi y/b) \end{Bmatrix} e^{-j\beta z}.$$

It is necessary to convert the boundary conditions from $\mathbf{E}$ to conditions on $\mathbf{A}$.

4.13   A coaxial cable has inner and outer radii of $\rho = a$ and $\rho = b$. Find ϕ-dependent field solutions.
Choose the potentials as $A_z = 0$ and $F_z = \psi(\phi)e^{-jkz}$ where ψ needs to be found and
$k = \omega\sqrt{\mu\epsilon}$.

4.14   Find a vector potential F_y to obtain the TE_y modes in a rectangular waveguide. That is, solve
$\nabla^2 F_y + k^2 F_y = 0$, and choose from the possibilities

$$F_y = \begin{Bmatrix} \cos(m\pi x/a) \\ \sin(m\pi x/a) \end{Bmatrix} \begin{Bmatrix} \cos(n\pi y/b) \\ \sin(n\pi y/b) \end{Bmatrix} e^{-j\beta z}.$$

It is necessary to convert the boundary conditions from $\mathbf{E}$ to conditions on $\mathbf{F}$.

4.15   A rectangular waveguide is partially filled with a dielectric, as in Figure 4.3. Find vector potentials
F_{y1} in the dielectric and F_{y0} in the air regions and the field components for the TE_y modes.
(a) From the continuity conditions for $\mathbf{E}_{tan}$ and $\mathbf{H}_{tan}$, obtain the equation relating k_{y1} and k_{y0}

$$k_{y1}\cot k_{y1}d = -k_{y0}\cot k_{y0}(b - d).$$

(b) Obtain the same result, using the transverse resonance method.

4.16   A rectangular waveguide as in Figure 4.3 is partially loaded with a dielectric that has $\epsilon_r = 2.45$,
$b/a = 0.45$ and $d/b = 0.50$. With $b = 0.3\lambda_0$, calculate β/k_0. Your result should be somewhere
between 0.8 and 1.2.

4.17   A rectangular waveguide as in Figure 4.3 is partially loaded with a dielectric that has $\epsilon_r = 2.45$,
$b/a = 0.45$ and $d/b = 0.50$.
(a) Find b/λ_0 at cutoff.
(b) Calculate and plot β/k_0 from cutoff up to $b = 0.6\lambda_0$.

4.18 An empty rectangular waveguide with $0 \le x \le a$, $0 \le y \le b$ has an electric field $\mathbf{E} = \hat{\mathbf{y}} E_0 \sin(\pi x / a) e^{-j\beta z}$.

  (a) Find an electric current sheet at $z = 0$ that produces this field for all $z > 0$.

  (b) Find a magnetic current sheet at $z = 0$ that produces this field for all $z > 0$.

  (c) Find electric and magnetic current sheets that produce this field for $z > 0$ and zero fields for $z < 0$.

4.19 For the dielectric half-space surface equivalents in Example 4.5, find $\mathbf{E}$ and $\mathbf{H}$ produced by the current sheets. You can use the results from Example 2.2 and duality. Show that the $z < 0$ and $z > 0$ equivalents produce the original $\mathbf{E}, \mathbf{H}$ in their respective regions.

4.20 A thin metal plate is at $z = 0$. The size of the plate is $-a/2 \le x \le a/2$, $-b/2 \le y \le b/2$ and the incident field is a plane wave $\mathbf{H}^i = \hat{\mathbf{x}} H_0 e^{jkz}$.

  (a) In the $\phi = 0$ plane, use PO to find the far-zone scattered field, H_θ^s.

  (b) Calculate and plot the radar cross section $\sigma(\theta)$ for $0 \le \theta \le 90°$. Assume that $\lambda = 1\text{m}$ and that $a = b = 2\,\text{m}$. In this problem $\mathbf{J}_s$ is being approximated but the way it radiates is known exactly.

4.21 Repeat the previous problem, but this time use an induction equivalent current $\mathbf{M}_s = -\mathbf{E}^i \times \hat{\mathbf{z}}$. The magnetic current acts in the presence of the plate. Use image theory (assume an infinite PEC plate) to approximately model the effect that a finite plate would have on the magnetic current. In this problem $\mathbf{M}_s$ is known exactly but the way it radiates is being approximated.

4.22 Show that for a perfectly conducting flat plate of area A and arbitrary edge shape, the monostatic radar cross section at broadside ($\theta = 0$) is

$$\sigma = \frac{4\pi A^2}{\lambda^2}.$$

Use PO. The plate is in the $x - y$ plane, and the incident field is $E_x^i = E_0 e^{jkz}$. Show that $\hat{\mathbf{r}}$ is perpendicular to $\mathbf{r}'$, which simplifies the radiation integral.

4.23 The RCS of a conducting plate was obtained in Example 4.6. Modify this to treat a dielectric plate. The question is limited to normal incidence $\mathbf{E}^i = \hat{\mathbf{x}} E_0 e^{jkz}$ and the backscatter direction ($\theta = 0$). Assume a plate thickness of $d = 1$ cm, with $\epsilon_r = 4$.

  (a) Using a surface equivalent, find $\mathbf{J}_s$ and $\mathbf{M}_s$ on the faces at $z = d$ and $z = 0$.

  (b) Develop expressions for $\mathbf{E}^s$ and the RCS σ in terms of these currents.

  (c) Assuming a 10 cm by 10 cm plate, calculate σ for 1–10 GHz. Modify PROGRAM mlslab to calculate the necessary reflection and transmission coefficients.

  (d) Check your code with test cases. You should find that the RCS is minimized when the plate is a half-wave thick. Also try a lossy material with $\epsilon_r = 4$ and $\tan \delta = 10$ (complex $\epsilon_c = \epsilon_0 \epsilon_r (1 - j \tan \delta)$) with a plate thickness of at least one skin depth. This should give an RCS result that is close to a conducting plate; $\sigma = 4\pi(ab)^2/\lambda^2$.

4.24 An infinite metal screen with an aperture is at $z = 0$. The aperture size is $-a/2 \le x \le a/2$, $-b/2 \le y \le b/2$. An incident plane wave is $\mathbf{H}^i = \hat{\mathbf{x}} H_0 e^{-jkz}$. In the $\phi = 0$ plane, find

  (a) the far-zone transmitted field for $z > 0$ and

  (b) the scattered field for $z < 0$.

4.25 An open-ended rectangular waveguide is mounted in an infinite ground plane. The aperture is along $-a/2 \leq x \leq a/2$, $-b/2 \leq y \leq b/2$ and the aperture field is $E_y = E_0 \cos(\pi x/a)$.

(a) What would be more accurate, the $\mathbf{J}_s$ or $\mathbf{M}_s$ formulation?

(b) Find the radiation pattern E_θ when $\phi = 90°$.

(c) Find the radiation pattern E_ϕ when $\phi = 0$.

4.26 A 2D perfectly conducting strip is at $y = 0$ and along $-a/2 \leq x \leq a/2$. Derive a simple expression for the monostatic echo width at broadside ($\phi_i = \phi = 90°$). The radiation integral can be simplified by noting that $\hat{\boldsymbol{\rho}} \cdot \boldsymbol{\rho}' = 0$.

4.27 A 2D perfectly conducting strip is at $y = 0$ and along $-a/2 \leq x \leq a/2$. Calculate and plot the bistatic echo width for width $a = 2\lambda$ and 10λ. Use an incidence angle $\phi_i = 120°$ and scattering angle $0 \leq \phi \leq 180°$. Assume that $\lambda = 1$ m. Do this for the TM case, and the TE case.

5

Canonical Problems

Canonical problems are those having the simplest form, such as a cylinder, wedge and sphere. For these problems the wave equation can be solved in closed form by the method of separation of variables. This chapter describes solution techniques for these 2D and 3D problems. In all cases an eigenfunction series solution of the wave equation is used.

Excitations considered are a plane wave, a 2D line source and a 3D point source. For problems in cylindrical coordinates, the point source field is found from a 2D line source solution and a Fourier transform with respect to the axial wavenumber k_z. The integral for the Fourier transform is evaluated approximately with good accuracy by using the method of stationary phase. Finding point source fields via the reciprocity theorem is also demonstrated.

5.1 Cylinder

In this section, 2D field solutions for a circular cylinder are obtained for several different types of excitations. The cylinder is assumed to be a perfect electric conductor, unless otherwise specified. In all cases we make use of the cylindrical wave equation

$$\nabla^2 \psi + k^2 \psi = 0. \tag{5.1}$$

From Section 3.3 the 2D solutions (those having $k_z = 0$) are

$$\psi(\rho, \phi) = \begin{Bmatrix} J_\nu(k\rho) \\ Y_\nu(k\rho) \\ H_\nu^{(1)}(k\rho) \\ H_\nu^{(2)}(k\rho) \end{Bmatrix} \begin{Bmatrix} \cos \nu\phi \\ \sin \nu\phi \\ e^{\pm j\nu\phi} \end{Bmatrix}. \tag{5.2}$$

The parameter ν is arbitrary, though $\nu = n$ an integer is chosen if a solution with a 2π periodicity in ϕ is required.

5.1.1 Plane Wave Incidence

Figure 5.1 shows an x-travelling plane wave illuminating a perfectly conducting cylinder. The field point (ρ, ϕ) is arbitrary. We will first consider the TM polarization. $\mathbf{E}$ has only a z component so we can let $\psi = E_z$. The field can be expressed as incident and scattered parts $E_z = E_z^i + E_z^s$. On the surface $\rho = a$

Applied Frequency-Domain Electromagnetics, First Edition. Robert Paknys.
© 2016 John Wiley & Sons, Ltd. Published 2016 by John Wiley & Sons, Ltd.
Companion Website: www.wiley.com/go/paknys9981

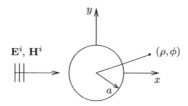

Figure 5.1   Plane wave incident on a perfectly conducting circular cylinder.

we require that $E_z = 0$. E_z^s will be in terms of cylindrical modes (5.2) so we should express E_z^i in the same way. This will make it easy to enforce the boundary condition at $\rho = a$.

Since a plane wave has a finite amplitude everywhere, only modes involving the J_n Bessel function are admissible. The incident field can be expanded as a Fourier series in ϕ:

$$E_z^i = e^{-jkx} = e^{-jk\rho\cos\phi} = \sum_{n=-\infty}^{\infty} c_n J_n(k\rho)e^{jn\phi}.$$

Multiplying both sides by $e^{-jm\phi}$ and integrating over $(0, 2\pi)$ gives

$$\int_0^{2\pi} e^{-j(m\phi+k\rho\cos\phi)} \, d\phi = \sum_{n=-\infty}^{\infty} c_n J_n(k\rho) \int_0^{2\pi} e^{j(n-m)\phi} \, d\phi.$$

The right-hand side is zero unless $m = n$ and the left-hand side is a well-known integral that can be evaluated with the help of (C.22). Therefore

$$2\pi j^{-m} J_m(k\rho) = 2\pi c_m J_m(k\rho)$$

from which $c_m = j^{-m}$ and hence

$$E_z^i = e^{-jkx} = \sum_{n=-\infty}^{\infty} j^{-n} J_n(k\rho)e^{jn\phi}. \qquad (5.3)$$

The scattered field will consist of outward-travelling cylindrical waves, so it should be expressed in terms of Hankel functions of the second kind:

$$E_z^s = \sum_{n=-\infty}^{\infty} a_n H_n^{(2)}(k\rho)e^{jn\phi}. \qquad (5.4)$$

Enforcing $E_z^i + E_z^s = 0$ at $\rho = a$ leads to

$$a_n = -j^{-n} \frac{J_n(ka)}{H_n^{(2)}(ka)} \qquad (5.5)$$

and the solution

$$E_z^s = - \sum_{n=-\infty}^{\infty} j^{-n} \frac{J_n(ka)}{H_n^{(2)}(ka)} H_n^{(2)}(k\rho)e^{jn\phi}. \qquad (5.6)$$

For the TE case $\mathbf{H}$ has only a z component, so

$$H_z^i = e^{-jkx} = \sum_{n=-\infty}^{\infty} j^{-n} J_n(k\rho) e^{jn\phi} \tag{5.7}$$

$$H_z^s = \sum_{n=-\infty}^{\infty} b_n H_n^{(2)}(k\rho) e^{jn\phi}. \tag{5.8}$$

Taking the curl of $\mathbf{H}$ gives $\mathbf{E}$, from which

$$E_\phi^i = \frac{jk}{\omega\epsilon} \sum_{n=-\infty}^{\infty} j^{-n} J_n(k\rho) e^{jn\phi} \tag{5.9}$$

$$E_\phi^s = \frac{jk}{\omega\epsilon} \sum_{n=-\infty}^{\infty} b_n H_n'^{(2)}(k\rho) e^{jn\phi}. \tag{5.10}$$

Enforcing the boundary condition $E_\phi^i + E_\phi^s = 0$ at $\rho = a$ leads to

$$b_n = -j^{-n} \frac{J_n'(ka)}{H_n'^{(2)}(ka)} \tag{5.11}$$

and the solution

$$H_z^s = -\sum_{n=-\infty}^{\infty} j^{-n} \frac{J_n'(ka)}{H_n'^{(2)}(ka)} H_n^{(2)}(k\rho) e^{jn\phi}. \tag{5.12}$$

In either case the far-zone field can be obtained by using the large-argument approximation in (C.13), repeated here as

$$H_n^{(2)}(k\rho) \sim \sqrt{\frac{2}{\pi k\rho}} e^{-j(k\rho - n\pi/2 - \pi/4)}; \ k\rho \to \infty. \tag{5.13}$$

The summation over negative n can be eliminated by using $J_{-n} = (-1)^n J_n$ and $Y_{-n} = (-1)^n Y_n$. For TM scattering, (5.6) becomes

$$E_z^s = -\sqrt{\frac{j2}{\pi k}} \frac{e^{-jk\rho}}{\sqrt{\rho}} \sum_{n=0}^{N} \epsilon_{0n} \frac{J_n(ka)}{H_n^{(2)}(ka)} \cos n\phi, \tag{5.14}$$

where $\epsilon_{00} = 1$ and $\epsilon_{0n} = 2; n \neq 0$ is Neumann's number. Equation (5.14) is coded in PROGRAM ezcyl.

The TM and TE echo widths σ as defined by (1.92) were calculated from E_z^s and H_z^s and plotted in Figure 5.2. The backscatter case ($\phi = 180°$) is assumed. For small ka, the backscatter for the TM case is much larger than for the TE case. This effect can be used to make a polarizing filter. By constructing a grid of thin parallel wires, a plane wave that is polarized parallel to the wires will get reflected, whereas the perpendicular polarization will pass through without much difficulty. This idea is used, for example, in polarizing sunglasses. These use long chains of conducting polymers that act like a grid of parallel wires.

For larger ka the backscatter can be given a physical explanation in terms of optics. Most of the backscatter is due to reflection by cylinder's metallic face. The ripples, which are more prominent in the TE case, are due to 'creeping waves' that propagate around the shadow side of the cylinder, and return to the observer. Depending on their phase they may reinforce or reduce the main reflection, which causes the ripples. In Section 5.5 we will obtain an approximate result $\sigma \approx \pi a$ that is valid for large ka and is quite close to the results in Figure 5.2(b).

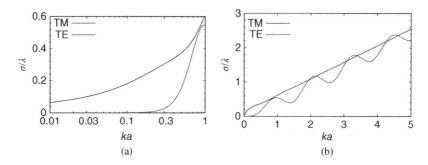

Figure 5.2 Perfectly conducting cylinder, normalized echo width σ/λ versus ka, for the backscatter case. (a) Small ka and (b) arbitrary ka.

5.1.2   Line Source Incidence

A perfectly conducting cylinder with line source illumination as in Figure 5.3 is now considered. It can be either an electric or magnetic source for the TM or TE case. We begin with the TM case, which has an electric line source of strength I_0 at (ρ', ϕ') and the field point at (ρ, ϕ). The distance between these points is R. A not so obvious but useful way to express the incident field is

$$E_z^i = -\frac{k^2 I_0}{4\omega\epsilon} H_0^{(2)}(kR) = -\frac{k^2 I_0}{4\omega\epsilon} \sum_{n=-\infty}^{\infty} J_n(k\rho_<)H_n^{(2)}(k\rho_>)e^{jn(\phi-\phi')}. \tag{5.15}$$

In (5.15) the larger of $\{\rho, \rho'\}$ is $\rho_>$ and the smaller one is $\rho_<$. This is a good way to write things because we get the same field if the source point (ρ', ϕ') and field point (ρ, ϕ) are interchanged. Therefore reciprocity is satisfied (as it must be) with one compact expression.

There are various ways to obtain Equation (5.15). One approach is to use the UT method for Green's function construction, which is explained in Section 12.8. Another method is to use the addition theorem; see for example, Abramowitz and Stegun (1965, Eq. 9.1.79); Harrington (2001, Section 5.8). For the moment, the lengthy derivation can be skipped, and it is sufficient to simply use (5.15) to solve the problem at hand.

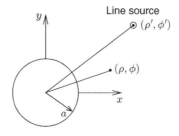

Figure 5.3   Line source illuminating a perfectly conducting circular cylinder.

The scattered field consists of outward-travelling waves so it should have a mode expansion of the form

$$E_z^s = \sum_{n=-\infty}^{\infty} c_n' H_n^{(2)}(k\rho) e^{jn(\phi-\phi')}.$$

The solution will proceed more readily if we rewrite E_z^s as

$$E_z^s = -\frac{k^2 I_0}{4\omega\epsilon} \sum_{n=-\infty}^{\infty} c_n H_n^{(2)}(k\rho_<) H_n^{(2)}(k\rho_>) e^{jn(\phi-\phi')}. \tag{5.16}$$

All we have done is define a new Fourier coefficient c_n in terms of the old one c_n'. They are related by

$$-\frac{k^2 I_0}{4\omega\epsilon} c_n H_n^{(2)}(k\rho_<) = c_n'.$$

Enforcing $E_z^i + E_z^s = 0$ at $\rho = a$ leads to

$$c_n = -\frac{J_n(ka)}{H_n^{(2)}(ka)}. \tag{5.17}$$

From (5.15)–(5.17)

$$E_z = -\frac{k^2 I_0}{4\omega\epsilon} \sum_{n=-\infty}^{\infty} \left[J_n(k\rho_<) - \frac{J_n(ka)}{H_n^{(2)}(ka)} H_n^{(2)}(k\rho_<) \right] H_n^{(2)}(k\rho_>) e^{jn(\phi-\phi')}. \tag{5.18}$$

The derivation for the TE case is similar. For this polarization a magnetic line source of strength M_0 produces an incident field

$$H_z^i = -\frac{k^2 M_0}{4\omega\mu} H_0^{(2)}(kR) = -\frac{k^2 M_0}{4\omega\mu} \sum_{n=-\infty}^{\infty} J_n(k\rho_<) H_n^{(2)}(k\rho_>) e^{jn(\phi-\phi')} \tag{5.19}$$

and the scattered field is

$$H_z^s = -\frac{k^2 M_0}{4\omega\mu} \sum_{n=-\infty}^{\infty} d_n H_n^{(2)}(k\rho_<) H_n^{(2)}(k\rho_>) e^{jn(\phi-\phi')}. \tag{5.20}$$

Taking the curl of $\mathbf{H}$ gives $\mathbf{E}$, allowing us to enforce the boundary condition $E_\phi^i + E_\phi^s = 0$ at $\rho = a$ which leads to

$$d_n = -\frac{J_n'(ka)}{H_n'^{(2)}(ka)}. \tag{5.21}$$

From (5.19)–(5.21)

$$H_z = -\frac{k^2 M_0}{4\omega\mu} \sum_{n=-\infty}^{\infty} \left[J_n(k\rho_<) - \frac{J_n'(ka)}{H_n'^{(2)}(ka)} H_n^{(2)}(k\rho_<) \right] H_n^{(2)}(k\rho_>) e^{jn(\phi-\phi')}. \tag{5.22}$$

By letting $\rho_< = \rho' = a$ and $\phi' = 0$, we can use (5.22) to represent a narrow axial slot of width $a\phi_0$ having a uniform electric field $E_\phi = E_0$ as in Figure 5.4. The source strength M_0 is in Volts and the surface equivalent current density is $M_z = -E_\phi$. The aperture width is $a\phi_0$ so that $M_0 = M_z a\phi_0 = -E_0 a\phi_0$. For a unit strength source, the product $E_0 a\phi_0 = 1$.

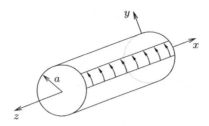

Figure 5.4   Slot with a ϕ-directed electric field in the wall of a perfectly conducting circular cylinder.

5.1.3   TE Slot

Although magnetic currents are nonphysical, they are useful for modelling aperture antennas. This section discusses a finite-size slot on a cylinder and explains its relationship to an infinitesimal magnetic current.

Suppose that the perfectly conducting cylinder has a narrow radiating slot at $(\rho, \phi) = (a, 0)$, with a width $a\phi_0$ and a uniform electric field $E_\phi = E_0$. This is shown in Figure 5.4. If the source has a unit strength, then the field strength times the aperture width is unity, so that $E_0 a\phi_0 = 1$. In the limit as $\phi_0 \to 0$, the source can be represented as a unit strength impulse $E_\phi(a, \phi) = \delta(\phi)/a$, which has the property $\int_{-\pi}^{\pi} E_\phi a\,d\phi = 1$.

Considering (5.10), we expect E_ϕ for $\rho \geq a$ to be of the form

$$E_\phi(\rho, \phi) = \sum_{n=-\infty}^{\infty} a_n H_n'^{(2)}(k\rho) e^{jn\phi}.$$

E_ϕ must take on the prescribed impulse behaviour at $\rho = a$ so that

$$\frac{\delta(\phi)}{a} = \sum_{n=-\infty}^{\infty} a_n H_n'^{(2)}(ka) e^{jn\phi}.$$

Multiplying both sides by $e^{-jm\phi}$ and integrating over $(-\pi, \pi)$ gives the Fourier coefficients

$$a_n = \frac{1}{2\pi a H_n'^{(2)}(ka)}$$

and a solution

$$E_\phi(\rho, \phi) = \frac{1}{2\pi a} \sum_{n=-\infty}^{\infty} \frac{H_n'^{(2)}(k\rho)}{H_n'^{(2)}(ka)} e^{jn\phi}. \tag{5.23}$$

It is instructive to note that we have specified the value of $\mathbf{E}_{tan}$ on the boundary at $\rho = a$, which ensures that the solution for $\rho > a$ is unique. Since a delta function can be written as

$$\sum_{n=-\infty}^{\infty} e^{jn\phi} = 2\pi\delta(\phi)$$

it follows that at $\rho = a$, Equation (5.23) reduces to $E_\phi(a, \phi) = \delta(\phi)/a$ as expected.

5.1.4 TM Dielectric Cylinder

A dielectric cylinder has a radius a and material properties ϵ_r, μ_r. Outside the cylinder $k_0 = \omega\sqrt{\mu_0\epsilon_0}$ and inside, $k_1 = k_0\sqrt{\mu_r\epsilon_r}$. An electric line source I_0 at (ρ', ϕ') is the excitation and $\rho' \geq a$ is assumed.

Only the $J_n(k_1\rho)$ solutions are allowed in the dielectric, because $Y_n(k_1\rho)$ is singular at $\rho = 0$. For $0 \leq \rho \leq a$ we can write the total field in the dielectric as

$$E_z^d = \sum_{n=-\infty}^{\infty} d_n' J_n(k_1\rho)e^{jn(\phi-\phi')}. \tag{5.24}$$

We need to make $\mathbf{E}_{tan}$ and $\mathbf{H}_{tan}$ continuous at $\rho = a$. The problem will be easier to solve if we use

$$E_z^d = \frac{-k_0^2 I_0}{4\omega\epsilon_0} \sum_{n=-\infty}^{\infty} d_n J_n(k_1\rho_<)H_n^{(2)}(k_0\rho_>)e^{jn(\phi-\phi')}. \tag{5.25}$$

d_n' and d_n are just different ways of expressing the unknown Fourier coefficient. The source is outside the cylinder so d_n' and d_n are related by

$$d_n' = \frac{-k_0^2 I_0}{4\omega\epsilon_0} d_n H_n^{(2)}(k_0\rho').$$

The incident field E_z^i is the same as in (5.15), and the scattered field E_z^s for $\rho \geq a$ is of the form in (5.16). We then enforce $E_z^i + E_z^s = E_z^d$ at $\rho = a$. Taking the curl of $\mathbf{E}$, we find the magnetic fields and enforce $H_\phi^i + H_\phi^s = H_\phi^d$ at $\rho = a$. The two equations allow us to find c_n and d_n, with the result that

$$c_n = \frac{J_n(k_0a)J_n'(k_1a) - \alpha J_n'(k_0a)J_n(k_1a)}{\alpha H_n'^{(2)}(k_0a)J_n(k_1a) - H_n^{(2)}(k_0a)J_n'(k_1a)} \tag{5.26}$$

$$d_n = \frac{J_n(k_0a) + c_n H_n^{(2)}(k_0a)}{J_n(k_1a)}, \tag{5.27}$$

where $\alpha = \sqrt{\mu_r/\epsilon_r}$. If $\epsilon_r \to \infty$ the perfectly conducting case is recovered.

Since μ_d was included in the formulation, duality can be readily applied to obtain the fields for the TE polarization and a magnetic line source.

5.2 Wedge

In this section, 2D field solutions for a perfectly conducting wedge are obtained for several different types of excitations. The wedge is shown in Figure 5.5. The angles ϕ and ϕ' are always positive, and between α and $2\pi - \alpha$. We begin with solutions of the cylindrical wave equation

$$\nabla 2\psi + k^2\psi = 0. \tag{5.28}$$

The boundary conditions on the wedge faces are either $\psi = 0$ or $\partial\psi/\partial\phi = 0$. Suitable modes are of the form

$$\psi(\rho, \phi) = \begin{Bmatrix} J_\nu(k\rho) \\ Y_\nu(k\rho) \\ H_\nu^{(1)}(k\rho) \\ H_\nu^{(2)}(k\rho) \end{Bmatrix} \begin{Bmatrix} \cos\nu(\phi - \alpha) \\ \sin\nu(\phi - \alpha) \end{Bmatrix}. \tag{5.29}$$

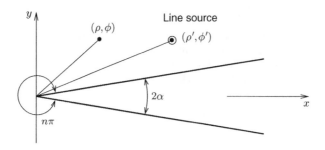

Figure 5.5 Line source illuminating a perfectly conducting wedge with faces at $\phi = \alpha$ and $\phi = 2\pi - \alpha$. The exterior wedge angle is $n\pi$.

The parameter ν is arbitrary. Since

$$\cos \nu(\phi - \alpha) = \cos \nu\phi \cos \nu\alpha + \sin \nu\phi \sin \nu\alpha$$

$$\sin \nu(\phi - \alpha) = \sin \nu\phi \cos \nu\alpha - \cos \nu\phi \sin \nu\alpha$$

and ν is a parameter, it can be seen that these modes are nothing more than a combination of the independent solutions $\cos \nu\phi$ and $\sin \nu\phi$ in (5.2) that were used earlier.

5.2.1 TM Case

An electric line source of strength I_0 is placed at (ρ', ϕ'), and the field point is at (ρ, ϕ). An efficient way to express the total field is

$$E_z = \sum_\nu a_\nu J_\nu(k\rho_<) H_\nu^{(2)}(k\rho_>) \sin \nu(\phi - \alpha) \sin \nu(\phi' - \alpha). \tag{5.30}$$

E_z must vanish on the wedge faces at $\phi = \alpha$ and $\phi = 2\pi - \alpha$ which implies that

$$\nu = \frac{m\pi}{2(\pi - \alpha)}; m = 1, 2, 3 \ldots . \tag{5.31}$$

For the moment, regarding the source as a current sheet J_{sz} at ρ', the tangential magnetic field must have a jump discontinuity in accordance with

$$J_{sz} = H_\phi(\rho = \rho' + \Delta) - H_\phi(\rho = \rho' - \Delta) \tag{5.32}$$

where Δ is small and positive. Taking the curl of $\mathbf{E}$ in (5.30) gives $\mathbf{H}$. When $\rho = \rho' + \Delta$ then $\rho_< = \rho'$, $\rho_> = \rho$ and so

$$H_\phi = \frac{k}{j\omega\mu} \sum_\nu a_\nu J_\nu(k\rho') H_\nu'^{(2)}(k\rho) \sin \nu(\phi - \alpha) \sin \nu(\phi' - \alpha). \tag{5.33}$$

When $\rho = \rho' - \Delta$ then $\rho_< = \rho$, $\rho_> = \rho'$ so that

$$H_\phi = \frac{k}{j\omega\mu} \sum_\nu a_\nu J_\nu'(k\rho) H_\nu^{(2)}(k\rho') \sin \nu(\phi - \alpha) \sin \nu(\phi' - \alpha). \tag{5.34}$$

Using (5.33) and (5.34) in (5.32) and the Wronskian relationship (C.44) gives

$$J_{sz} = -\frac{2}{\pi\omega\mu\rho'} \sum_\nu a_\nu \sin\nu(\phi-\alpha)\sin\nu(\phi'-\alpha). \tag{5.35}$$

We now think of the J_{sz} current sheet at ρ' as having a narrow angular extent of $\phi = \phi' \pm \Delta$. Since the source strength is I_0, this requires

$$\int_{\phi-\Delta}^{\phi+\Delta} J_{sz}(\phi')\rho'd\phi' = I_0. \tag{5.36}$$

A surface current satisfying this requirement is

$$J_{sz} = \frac{I_0}{\rho'}\delta(\phi-\phi'). \tag{5.37}$$

The delta function can be expanded as a Fourier series[1]

$$\delta(\phi-\phi') = \frac{1}{\pi-\alpha}\sum_\nu \sin\nu(\phi-\alpha)\sin\nu(\phi'-\alpha). \tag{5.38}$$

Using (5.38) in (5.37) and equating it to (5.35) gives the unknown coefficient a_ν

$$a_\nu = \frac{-\pi\omega\mu I_0}{2(\pi-\alpha)} \tag{5.39}$$

and hence

$$E_z = \frac{-\pi\omega\mu I_0}{2(\pi-\alpha)} \sum_{m=1}^\infty J_{\nu_m}(k\rho_<)H_{\nu_m}^{(2)}(k\rho_>)\sin\nu_m(\phi-\alpha)\sin\nu_m(\phi'-\alpha), \tag{5.40}$$

where ν_m is related to m by (5.31).

In a cylindrical coordinate system, the wedge has its metal faces at $\phi = \alpha$ and $\phi = 2\pi - \alpha$. It is customary in the wedge problem to define an exterior wedge angle $n\pi = 2(\pi-\alpha)$ that is in terms of the real number n. Then, (5.40) becomes

$$E_z = \frac{-\omega\mu I_0}{n} \sum_{m=1}^\infty J_{m/n}(k\rho_<)H_{m/n}^{(2)}(k\rho_>)\sin\left[\frac{m}{n}(\phi'-\alpha)\right]\sin\left[\frac{m}{n}(\phi-\alpha)\right]. \tag{5.41}$$

To obtain plane wave incidence, we let the source distance ρ' become large. With $\rho' \gg \rho$, $R \approx \rho' - \rho\cos(\phi-\phi')$ and using the large argument Hankel function approximation (C.13),

$$E_z^i = -\frac{k^2 I_0}{4\omega\epsilon}H_0^{(2)}(kR) \approx -\frac{\omega\mu I_0}{4}\sqrt{\frac{j2}{\pi k\rho'}}e^{-jk\rho'}e^{jk\rho\cos(\phi-\phi')} \equiv E_0 e^{jk\rho\cos(\phi-\phi')}. \tag{5.42}$$

---

[1] The angular part of the separated wave equation is $\Phi''(\phi) + \nu^2\Phi(\phi) = 0$ with boundary conditions $\Phi(0) = 0$, $\Phi(2\pi-\alpha) = 0$. The eigenfunctions are $\Phi_m(\phi) = \sin\nu_m(\phi-\alpha)$ and the eigenvalues are $\nu_m = \frac{m\pi}{2(\pi-\alpha)}$. The eigenfunctions are orthogonal on $\alpha \le \phi \le 2\pi - \alpha$ and obey

$$\int_\alpha^{2\pi-\alpha} \Phi_m(\phi)\Phi_n(\phi)\,d\phi = \begin{cases} (\pi-\alpha); n = m \\ 0; n \ne m \end{cases}.$$

These can be used to obtain a Fourier series.

The term $e^{jk\rho\cos(\phi-\phi')}$ is an incident plane wave in the direction $-\hat{\mathbf{x}}\cos\phi' - \hat{\mathbf{y}}\sin\phi'$. From the definition in (5.42), we can write I_0 in terms of E_0, which allows us to replace I_0 in (5.41). Using (C.13) to replace the Hankel function in (5.41) then leads to

$$E_z = \frac{4E_0}{n}\sum_{m=1}^{\infty} j^{m/n} J_{m/n}(k\rho)\sin\left[\frac{m}{n}\nu(\phi-\alpha)\right]\sin\left[\frac{m}{n}(\phi'-\alpha)\right]. \tag{5.43}$$

For the special case of a half plane, $n = 2$ and (5.43) becomes

$$E_z = 2E_0\sum_{m=1}^{\infty} j^{m/2} J_{m/2}(k\rho)\sin\frac{m\phi}{2}\sin\frac{m\phi'}{2}. \tag{5.44}$$

5.2.2   TE Case

In this case a magnetic line source of strength M_0 is placed at (ρ',ϕ') and the field point is at (ρ,ϕ). The total field is

$$H_z = \sum_{\nu} b_{\nu} J_{\nu}(k\rho_<)H_{\nu}^{(2)}(k\rho_>)\cos\nu(\phi-\alpha)\cos\nu(\phi'-\alpha). \tag{5.45}$$

Requiring $E_{\rho} = 0$ on the metal implies that $\partial H_z/\partial\phi = 0$ so that

$$\nu = \frac{m\pi}{2(\pi-\alpha)}; m = 0, 1, 2, \ldots . \tag{5.46}$$

Following a procedure similar to the TM case and imposing a jump condition on E_{ϕ} at $\rho = \rho'$ gives

$$b_{\nu} = \frac{\pi\omega\epsilon M_0}{4(\pi-\alpha)}\epsilon_{0m} \tag{5.47}$$

where ϵ_{0m} is Neumann's number. For plane wave incidence $H_z^i = H_0 e^{jk\rho\cos(\phi-\phi')}$, and with $n\pi = 2(\pi-\alpha)$,

$$H_z = \frac{2H_0}{n}\sum_{m=0}^{\infty}\epsilon_{0m} j^{m/n} J_{m/n}(k\rho)\cos\left[\frac{m}{n}(\phi-\alpha)\right]\cos\left[\frac{m}{n}(\phi'-\alpha)\right]. \tag{5.48}$$

For a half plane, $n = 2$ and (5.48) becomes

$$H_z = H_0\sum_{m=0}^{\infty}\epsilon_{0m} j^{m/2} J_{m/2}(k\rho)\cos\frac{m\phi}{2}\cos\frac{m\phi'}{2}. \tag{5.49}$$

For plane wave incidence and a half plane, the fields in (5.44) and (5.49) can be used to get the surface currents; see Problems 5.13 and 5.14. The fractional-order Bessel functions can be computed using MODULE amos_mod. PROGRAM jxjz was used to obtain the results in Figure 5.6. The incident plane wave strength is $E_0 = 1$ V/m, the incidence angle is $\phi' = 90°$ (broadside from above) and the wavelength is $\lambda = 1$ m. The lit-side currents are strong, and for large distances from the edge they are in good agreement with the geometrical optics approximation, which gives $J_s \approx 2E_0/\eta$. On the shadow side the currents are weak but nonzero, whereas in the geometrical optics approximation they are zero.

The ripples are caused by 'edge diffraction' which appears as a cylindrical wave that emanates from the edge and interferes with the incident wave. The field behaviour near the edge is discussed in Problems 5.13 and 5.14.

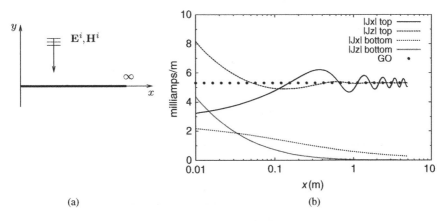

(a)   (b)

Figure 5.6 (a) Half plane. A plane wave is broadside incident from above, has a strength of $E_0 = 1$ V/m and a wavelength of $\lambda = 1$ m. (b) Surface currents and geometrical optics (GO) approximation.

5.3 The Relation Between 2D and 3D Solutions

The 3D solution of the inhomogeneous Helmholtz equation can be found from a 2D cylindrical solution. This is accomplished by using a Fourier transform.

For a point source at $z' = 0$, the inhomogeneous Helmholtz equation is

$$(\nabla^2 + k^2)\psi(\rho, \phi, z) = -C_0 \frac{\delta(\rho - \rho')}{\rho'}\delta(\phi - \phi')\delta(z) \tag{5.50}$$

where $k = 2\pi/\lambda$, C_0 is a constant and ψ represents a potential function such as A_z or F_z.

Let us define a Fourier transformation with respect to z:

$$\bar{\psi}(\rho, \phi, k_z) = \int_{-\infty}^{\infty} \psi(\rho, \phi, z)e^{-jk_z z}dz \tag{5.51}$$

and its inverse

$$\psi(\rho, \phi, z) = \frac{1}{2\pi}\int_{-\infty}^{\infty} \bar{\psi}(\rho, \phi, k_z)e^{jk_z z}dk_z. \tag{5.52}$$

Taking the Fourier transform of (5.50) gives

$$(\nabla_t^2 + k_t^2)\bar{\psi}(\rho, \phi; k_z) = -C_0 \frac{\delta(\rho - \rho')}{\rho'}\delta(\phi - \phi') \tag{5.53}$$

where the transverse Laplacian is $\nabla_t^2 = \nabla^2 - \partial^2/\partial z^2$, and $k_t^2 = k^2 - k_z^2$.

Equation (5.53) is a 2D problem, where k_z is treated as a parameter. In many instances, it is easier to solve than (5.50). Then, the inverse Fourier transform of $\bar{\psi}$ gives us ψ. In general, the inversion integral is difficult. However, the method of stationary phase (see Section 5.5) can be used to obtain a good approximation. One particularly useful integral for problems in cylindrical coordinates is

$$\int_{-\infty}^{\infty} \bar{F}(k_z)H_n^{(2)}(\rho\sqrt{k^2 - k_z^2})e^{jk_z z}dk_z \sim 2\frac{e^{-jkr}}{r}j^{n+1}\bar{F}(-k\cos\theta), \tag{5.54}$$

where $kr\sin\theta$ is large, and $\bar{F}$ is any function.

Table 5.1 TM cylindrical field transforms.

Field	z domain	k_z domain
E_ρ	$-j/(\omega\mu\epsilon)\,\partial^2 A_z/\partial\rho\partial z$	$-j/(\omega\mu\epsilon)\,jk_z\partial\bar{A}_z/\partial\rho$
E_ϕ	$-j/(\omega\mu\epsilon\rho)\,\partial^2 A_z/\partial\phi\partial z$	$-j/(\omega\mu\epsilon\rho)\,jk_z\partial\bar{A}_z/\partial\phi$
E_z	$-j/(\omega\mu\epsilon)\,(\partial^2/\partial z^2 + k^2)A_z$	$-j/(\omega\mu\epsilon)\,k_t^2\bar{A}_z$
H_ρ	$1/(\mu\rho)\,\partial A_z/\partial\phi$	$1/(\mu\rho)\,\partial\bar{A}_z/\partial\phi$
H_ϕ	$-1/\mu\partial A_z/\partial\rho$	$-1/\mu\partial\bar{A}_z/\partial\rho$

Table 5.2 TE cylindrical field transforms.

Field	z domain	k_z domain
H_ρ	$-j/(\omega\mu\epsilon)\,\partial^2 F_z/\partial\rho\partial z$	$-j/(\omega\mu\epsilon)\,jk_z\partial\bar{F}_z/\partial\rho$
H_ϕ	$-j/(\omega\mu\epsilon\rho)\,\partial^2 F_z/\partial\phi\partial z$	$-j/(\omega\mu\epsilon\rho)\,jk_z\partial\bar{F}_z/\partial\phi$
H_z	$-j/(\omega\mu\epsilon)\,(\partial^2/\partial z^2 + k^2)F_z$	$-j/(\omega\mu\epsilon)\,k_t^2\bar{F}_z$
$-E_\rho$	$1/(\epsilon\rho)\,\partial F_z/\partial\phi$	$1/(\epsilon\rho)\,\partial\bar{F}_z/\partial\phi$
$-E_\phi$	$-1/\epsilon\partial F_z/\partial\rho$	$-1/\epsilon\partial\bar{F}_z/\partial\rho$

If needed, a formula similar to (5.54) involving $H_n'^{(2)}$ can be obtained. From the large argument asymptotic approximation of $H_n^{(2)}$, it follows that $H_n'^{(2)}(x) \sim -jH_n^{(2)}(x)$, which can be used in (5.54) to obtain the desired result.

The potential functions A_z and F_z provide the TM and TE fields via Equations (4.43) and (4.44). In the 2D case, one replaces $\partial/\partial z \to jk_z$. The resulting relations between potentials and fields are summarized in Tables 5.1 and 5.2.

For the TM case, $\bar{A}_z$ can be found from a 2D solution for $\bar{E}_z$ by using $\bar{E}_z = -j\omega\bar{A}_z$. Here, $\bar{E}_z$ is a 2D solution having $k = k_t$. The 3D far-field component E_θ can be found from E_z by noting that

$$\mathbf{E} = \hat{\rho}E_\rho + \hat{\phi}E_\phi + \hat{z}E_z = \hat{r}E_r + \hat{\theta}E_\theta + \hat{\phi}E_\phi. \tag{5.55}$$

Taking the dot product of (5.55) with $\hat{z}$ and noting that E_r vanishes to $O(1/r)$ in the far field,

$$E_z = \hat{z}\cdot\hat{\theta}E_\theta = -\sin\theta E_\theta. \tag{5.56}$$

Similarly, the TE fields are obtained from $\bar{H}_z = -j\omega\bar{F}_z$ and the 3D far-field component H_θ follows from $H_z = -\sin\theta H_\theta$.

5.3.1 Magnetic Dipole on a Cylinder

The Fourier transform method will now be used to find the radiation pattern due to a magnetic dipole on a perfectly conducting cylinder of radius a. The dipole can be either z- or ϕ-directed and in source coordinates is located at $(\rho', \phi', 0)$.

We will start with a z-directed dipole. To construct the field using vector potentials, it is noted that a z-directed magnetic line source (in 2D) or a magnetic dipole (in 3D) produces a TE field. Therefore the fields can be obtained from F_z. Denoting the source strength as p_m, the 3D vector potential satisfies

$$(\nabla^2 + k^2)F_z = -\epsilon p_m \frac{\delta(\rho - \rho')}{\rho'} \delta(\phi - \phi')\delta(z). \tag{5.57}$$

Taking the Fourier transform turns it into the 2D wave equation

$$(\nabla_t^2 + k_t^2)\bar{F}_z = -\epsilon p_m \frac{\delta(\rho - \rho')}{\rho'} \delta(\phi - \phi'). \tag{5.58}$$

Earlier we obtained the 2D magnetic line source solution (5.22) which satisfies

$$(\nabla_t^2 + k^2)H_z = -\epsilon M_0 \frac{\delta(\rho - \rho')}{\rho'} \delta(\phi - \phi'). \tag{5.59}$$

Comparing (5.58) and (5.59) shows that the solution of (5.58) can be obtained by replacing $k \to k_t$, $M_0 \to p_m$ and $H_z \to \bar{H}_z$ in (5.22). Since $\bar{H}_z = -j\omega\bar{F}_z$ in 2D, (5.22) gives us

$$\bar{F}_z = \frac{-j\epsilon p_m}{4} \sum_{n=-\infty}^{\infty} \left[J_n(k_t\rho_<) - \frac{J_n'(k_t a)}{H_n'^{(2)}(k_t a)} H_n^{(2)}(k_t\rho_<) \right] H_n^{(2)}(k_t\rho_>) e^{jn(\phi-\phi')}. \tag{5.60}$$

The magnetic source is on the surface, so $\rho_< = \rho' = a$ and $\rho_> = \rho$ so that

$$\bar{F}_z = \frac{-\epsilon p_m}{2\pi k_t a} \sum_{n=-\infty}^{\infty} \frac{H_n^{(2)}(k_t\rho)}{H_n'^{(2)}(k_t a)} e^{jn(\phi-\phi')}. \tag{5.61}$$

Taking the inverse Fourier transform,

$$F_z(\rho, \phi, z) = \frac{\epsilon}{2\pi} \int_{-\infty}^{\infty} \frac{-p_m}{2\pi k_t a} \sum_{n=-\infty}^{\infty} \frac{H_n^{(2)}(k_t\rho)}{H_n'^{(2)}(k_t a)} e^{jn(\phi-\phi')} e^{jk_z z} dk_z. \tag{5.62}$$

Referring to Table 5.2,

$$E_\phi = (1/\epsilon)\partial F_z/\partial\rho \tag{5.63}$$

and evaluating the integral with the stationary phase method (where $k_t = k\sin\theta$ and $k_t^2 + k_z^2 = k^2$),

$$E_\phi(\rho, \phi, z) = \frac{-p_m}{2\pi^2 ar} e^{-jkr} \sum_{n=-\infty}^{\infty} \frac{j^n e^{jn(\phi-\phi')}}{H_n'^{(2)}(ka\sin\theta)}. \tag{5.64}$$

It is useful to examine E_ϕ at the surface. From (5.62) and (5.63) at $\rho = a$,

$$E_\phi(a, \phi, z) = \frac{-p_m}{2\pi a} \sum_{n=-\infty}^{\infty} e^{jn(\phi-\phi')} \frac{1}{2\pi} \int_{-\infty}^{\infty} e^{jk_z z} dk_z = -p_m \frac{\delta(\phi - \phi')}{a} \delta(z).$$

From the relation $\mathbf{E} \times \hat{\mathbf{n}} = \mathbf{M}_s$,

$$M_{sz} = -E_\phi = p_m \frac{\delta(\phi - \phi')}{a} \delta(z)$$

which confirms that our solution produces the correct $\mathbf{E}_{tan}$ for a perfectly conducting surface with a magnetic dipole. These observations also bring to light that the solution (5.62) is a Fourier series in ϕ

and a Fourier integral in z. These are natural consequences of the fact that the ϕ domain is 2π periodic, whereas the z domain is infinite.

A similar procedure will now be outlined for a magnetic dipole p_m^c on the surface, oriented in the circumferential $(+\phi)$ direction and located at $(\rho', \phi', z') = (a, \phi', 0)$. The fields will have both H_z and E_z components (this is true even if there is just a dipole and no cylinder). Therefore a combination of A_z and F_z potentials is needed. These are found to be

$$A_z(\rho, \phi, z) = \frac{\mu}{2\pi} \int_{-\infty}^{\infty} \frac{jkp_m^c}{2\pi\eta k_t^2 a} \sum_{n=-\infty}^{\infty} \frac{H_n^{(2)}(k_t\rho)}{H_n^{(2)}(k_t a)} e^{jn(\phi-\phi')} e^{jk_z z} dk_z \tag{5.65}$$

and

$$F_z(\rho, \phi, z) = \frac{\epsilon}{2\pi} \int_{-\infty}^{\infty} \frac{k_z p_m^c}{2\pi k_t^3 a^2} \sum_{n=-\infty}^{\infty} n \frac{H_n^{(2)}(k_t\rho)}{H_n^{\prime(2)}(k_t a)} e^{jn(\phi-\phi')} e^{jk_z z} dk_z. \tag{5.66}$$

All the field components can be obtained from these potentials; the derivation is left as an exercise in Problem 5.17. At $\rho = a$ the potentials provide the required field behaviour which is

$$M_{s\phi} = E_z = p_m^c \frac{\delta(\phi - \phi')}{a} \delta(z).$$

5.3.2   Electric Dipole Near a Wedge

In this configuration, the source is a z-directed electric dipole p_e, which produces a TM field. The 3D vector potential satisfies the wave equation

$$(\nabla^2 + k^2)A_z = -\mu p_e \frac{\delta(\rho - \rho')}{\rho'} \delta(\phi - \phi')\delta(z). \tag{5.67}$$

Taking the Fourier transform, we obtain

$$(\nabla_t^2 + k_t^2)\bar{A}_z = -\mu p_e \frac{\delta(\rho - \rho')}{\rho'} \delta(\phi - \phi'). \tag{5.68}$$

Equation (5.41) gives the 2D solution $\bar{E}_z$ for a line source I_0. Replacing $k \to k_t$ and $I_0 \to p_e$,

$$\bar{E}_z = \frac{-\omega\mu p_e}{n} \sum_{m=1}^{\infty} J_{m/n}(k_t\rho_<)H_{m/n}^{(2)}(k_t\rho_>) \sin\left[\frac{m}{n}(\phi' - \alpha)\right] \sin\left[\frac{m}{n}(\phi - \alpha)\right]. \tag{5.69}$$

In 2D, $\bar{E}_z = -j\omega\bar{A}_z$ so that

$$\bar{A}_z = \frac{-j\mu p_e}{n} \sum_{m=1}^{\infty} J_{m/n}(k_t\rho_<)H_{m/n}^{(2)}(k_t\rho_>) \sin\left[\frac{m}{n}(\phi' - \alpha)\right] \sin\left[\frac{m}{n}(\phi - \alpha)\right]. \tag{5.70}$$

Equation (5.70) satisfies the 2D wave equation (5.68), the boundary conditions at $\phi = \alpha$ and $\phi = 2\pi - \alpha$, and it has the proper jump discontinuity at the line source.

A stationary phase evaluation of the inverse Fourier transform gives

$$E_z = \frac{e^{-jkr}}{r} \frac{\omega\mu p_e \sin^2\theta}{n\pi} \sum_{m=1}^{\infty} j^{(m/n)-1} J_{m/n}(k\rho_< \sin\theta) \sin\left[\frac{m}{n}(\phi' - \alpha)\right] \sin\left[\frac{m}{n}(\phi - \alpha)\right]. \quad (5.71)$$

The far-field component E_θ can be found from (5.71) by noting that $E_z = -\sin\theta E_\theta$ so that

$$E_\theta = \frac{e^{-jkr}}{r} \frac{\omega\mu p_e \sin\theta}{n\pi} \sum_{m=1}^{\infty} j^{(m/n)+1} J_{m/n}(k\rho_< \sin\theta) \sin\left[\frac{m}{n}(\phi' - \alpha)\right] \sin\left[\frac{m}{n}(\phi - \alpha)\right]. \quad (5.72)$$

The case of a z-directed magnetic dipole source is discussed in Problem 5.18.

5.3.3   Reciprocity-Based Solutions

An alternative way to find the point source radiation is to first solve the electromagnetic problem using plane wave incidence. Then an application of the Lorentz reciprocity theorem gives the point source result. This is best demonstrated as an example.

Example 5.1 (Radial Dipole on a Cylinder) A metal cylinder of radius a is illuminated by a TE plane wave, and it is assumed that the near field is known. Use reciprocity to find the far-zone radiated field due to an electric dipole that is normal to the cylinder. The source and field points are assumed to be in the $z = 0$ plane.

Solution: We can solve this problem using (1.79) with electric current sources, so that

$$\int_V \mathbf{E}_1 \cdot \mathbf{J}_2 \, dV = \int_V \mathbf{E}_2 \cdot \mathbf{J}_1 \, dV.$$

If the currents are point sources $\mathbf{J}_1$ at $\mathbf{r}_1$ and $\mathbf{J}_2$ at $\mathbf{r}_2$, then

$$\mathbf{E}_1(\mathbf{r}_2) \cdot \mathbf{J}_2(\mathbf{r}_2) = \mathbf{E}_2(\mathbf{r}_1) \cdot \mathbf{J}_1(\mathbf{r}_1).$$

The situation is shown in Figure 5.7.
   Let source #1 be the incident plane wave:

$$\mathbf{E}^i = \hat{\phi} E_0 e^{jk(x\cos\phi_1 + y\sin\phi_1)}.$$

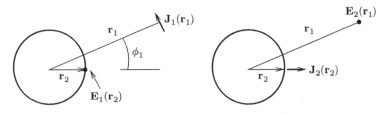

Figure 5.7   Reciprocity and cylinder.

The plane wave source can be replaced by a current element at $(\rho, \phi) = (r_1, \phi_1)$, given by

$$\mathbf{J}_1(\mathbf{r}_1) = \hat{\phi}(\phi_1)\, p_{e1}\, \delta(\mathbf{r} - \mathbf{r}_1); \quad p_{e1} = \frac{4\pi r_1 E_0}{-jk\eta e^{-jkr_1}}$$

where $r_1 \gg a$. The cylinder near field $\mathbf{E}_1(\rho, \phi)$ under plane wave incidence is assumed to be known. Source #2 is a radially directed electric dipole of strength p_{e2} located at $(\rho, \phi) = (a, 0)$ and is given by

$$\mathbf{J}_2(\mathbf{r}_2) = \hat{\rho}\, p_{e2}\, \delta(\mathbf{r} - \mathbf{r}_2).$$

Applying the reciprocity theorem gives

$$p_{e2}\, \hat{\rho} \cdot \mathbf{E}_1(a, 0) = p_{e1}\, \hat{\phi} \cdot \mathbf{E}_2(r_1, \phi_1)$$

and $E_{2\phi}$ due to source #2 is then

$$E_{2\phi}(r_1, \phi_1) = \frac{-jk\eta}{4\pi E_0} \frac{e^{-jkr_1}}{r_1} p_{e2} E_{1\rho}(a, 0).$$

More generally, if $z \neq 0$ there will be both E_θ and E_ϕ components. ∎

5.4   Spherical Waves

Let us begin with the scalar wave equation in a source-free region:

$$\nabla^2 \psi + k^2 \psi = 0. \tag{5.73}$$

Using the separation of variables method, the proposed solution is a product of three functions $\psi(r, \theta, \phi) = R(r)\Theta(\theta)\Phi(\phi)$. Substituting this into (5.73) and multiplying by r^2/ψ,

$$\underbrace{\frac{1}{R}\frac{d}{dr}\left(r^2\frac{dR}{dr}\right) + k^2 r^2}_{n(n+1)} + \frac{1}{\Theta \sin\theta}\frac{d}{d\Theta}\left(\sin\theta\frac{d\Theta}{d\theta}\right) + \frac{1}{\sin^2\theta}\underbrace{\frac{1}{\Phi}\frac{d^2\Phi}{d\phi^2}}_{-m^2} = 0. \tag{5.74}$$

Equation (5.74) breaks down to

$$\frac{1}{R}\frac{d}{dr}\left(r^2\frac{dR}{dr}\right) + k^2 r^2 = n(n+1)$$

$$\frac{1}{\Theta \sin\theta} + \frac{d}{d\Theta}\left(\sin\theta\frac{d\Theta}{d\theta}\right) - \frac{m^2}{\sin^2\theta} = -n(n+1)$$

$$\frac{1}{\Phi}\frac{d^2\Phi}{d\phi^2} = -m^2.$$

The solution is a product of spherical Bessel functions in r, Legendre functions in θ and trigonometric functions in ϕ. The seemingly odd choice of separation constants is desirable because it leads to easy notation for the standard functions in the solution. The resulting product $\psi(r, \theta, \phi) = R(r)\Theta(\theta)\Phi(\phi)$ becomes

$$\psi(r, \theta, \phi) = \begin{Bmatrix} \hat{\jmath}_n(kr) \\ y_n(kr) \\ h_n^{(1)}(kr) \\ h_n^{(2)}(kr) \end{Bmatrix} \begin{Bmatrix} P_n^m(\cos\theta) \\ Q_n^m(\cos\theta) \end{Bmatrix} \begin{Bmatrix} \cos m\phi \\ \sin m\phi \end{Bmatrix}. \tag{5.75}$$

The spherical Bessel functions use a notation that is analogous to the cylindrical case. That is, $j_n(kr)$ is finite and $y_n(kr)$ is singular at the origin. $h_n^{(1)}(kr)$ and $h_n^{(2)}(kr)$ are inward-and outward-travelling waves. Spherical Bessel functions are described in Appendix C and Legendre functions are in Appendix D.

For an electromagnetic problem we need to solve

$$\nabla^2 \mathbf{A} + k^2 \mathbf{A} = 0 \tag{5.76}$$

$$\nabla^2 \mathbf{F} + k^2 \mathbf{F} = 0. \tag{5.77}$$

It is useful to construct solutions that are TM and TE with respect to r. This is done by using $\mathbf{A} = \hat{\mathbf{r}} A_r$ and $\mathbf{F} = \hat{\mathbf{r}} F_r$. From the vector potentials the fields can then be obtained.

Given the similarity of (5.73) and (5.76), it is tempting to write $\mathbf{A} = \hat{\mathbf{r}}\psi$. However, this idea would not work because vector and scalar Laplacians are different, that is, $\nabla^2 \mathbf{A} = \nabla^2 \hat{\mathbf{r}} A_r \neq \hat{\mathbf{r}} \nabla^2 A_r$. It is not at all obvious, but it turns out that A_r and F_r satisfy the scalar equations

$$(\nabla^2 + k^2)\frac{A_r}{r} = 0 \tag{5.78}$$

$$(\nabla^2 + k^2)\frac{F_r}{r} = 0. \tag{5.79}$$

Comparing (5.73) with (5.78), (5.79) shows that $\psi = A_r/r$ and $\psi = F_r/r$. The validity of (5.78) and (5.79) can be confirmed by substituting $A_r = r\psi$ and $F_r = r\psi$ into (5.76) and (5.77). Alternatively, a derivation can be found in Harrington (2001, p. 267).

We can now construct solutions for A_r and F_r from (5.75). Since the solutions for A_r and F_r are always the ψ solutions multiplied by r, it is customary to make use of the Schelkunoff spherical Bessel functions $\hat{B}_n(kr)$ defined by

$$\hat{B}_n(kr) = kr\, b_n(kr)$$

where $b_n(kr)$ is any one of $j_n(kr)$, $y_n(kr)$, $h_n^{(1)}(kr)$ or $h_n^{(2)}(kr)$. Then, the solutions of (5.76), (5.77) become

$$\begin{Bmatrix} A_r \\ F_r \end{Bmatrix} = \begin{Bmatrix} \hat{J}_n(kr) \\ \hat{Y}_n(kr) \\ \hat{H}_n^{(1)}(kr) \\ \hat{H}_n^{(2)}(kr) \end{Bmatrix} \begin{Bmatrix} P_n^m(\cos\theta) \\ Q_n^m(\cos\theta) \end{Bmatrix} \begin{Bmatrix} \cos m\phi \\ \sin m\phi \end{Bmatrix}. \tag{5.80}$$

To complete the solution we need to find the fields from the potentials. This follows from (4.22) and (4.23). For the special case having $\mathbf{A} = \hat{\mathbf{r}}A_r$ and $\mathbf{F} = \hat{\mathbf{r}}F_r$,

$$\mathbf{E} = \hat{\mathbf{r}}\frac{1}{j\omega\mu\epsilon}\left(\frac{\partial^2}{\partial r^2} + k^2\right)A_r + \hat{\boldsymbol{\theta}}\left(\frac{1}{j\omega\mu\epsilon r}\frac{\partial^2 A_r}{\partial r\partial\theta} - \frac{1}{\epsilon r\sin\theta}\frac{\partial F_r}{\partial\phi}\right)$$

$$+ \hat{\boldsymbol{\phi}}\left(\frac{1}{j\omega\mu\epsilon r\sin\theta}\frac{\partial^2 A_r}{\partial r\partial\phi} + \frac{1}{\epsilon r}\frac{\partial F_r}{\partial\theta}\right) \tag{5.81}$$

$$\mathbf{H} = \hat{\mathbf{r}}\frac{1}{j\omega\mu\epsilon}\left(\frac{\partial^2}{\partial r^2} + k^2\right)F_r + \hat{\boldsymbol{\theta}}\left(\frac{1}{\mu r\sin\theta}\frac{\partial A_r}{\partial\phi} + \frac{1}{j\omega\mu\epsilon r}\frac{\partial^2 F_r}{\partial r\partial\theta}\right)$$

$$+ \hat{\boldsymbol{\phi}}\left(\frac{-1}{\mu r}\frac{\partial A_r}{\partial\theta} + \frac{1}{j\omega\mu\epsilon r\sin\theta}\frac{\partial^2 F_r}{\partial r\partial\phi}\right). \tag{5.82}$$

These expressions show that if $F_r = 0$ the fields are TM_r, and if $A_r = 0$ they are TE_r.

5.4.1   Scattering by a Sphere

The electromagnetic solution for scattering by a sphere has many uses. For instance, it is an important benchmark for the validation of numerical solutions. It is used as a standard target for the calibration of radars. A lossy sphere is used to model radio propagation over the earth. Light scattering by small spheres is used as a diagnostic tool in many branches of science.

We will now find the scattered field for a perfectly conducting metal sphere and a dielectric sphere, illuminated by an incident plane wave. We begin with the PEC sphere. The radius is a and the incident field is

$$\mathbf{E}^i = \hat{\mathbf{x}}E_x^i = \hat{\mathbf{x}}E_0 e^{-jkz}; \quad H_y^i = \frac{E_x^i}{\eta} \tag{5.83}$$

where $k = \omega\sqrt{\mu\epsilon}$ and $\eta = \sqrt{\mu/\epsilon}$ having $\mu = \mu_0$ and $\epsilon = \epsilon_0$. We want to find the scattered field. To solve this problem we have to enforce the boundary condition $\mathbf{E}_{tan} = 0$ at $r = a$.

In order to enforce boundary conditions in spherical coordinates, we need to find a way to express the incident plane wave in terms of spherical modes. We start with a Fourier Legendre series

$$e^{-jkz} = e^{-jkr\cos\theta} = \sum_{n=0}^{\infty} a_n j_n(kr)P_n(\cos\theta).$$

Out of all the possible solutions, we have selected $j_n(kr)$ because it is finite at $r = 0$ and $P_n(\cos\theta)$ because it is finite at $\theta = 0, \pi$. Multiplying both sides by $P_m(\cos\theta)\sin\theta$ and integrating from 0 to π, the orthogonality of Legendre polynomials gives

$$\int_o^\pi e^{-jkr\cos\theta}P_m(\cos\theta)\sin\theta d\theta = a_m j_m(kr)\frac{2}{2m+1}.$$

Using Gegenbauer's generalization (Abramowitz and Stegun 1965, Eq. 10.1.14), the integral on the left-hand side can be evaluated as

$$\int_o^\pi e^{-jkr\cos\theta}P_m(\cos\theta)\sin\theta d\theta = 2j^{-m}j_m(kr)$$

so that

$$a_m = j^{-m}(2m+1).$$

This gives us the spherical mode representation of e^{-jkz}:

$$e^{-jkr\cos\theta} = \sum_{n=0}^{\infty} j^{-n}(2n+1)j_n(kr)P_n(\cos\theta).$$

(5.84)

By taking $\partial/\partial\theta$ of (5.84), another useful relation is obtained[2]

$$jkr\sin\theta e^{-jkr\cos\theta} = \sum_{n=1}^{\infty} j^{-n}(2n+1)j_n(kr)P_n^1(\cos\theta).$$

(5.85)

Returning to the electromagnetic problem, the incident plane wave can be written as a spherical mode expansion in terms of $\hat{J}_n(kr)P_n^1(\cos\theta)$ that is finite at the origin. The scattered field is in terms of outward-propagating modes $\hat{H}_n^{(2)}(kr)P_n^1(\cos\theta)$. We will see that potentials chosen in the following way,

$$A_r = \frac{E_0}{\omega}\cos\phi\sum_{n=1}^{\infty}(a_n\hat{J}_n(kr)+b_n\hat{H}_n^{(2)}(kr))P_n^1(\cos\theta)$$

(5.86)

$$F_r = \frac{\epsilon E_0}{k}\sin\phi\sum_{n=1}^{\infty}(a_n\hat{J}_n(kr)+c_n\hat{H}_n^{(2)}(kr))P_n^1(\cos\theta)$$

(5.87)

lead to fields that meet the boundary condition. The a_n terms correspond to a spherical mode expansion of the incident field, and b_n, c_n are associated with the scattered field. Only the P_n^1 modes are used.

Using (5.86), (5.87) in (5.81), (5.82) gives the fields

$$E_r = \frac{1}{j\omega\epsilon}\left(\frac{\partial^2}{\partial r^2}+k^2\right)A_r$$

$$= \frac{E_0\cos\phi}{j(kr)^2}\sum_{n=1}^{\infty}n(n+1)\left\{a_n\hat{J}_n(kr)+b_n\hat{H}_n^{(2)}(kr)\right\}P_n^1(\cos\theta)$$

(5.88)

$$E_\theta = \frac{-1}{r\sin\theta}\frac{\partial F_r}{\partial\phi}+\frac{1}{j\omega\epsilon r}\frac{\partial^2 A_r}{\partial r\partial\theta}$$

$$= \frac{-E_0}{kr\sin\theta}\cos\phi\sum_{n=1}^{\infty}\left\{a_n\hat{J}_n(kr)+c_n\hat{H}_n^{(2)}(kr)\right\}P_n^1(\cos\theta)$$

$$+\frac{E_0}{jkr}\cos\phi\sum_{n=1}^{\infty}\left\{a_n\hat{J}_n'(kr)+b_n\hat{H}_n'^{(2)}(kr)\right\}\frac{dP_n^1(\cos\theta)}{d\theta}$$

(5.89)

$$E_\phi = \frac{1}{r}\frac{\partial F_r}{\partial\theta}+\frac{1}{j\omega\epsilon r\sin\theta}\frac{\partial^2 A_r}{\partial r\partial\phi}$$

$$= \frac{E_0}{kr}\sin\phi\sum_{n=1}^{\infty}\left\{a_n\hat{J}_n(kr)+c_n\hat{H}_n^{(2)}(kr)\right\}\frac{dP_n^1(\cos\theta)}{d\theta}$$

$$+\frac{-E_0}{jkr\sin\theta}\sin\phi\sum_{n=1}^{\infty}\left\{a_n\hat{J}_n'(kr)+b_n\hat{H}_n'^{(2)}(kr)\right\}P_n^1(\cos\theta)$$

(5.90)

---

[2] The $n=0$ term is unnecessary because $P_0^1=0$.

$$H_r = \frac{1}{j\omega\mu}\left(\frac{\partial^2}{\partial r^2} + k^2\right)F_r$$

$$= \frac{E_0\sin\phi}{j\eta(kr)^2}\sum_{n=1}^{\infty}n(n+1)\left\{a_n\hat{J}_n(kr) + c_n\hat{H}_n^{(2)}(kr)\right\}P_n^1(\cos\theta) \qquad (5.91)$$

$$H_\theta = \frac{1}{r\sin\theta}\frac{\partial A_r}{\partial\phi} + \frac{1}{j\omega\mu r}\frac{\partial^2 F_r}{\partial r\partial\theta}$$

$$= \frac{-E_0}{\eta kr\sin\theta}\sin\phi\sum_{n=1}^{\infty}\left\{a_n\hat{J}_n(kr) + b_n\hat{H}_n^{(2)}(kr)\right\}P_n^1(\cos\theta)$$

$$+ \frac{E_0}{j\eta kr}\sin\phi\sum_{n=1}^{\infty}\left\{a_n\hat{J}'_n(kr) + c_n\hat{H}_n'^{(2)}(kr)\right\}\frac{dP_n^1(\cos\theta)}{d\theta} \qquad (5.92)$$

$$H_\phi = \frac{-1}{r}\frac{\partial A_r}{\partial\theta} + \frac{1}{j\omega\mu r\sin\theta}\frac{\partial^2 F_r}{\partial r\partial\phi}$$

$$= \frac{-E_0}{\eta kr}\cos\phi\sum_{n=1}^{\infty}\left\{a_n\hat{J}_n(kr) + b_n\hat{H}_n^{(2)}(kr)\right\}\frac{dP_n^1(\cos\theta)}{d\theta}$$

$$+ \frac{E_0}{j\eta kr\sin\theta}\cos\phi\sum_{n=1}^{\infty}\left\{a_n\hat{J}'_n(kr) + c_n\hat{H}_n'^{(2)}(kr)\right\}P_n^1(\cos\theta). \qquad (5.93)$$

It was very helpful that we made the right choices early on for the basic form of A_r and F_r in (5.86) and (5.87). The motivation behind those choices is now explained. First, consider the incident field. If we make $b_n = 0$ and $c_n = 0$, we are left with the spherical mode expansion of the incident plane wave. This must equal (5.83). The radial part of (5.83) is $E_r^i = E_x^i\sin\theta\cos\phi$. If we equate this with the incident field part of (5.88) and then use (5.84), we can obtain a_n. The form of (5.86) was chosen to give a $\cos\phi$ dependence that matches the behaviour of E_r^i. Similar considerations apply to $H_r^i = H_y^i\sin\theta\sin\phi$, which must equal the incident field part of (5.91). This in turn determined the required form of (5.87). Once the radial field components are determined, the other components automatically follow because they all depend on the same a_n.

The boundary condition $\mathbf{E}_{tan} = 0$ at $r = a$ requires that (5.89) and (5.90) must be zero, which then gives b_n and c_n. The results are

$$a_n = j^{-n}\frac{(2n+1)}{n(n+1)}; \quad b_n = -a_n\frac{\hat{J}'_n(ka)}{\hat{H}_n'^{(2)}(ka)}; \quad c_n = -a_n\frac{\hat{J}_n(ka)}{\hat{H}_n^{(2)}(ka)}. \qquad (5.94)$$

For a perfectly conducting sphere, the backscatter radar cross section can be found by calculating E_θ in (5.89) at $(\theta,\phi) = (\pi, 0)$. Using the relationships

$$\lim_{\theta\to\pi}\frac{P_n^1(\cos\theta)}{\sin\theta} = \frac{(-1)^n}{2}n(n+1)$$

$$\lim_{\theta\to\pi}\frac{dP_n^1(\cos\theta)}{d\theta} = -\frac{(-1)^n}{2}n(n+1)$$

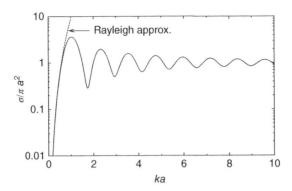

Figure 5.8   Normalized radar cross section $\sigma/\pi a^2$ versus ka for a perfectly conducting sphere.

and Wronskians for spherical Hankel functions, one can obtain the RCS

$$\sigma = \frac{\lambda^2}{4\pi} \left| \sum_{n=1}^{\infty} \frac{(-1)^n (2n+1)}{\hat{H}_n^{(2)}(ka)\, \hat{H}_n^{\prime(2)}(ka)} \right|^2. \tag{5.95}$$

Equation (5.95) was coded in PROGRAM sphrcs and used to calculate the RCS. The spherical Hankel functions of order n are related to cylindrical Hankel functions of order $n + 1/2$ and were calculated using MODULE amos_mod.

The results are shown in Figure 5.8. Similar to the scattering by a cylinder, for larger ka the backscatter can be given a physical explanation in terms of optics. Most of the backscatter is due to reflection by the sphere's metallic face. The ripples are due to creeping waves that propagate around the shadow side of the sphere and return to the observer. Depending on their phase they may reinforce or reduce the main reflection, which causes the ripples. In Chapter 8 we will use geometrical optics to show that for a sphere $\sigma \approx \pi a^2$ when ka is large.

For small ka, using just the $n = 1$ term in (5.95) leads to

$$\sigma = \frac{9\lambda^2}{4\pi} (ka)^6. \tag{5.96}$$

This result was obtained by Lord Rayleigh. The Rayleigh approximation is good to about $ka = 0.6$. It shows that the scattering is proportional to $1/\lambda^4$. Dust in the atmosphere scatters most strongly at short wavelengths and his result was used to explain why the sky is blue. The term *Rayleigh scattering* is used whenever scattering by small objects (that is, small in terms of the wavelength) is being considered.

With minor modifications we can find the fields for a dielectric sphere. We need to impose the continuity of E_θ, E_ϕ, H_θ and H_ϕ at $r = a$. For $r \geq a$ the incident field is a plane wave and the scattered field is an outgoing spherical wave, so we keep the same a_n and potentials (5.86), (5.87) that we had before, with $k = k_0$ and $\epsilon = \epsilon_0$. The total field inside a material having $\epsilon_d = \epsilon_r \epsilon_0$, $\mu_d = \mu_r \mu_0$ can be expressed in terms of the potentials

$$A_r^d = \frac{E_0}{\omega} \cos \phi \sum_{n=1}^{\infty} d_n \hat{J}_n(k_d r) P_n^1(\cos \theta) \tag{5.97}$$

$$F_r^d = \frac{\epsilon_d E_0}{k_d} \sin \phi \sum_{n=1}^{\infty} e_n \hat{J}_n(k_d r) P_n^1(\cos \theta). \tag{5.98}$$

These potentials lead to all the field components inside the sphere. Enforcing tangential field continuity for E_θ, E_ϕ, H_θ and H_ϕ at $r = a$ leads to

$$b_n = \frac{-\alpha \hat{J}'_n(k_0 a) \hat{J}_n(k_d a) + \hat{J}_n(k_0 a) \hat{J}'_n(k_d a)}{\alpha \hat{H}^{(2)'}_n(k_0 a) \hat{J}_n(k_d a) - \hat{H}^{(2)}_n(k_0 a) \hat{J}'_n(k_d a)} a_n \qquad (5.99)$$

$$c_n = \frac{-\alpha \hat{J}_n(k_0 a) \hat{J}'_n(k_d a) + \hat{J}'_n(k_0 a) \hat{J}_n(k_d a)}{\alpha \hat{H}^{(2)}_n(k_0 a) \hat{J}'_n(k_d a) - \hat{H}^{(2)'}_n(k_0 a) \hat{J}_n(k_d a)} a_n \qquad (5.100)$$

$$d_n = \frac{-j\alpha \mu_r}{\alpha \hat{H}^{(2)'}_n(k_0 a) \hat{J}_n(k_d a) - \hat{H}^{(2)}_n(k_0 a) \hat{J}'_n(k_d a)} a_n \qquad (5.101)$$

$$e_n = \frac{j k_d / k_0}{\alpha \hat{H}^{(2)}_n(k_0 a) \hat{J}'_n(k_d a) - \hat{H}^{(2)'}_n(k_0 a) \hat{J}_n(k_d a)} a_n \qquad (5.102)$$

where $\alpha = \sqrt{\epsilon_r / \mu_r}$ and $k_d = k_0 \sqrt{\mu_r \epsilon_r}$.

5.5   Method of Stationary Phase

Consider the integral

$$I(\Omega) = \int_a^b f(x) e^{j\Omega \phi(x)} \, dx \qquad (5.103)$$

where x is a real variable and Ω is real, positive and large. $\phi(x)$ is real and continuous on $a \leq x \leq b$. The 'stationary phase' point is defined by $\phi'(x_s) = 0$, where $a \leq x_s \leq b$. The function $f(x)$ may be complex, but it is slowly varying and well behaved in the range of integration.

The *method of stationary phase* (Felsen and Marcuvitz 1994, Section 4.2c) gives

$$I(\Omega) \sim f(x_s) \sqrt{\frac{2\pi}{\Omega |\phi''(x_s)|}} e^{j[\Omega \phi(x_s) + \frac{\pi}{4} \operatorname{sgn} \phi''(x_s)]}$$

$$+ \frac{1}{\Omega} \frac{f(b)}{\phi'(b)} e^{j[\Omega \phi(b) - \frac{\pi}{2}]} - \frac{1}{\Omega} \frac{f(a)}{\phi'(a)} e^{j[\Omega \phi(a) - \frac{\pi}{2}]} + O(\Omega^{-3/2}). \qquad (5.104)$$

$\operatorname{sgn}(t)$ is the *signum function*; $\operatorname{sgn}(t) = \pm 1$ when $t \gtrless 0$. The first term in (5.104) represents the stationary phase contribution, and the other two terms are endpoint contributions. Equation (5.104) is the first term of an asymptotic series, and the higher-order terms of $O(\Omega^{-3/2})$ and beyond are neglected here. The formula (5.104) was obtained by making suitable approximations of the integrand in the vicinity of x_s which can then be evaluated analytically. It will not be derived here, but an explanation of how it works will be given via an example.

An example integrand is plotted in Figure 5.9 using $a = -1$, $b = 1$, $f(x) = 1$, $\Omega = 100$ and $\phi(x) = \cos x$. The stationary phase point is at $x_s = 0$. The real and imaginary parts of the integrand are plotted, and it shows intuitively why the method of stationary phase works. That is, the integrand oscillates rapidly when Ω is large. As a result, the positive and negative areas cancel, *except* in the vicinity of the stationary phase point x_s. Also, the cancellations do not generally occur at the endpoints, where uncancelled areas remain. Hence, there are also two endpoint contributions in (5.104).

The integrand illustrated in Figure 5.9 shows a stationary point x_s that is not too close to an endpoint a or b. That is, the stationary point is 'isolated'. If x_s is close to an endpoint then the stationary phase argument which is based on area cancellation is no longer valid and (5.104) cannot be used.

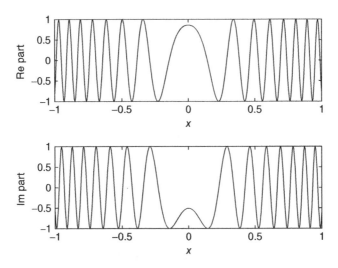

Figure 5.9 Behaviour of the integrand, when $f(x) = 1$, $\Omega = 100$ and $\phi(x) = \cos x$. The stationary phase point is at $x_s = 0$. The areas cancel when the oscillation is rapid, away from the stationary phase point.

Example 5.2 A z-polarized plane wave of strength E_0 is travelling in the negative x direction and is incident on a perfectly conducting cylinder of radius a. Find the scattered field by using the method of stationary phase and obtain the echo width.

Solution: The incident field is

$$\mathbf{E}^i = \hat{\mathbf{z}} E_0 e^{jkx}; \; \mathbf{H}^i = \hat{\mathbf{y}} \frac{E_0}{\eta} e^{jkx}.$$

The surface current is $\mathbf{J}_s = \hat{\boldsymbol{\rho}} \times \mathbf{H}$ where $\mathbf{H}$ is the total magnetic field at the surface. If we use the geometrical optics approximation, then $\mathbf{J}_s = 0$ when $|\phi| > \pi/2$ on the shadow side and

$$\mathbf{J}_s = 2\hat{\boldsymbol{\rho}} \times \mathbf{H}^i = \hat{\mathbf{z}} \frac{2E_0}{\eta} \cos \phi \, e^{jka\cos\phi}$$

when $|\phi| < \pi/2$ on the lit side.

The radiation integral (4.107) gives the scattered field

$$E_z^s = -e^{j\pi/4} \eta \sqrt{\frac{k}{8\pi}} \frac{e^{-jk\rho}}{\sqrt{\rho}} \int_{C_0} e^{jk\hat{\boldsymbol{\rho}}\cdot\boldsymbol{\rho}'} J_z(\boldsymbol{\rho}') \, d\ell'.$$

With $\hat{\boldsymbol{\rho}} = \hat{\mathbf{x}}\cos\phi + \hat{\mathbf{y}}\sin\phi$, $\boldsymbol{\rho}' = \hat{\mathbf{x}}a\cos\phi' + \hat{\mathbf{y}}a\sin\phi'$ and $d\ell' = a\,d\phi'$, this becomes

$$E_z^s(\phi) = -E_0 \sqrt{\frac{k}{2\pi}} e^{j\pi/4} \frac{e^{-jk\rho}}{\sqrt{\rho}} \int_{-\pi/2}^{\pi/2} e^{jka\cos(\phi-\phi')} \cos\phi' \, e^{jka\cos\phi'} \, a\,d\phi'.$$

In the backscatter direction, $\phi = 0$ and we obtain

$$E_z^s = -E_0 \sqrt{\frac{k}{2\pi\rho}} e^{j\pi/4} e^{-jk\rho} I(ka)\,a$$

where

$$I(ka) = \int_{-\pi/2}^{\pi/2} \cos\phi'\, e^{j2ka\cos\phi'}\, d\phi'$$

is of the form in (5.103). Replacing $x \to \phi'$, $f(x) \to \cos\phi'$, $\Omega \to ka$ and $\phi(x) \to 2\cos\phi'$ allows evaluation at the stationary phase point $\phi' = 0$ so that

$$I(ka) = \sqrt{\frac{\pi}{ka}}\, e^{-j\pi/4}\, e^{j2ka}$$

and

$$E_z^s = -E_0\sqrt{\frac{a}{2\rho}}\, e^{-jk\rho}\, e^{j2ka}.$$

The echo width for large ka is thus

$$\sigma = \lim_{\rho\to\infty} 2\pi\rho\left|\frac{E_z^s}{E_z^i}\right|^2 = 2\pi\left(\frac{a}{2}\right) = \pi a.$$
               ∎

Example 5.3 Use the stationary phase method to show that

$$\int_{-\infty}^{\infty} \bar{F}(k_z) H_n^{(2)}\!\left(\rho\sqrt{k^2 - k_z^2}\right) e^{jk_z z}\, dk_z \sim 2\frac{e^{-jkr}}{r} j^{n+1}\bar{F}(-k\cos\theta)$$

where $kr\sin\theta$ is large, and $\bar{F}$ is any function. This integral is useful for problems in cylindrical coordinates.

Solution: Make a change of variables $k_z = -k\cos\theta$ and in (5.103) replace $\Omega \to k$, $x \to \theta$, with the phase function $\phi(x) \to \phi(\theta) = -\rho\sin\theta - z\cos\theta$. The stationary phase point defined by $\phi'(\theta) = 0$ occurs at

$$z\sin\theta = \rho\cos\theta.$$

Since $r = \sqrt{\rho^2 + z^2}$ it can be recognized that $z/r = \cos\theta$, which is used to obtain $\phi''(\theta) = r$. Application of the stationary phase formula then gives the final result.   ∎

5.6   Further Reading

Analytical solutions for canonical problems are covered in many books. The cylinder, wedge and sphere are discussed by Harrington (2001). Like this book, he uses **A** and **F** to construct the solutions. Ishimaru (1991) offers similar developments using Hertz potentials. The book by Wait (1959) treats cylinders and wedges also using Hertz potentials. An encyclopaedic collection of conducting bodies is described in Bowman et al. (1969). Their coverage includes spheroids, cones, elliptic cylinders and many other shapes having analytical solutions. The electromagnetics of light scattering is discussed in depth in the classic book by Kerker (1969). As a professor of chemistry, he offers interesting science-motivated applications.

References

Abramowitz M and Stegun I (1965) *Handbook of Mathematical Functions*. Dover Publications.

Bowman JJ, Senior TBA and Uslenghi PLE (1969) *Electromagnetic and Acoustic Scattering by Simple Shapes*. North Holland.

Felsen LB and Marcuvitz N (1994) *Radiation and Scattering of Waves*. IEEE Press.

Harrington RF (2001) *Time-Harmonic Electromagnetic Fields*. IEEE Press.

Ishimaru A (1991) *Electromagnetic Wave Propagation, Radiation and Scattering*. Prentice-Hall.

Kerker M (1969) *The Scattering of Light and Other Electromagnetic Radiation*. Academic Press.

Wait JR (1959) *Electromagnetic Radiation from Cylindrical Structures*. Pergamon Press.

Problems

5.1 Equation (5.14) is coded in PROGRAM ezcyl, which calculates the TM scattered field and echo width σ for a PEC cylinder.

(a) Use the program to calculate the backscatter echo width for a cylinder with $ka = 0.05$ and $ka = 5$. How many summation terms are needed for a good result?

(b) Using just the $n = 0$ term, derive the small ka approximation

$$\frac{\sigma}{\lambda} = \frac{\pi}{2} \left| \frac{1}{\ln(0.89ka)} \right|^2.$$

(c) From geometrical optics, the backscatter for large ka is approximately $\sigma = \pi a$. Calculate the exact backscatter echo width for $0.05 \le ka \le 5$. Compare it with the small ka and large ka approximations.

5.2 For the TE scattering by a PEC cylinder, develop an expression for H_z^s that is similar to Equation (5.14). Modify PROGRAM ezcyl for this case. Calculate and plot the forward scatter ($\phi = 0$) for $0.01 \le ka \le 1$ for both the TE and TM polarizations.

5.3 From Equation (5.6) E_z^s is known for a metal cylinder that is illuminated by a plane wave $E_z^i = E_0 e^{-jkx}$. From $\mathbf{E} = \hat{\mathbf{z}}(E_z^i + E_z^s)$, find $\mathbf{H}$ and $\mathbf{J}_s = \hat{\boldsymbol{\rho}} \times \mathbf{H}$. Modify your program to compute the surface current J_{sz}.

(a) Calculate and plot $|J_{sz}|$ for $0 \le \phi \le 360°$ and $ka = 5$. Assume $\lambda = 1$ m and $E_0 = 1$ V/m.

(b) Find the geometrical optics approximation of the surface current. Plot it and compare with your result in part (a).

5.4 From Equation (5.6) E_z^s is known for a metal cylinder that is illuminated by a plane wave $E_z^i = E_0 e^{-jkx}$. The solution depends on the angular separation between the incident wave direction and ϕ. Using this observation, generalize the solution to the case of a plane wave incident at any angle ϕ' and a field point at ϕ.

5.5 An electric line source I_0 at $(\rho', \phi') = (b, 0)$ illuminates a PEC cylinder of radius a.

(a) Show that the far-zone radiated field is

$$E_z = -I_0 \eta \sqrt{\frac{jk}{8\pi}} \frac{e^{-jk\rho}}{\sqrt{\rho}} \sum_{n=0}^{N} \epsilon_{0n} \left[J_n(kb) - \frac{J_n(ka)}{H_n^{(2)}(ka)} H_n^{(2)}(kb) \right] \cos(n(\phi + \pi/2)).$$

(b) Calculate and plot the radiation pattern for a cylinder with $ka = 5$, $I_0 = 1\,\mathrm{A}$ and $\lambda = 1\,\mathrm{m}$. The source is at a height of $b - a = \lambda/4$ above the cylinder.

(c) For an electric line source that is $\lambda/4$ above a flat ground plane, make a hand calculation of $|E_z|$. This should be close to the electric field level from part (b) when ϕ is near zero.

5.6 A magnetic line source of strength M_0 at (ρ', ϕ') illuminates a cylinder with a radius a, relative permittivity ϵ_r and relative permeability μ_r. Find an expression for the scattered field $H_z^s(\rho, \phi)$.

5.7 Find the scattered field produced by a TM plane wave that is incident on a dielectric cylinder with relative permittivity ϵ_r and radius a. When $\rho \geq a$ use (5.3) for the incident field, and assume a scattered field of the form

$$E_z^s = \sum_{n=-\infty}^{\infty} a_n H_n^{(2)}(k_0\rho)e^{jn\phi}.$$

Inside the dielectric, $k_d = k_0\sqrt{\epsilon_r}$ and the total field is of the form

$$E_z^d = \sum_{n=-\infty}^{\infty} b_n J_n(k_d\rho)e^{jn\phi}.$$

By enforcing $E_z^i + E_z^s = E_z^d$ and $H_\phi^i + H_\phi^s = H_\phi^d$ at $\rho = a$, obtain the coefficients a_n and b_n.

5.8 Repeat the previous problem for the TE polarization. The field expressions are readily obtained from duality.

5.9 Modify PROGRAM ezcyl to compute the TM echo width $\sigma(\phi)$ of a dielectric cylinder. Calculate the backscatter echo width for $0.01 \leq k_0a \leq 5$ and $\epsilon_r = 4$; here k_0 is the wavenumber in free space. Assume $\lambda = 1\,\mathrm{m}$.

5.10 A metal cylinder with radius a has a material coating with parameters μ_d, ϵ_d. The coating thickness is d, so that the outer radius is $b = a + d$. Assuming an incident plane wave $E_z^i = E_0 e^{-jk_0x}$, find E_z^s.

When $\rho \geq b$ use (5.3) for the incident field, and assume a scattered field of the form

$$E_z^s = \sum_{n=-\infty}^{\infty} a_n H_n^{(2)}(k_0\rho)e^{jn\phi}.$$

Inside the dielectric, $k_d = k_0\sqrt{\epsilon_r}$ and the total field is of the form

$$E_z^d = \sum_{n=-\infty}^{\infty} b_n (H_n^{(1)}(k_d\rho) + c_n H_n^{(2)}(k_d\rho))e^{jn\phi}.$$

$E_z = 0$ at $\rho = a$ gives c_n. Then, enforcing $E_z^i + E_z^s = E_z^d$ and $H_\phi^i + H_\phi^s = H_\phi^d$ at $\rho = b$ gives the coefficients a_n and b_n.

5.11 Convert the wave $E_z = E_0 e^{jk\rho\cos(\phi-\phi')}$ to rectangular coordinates, and identify the direction of propagation.

5.12 For a wedge region $\alpha \leq \phi \leq 2\pi - \alpha$, use a Fourier series with basis functions $\sin[\nu_m(\phi - \alpha)]$ to show that

$$\delta(\phi - \phi') = \frac{1}{\pi - \alpha} \sum_{m=1}^{\infty} \sin[\nu_m(\phi' - \alpha)] \sin[\nu_m(\phi - \alpha)].$$

5.13   A TM plane wave with an amplitude E_0 is incident at an angle ϕ' on a perfectly conducting half plane. The coordinates are as in Figure 5.5. Show that the surface currents on the top/bottom faces are

$$J_{sz} = \frac{\pm E_0}{j\omega\mu\rho} \sum_{m=1}^{\infty} m j^{m/2} J_{m/2}(k\rho)(\pm 1)^m \sin\frac{m\phi'}{2}.$$

Near the edge, show that on either face

$$J_{sz} = \frac{E_0}{\eta} \sqrt{\frac{2}{j\pi k\rho}} \sin\frac{\phi'}{2}.$$

This $1/\sqrt{\rho}$ *edge singularity* of the current always occurs when an electric field is parallel to a sharp metal edge.

5.14   A TE plane wave with an amplitude H_0 is incident at an angle ϕ' on a perfectly conducting half plane. The coordinates are as in Figure 5.5. Show that the surface currents on the top/bottom faces are

$$J_{s\rho} = \pm H_0 \sum_{m=0}^{\infty} \epsilon_{0m} j^{m/2} J_{m/2}(k\rho)(\pm 1)^m \cos\frac{m\phi'}{2}.$$

Near the edge, show that on the top/bottom faces

$$J_{s\rho} = \pm H_0.$$

The surface current is finite and 'wraps around' the edge as it flows.

5.15   PROGRAM jxjz was used to compute the surface currents on the half plane in Figure 5.6. Modify the program to calculate $E_z(\rho,\phi)$ for a half plane illuminated by a plane wave

$$E_z^i = E_0 e^{jk(x\cos\phi' + y\sin\phi')}.$$

(a)   Compute and plot $|E_z(\rho,\phi)|$ in dB at $(\rho,\phi) = (1\text{ m}, 30°)$ for a range of incident angles $0 \le \phi' \le 360°$. Assume $\lambda = 1\text{ m}$ and $E_0 = 1$ V/m.

(b)   Replace the half plane by an infinite ground. We now have two plane waves—an incident wave and its reflection. Given that $(\rho,\phi) = (1\text{ m}, 30°)$, at what angle ϕ would $|E_z|$ be maximum?

5.16   The far-zone radiation pattern E_ϕ of an axial magnetic dipole p_m on a cylinder is given by Equation (5.64). Write a program that calculates $|E_\phi|$. The pattern should not be normalized, but the far-field factor e^{-jkr}/r is suppressed.

(a)   Compute and plot the pattern for $0 \le \phi \le 360°$ and $\theta = 90°$, for the cases $ka = 0.1$ and $ka = 10$. Assume that $\lambda = 1\text{ m}$ and $p_m = 1$ V-m.

(b)   The $ka = 10$ cylinder is large enough so that E_ϕ near $\phi = 0$ should be almost the same as for a magnetic dipole on a flat ground plane. Use this idea and a hand calculation to check if your result from part (a) is sensible.

5.17   A perfectly conducting circular cylinder of radius a has a ϕ-directed magnetic dipole p_m^c at $(\rho,\phi,z) = (a,0,0)$.

(a)   Find **E** in cylindrical coordinates from the vector potentials (5.65) and (5.66).

(b)   Use the stationary phase method to find the far-zone E_ϕ in the $\theta = 90°$ plane. At other values of θ, there will be both E_θ and E_ϕ components.

5.18 For a perfectly conducting wedge, find the far-field H_θ due to a z-directed magnetic dipole of strength p_m. The result resembles the electric dipole case in Equation (5.72).

5.19 Find an expression for the scattered field $E_z^s(\rho, \phi)$ for a PEC TM cylinder illuminated by an incident plane wave

$$E_z^i = E_0 e^{jk(x\cos\phi' + y\sin\phi')}.$$

Start with the solution for line source illumination. Eliminate the line source strength I_0 by letting the line source recede to infinity, that is,

$$E_z^i = -\frac{k^2 I_0}{4\omega\epsilon} H_0^{(2)}(kR) \sim -\frac{\omega\mu I_0}{4}\sqrt{\frac{j2}{\pi k\rho'}}\, e^{-jk\rho'} e^{jk\rho\cos(\phi-\phi')} \equiv E_0 e^{jk\rho\cos(\phi-\phi')}.$$

5.20 Use reciprocity in 2D to find the E_z produced by a ϕ-directed magnetic line dipole M_0 that illuminates a PEC cylinder.

   Start with a cylinder and line source illumination I_0. From Equation (5.18), E_z is known and can be used to find H_ϕ in Figure 5.10(a). Then, reciprocity can be used to find the field produced by the magnetic line dipole in Figure 5.10(b) via

$$-M_0 H_\phi(\rho^b, \phi^b) = I_0 E_z(\rho^a, \phi^a).$$

Assuming $\rho > \rho'$ with M_0 at (ρ', ϕ'), obtain a result that is of the following form and determine K_0 in the expression

$$E_z(\rho, \phi) = M_0 K_0 \sum_{n=-\infty}^{\infty} \left[J_n'(k\rho') - \frac{J_n(ka)}{H_n^{(2)}(ka)} H_n'^{(2)}(k\rho') \right] H_n^{(2)}(k\rho) e^{jn(\phi-\phi')}.$$

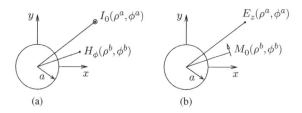

Figure 5.10 Perfectly conducting circular cylinder. (a) Line source I_0 produces H_ϕ. (b) Line dipole M_0 produces E_z.

5.21 In the previous problem let the magnetic line dipole reside on the surface, that is, let $(\rho', \phi') = (a, 0)$. Use the Wronskian to simplify your result. This can be used to model a narrow slot of width ϕ_0 and an aperture field E_z^a such that $a\phi_0 E_z^a = M_0$.

5.22 A cylinder of radius a has a radial monopole of strength p_e at $(\rho', \phi', z') = (a, 0, 0)$. Use the reciprocity method to find an expression for the radiation pattern in the $z = 0$ plane. The final result should be of the form

$$E_\phi = f(r) \sum_{n=1}^{\infty} \frac{nj^n \sin n\phi}{H_n'^{(2)}(ka)}.$$

Find $f(r)$. In other planes the field will have θ and ϕ components.

5.23 A conducting sphere of radius a is illuminated by an incident plane wave $E_x^i = E_0 e^{-jkz}$. Show that the E_θ component of the far-zone scattered field is

$$E_\theta^s = \frac{jE_0}{k} \frac{e^{-jkr}}{r} \cos\phi \sum_{n=1}^{\infty} -c_n j^n \frac{P_n^1(\cos\theta)}{\sin\theta} + b_n j^n \sin\theta P_n'^1(\cos\theta).$$

5.24 A conducting sphere of radius a is illuminated by an incident plane wave $E_x^i = E_0 e^{-jkz}$. The E_r field at the surface is known from (5.88), and for compactness we may write it as $E_r = f(\theta, \phi)$.

   (a) Find out what kind of dipole on the negative-z axis can be used to approximately generate the incident plane wave. Assume that the dipole is far away from the sphere so that suitable approximations can be made.
   (b) Use the reciprocity theorem to find the far-zone E_x on the negative-z axis, when the source is a radially directed short electric monopole p_e, located at $(r, \theta, \phi) = (a, \theta_0, 0)$. Your result should be in terms of f.
   (c) Use the result of part (b) to find the radiation pattern $E_\theta(\theta)$ in the $\phi = 0$ plane for a short monopole at $(r, \theta, \phi) = (a, 0, 0)$.

5.25 Derive Equation (5.95) which gives the RCS of a perfectly conducting sphere.

5.26 For a dielectric sphere
   (a) Obtain an expression for the RCS that is similar to Equation (5.95).
   (b) Find the Rayleigh approximation from the leading term of the series.

5.27 PROGRAM sphrcs calculates the RCS for a perfectly conducting sphere. Modify the program for the dielectric sphere and calculate the RCS for $0 \le k_0 a \le 5$ using a relative permittivity of $\epsilon_r = 4$.

When $k_0 a$ is large the RCS can be approximated by geometrical optics, as $\sigma = \pi a^2 |\Gamma|^2$ where Γ is the Fresnel reflection coefficient

$$\Gamma = \frac{1/\sqrt{\epsilon_r} - 1}{1/\sqrt{\epsilon_r} + 1}.$$

Compare this with your results from the program.

6

Method of Moments

6.1   Introduction

The method of moments (MoM), which is also known as the moment method, is a way to solve an integral equation. The general concepts are first discussed, and a simple example is used to illustrate the procedure. The MoM is then used to find the surface currents on a perfectly conducting two-dimensional (2D) strip that are induced by an incident plane wave. A source on the surface (as in an antenna) is also considered.

The special case of a 2D thin wire (infinitely long) is treated next. This is readily extended to an array of wires. The use of thin wires to approximate a conducting surface is also described. An infinite array of wires (periodic array) is considered next. Special solution techniques for periodic arrays involving Poisson summation and Floquet harmonics are explained.

The problem of plane wave scattering by a straight thin wire is used to demonstrate some 3D techniques. With minor modifications it permits the analysis of a dipole antenna's current distribution, input impedance and radiation pattern.

There are several ways to obtain an MoM formulation for electromagnetic boundary value problems and these are discussed next. The MoM can be formulated as an electric field integral equation (EFIE), which is based on meeting a boundary condition involving $\mathbf{E}_{tan}$. There is also the magnetic field integral equation (MFIE) which is based on meeting a boundary condition involving $\mathbf{H}_{tan}$. There are combinations of the two, leading to the combined field integral equation (CFIE) and the formulation of Poggio et al. (PMCHWT) for dielectrics.

The MoM expands the unknown current as a weighted sum of N *basis functions* with unknown coefficients, and the MoM procedure allows us to find those coefficients. A few possibilities for basis functions for currents on wires, surfaces and in volumes are reviewed.

6.2   General Concepts

In its simplest form, the MoM is a way to solve an integral equation of the form

$$\int_{x'=a}^{b} f(x')g(x,x')\,dx' = h(x); \ a \le x \le b. \tag{6.1}$$

The functions $g(x,x')$ and $h(x)$ are given, and $f(x)$ is an unknown function to be determined. Two methods of solution will be described: point matching and Galerkin's method.

Applied Frequency-Domain Electromagnetics, First Edition. Robert Paknys.
© 2016 John Wiley & Sons, Ltd. Published 2016 by John Wiley & Sons, Ltd.
Companion Website: www.wiley.com/go/paknys9981

The unknown is approximated by an expansion in terms of N *basis functions* $f_n(x)$ so that

$$f(x) \approx \sum_{n=1}^{N} a_n f_n(x). \tag{6.2}$$

Substituting (6.2) in (6.1) and interchanging the order of summation and integration gives

$$\sum_{n=1}^{N} a_n \int_{x'=a}^{b} f_n(x')g(x,x') \, dx' = h(x). \tag{6.3}$$

Ideally, this equation must hold true for all $a \le x \le b$. In addition, it represents one equation with N unknowns.

6.2.1 Point Matching

Equation (6.3) can be solved by forcing the equality to hold at N equally spaced points on the interval (a, b):

$$x_m = x_1, x_2, x_3, \ldots x_N. \tag{6.4}$$

This is called *point matching*. (It is also known as *collocation*.) Then (6.3) becomes

$$\sum_{n=1}^{N} a_n \int_{x'=a}^{b} f_n(x')g(x_m,x') \, dx' = h(x_m); \; m = 1, 2, \ldots, N. \tag{6.5}$$

This gives us N equations and N unknowns. Often this is written in a matrix format $\mathbf{V} = \mathbf{ZI}$ where $\mathbf{V}$ and $\mathbf{I}$ are column vectors and $\mathbf{Z}$ is an $N \times N$ matrix:

$$V_m = h(x_m) \tag{6.6}$$

$$I_n = a_n \tag{6.7}$$

$$Z_{mn} = \int_{x'=a}^{b} f_n(x')g(x_m,x') \, dx'. \tag{6.8}$$

6.2.2 Galerkin's Method

More generally, (6.3) is an equation of the form

$$u(x) = h(x) \tag{6.9}$$

where

$$u(x) = \sum_{n=1}^{N} a_n \int_{x'=a}^{b} f_n(x')g(x,x') \, dx'.$$

Equation (6.9) was solved earlier by making it hold at N discrete match points x_m. This is not the only way to get N conditions. For example, another possibility could be to enforce

$$\int_{x=a}^{b} u(x)w_m(x) \, dx = \int_{x=a}^{b} h(x)w_m(x) \, dx; \; m = 1, 2, \ldots, N. \tag{6.10}$$

The w_m here are *testing functions* (also called *weighting functions*). If $w_m(x) = \delta(x - x_m)$, then (6.10) reduces to point matching. If the weighting functions are pulses $w_m(x) = p(x - x_m)$, then the average of $u(x)$ equals the average of $h(x)$ within each pulse interval. If the weighting functions are chosen to be the same as the basis functions, then $w_m = f_m$ and this is known as *Galerkin's method*.

Applying this idea to (6.9), we multiply both sides by $w_m(x)$ and integrate over (a, b) to obtain

$$\sum_{n=1}^{N} a_n \int_{x=a}^{b} \int_{x'=a}^{b} f_n(x')g(x, x')w_m(x)\,dx'dx =$$

$$\int_{x=a}^{b} h(x)w_m(x)\,dx; \; m = 1, 2, \ldots, N. \tag{6.11}$$

As before, this leads to an $N \times N$ system of equations $\mathbf{V} = \mathbf{ZI}$ in which

$$V_m = \int_{x=a}^{b} h(x)w_m(x)\,dx \tag{6.12}$$

$$I_n = a_n \tag{6.13}$$

$$Z_{mn} = \int_{x=a}^{b} \int_{x'=a}^{b} f_n(x')g(x, x')w_m(x)\,dx'dx \tag{6.14}$$

and letting $w_m = f_m$ gives Galerkin's method.

6.2.3 Fredholm Integral Equation

An integral equation of the form

$$\int_{x'=a}^{b} f(x')\,K(x, x')\,dx' = v(x) \tag{6.15}$$

is called a *Fredholm integral equation of the first kind*. The function $K(x, x')$ is called the *kernel*. Both $K(x, x')$ and the right-hand side $v(x)$ are known functions. Equation (6.15) has to be solved for the unknown function $f(x)$. The defining characteristics of this equation are that the limits of integration are constant and that the unknown function $f(x)$ only appears inside the integral. If the unknown function appears both inside and outside the integral, then it is called a Fredholm integral equation of the *second kind*.

As an example, let us solve the following integral equation using the MoM

$$\int_{x'=0}^{\pi/2} f(x')\,|x - x'|\,dx' = 1 + \frac{\pi}{2} - 2\cos x - x. \tag{6.16}$$

It has an exact solution, $f(x) = \cos x$.

We can use pulse bases and point matching. The pulses have a width Δ and are centred at $x = x_n$:

$$p(x) = \begin{cases} 1; \; |x| < \Delta/2 \\ 0; \; |x| > \Delta/2 \end{cases} \tag{6.17}$$

$$f(x) = \sum_{n=1}^{N} f_n p(x - x_n) \tag{6.18}$$

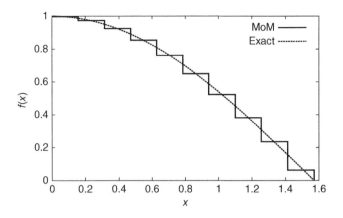

Figure 6.1 MoM and exact solutions of the Fredholm integral equation (6.16), with $N = 10$ pulses.

and the match points x_m are at the pulse centres. This leads to

$$\sum_{n=1}^{N} f_n K_{mn} = v_m \tag{6.19}$$

where

$$v_m = v(x_m) = 1 + \frac{\pi}{2} - 2\cos x_m - x_m \tag{6.20}$$

$$K_{mn} = \begin{cases} |m - n|\Delta^2; & m \neq n \\ \Delta^2/4; & m = n. \end{cases} \tag{6.21}$$

For $N = 3$, K looks like this

$$K = \Delta^2 \begin{bmatrix} 1/4 & 1 & 2 \\ 1 & 1/4 & 1 \\ 2 & 1 & 1/4 \end{bmatrix}.$$

This leads to a very nice numerical solution that well approximates the exact solution $f(x) = \cos x$. The results are shown in Figure 6.1 for $N = 10$ pulses. Notice that the approximate and true solutions are not equal at the match points $x = x_m$ and that is correct. It is the left- and right-hand sides of (6.16) that are equal.

6.3   2D Conducting Strip

The TM and TE MoM solution for a 2D perfectly conducting strip will now be developed. The configuration is shown in Figure 6.2. The strip is at $y = 0$ and along $-a/2 \leq x \leq a/2$. It may be illuminated by a plane wave with an incidence angle ϕ_i. Line source excitation and magnetic sources on the surface will also be discussed.

6.3.1   TM Case

For the TM polarization the surface current on the strip is $J_z(x)$ and the nonzero field components are E_z, H_x and H_y. Assuming a 1 V/m incident wave, the incident and scattered fields are given by (see (4.103))

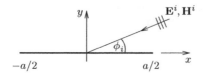

Figure 6.2   2D perfectly conducting strip of width a and plane wave incidence.

$$E_z^i = e^{jk(x\cos\phi_i + y\sin\phi_i)} \tag{6.22}$$

$$E_z^s = -\frac{\omega\mu}{4}\int_{x'=-a/2}^{a/2} J_z(x')H_0^{(2)}(kR)\,dx', \tag{6.23}$$

where $R = |\boldsymbol{\rho} - \boldsymbol{\rho}'|$, $\boldsymbol{\rho} = \hat{x}\rho\cos\phi + \hat{y}\rho\sin\phi$ and $\boldsymbol{\rho}' = \hat{x}x'$. Enforcing $\mathbf{E}_{tan} = 0$ on the metal strip gives an integral equation

$$E_z^i(x,0) = -E_z^s(x,0). \tag{6.24}$$

The above formulation is in free space. This is explained in Figure 6.3. Case (a) shows the metal strip and its electric surface current $\mathbf{J}_s = \hat{z}J_z(x)$. This is replaced by a surface equivalent (b) in free space. The equivalent has zero fields inside S and the same fields as (a) outside S. Therefore, the actual physical currents in (a) and the surface-equivalent currents in (b) are the same.

The surface currents on the two sides of the strip are different. However, if the strip thickness shrinks to zero (i.e. $t \to 0$) as in Figure 6.3(c), the top and bottom currents which are in free space will merge, so the distinction between a top and bottom current is lost. Nevertheless, the merging does not affect the fields, which are different on the two sides of the strip. It is possible to separate out the top and bottom currents; this is discussed in Problem 6.3.

In Figure 6.3(d) the unknown current (which is the sum of the top and bottom J_z) is represented as a pulse-basis point-matching formulation. The point matching can be thought of as enforcing $\mathbf{E}_{tan} = 0$ on the top surface or on the bottom surface. It does not matter, because this is an electric current sheet,

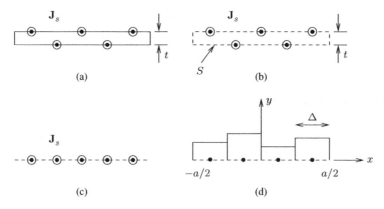

Figure 6.3   (a) Metal strip with top and bottom surface currents. (b) Same currents, on a surface equivalent in free space. (c) Strip thickness $t \to 0$. (d) $J_z(x)$ using pulse bases. The match points are at the dots •.

and $\mathbf{E}_{tan}$ is continuous across the sheet. Therefore, enforcing $\mathbf{E}_{tan} = 0$ on the top surface automatically ensures that $\mathbf{E}_{tan} = 0$ on the bottom surface (and vice versa).

Enforcing $\mathbf{E}_{tan} = 0$ at the match points $(x, y) = (x_m, 0)$ gives

$$E_z^i(x_m, 0) = -E_z^s(x_m, 0) \tag{6.25}$$

where the match points x_m are at

$$x_m = -(N/2)\,\Delta + \Delta/2 + (m-1)\Delta; \quad m = 1, 2, \ldots, N.$$

The unknown current is expanded with pulse bases

$$J_z(x) = \sum_{n=1}^{N} a_n p(x - x_n) \tag{6.26}$$

where $p(x)$ is a unit pulse of width Δ, centred at $x = 0$. As shown in Figure 6.3(d), the pulses are centred at the match points so that

$$x_n = -\left(\frac{N}{2}\right)\Delta + \frac{\Delta}{2} + (n-1)\Delta; \quad n = 1, 2, \ldots, N.$$

Using (6.22), (6.23) and (6.26) in (6.24), we obtain

$$e^{jkx_m \cos\phi_i} = \frac{\omega\mu}{4} \int_{x'=-a/2}^{a/2} \sum_{n=1}^{N} a_n p(x' - x_n) H_0^{(2)}(k|x_m - x'|)\,dx' \tag{6.27}$$

which is a matrix equation of the form $V = ZI$, with the mth row being

$$V_m = \sum_{n=1}^{N} a_n Z_{mn} \tag{6.28}$$

and matrix elements

$$V_m = e^{jkx_m \cos\phi_i} \tag{6.29}$$

$$I_n = a_n \tag{6.30}$$

$$Z_{mn} = \frac{\omega\mu}{4} \int_{x'=-a/2}^{a/2} p(x' - x_n) H_0^{(2)}(k|x_m - x'|)\,dx'. \tag{6.31}$$

Since the pulse equals unity when $|x' - x_n| \le \Delta/2$, we get

$$Z_{mn} = \frac{\omega\mu}{4} \int_{x'=x_n-\Delta/2}^{x_n+\Delta/2} H_0^{(2)}(k|x_m - x'|)\,dx'. \tag{6.32}$$

Z_{mn} is the negative scattered field at x_m produced by a unit current pulse at x_n. When $m \ne n$, a possible approximation for Z_{mn} is

$$Z_{mn} \approx \frac{\omega\mu}{4} \Delta H_0^{(2)}(k|x_m - x_n|). \tag{6.33}$$

However, if the match point x_m and current pulse at x_n are not far apart (this happens with narrow pulses), a numerical integration of (6.32) may result in improved accuracy.

When $m = n$, the integral (6.32) represents the negative scattered field at x_m due to a pulse at x_m. It has an integrable singularity. Since all the Z_{mm} are the same, it is sufficient to consider the case $m = n = 0$, and so

$$Z_{mm} = \frac{\omega\mu}{2}\int_{x'=0}^{\Delta/2} H_0^{(2)}(kx')\,dx'. \tag{6.34}$$

The Hankel function can be approximated as

$$H_0^{(2)}(x) \approx 1 - j\frac{2}{\pi}\ln(\gamma x/2)$$

where $\gamma = 1.781072416\cdots$. Using

$$\int \ln ax\,dx = x\ln ax - x$$

and

$$\lim_{x\to 0} x\ln ax = 0$$

leads to

$$Z_{mm} \approx \frac{\omega\mu}{4}\Delta\left(1 - j\frac{2}{\pi}[\ln(\gamma k\Delta/4) - 1]\right). \tag{6.35}$$

As a side note, singular integrals of the form

$$\int_0^1 f(x)\ln x\,dx$$

where $f(x)$ is slowly varying on $(0,1)$ can also be evaluated numerically with Gaussian quadrature formulas developed by Ma et al. (1996); this is known as the MRW method. It can be useful for implementation in general-purpose codes.

6.3.2   TE Case

A 2D perfectly conducting strip is at $y = 0$ and along $-a/2 \le x \le a/2$. It is illuminated by a 1 V/m plane wave. As before, we require that $\mathbf{E}_{tan} = 0$ on the strip, and the strip is represented by a surface equivalent in free space. For this polarization the surface current is $J_x(x)$ and the nonzero field components are H_z, E_x and E_y. The integral equation is

$$E_x^i(x,0) = -E_x^s(x,0). \tag{6.36}$$

The incident field is assumed to be 1 V/m so that

$$H_z^i = \frac{1}{\eta}e^{jk(x\cos\phi_i + y\sin\phi_i)} \tag{6.37}$$

from which

$$E_x^i(x,0) = \sin\phi_i\,e^{jkx\cos\phi_i}. \tag{6.38}$$

In order to find the scattered field, we must first know the near field of a 2D x-directed dipole (sometimes called a 'current blade'). If the dipole strength is p_e Amps and is located at $(0,0)$, then from (4.101),

$$H_z = -\frac{jkp_e}{4} H_1^{(2)}(k\rho) \frac{y}{\rho}$$

(6.39)

and it follows that

$$E_x = \frac{1}{jwe} \frac{\partial H_z}{\partial y} = -p_e \frac{k\eta}{8} \left(H_0^{(2)}(k\rho) + H_2^{(2)}(k\rho) \cos 2\phi \right).$$

(6.40)

Here, $\cos 2\phi = 1 - 2\sin^2\phi = 1 - 2(y/\rho)^2$; in other words, ϕ is the angle between the vectors $\hat{\rho}$ and $\hat{x}$. If the field point is on the strip, then $y = 0$ and $\cos 2\phi = 1$. Equation (6.40) can be used to find E_x^s due to J_x on the strip, and so (6.36) becomes

$$\sin\phi_i \, e^{jkx\cos\phi_i} = \frac{k\eta}{8} \int_{x'=-a/2}^{a/2} J_x(x')[H_0^{(2)}(k|x-x'|) + H_2^{(2)}(k|x-x'|)] \, dx'.$$

(6.41)

Using $xH_0 - H_1 = H_1 - xH_2$, (6.41) becomes

$$\sin\phi_i \, e^{jkx\cos\phi_i} = \frac{k\eta}{4} \int_{x'=-a/2}^{a/2} J_x(x') \frac{H_1^{(2)}(k|x-x'|)}{k|x-x'|} \, dx'.$$

(6.42)

Next, the current is expanded in terms of pulse bases

$$J_x(x) = \sum_{n=1}^{N} a_n p(x - x_n)$$

(6.43)

and point matching is applied

$$E_x^i(x_m, 0) = -E_x^s(x_m, 0)$$

(6.44)

at the match points

$$x_m = -(N/2)\Delta + \Delta/2 + (m-1)\Delta; \, m = 1, 2, \ldots, N$$

(6.45)

so that (6.42) becomes

$$\sin\phi_i \, e^{jkx_m\cos\phi_i} = \frac{k\eta}{4} \int_{x'=x_n-\Delta/2}^{x_n+\Delta/2} \sum_{n=1}^{N} a_n \frac{H_1^{(2)}(k|x_m-x'|)}{k|x_m-x'|} \, dx'.$$

(6.46)

This can be written in the matrix format $V = ZI$ with the mth row being

$$V_m = \sum_{n=1}^{N} a_n Z_{mn}$$

(6.47)

and matrix elements

$$V_m = \sin\phi_i \, e^{jkx_m\cos\phi_i}$$

(6.48)

$$I_n = a_n \tag{6.49}$$

$$Z_{mn} = \frac{k\eta}{4} \int_{x'=x_n-\Delta/2}^{x_n+\Delta/2} \frac{H_1^{(2)}(k|x_m - x'|)}{k|x_m - x'|} \, dx'. \tag{6.50}$$

6.3.3   Self-Impedance Term for the TE Strip

$E_x(x, 0)$ due to a surface current J_x was obtained in (6.40), (6.41). It can also be written in an alternate form, as[1]

$$E_x = -\frac{\eta}{4k} \int_{-\infty}^{\infty} J_x(x') \left[\frac{d^2}{dx^2} + k^2 \right] H_0^{(2)}(k|x - x'|) \, dx'. \tag{6.51}$$

This requires the second derivative of the Hankel function. The singularity at $x = x'$ cannot be integrated. This can be avoided by using another representation, obtainable from $\mathbf{E} = -j\omega\mathbf{A} - \nabla V$. In 2D, the potentials obey

$$(\nabla^2 + k^2)\mathbf{A}(\boldsymbol{\rho}) = -\mu\mathbf{J}(\boldsymbol{\rho})$$

$$\mathbf{A}(\boldsymbol{\rho}) = \frac{-j\mu}{4} \int \mathbf{J}(\boldsymbol{\rho}\,')H_0^{(2)}(k|\boldsymbol{\rho} - \boldsymbol{\rho}\,'|) \, dS'$$

$$(\nabla^2 + k^2)V(\boldsymbol{\rho}) = \frac{-1}{\epsilon}\rho_s(\boldsymbol{\rho})$$

$$V(\boldsymbol{\rho}) = \frac{-j}{4\epsilon} \int \rho_s(\boldsymbol{\rho}\,')H_0^{(2)}(k|\boldsymbol{\rho} - \boldsymbol{\rho}\,'|) \, dS'.$$

For a unit current pulse $J_x(x)$ on $-\Delta/2 \le x \le \Delta/2$, let us denote the field contributions from J as E^J and from ρ_s as E^ρ. The J part is $E_x = -j\omega A_x$, or

$$E_x^J(x = 0) = -\frac{k\eta}{4} \int_{-\Delta/2}^{\Delta/2} H_0^{(2)}(k|x'|) \, dx'. \tag{6.52}$$

The x-directed surface current is related to the surface charge density by the continuity equation

$$\frac{dJ_x}{dx} = -j\omega\rho_s$$

so that

$$V(x) = \frac{1}{4\omega\epsilon} \int_{-\Delta/2}^{\Delta/2} J_x'(x')H_0^{(2)}(k(|x - x'|) \, dx'.$$

Since we are using a unit pulse basis centred around $x = 0$, we have $J_x'(x) = \delta(x + \Delta/2) - \delta(x - \Delta/2)$. (Note that $f'(|t|) = \pm f'(t)$ for $t \gtrless 0$.) The electric field due to ρ_s is then

$$E_x^\rho(x = 0) = -\frac{dV}{dx} = \frac{\eta}{2} H_1^{(2)}(k\Delta/2). \tag{6.53}$$

---

[1] One way to find E_x in this form is to use (6.40) or (6.41), with $H_0''(x) = -H_1'(x) = (H_0(x) - H_2(x))/2$ to eliminate H_2. Another way is to obtain E_x from a potential $A_x = (\mu/j4)H_0^{(2)}(kR)J_x \, dx'$.

The total field comes from (6.52) plus (6.53), that is, $E_x = E_x^J + E_x^\rho$. Noting that $Z_{mn} = -E_x$ for pulse basis n at match point m, we get

$$Z_{mm} = \frac{k\eta}{4} \int_{-\Delta/2}^{\Delta/2} H_0^{(2)}(k|x'|)\,dx' - \frac{\eta}{2} H_1^{(2)}(k\Delta/2). \qquad (6.54)$$

Since Δ is small, we can use the small argument approximations

$$J_0(x) \approx 1 - \frac{1}{4}x^2 + \frac{1}{64}x^4 + \cdots$$

$$\pi Y_0(x) \approx 2J_0(x)\{\ln(x/2) + \gamma\} + \frac{1}{2}x^2 - \frac{3}{64}x^4 + \cdots .$$

By using $J_0' = -J_1$, $Y_0' = -Y_1$, we also have

$$J_1(x) \approx \frac{1}{2}x - \frac{1}{16}x^3 + \cdots$$

$$\pi Y_1(x) \approx x \ln(x/2) - \frac{2}{x} + \left(\gamma - \frac{1}{2}\right)x + \frac{6}{32}x^3 + \cdots .$$

Using the leading term for H_0 in (6.52) gives

$$\int_{-\Delta/2}^{\Delta/2} H_0^{(2)}(k|x'|)\,dx' \approx \Delta \left[1 - \frac{j2}{\pi}\ln\frac{\gamma_0 k\Delta}{4e}\right].$$

Here, $\gamma_0 = \exp\gamma$ where $\gamma = 0.577215665\cdots$ is Euler's constant. Using this in (6.54), Z_{mm} becomes

$$Z_{mm} = \frac{\eta k\Delta}{8}\left\{1 - \frac{j}{\pi}\left[-1 + 2\ln\left(\frac{\gamma_0 k\Delta}{4e}\right)\right]\right\} - \frac{\eta}{2}H_1^{(2)}(k\Delta/2). \qquad (6.55)$$

Since H_1 is not being integrated, it is consistent to retain terms to order x. Then,

$$Z_{mm} = \frac{\eta k\Delta}{8}\left\{1 - \frac{j}{\pi}\left[-1 + 2\ln\left(\frac{\gamma_0 k\Delta}{4e}\right) + \frac{16}{(k\Delta)^2}\right]\right\}. \qquad (6.56)$$

Although (6.56) avoids a Hankel function evaluation, in practice it is much better to use the more accurate (6.55).

6.3.4 Other Source Types

It is easy to replace the incident plane wave with a line source illumination, by using (4.96) for E_z^i or (4.97) for H_z^i. If the source is at a finite height h above the strip, no complications arise. The special case of $h \to 0$ for a magnetic current right on the surface is useful for modelling aperture antennas; however, this limit requires some special care.

For the TE case, suppose that a z-directed magnetic line source $\mathbf{M} = \hat{z}p_m\delta(x)\delta(y)$ is placed slightly above the strip, at $x = 0$, $y = 0^+$. Since this source produces the incident field, it is by definition in free space. If we tried to evaluate the incident field at the match point right underneath the source, it would be infinite. The solution to this difficulty is to not use an infinitesimal source but instead use a thin ribbon of current of width Δ and strength $p_m = M_0\Delta$. (Here, Δ is the width of the J_x pulse basis under the source.) From symmetry considerations and the boundary condition for a magnetic current ribbon, the electric field just above/below the current ribbon is $\mathbf{E} = \pm\hat{\mathbf{x}}M_0/2$. The field just below the ribbon is the incident field on the strip, so that $E_x^i = -M_0/2$ and

$$V_m = E_x^i = -\frac{p_m}{2\Delta} \tag{6.57}$$

on the source segment. On the other segments, $V_m = 0$. This is because the source only produces an E_ϕ component, which implies that $E_x^i = 0$. (This argument only holds for a flat strip; the match points on a non-planar strip would have $E_x^i \neq 0$.)

A similar development holds for the TM case with an x-directed magnetic current $\mathbf{M} = \hat{\mathbf{x}}p_m\delta(x)\delta(y)$ at $x = 0, y = 0^+$. In this instance, the equivalent ribbon has $p_m = M_0\Delta$ and $\mathbf{E} = \mp\hat{\mathbf{z}}M_0/2$ above/below the ribbon, so that

$$V_m = E_z^i = \frac{p_m}{2\Delta} \tag{6.58}$$

on the source segment, and $V_m = 0$ elsewhere on a flat strip.

Example 6.1 (Strip Currents, MoM) Calculate the TM and TE surface currents on a metal strip of width 2 m with $\lambda = 1$ m. The incident field is a 1 V/m plane wave at an angle of $\phi = \phi_i = 90°$. Compare with geometrical optics.

Solution: Equations (6.26) and (6.43) give the TM and TE surface currents. The MoM solutions have been coded in PROGRAM mmtmstrip and PROGRAM mmtestrip. Using $N = 100$ pulses, the results are shown in Figure 6.4.

For a plane wave of strength E_0, the amplitude of the geometrical optics current is $J_s = 2E_0/\eta$. Since $E_0 = 1$ V/m, this gives $J_s = 5.31$ mA/m.

The GO current is for the lit side, whereas the MoM current is for the lit plus shadow sides. Therefore, the comparison is only fair if the true shadow-side current is very weak. This is somewhat true for J_z but not for J_x. The results also show the correct edge behaviour; that is, J_z is singular whereas J_x tends towards zero. (J_x on the top and bottom surfaces are equal and opposite near the edge, so in the free-space equivalent they cancel.) ∎

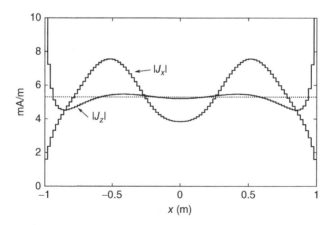

Figure 6.4 TM and TE surface currents on a strip, for a 1 V/m plane wave incident at $\phi_i = 90°$. The strip width is 2 m and the wavelength is $\lambda = 1$ m. $N = 100$ pulses. Also shown (dashed) is the geometrical optics approximation.

6.4   2D Thin Wire MoM

In this section the 2D scattering by a thin perfectly conducting wire and wire array will be developed. These are good problems for demonstrating MoM techniques. The structures have a practical importance in devices such as polarizers and waveguides and in approximate modelling techniques (Ozturk and Paknys 2012; Paknys 1991). By 'thin' it is meant that the wire radius is $a \ll \lambda$. Consequently, the surface current does not vary around the wire circumference. This is the so-called thin wire assumption. If the wire is z-directed and carries a total current I_0, then the surface current density is $\mathbf{J}_s = \hat{\mathbf{z}} I_0 / 2\pi a$.

First, a single wire with plane wave illumination will be considered. Second, an array (forming a grid of wires) will be discussed. Third, special techniques for an infinite array (i.e. a periodic structure) of wires will be described.

6.4.1   One Wire

A single perfectly conducting wire is shown in Figure 6.5(a). It is illuminated by a 1 V/m TM plane wave with an incidence angle $\phi = \phi_i$

$$E_z^i = e^{jk(x \cos \phi_i + y \sin \phi_i)}. \tag{6.59}$$

A uniform surface current having $\mathbf{J}_s = \hat{\mathbf{z}} J_{so}$ is induced on the wire. The wire is regarded as a hollow shell that supports the current. The fields due to $\mathbf{J}_s$ outside the wire ($\rho \geq a$) are

$$E_{1z} = A_0 H_0^{(2)}(k\rho) \tag{6.60}$$

$$H_{1\phi} = \frac{-jA_0}{\eta} H_0'^{(2)}(k\rho) \tag{6.61}$$

where $k = \omega \sqrt{\mu_0 \epsilon_0} = 2\pi/\lambda$ and $\eta = \sqrt{\mu_0/\epsilon_0}$. Inside the wire ($\rho \leq a$) the fields due to $\mathbf{J}_s$ are

$$E_{2z} = A_0 \frac{H_0^{(2)}(ka)}{J_0(ka)} J_0(k\rho) \tag{6.62}$$

$$H_{2\phi} = \frac{-jA_0}{\eta} \frac{H_0^{(2)}(ka)}{J_0(ka)} J_0'(k\rho). \tag{6.63}$$

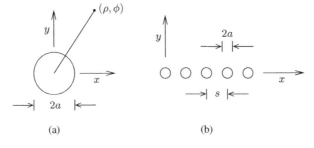

(a)   (b)

Figure 6.5   (a) Perfectly conducting thin wire with radius a. (b) Array of thin wires, with spacing s.

There is no $\mathbf{M}_s$, only a $\mathbf{J}_s$, so E_z is continuous at $\rho = a$. From the boundary conditions at $\rho = a$, a relation between the current strength J_{so} and the arbitrary constant A_0 can be found:

$$J_{so} = \frac{A_0}{\eta}\left[Y_1(ka) - \frac{Y_0(ka)}{J_0(ka)}J_1(ka)\right]. \tag{6.64}$$

The current strength J_{so} (or, equivalently, A_0) is unknown and can be found from the MoM. Since $J_{so} = \text{const.}$, it can be thought of as a pulse basis. The condition $E_z^i + E_z^s = 0$ holds true at $\rho = a$ but also anywhere inside; that is, $\rho < a$. It is convenient to impose this at $(x, y) = (0, 0)$; this is point matching. From (6.59), $E_z^i = 1$. From (6.62) the form of scattered field at $(0, 0)$ is known. Therefore,

$$E_z^i + E_z^s = 1 + A_0\frac{H_0^{(2)}(ka)}{J_0(ka)} = 0 \tag{6.65}$$

which gives the solution $A_0 = -J_0(ka)/H_0^{(2)}(ka)$ and the scattered field for $\rho \geq a$ as

$$E_z^s = -\frac{J_0(ka)}{H_0^{(2)}(ka)}H_0^{(2)}(k\rho). \tag{6.66}$$

6.4.2   Wire Array

It is a small step to extend the previous results to the wire array in Figure 6.5(b). The number of wires is N and they are indexed as $n = 0, 1, 2, \ldots, N - 1$. In this case the match points are at the wire centres $(x_m, y_m) = (ms, 0)$ and each wire radiates a field $E_z^s = A_n H_0^{(2)}(k\rho_n)$ where the A_n are unknown strengths and

$$\rho_n = \sqrt{(x - x_n)^2 + (y - y_n)^2}$$

is the distance from the nth wire at $(x_n, y_n) = (ns, 0)$ to the field point (x, y). Imposing point matching at the wire centres $(x, y) = (x_m, y_m)$,

$$E_z^i + E_z^s = e^{jk(x_m \cos\phi_i + y_m \sin\phi_i)} + \sum_{n=0}^{N-1} A_n H_0^{(2)}(k\rho_{mn})$$

$$= e^{jkms\cos\phi_i} + \sum_{n=0}^{N-1} A_n H_0^{(2)}(k\rho_{mn}) = 0, \tag{6.67}$$

where ρ_{mn} is the distance between the mth match point and nth wire

$$\rho_{mn} = \sqrt{(x_m - x_n)^2 + (y_m - y_n)^2} = |m - n|s.$$

The evaluation of $H_0^{(2)}(k\rho_{mn})$ is easy when $m \neq n$. For $m = n$ extra care is needed because the test point is on the wire's centre and the current is on the wire surface. Such was the case in the previous section when we had only one wire. Therefore, (6.67) becomes

$$e^{jkms\cos\phi_i} + \sum_{n=0}^{N-1} A_n Z_{mn} = 0, \tag{6.68}$$

where

$$Z_{mn} = \begin{cases} H_0^{(2)}(k\rho_{mn}); & m \neq n \\ H_0^{(2)}(ka)/J_0(ka); & m = n \end{cases} \tag{6.69}$$

and $m = 0, 1, \ldots, N-1$. This is an $N \times N$ matrix equation that can be easily solved by standard techniques. It is of the form $V = ZI$ in which $V_m = -e^{jkms\cos\phi_i}$ and $I_n = A_n$. Once the A_n are known, E_z^s readily follows from

$$E_z^s(x, y) = \sum_{n=0}^{N-1} A_n H_0^{(2)}(k\rho_n). \tag{6.70}$$

If only the far field is needed, then at $\phi = \phi_s$,

$$\rho_n \approx \rho - ns\cos\phi_s$$

and using the large-argument Hankel function approximation (C.13) gives

$$E_z^s(\rho, \phi_s) = e^{j\pi/4}\sqrt{\frac{2}{\pi k}} \sum_{n=0}^{N-1} A_n e^{jkns\cos\phi_s} \frac{e^{-jk\rho}}{\sqrt{\rho}}. \tag{6.71}$$

Example 6.2 (Strip Scattering, MoM) Calculate the TM bistatic echo width for a metal strip of width $w = 2$ m with $\lambda = 1$ m. The incident field is a 1 V/m plane wave at an angle of $\phi_i = 90°$. Compare with an approximate wire- grid model. Use 20 wires with a spacing of $s = 0.1$ m and radius 0.0159 m.

Solution: Using PROGRAM mmtmstrip the pulse-basis MoM current in Equation (6.26) can be calculated. By adding up the far-zone radiation from each pulse, one can obtain

$$E_z^s(\rho, \phi_s) = -e^{j\pi/4}\eta\Delta\sqrt{\frac{k}{8\pi}} \sum_{n=1}^{N} a_n e^{jkx_n\cos\phi_s} \frac{e^{-jk\rho}}{\sqrt{\rho}}$$

where a_n are the pulse amplitudes, Δ is the pulse width and x_n are the pulse-basis locations. From (1.92) the echo width is

$$\sigma = \lim_{\rho\to\infty} 2\pi\rho \frac{|\mathbf{E}^s|^2}{|\mathbf{E}^i|^2} = \frac{k}{4}(\eta\Delta)^2 \left|\sum_{n=1}^{N} a_n e^{jkx_n\cos\phi_s}\right|^2.$$

For the wire grid, solving the matrix equation (6.68) with a small computer program gives the coefficients A_n. The scattered field (6.71) can be used to obtain

$$\sigma = \frac{4}{k}\left|\sum_{n=0}^{N-1} A_n e^{jkns\cos\phi_s}\right|^2.$$

It is noted that a_n is a current density in A/m, whereas A_n is an electric field strength in V/m. Had we assumed an incident field with $E_0 \neq 1$ a factor $|E_0|^2$ would be present in the denominators. Therefore, in either case it works out that σ is in metres.

Figure 6.6 shows the echo width from both techniques. The peak at $\phi = 90°$ is the broadside backscatter case, and $\sigma = 13.8$ dB. This is in excellent agreement with the PO prediction of

$$\sigma = 2\pi w^2/\lambda = 25.1 \text{ m} = 14.0 \text{ dB}.$$

■

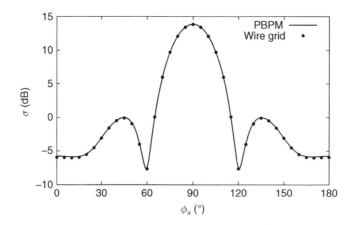

Figure 6.6 TM bistatic echo width for a strip of width $w = 2$ m and $\lambda = 1$ m (reference is 0 dB = 1 m). Plane wave incident at $\phi_i = 90°$. Comparison of pulse-basis point-matching solution ($N = 20$) and wire-grid equivalent ($N = 20$).

In the above example the wire radii a were chosen so that their total surface area is $2\pi a \times 20 = 2$ m. That is, the surface area of the wires is the same as the 2 m strip. In wire-grid models this gives the most accurate result and this is known as the *same surface area rule* (Ozturk and Paknys 2012, Paknys 1991).

6.5 Periodic 2D Wire Array

A periodic 2D wire array resembles the structure shown in Figure 6.5(b). The only new thing now is that there is an infinite number of wires on $-\infty < x < \infty$. The wires are located at $(x, y) = (ns, 0)$, where n is an integer and the spacing is s. Since the structure repeats endlessly with a period s, it is said to be 'periodic'.

In general, the electromagnetic problem of scattering by a periodic structure involves special techniques using Poisson summation and Floquet harmonics. In the following sections we will gain some familiarity with these techniques by analysing the TM scattering by an infinite array of z-directed thin wires.

6.5.1 Poisson Summation

When calculating the scattered field for a periodic array of wires, we encounter a summation of the form

$$I_1 = \sum_{n=-\infty}^{\infty} H_0^{(2)}[k\sqrt{(x - ns)^2 + y^2}]e^{j\alpha ns} \tag{6.72}$$

where α is a real parameter. The convergence of this sum is slow and can be accelerated by using the Poisson summation formula. To accomplish this, we use the following integral representation of the Hankel function

$$H_0^{(2)}(k|\boldsymbol{\rho} - \boldsymbol{\rho}'|) = \frac{1}{\pi} \int_{-\infty}^{\infty} \frac{e^{-j\nu(x-x')}e^{-\gamma|y-y'|}}{-j\gamma} d\nu \tag{6.73}$$

where $\gamma = j\sqrt{k^2 - \nu^2}$. The sign of the square root is chosen so that $\text{Re }\gamma \geq 0$. Using (6.73) in (6.72) gives

$$I_1 = \sum_{n=-\infty}^{\infty} \frac{1}{\pi} \int_{-\infty}^{\infty} \frac{e^{-j\nu(x-ns)}e^{-\gamma|y|}}{-j\gamma} \, d\nu \, e^{j\alpha ns}. \tag{6.74}$$

Next, we use the Poisson summation formula (Papoulis 1962, p. 47)

$$\sum_{m=-\infty}^{\infty} f(t + mT) = \frac{1}{T} \sum_{n=-\infty}^{\infty} e^{jn\omega_0 t} F(n\omega_0),$$

where $\omega_0 T = 2\pi$ and $f(t) \leftrightarrow F(\omega)$ is a Fourier transform pair

$$F(\omega) = \int_{-\infty}^{\infty} f(t)e^{-j\omega t} \, dt; \quad f(t) = \frac{1}{2\pi} \int_{-\infty}^{\infty} F(\omega)e^{j\omega t} \, d\omega.$$

By letting $f(t) = \delta(t)$, $T = 2\pi$ and $t = s\alpha$, we can obtain

$$\frac{1}{2\pi} \sum_{n=-\infty}^{\infty} e^{jns\alpha} = \sum_{m=-\infty}^{\infty} \delta(s\alpha - m2\pi) \tag{6.75}$$

where $s\alpha$ is an arbitrary parameter. Using (6.75) in (6.74) gives

$$I_1 = 2 \int_{-\infty}^{\infty} \sum_{m=-\infty}^{\infty} \delta(s(\nu + \alpha) - m2\pi) \frac{e^{-j\nu x}e^{-\gamma|y|}}{-j\gamma} \, d\nu. \tag{6.76}$$

Integration with the delta function gives

$$\int_{-\infty}^{\infty} \delta(s(\nu + \alpha) - m2\pi)F(\nu) \, d\nu = \frac{1}{s}F(-\alpha + m2\pi/s) \tag{6.77}$$

where F is any function. Using (6.77) in (6.76) yields

$$I_1 = \frac{2}{s} \sum_{m=-\infty}^{\infty} e^{jx(\alpha - m2\pi/s)} \frac{e^{-\gamma|y|}}{-j\gamma} \tag{6.78}$$

where $\gamma = \sqrt{\nu^2 - k^2} = \sqrt{(\alpha - m2\pi/s)^2 - k^2}$. The Poisson summation formula is often useful in accelerating the convergence of a sum. In this instance the expression (6.78) is much more rapidly converging than (6.72).

6.5.2   Scattering Formulation

Returning now to the structure in Figure 6.5(b), a plane wave is incident at an angle ϕ_i:

$$E_z^i(x, y) = E_0 e^{jk(x \cos \phi_i + y \sin \phi_i)}. \tag{6.79}$$

The wavenumber is $k = 2\pi/\lambda$, and λ is the free-space wavelength. The wire radius is a, and it is assumed that the wires are electrically thin and not too close together, so that the current is uniform around each wire. Under these conditions the scattered field can be expressed as[2]

---

[2] For $\rho \geq a$, a thin wire of radius a centred at $(x, y) = (0, 0)$ and carrying J_{sz} will radiate the same field as an infinitesimal filament at $(0, 0)$ carrying $I_0 = 2\pi a J_{sz}$. In the present development it is convenient to use the filament.

$$E_z^s(x, y) = \sum_{n=-\infty}^{\infty} A_n H_0^{(2)}(k R_n). \tag{6.80}$$

Here, $R_n = \sqrt{(x - ns)^2 + y^2}$ is the distance from the nth wire to the field point (x, y).

Floquet's theorem indicates that since the structure is periodic and the excitation is periodic, the wire currents will be periodic, so that

$$A_n = A_0 e^{jkns \cos \phi_i}. \tag{6.81}$$

Using (6.81) in (6.80),

$$E_z^s(x, y) = A_0 \sum_{n=-\infty}^{\infty} e^{jkns \cos \phi_i} H_0^{(2)}(k R_n). \tag{6.82}$$

Since the sum over n is slowly convergent, it is useful to apply the Poisson summation formula (6.78), with the result that

$$E_z^s(x, y) = \frac{j2A_0}{s} \sum_{m=-\infty}^{\infty} e^{jx(k \cos \phi_i - 2m\pi/s)} \frac{e^{-\gamma_m |y|}}{\gamma_m} \tag{6.83}$$

where

$$\gamma_m = \sqrt{(k \cos \phi_i - 2m\pi/s)^2 - k^2}. \tag{6.84}$$

To find A_0 we enforce $\mathbf{E}_{tan} = 0$ at any point (x_p, y_p) on a wire surface

$$E_z^i(x_p, y_p) = -E_z^s(x_p, y_p). \tag{6.85}$$

This amounts to the point-matching MoM, with just one unknown A_0. Using (6.79) and (6.83) in (6.85) leads to

$$A_0 = \frac{j s E_0 e^{jky_p \sin \phi_i}}{2 \sum\limits_{m=-\infty}^{\infty} \frac{1}{\gamma_m} e^{-jx_p(2m\pi/s)} e^{-\gamma_m |y_p|}}. \tag{6.86}$$

For the scattered field (6.83), the terms in the summation over m are called *space harmonics*. Most of them will be evanescent in y, but some may propagate. From the propagation implied by (6.83), we can denote[3]

$$\beta_{x,m} = -k \cos \phi_i + 2m\pi/s. \tag{6.87}$$

With $\gamma_m = j\beta_{y,m}$,

$$\beta_{y,m} = \sqrt{k^2 - \beta_{x,m}^2}. \tag{6.88}$$

Table 6.1 shows the propagation constants of several space harmonics m, for $k = 1$, an incidence angle of $\phi_i = 120°$ and a wire spacing of $s = \lambda/2$. In this instance, only the $m = 0$ space harmonic radiates; all the rest are evanescent in y. Furthermore, the $m = 0$ harmonic has a phase velocity along x that is $u_p = ck/\beta_x = 2c$ which is greater than the speed of light, so this is a fast wave. In general, any fast wave will radiate. Also, any $m \neq 0$ space harmonic that can radiate is called a *grating lobe*.

The $m = 0$ space harmonic is of special interest because its propagation direction corresponds to specular reflection. Since the specular reflection is in the direction $\phi_s = \pi - \phi_i$, we can rewrite the $m = 0$ term of (6.83) as

$$E_z^s(m = 0) = \frac{2A_0}{ks \sin \phi_i} e^{-jkx \cos \phi_s} e^{-jk|y| \sin \phi_s}. \tag{6.89}$$

---

[3] Since $\beta_{x,0} = -k \cos \phi_i$, (6.87) can also be written as $\beta_{x,m} = \beta_{x,0} + 2m\pi/s$. The sign of m does not affect the summations, so replacing $m \rightarrow -m$ gives an equally valid alternative $\beta_{x,m} = \beta_{x,0} - 2m\pi/s$.

Table 6.1 Propagation constants for the space harmonics.

m	$\beta_{x,m}$	γ_m
-2	-3.5	3.354
-1	-1.5	1.118
0	0.5	$j0.866$
1	1.5	1.118
2	4.5	4.388

for $k = 1$, $\phi_i = 120°$ and $s = \pi = \lambda/2$.

From the scattered field we can deduce the reflection coefficient

$$\Gamma = \left.\frac{E_z^s}{E_z^i}\right|_{y=0} = \frac{j\,e^{jky_p\sin\phi_i}}{k\sin\phi_i\,\sum\limits_{m=-\infty}^{\infty}\frac{1}{\gamma_m}e^{-jx_p(2m\pi/s)}e^{-\gamma_m|y_p|}}. \tag{6.90}$$

6.5.3 *Numerical Considerations*

We are free to choose a match point anywhere on a wire surface. With $(x_p, y_p) = (0, a)$, the convergence of (6.86) is monotonic. If $(x_p, y_p) = (a, 0)$, the convergence is oscillatory. Both will give the correct result, though using $(0, a)$ might be preferred.

It is tempting to try avoiding the Poisson summation by calculating the scattered field from (6.80). By replacing the Hankel function with its large-argument approximation and using $R \approx \rho - ns\cos\phi_s$, one obtains

$$E_z^s = \sqrt{\frac{j2}{\pi k\rho}}e^{-jk\rho}A_0\sum_{n=-\infty}^{\infty}e^{jkns(\cos\phi_i+\cos\phi_s)}.$$

It is seen that the summation diverges in the specular direction where $\cos\phi_i = -\cos\phi_s$. So, this approach cannot be used to find the reflection coefficient.

As a point of interest, it is possible to calculate A_0 from the slowly converging partial sum

$$S(N) = \sum_{n=-N}^{N}e^{jkns\cos\phi_i}H_0^{(2)}(k|a - ns|).$$

If S is plotted versus N in the complex S plane, the resulting trajectory is almost a circle. It takes a very large number of terms to achieve convergence; however, the centre of the circle is a good approximation for the converged value.[4]

In this periodic structure, the region $-s/2 \leq x \leq s/2$ occupied by the $n = 0$ wire is called the *unit cell*. If the current in the unit cell is known (in this case, A_0), then the currents on the entire structure (all the other A_n) are automatically known via (6.81); this is an important characteristic of any periodic solution. In the present example the wire is thin so there is only one unknown A_0 in the unit cell. More generally, a unit cell might contain some other more complicated shape, for example, a strip or a triangular cylinder. In that case an MoM expansion using several modes might be needed in the $n = 0$ cell. However,

---

[4] There are other Poisson summation formulas that exclude the $n = 0$ term. They can be found in Gradshteyn and Ryzhik (1980, Section 8.522).

it remains true that once the current in the $n = 0$ cell is known, then the currents on the entire structure are automatically known from Floquet's theorem. This will be further discussed in Chapter 12 when a strip grating on a dielectric slab is solved.

6.6   3D Thin Wire MoM

In this section we will apply the MoM to a straight thin wire. By 'thin' we mean that the wire radius is $a \ll \lambda$. Therefore, the current flows mainly along the wire axis and this axial current does not vary around the wire circumference.

We will consider both plane wave incidence (the scattering problem) and excitation by a voltage source (the antenna problem). Before doing this we need to develop the expression for the near-zone electric field due to the wire current. For a z-directed thin wire, the surface current $\mathbf{J}_s$ and total current I are related by

$$\mathbf{J}_s(z) = \hat{\mathbf{z}} J_z(z) = \hat{\mathbf{z}} \frac{I(z)}{2\pi a}. \tag{6.91}$$

$I(z)$ is the total current in Amps that pierces a $z = $ const. plane.

For our purposes it will suffice to know the near-zone E_z on the wire axis. The wire length is ℓ, and $-\ell/2 \leq z \leq \ell/2$. The configuration is shown in Figure 6.7.

From vector potential theory we can find the exact near field at $\rho = 0$ that is radiated by $\mathbf{J}_s(z)$ on the surface. First, (4.13) can be used to find $A_z(z)$ from J_z. Then, (4.43) gives E_z from A_z. This takes quite a bit of work; the end result (Richmond 1965) is

$$E_z(z) = \int_{z'=-\ell/2}^{\ell/2} \int_{\phi'=0}^{2\pi} J_z(z') g(z, z') a \, d\phi' dz'$$

where $R = \sqrt{a^2 + (z - z')^2}$ and

$$g(z, z') = \frac{-j}{\omega\epsilon} \frac{e^{-jkR}}{4\pi R^5} [(1 + jkR)(2R^2 - 3a^2) + (kaR)^2]. \tag{6.92}$$

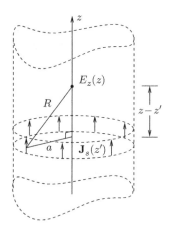

Figure 6.7 $\mathbf{J}_s$ on wire surface and field point on the z axis.

Performing the integration over $d\phi'$,

$$E_z(z) = \int_{z'=-\ell/2}^{\ell/2} 2\pi a J_z(z') g(z, z') \, dz'$$

and using (6.91) for J_z gives

$$E_z(z) = \int_{z'=-\ell/2}^{\ell/2} I(z') g(z, z') \, dz'. \tag{6.93}$$

Finding E_z at $\rho = a$ is more difficult and will not be done here. One issue is that the $d\phi'$ integration no longer benefits from symmetry. Also, the integrand is singular because $R \to 0$ when the integration point coincides with the field point.

6.6.1 The Scattering Problem

We now want to find the scattered field when the wire is illuminated by a θ-polarized plane wave in the $\phi = 0$ plane

$$E_\theta^i = E_0 e^{jk(x \sin \theta_i + z \cos \theta_i)}.$$

On the wire surface we require $\mathbf{E}_{tan} = 0$ so that

$$E_z^i(z) = -E_z^s(z); \quad -\ell/2 \le z \le \ell/2 \tag{6.94}$$

and we note that $E_z^i = -E_\theta^i \sin \theta_i$.

The wire is replaced by a surface S in free space that carries an unknown surface current $J_z(z)$ at $\rho = a$ (currents on the endcaps are neglected in this approximation). The surface-equivalent current on S together with the incident field will generate the original $\mathbf{E}$, $\mathbf{H}$ outside S and $\mathbf{E} = 0$, $\mathbf{H} = 0$ inside S. In other words, the surface current generates the negative incident field inside S and the scattered field outside S.

Instead of imposing $E_z = 0$ on S (at $\rho = a$), we will impose $E_z = 0$ on the wire axis. This alternative is acceptable because $\mathbf{E} = 0$ inside the surface equivalent. It also simplifies the evaluation of the integral for E_z, as mentioned in the previous section.

Using (6.93) for E_z^s, (6.94) becomes

$$E_z^i(z) = -E_z^s(z) = -\int_{z'=-\ell/2}^{\ell/2} I(z') g(z, z') \, dz'. \tag{6.95}$$

Using pulse bases for $I(z)$

$$I(z) = \sum_{n=1}^{N} a_n p(z - z_n) \tag{6.96}$$

and match points at z_m on the wire axis, (6.95) becomes a matrix equation of the form $\mathbf{V} = \mathbf{ZI}$ where row m is

$$V_m = E_z^i(z_m) = \sum_{n=1}^{N} a_n Z_{mn} \tag{6.97}$$

with

$$V_m = -E_0 \sin \theta_i \, e^{jk z_m \cos \theta_i}; \quad I_n = a_n \tag{6.98}$$

and

$$Z_{mn} = -\int_{z'=z_n-\Delta/2}^{z_n+\Delta/2} g(z_m, z') \, dz'. \qquad (6.99)$$

Z_{mn} represents the negative scattered field at z_m due to a unit current pulse at z_n.

6.6.2   A Reciprocal Equivalent

We solved the scattering problem by using pulse bases on S and point matching on the z axis. In some other treatments of this topic, the formulation replaces the surface current by an equivalent filament on the z axis and does the testing on S. This alternative is allowed because both problems have the same E_z. The equivalence will now be shown by using reciprocity. (This equivalent is provided for completeness, but it will not be used in this book.)

In Figure 6.8(a), a current I is uniformly spread out over an annular surface of radius a and height Δ, so it is given by $J_{z1} = I/2\pi a$; at point P_1 at $z = h$, the z component of the electric field is $E_z(P_1)$. In case (b), a dipole of strength $p_{e2} = I\Delta$ produces $E_z(P_2)$ at $\rho = a$. From reciprocity,

$$\int_{z=-\Delta/2}^{\Delta/2} \int_{\phi=0}^{2\pi} J_{z1} E_{z2}(\rho = a) \, a \, d\phi \, dz = p_{e2} E_z(P_1).$$

Substituting $J_{z1} = I/2\pi a$ and $p_{e2} = I\Delta$ gives $E_{z2}(P_2) = E_{z1}(P_1)$. That is, (a) and (b) have the same E_z. Next, we note that a dipole's E_z is the same, both above and below the dipole, so if $p_{e2} = p_{e3}$ then $E_z(P_2) = E_z(P_3)$. Therefore, E_z is the same for all three cases. Comparing cases (a) and (c), we conclude that E_z on the wire axis due to J_z on the wire surface is equivalent to the E_z at $\rho = a$ due to an infinitely thin wire carrying $I = 2\pi a J_z$ on the z axis.

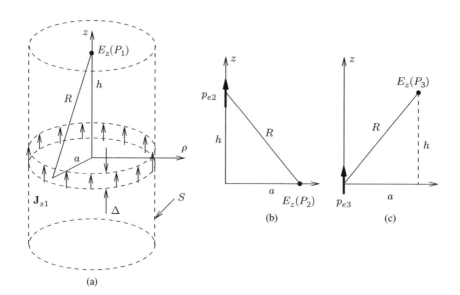

Figure 6.8   (a) $\mathbf{J}_{s1}$ on wire surface; (b) p_{e2} on z axis; (c) equivalent configuration with p_{e3}.

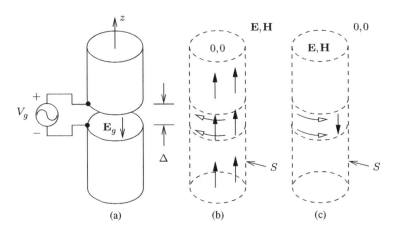

Figure 6.9   (a) Dipole antenna with delta-gap feed. (b) External equivalent, with surface currents on the wire and gap. (c) Internal equivalent, with surface currents on the gap.

6.6.3   The Antenna Problem

A dipole antenna can be modelled with a *delta-gap feed*. This is shown in Figure 6.9(a). A small gap of height Δ is at $z = 0$. A voltage source V_g is connected with the positive terminal at $z = +\Delta/2$ and negative terminal at $z = -\Delta/2$. In this model it is assumed that a uniform z-directed electric field exists in the gap, like in a capacitor. There is also a magnetic field in the gap and it is unknown. The gap electric field is

$$\mathbf{E}_g = -\hat{\mathbf{z}} E_0$$

where $E_0 = V_g/\Delta$. In the antenna problem, we require

$$\mathbf{E}_{tan} = \begin{cases} 0; \text{ on the wire} \\ \mathbf{E}_g; \text{ at the gap.} \end{cases} \tag{6.100}$$

Here, $\mathbf{E}_{tan}$ is the tangential part of the total field that is radiated by the antenna.

The gap surface at $\rho = a$ can be thought of as an aperture antenna. Figure 6.9(b) shows the external equivalent. The electric currents are shown as $\longrightarrow$ and the magnetic currents are $\longrightarrow\!\!\!\!\triangleright$. The internal fields are zero and the external fields are $\mathbf{E}$, $\mathbf{H}$. An electric surface current J_z flows on the antenna and on the gap. A magnetic surface current $M_\phi = -E_0$ is at the gap. Since $\mathbf{E}_g$ is known, M_ϕ is known; however, J_z is not known.

Figure 6.9(c) shows the internal equivalent. The external fields are zero and the internal fields are $\mathbf{E}$, $\mathbf{H}$. A magnetic surface current $M_\phi = +E_0$ is at the gap. As before, M_ϕ is known but J_z is not. (The gap M_ϕ are equal and opposite in cases (b) and (c) but the J_z are not.)

In cases (b) and (c) the exterior/interior fields are found from their sources $\mathbf{J}_s$ and $\mathbf{M}_s$ on S. This is Love's equivalence. We could solve this problem using the exterior equivalent. This would lead to an integral equation of the form

$$\mathbf{E}_{tan}(J_z) + \mathbf{E}_{tan}(M_\phi) = \begin{cases} 0; \text{ on the wire} \\ \mathbf{E}_g; \text{ at the gap.} \end{cases}$$

That is, we would have to find the near-field contributions to $\mathbf{E}_{tan}$ just outside S, from *both* J_z and M_ϕ, which is not so desirable.

A better approach is to make a new equivalent by adding together the sources and fields shown in (b) and (c). The sources from (c) produce no fields outside S, so the combined sources produce the same $\mathbf{E}, \mathbf{H}$ outside S that we had before. Similarly, the combined sources produce the same $\mathbf{E}, \mathbf{H}$ inside S that we had before. In combining the two cases, the M_ϕ contributions cancel, leaving a formulation purely in terms of a new J_z, which is now the sum of the outside and inside equivalent currents

$$J_z = J_z^b + J_z^c$$

which satisfies

$$\mathbf{E}_{tan}(J_z) = \begin{cases} 0; \text{ on the wire} \\ \mathbf{E}_g; \text{ at the gap.} \end{cases}$$

Although we lost the distinction between the inside and outside currents, this new J_z still gives the correct fields everywhere.[5] Since $\mathbf{H} = 0$ inside the dipole, J_z gives the actual physical current on the dipole's metal surface. The interior field comes from $-J_z$.

The total tangential field $\mathbf{E}_{tan}$ produced by the antenna can be found from (6.93). Therefore, (6.100) becomes

$$\int_{z'=-\ell/2}^{\ell/2} I(z')g(z, z')\,dz' = \mathbf{E}_{tan} = \begin{cases} 0; \text{ on the wire} \\ \mathbf{E}_g; \text{ at the gap.} \end{cases} \tag{6.101}$$

At the gap, $\mathbf{E}_g = -\hat{z}V_g/\Delta$. As before, we can solve (6.101) with pulse bases and point matching. The result is again a matrix equation of the form

$$V = ZI$$

where

$$V_m = -E_z(z_m) = \sum_{n=1}^{N} a_n Z_{mn}. \tag{6.102}$$

The previous definitions (6.96) for $I(z)$ and (6.99) for Z_{mn} have been retained, and the excitation vector V is given by

$$V_m = \frac{V_g}{\Delta} \tag{6.103}$$

on the generator segment, and $V_m = 0$ on all other segments. The antenna input impedance can be found from the current $I(0)$ on the generator segment using

$$Z_{in} = \frac{V_g}{I(0)}.$$

An actual physical dipole gap will have nonideal aspects such as fringing fields. The relationship between this reality and an idealized delta gap can be a bit nebulous. In spite of this, the delta-gap model can give surprisingly good results for Z_{in}. Nevertheless, one can rightfully still expect that a model with a closer resemblance to an actual physical feed would give better results. One such case is the well-studied monopole on a ground plane, in Figure 6.10. This is much better defined in physical terms than the delta gap. The cable's centre conductor is extended above the ground to form the monopole. The shield is connected to the ground under the ground plane. On the top side, the cable's dielectric is visible in a small aperture on the ground plane and this aperture electric field can be modelled with a small disc of

---

[5] The aperture equivalent has been described by Wallenberg and Harrington (1969, see Fig. 1c).

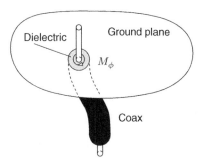

Figure 6.10   Monopole on a ground plane and magnetic frill excitation M_ϕ in the coaxial aperture.

magnetic current, a so-called magnetic frill. With $\hat{\mathbf{n}} = \hat{\mathbf{z}}$ upwards, the magnetic current in the coaxial aperture is $\mathbf{M} = \mathbf{E} \times \hat{\mathbf{z}}$ so that $M_\phi = -E_\rho$; in the figure E_ρ in the aperture is assumed to be radially outward from the monopole so that M_ϕ is negative and hence clockwise directed. Further details of this model can be found in Tsai (1972) or Stutzman and Thiele (2012, Section 14.5).

6.6.4   Numerical Considerations

Z_{mn} has to be found by a numerical integration of (6.99) over z' on segment n. Extra care must be taken when $m = n$ because R can be small and the integrand is nearly singular. The change of variables $z_m - z' = a \tan \theta$ can be helpful.

For a straight wire, the $m + 1$th row of the impedance matrix $\mathbf{Z}$ is the same as the mth row shifted to the right by one. This is known as a *Toeplitz matrix*, and it can be solved more quickly than the general case.

6.7   EFIE and MFIE

There are several different types of integral equations that will now be discussed. They all make use of a surface equivalent, and we will apply it to the perfectly conducting body in Figure 6.11. The impressed source $\mathbf{J}_0$ is a dipole but could just as well be an incident plane wave. This source produces $\mathbf{E}$, $\mathbf{H}$. Surface currents are introduced on S such that they 'turn off' the $\mathbf{E}$, $\mathbf{H}$ inside S. That is, in case (a) they

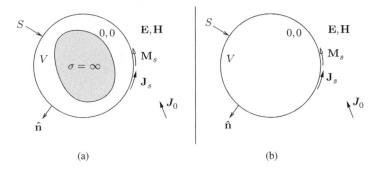

Figure 6.11   (a) Perfect electric conductor surrounded by surface S. (b) Same currents and fields in free space.

radiate nothing outside S and $-\mathbf{E}$, $-\mathbf{H}$ inside S. The surface currents follow from the usual continuity conditions $\mathbf{J}_s = \hat{\mathbf{n}} \times \mathbf{H}$ and $\mathbf{M}_s = \mathbf{E} \times \hat{\mathbf{n}}$. In case (b) the PEC is removed. The fields are unchanged, but the contribution by each current to $\mathbf{E}$, $\mathbf{H}$ will be different from what it was in case (a).

Often, fields are written as incident and scattered parts $\mathbf{E} = \mathbf{E}^i + \mathbf{E}^s$ and $\mathbf{H} = \mathbf{H}^i + \mathbf{H}^s$. In Figure 6.11(a), the impressed source $\mathbf{J}_0$ generates the total $\mathbf{E}$, $\mathbf{H}$. In case (b) the same $\mathbf{J}_0$ generates $\mathbf{E}^i$, $\mathbf{H}^i$ and the currents $\mathbf{J}_s$, $\mathbf{M}_s$ produce the rest. In case (a) there really is no incident field, just the total – and it is incorrect to try and associate $\mathbf{J}_0$ with any kind of 'incident' $\mathbf{E}^i$, $\mathbf{H}^i$.

We can shrink S down to the metal surface. Since $\mathbf{E}_{tan} = 0$, it automatically makes $\mathbf{M}_s = 0$. Then we only need to worry about the radiation from $\mathbf{J}_s$. If we are in free space as in Figure 6.11(b), the fields can be found from (4.13) and (4.21)–(4.23). These can be compactly expressed with operators $\mathcal{L}$ and $\mathcal{K}$ such that[6]

$$\mathbf{E}(\mathbf{r}) = \mathcal{L}(\mathbf{J}_s) \tag{6.104}$$

$$\mathbf{H}(\mathbf{r}) = \mathcal{K}(\mathbf{J}_s) \tag{6.105}$$

where

$$\mathcal{L}(\mathbf{J}_s) = \left(-j\omega\mu + \frac{\nabla\nabla\cdot}{j\omega\epsilon} \right) \int_S \mathbf{J}_s(\mathbf{r}') \frac{e^{-jk|\mathbf{r}-\mathbf{r}'|}}{4\pi|\mathbf{r}-\mathbf{r}'|} dS' \tag{6.106}$$

$$\mathcal{K}(\mathbf{J}_s) = \nabla \times \int_S \mathbf{J}_s(\mathbf{r}') \frac{e^{-jk|\mathbf{r}-\mathbf{r}'|}}{4\pi|\mathbf{r}-\mathbf{r}'|} dS'. \tag{6.107}$$

To get an integral equation, we will use the free-space equivalent of Figure 6.11(b). Since the fields are zero inside S and $\mathbf{E}$, $\mathbf{H}$ outside,

$$\left. \begin{array}{l} \text{outside } S; \ \mathbf{E} \\ \text{inside } S; \ 0 \end{array} \right\} = \mathbf{E}^i + \mathcal{L}(\mathbf{J}_s) \tag{6.108}$$

and

$$\left. \begin{array}{l} \text{outside } S; \ \mathbf{H} \\ \text{inside } S; \ 0 \end{array} \right\} = \mathbf{H}^i + \mathcal{K}(\mathbf{J}_s). \tag{6.109}$$

In the MoM, $\mathbf{J}_s$ will be an unknown. The boundary conditions are to be enforced on S, so $\mathbf{r} \in S$. They may involve $\mathbf{E}_{tan}$ and/or $\mathbf{H}_{tan}$ but never the normal components, so we should write (6.108), (6.109) as

$$\left. \begin{array}{l} \mathbf{r} \in S^+; \ \hat{\mathbf{n}} \times \mathbf{E} \\ \mathbf{r} \in S^-; \ 0 \end{array} \right\} = \hat{\mathbf{n}} \times \mathbf{E}^i + \hat{\mathbf{n}} \times \mathcal{L}(\mathbf{J}_s) \tag{6.110}$$

and

$$\left. \begin{array}{l} \mathbf{r} \in S^+; \ \hat{\mathbf{n}} \times \mathbf{H} \\ \mathbf{r} \in S^-; \ 0 \end{array} \right\} = \hat{\mathbf{n}} \times \mathbf{H}^i + \hat{\mathbf{n}} \times \mathcal{K}(\mathbf{J}_s). \tag{6.111}$$

In the above, we are on the surface so 'inside S' has become the inner face of S (call it S^-) and 'outside S' has become the outer face (call it S^+). The outer surface has to be far enough 'out' so that the surface currents are on the inside. Then,

$$-\hat{\mathbf{n}} \times \mathbf{E}^i(\mathbf{r}) = \hat{\mathbf{n}} \times \mathcal{L}(\mathbf{J}_s); \ \mathbf{r} \in S^- \tag{6.112}$$

---

[6] The fields from magnetic currents are readily obtained by applying duality to (6.104) and (6.105), that is, $\mathbf{H}(\mathbf{r}) = \mathcal{L}(\mathbf{M}_s)$ and $-\mathbf{E}(\mathbf{r}) = \mathcal{K}(\mathbf{M}_s)$. In addition, the operator $\mathcal{L}$ requires the interchange $\mu \leftrightarrow \epsilon$.

and

$$-\hat{\mathbf{n}} \times \mathbf{H}^i(\mathbf{r}) = \hat{\mathbf{n}} \times \mathcal{K}(\mathbf{J}_s); \mathbf{r} \in S^-. \tag{6.113}$$

If we expand out (6.112), it becomes

$$-\hat{\mathbf{n}} \times \mathbf{E}^i(\mathbf{r}) = \hat{\mathbf{n}} \times \left(-j\omega\mu + \frac{\nabla\nabla\cdot}{j\omega\epsilon}\right) \int_S \mathbf{J}_s(\mathbf{r}') \frac{e^{-jk|\mathbf{r}-\mathbf{r}'|}}{4\pi|\mathbf{r}-\mathbf{r}'|} dS'; \mathbf{r} \in S. \tag{6.114}$$

There is no $\mathbf{M}_s$ on S so $\mathbf{E}_{tan}$ is continuous across the surface and it doesn't really matter if we use S^-, S or S^+ for the domain of $\mathbf{r}$.

The result (6.114) is called the *electric field integral equation*, or EFIE. It amounts to enforcing the electric field boundary condition $-\mathbf{E}^i_{tan}(\mathbf{r}) = \mathbf{E}^s_{tan}(\mathbf{r})$ on the surface S. In the MoM, one can then expand $\mathbf{J}_s$ with basis functions and perform testing on the surface, that is, for field points at $\mathbf{r} \in S$. We have been using the EFIE all along, for the MoM formulations developed in earlier sections. Because the field point $\mathbf{r}$ is in the null field region, it is sometimes called the *null field integral equation*. The cancellation of incident and scattered parts to produce a null field is referred to as the *extinction theorem*.

Similarly, we can expand out (6.113) which becomes

$$-\hat{\mathbf{n}} \times \mathbf{H}^i(\mathbf{r}) = \hat{\mathbf{n}} \times \nabla \times \int_{S^+} \mathbf{J}_s(\mathbf{r}') \frac{e^{-jk|\mathbf{r}-\mathbf{r}'|}}{4\pi|\mathbf{r}-\mathbf{r}'|} dS'; \mathbf{r} \in S^-. \tag{6.115}$$

There is a $\mathbf{J}_s$ on S, so $\mathbf{H}_{tan}$ will be discontinuous across the surface. The integral equation is based on having a null field in V, so it is crucial for the field point to be inside, that is, that $\mathbf{r} \in S^-$. Equation (6.115) is called the *magnetic field integral equation*, or MFIE. It amounts to enforcing the magnetic field boundary condition $-\mathbf{H}^i_{tan}(\mathbf{r}) = \mathbf{H}^s_{tan}(\mathbf{r})$ on the surface S^-.

An alternative form of the MFIE is (Chew et al. 2009, p. 56)

$$-\hat{\mathbf{n}} \times \mathbf{H}^i(\mathbf{r}) = -\frac{1}{2}\mathbf{J}_s(\mathbf{r}) + \hat{\mathbf{n}} \times PV \int_S \nabla \frac{e^{-jk|\mathbf{r}-\mathbf{r}'|}}{4\pi|\mathbf{r}-\mathbf{r}'|} \times \mathbf{J}_s(\mathbf{r}') dS'; \mathbf{r} \in S \tag{6.116}$$

and the integral is taken in the principal value sense. The integral has the same value on S^+ and S^- so the distinction does not matter and S can be used. The unknown $\mathbf{J}_s$ appears both inside and outside the integral so this is called a Fredholm integral equation of the 'second kind'. The EFIE has the unknown $\mathbf{J}_s$ only inside the integral, so it is of the 'first kind'.

The EFIE and MFIE involve differentiations and singular integrals. There are still many steps between these equations and having a working computer program. Reference books on these topics are mentioned at the end of the chapter.

The EFIE is suitable for both closed and open surfaces, for example, a sphere or a disc. In our earlier 2D example of strip scattering, it was seen that with the EFIE, the MoM currents on the top and bottom sides of the strip add up. This gives half the number of unknowns that we had before. It also prevents us from knowing $\mathbf{J}_s$ on the top and bottom surfaces separately.

A well-known and important limitation of the MFIE is that it only works for closed surfaces, for example, spheres, cubes and so on. It cannot be used with an open surface, for example, a disc or plate. The MFIE depends on having a null field on the inside surface S^-. If the volume takes the form of, say a cuboid, the impedance matrix will become ill conditioned as the cuboid's thickness shrinks.

6.8 Internal Resonances

The EFIE fails for closed bodies that are resonant. For example, this happens when using the EFIE to calculate the plane wave scattering by a PEC sphere; call this the 'source-excited solution'. It does not matter if the sphere is hollow or solid – the surface currents and scattered field from the MoM will be the same in either case. However, at some discrete frequencies there will be a second source-free solution, corresponding to a resonant mode of a hollow sphere. This, superimposed on the original source-excited solution, will make the total incorrect. In theory this is a cavity mode, so it should not contribute radiation. However, the minor imperfections of an MoM mesh will lead to some weakly radiating currents. Also, the mode will get excited by the incident wave because of minor leakage through the mesh itself.

One can use the EFIE for a closed body and simply 'skip over' the problematic resonances. However, if the body is electrically large, then the resonances will become close together in frequency and hard to avoid.

The MFIE also has internal resonances but they are of a different nature. One remedy to the resonance problem is to use the *combined field integral equation* (CFIE)

$$\text{CFIE} = \alpha\text{EFIE} + \eta(1 - \alpha)\text{MFIE}. \tag{6.117}$$

α is usually around 0.5 and $\eta = \sqrt{\mu/\epsilon}$. The EFIE and MFIE have real resonant frequencies, but when combined as in (6.117), they become complex and are thus not encountered. As a result, the internal resonance problem does not occur.

6.9 PMCHWT Formulation

It is possible to formulate a surface integral equation for a homogeneous dielectric. The method is named after its inventors Poggio, Miller, Chang, Harrington, Wu and Tsai, and a good account can be found in Mautz and Harrington (1983).

Figure 6.12(a) shows a dielectric body illuminated by an impressed source $\mathbf{J}_0$. Two surface equivalents are developed. In case (b), $\mathbf{J}_s = \hat{\mathbf{n}} \times \mathbf{H}$, $\mathbf{M}_s = \mathbf{E} \times \hat{\mathbf{n}}$, plus $\mathbf{J}_0$ generate $\mathbf{E}, \mathbf{H}$ outside S and the null fields inside S. The fields are also zero on S^-, the inner face of S.

Case (c) has $\mathbf{E}, \mathbf{H}$ inside S and the null fields outside S. The fields are also zero on S^+, the outer face of S. Since the normal for case (c) is reversed and $\mathbf{E}_{tan}$, $\mathbf{H}_{tan}$ are continuous at the boundary, the surface currents $\mathbf{J}_s$ and $\mathbf{M}_s$ turn out to be the same as in case (b).

Applying the extinction theorem to the inside for case (b),

$$- \mathbf{E}^i(\mathbf{r}) = \mathcal{L}_{1E}^+(\mathbf{J}_s) + \mathcal{K}_{1E}^+(\mathbf{M}_s); \, \mathbf{r} \in S^-, \mathbf{r}' \in S^+ \tag{6.118}$$

and for the outside in case (c),

$$0 = \mathcal{L}_{2E}^-(\mathbf{J}_s) + \mathcal{K}_{2E}^-(\mathbf{M}_s); \, \mathbf{r} \in S^+, \mathbf{r}' \in S^-. \tag{6.119}$$

In the above notation, $\mathcal{L}_{1E}^+$ means medium 1, the E field, due to the current on S^+, with similar interpretations for the other terms. Only the tangential parts of these equations are needed so they should be used

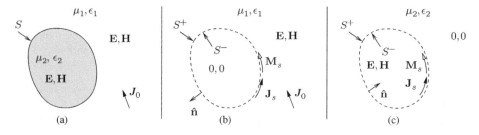

Figure 6.12   (a) Dielectric body and source $\mathbf{J}_0$. (b) External equivalent, μ_1, ϵ_1 everywhere. Fields are zero on S^-. (c) Internal equivalent, μ_2, ϵ_2 everywhere. Fields are zero on S^+.

with an overall $\hat{\mathbf{n}} \times$ factor. Equations (6.118) and (6.119) provide two equations, which can be solved for the two unknowns $\mathbf{J}_s$ and $\mathbf{M}_s$. Since they are based on extinction of the electric field, it is an EFIE.

By applying duality to (6.118) and (6.119) we obtain

$$- \mathbf{H}^i(\mathbf{r}) = \mathcal{L}^+_{1H}(\mathbf{M}_s) + \mathcal{K}^+_{1H}(\mathbf{J}_s); \, \mathbf{r} \in S^-, \mathbf{r}' \in S^+ \qquad (6.120)$$

and

$$0 = \mathcal{L}^-_{2H}(\mathbf{M}_s) + \mathcal{K}^-_{2H}(\mathbf{J}_s); \, \mathbf{r} \in S^+, \mathbf{r}' \in S^-. \qquad (6.121)$$

These provide an MFIE that is based on the extinction of the magnetic field.

The EFIE and MFIE are usable as is, but they do have the problem of failing at the internal resonances of a closed body S that is filled with a material having μ_1, ϵ_1. This problem can be solved. Adding (6.118) and (6.119) gives

$$- \mathbf{E}^i(\mathbf{r}) = \mathcal{L}^+_{1E}(\mathbf{J}_s) + \mathcal{L}^-_{2E}(\mathbf{J}_s) + \mathcal{K}^+_{1E}(\mathbf{M}_s) + \mathcal{K}^-_{2E}(\mathbf{M}_s); \, \mathbf{r} \in S^-. \qquad (6.122)$$

Adding (6.120) and (6.121) gives

$$- \mathbf{H}^i(\mathbf{r}) = \mathcal{L}^+_{1H}(\mathbf{M}_s) + \mathcal{L}^-_{2H}(\mathbf{M}_s) + \mathcal{K}^+_{1H}(\mathbf{J}_s) + \mathcal{K}^-_{2H}(\mathbf{J}_s); \, \mathbf{r} \in S^-. \qquad (6.123)$$

This provides two equations (6.122) and (6.123) which can be solved for the two unknowns $\mathbf{J}_s$ and $\mathbf{M}_s$. This forces the extinction of *both* $\mathbf{E}$ and $\mathbf{H}$ in the null field region and is free of the internal resonance problem. It is known as the PMCHWT formulation.

In the above, we have added the equations in equal amounts. It is possible to use different weights and arrive at the *Muller formulation* (Chew et al. 2009, Section 4.2). The PMCHWT method is better suited for a high-contrast material, and the Muller formulation is better for low contrasts.

6.10 Basis Functions

So far we have used pulse bases for electric currents and point matching to solve simple MoM problems. More generally, basis functions are also used to represent magnetic currents. These are needed, for

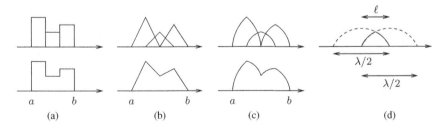

Figure 6.13 Different types of subdomain basis functions for wires: (a) pulse, (b) triangle, (c) piecewise sinusoidal. (d) One piecewise sinusoid.

instance, for the dielectric problems in Section 6.9 and can be useful for aperture problems as well. Going beyond pulse bases, there are numerous other possibilities and a few of these will now be reviewed.

6.10.1 Wires

Some possibilities for wire basis functions are shown in Figure 6.13. Case (a) shows the pulse basis that we used in Section 6.6. Case (b) shows triangular functions. They add up as shown in the lower part of the figure. They are *interpolatory*, as at any given peak, all other triangles are zero. Case (c) shows *piecewise sinusoidal* (PWS) functions. They are formed from portions of two half-wave sinusoids; see (d). If the segment length is exactly $\ell = \lambda/2$, then the PWS will be smooth. Otherwise, if $\ell < \lambda/2$, there will be a kink, as shown in (d).

All of the cases shown in Figure 6.13 are called *subdomain basis functions*. Each basis function is zero over all but a small part of the domain (a, b). One could use a Fourier series approach whereby each basis function (mode) is nonzero over (a, b). These are called *entire domain basis functions*.

Triangular functions with the Galerkin method are well suited for implementation in general-purpose computer codes. On the other hand, piecewise sinusoids have an advantage in that their near field can be obtained in closed form, making numerical integration unnecessary. When the segment length is small, the triangle and piecewise sinusoid currents look very similar.

6.10.2 Surfaces

For modelling the surface currents on complex-shaped and highly realistic objects, the most widely used MoM basis function is the so-called RWG basis, named after Rao et al. (1982). It provides a piecewise-linear approximation of a surface current, over a triangular domain.

The ideas behind the RWG basis function are illustrated in Figure 6.14. The objective is to represent a vector surface current $\mathbf{J}_s(u, v)$ over a 2D domain (u, v). If this were a scalar function $f(u, v)$, it would be straightforward to represent f as a mesh of triangles. However, the goal here is to represent a *vector* function and that is a little more complicated.

Figure 6.14(a) shows one basis function. It is important to recognize (but hard to draw) that in general none of the triangles need be coplanar. The triangles can conform to an arbitrary-shaped physical surface such as a plate, sphere and so forth (the domain). Within that domain, the surface current is represented in a piecewise-linear fashion; it can be thought of as a generalization of the triangular basis function on a wire. The vector surface current emanates from node 1 and linearly increases with distance until

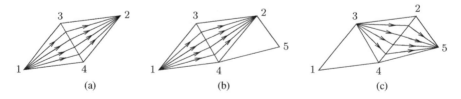

Figure 6.14   (a) One RWG basis function covers a domain with two triangles. (b) Three triangles showing current across edge 3–4. (c) Three triangles showing current across edge 2–4.

the edge defined by nodes 3 and 4; call it edge 3–4. It then linearly decreases towards node 2, where it becomes zero. Therefore, its value at edge 3–4 completely determines the current, and everywhere else it follows a linear interpolation, in a rooftop fashion. Notice also that there is no current flow across edges 1–3, 1–4, 3–2 and 4–2. An RWG function is of the subdomain type, as it is zero outside the triangle pair.

Figures 6.14(b) and (c) show a mesh of five nodes having two basis functions. In (b), a basis function covers the triangle pair 1–3–4 and 2–3–4. In (c), a second basis function covers triangles 2–3–4 and 2–4–5. The two basis functions overlap in triangle 2–3–4, where the total current comes from the vector sum of (b) and (c).

In Figure 6.14(a) the normal part of the surface current (the part that crosses edge 3–4) is continuous. The tangential part (the part that is parallel to edge 3–4) is not necessarily continuous. Because of these properties, the divergence is finite and the RWG function is called a *divergence-conforming* basis function. Unlike pulse bases, there is no step discontinuity and consequently no charge accumulation at the junction between triangles at 3–4. This is a desirable property which leads to improved accuracy when using the EFIE (Peterson et al. 1998, Section 9.13).

The evaluation of self and mutual impedances is usually done with Gaussian quadrature. Self-impedance terms require a careful treatment of singular integrals (Khayat and Wilton 2005). The implementation details are beyond the scope of this discussion, and the reader is referred to references at the end of the chapter for further information.

6.10.3   Volumes

So far we have considered wire and surface integral equations. It is also possible to formulate a volume integral equation. These are based on a volume equivalent as discussed in Section 4.11 with unknown volume currents $\mathbf{J}, \mathbf{M}$. If the material is non-magnetic, then there is only $\mathbf{J}$. The volume currents can be approximated in a staircase manner using cuboids. This is very good for inhomogeneous media. A disadvantage is that the number of unknowns is much larger than for a surface integral equation; the latter is preferred for homogeneous and piecewise homogeneous media. Again, the reader is referred to the references at the end of the chapter for further details.

6.11   Further Reading

The book by Peterson et al. (1998) covers the MoM and the finite element method and considers a wide range of theoretical and computational issues. Kolundzija and Djordjevic (2002) describe modern MoM implementations for wires, plates and dielectrics. Chew et al. (2009) is more on the theoretical side, yet very readable.

A collection of key MoM papers and a historical perspective to 1992 can be found in Miller et al. (1992). The paper by Jakobus (2000) compactly describes surface and volume integral equations for conducting, dielectric and magnetic bodies, along with some numerical considerations. Harrington (1989) reviews surface integral equations and resonance issues. At an introductory level, the articles by Newman (1988) and Wheless and Wurtz (1995) are helpful.

So-called fast methods use iterative solvers for very large matrix equations involving millions of unknowns to model structures such as aircraft at microwave frequencies. The multilevel fast multipole method (MLFMM) is a central and indispensable technique used in commercial codes today. The intricate details can be found in the books by Chew et al. (2001), Ergül and Gürel (2014) and other sources.

References

Chew WC, Michielssen E, Song JM and Jin JM (2001) *Fast and Efficient Algorithms in Computational Electromagnetics*. Artech House.

Chew WC, Tong MS and Hu B (2009) *Integral Equation Methods for Electromagnetic and Elastic Waves*. Morgan & Claypool.

Ergül Ö and Gürel L (2014) *The Multilevel Fast Multipole Algorithm (MLFMA) for Solving Large-Scale Computational Electromagnetics Problems*. John Wiley & Sons, Inc.

Gradshteyn IS and Ryzhik IM (1980) *Table of Integrals, Series, and Products*. Academic Press.

Harrington RF (1989) Boundary integral formulations for homogeneous material bodies. *J. Electromagn. Waves Appl.* **3**(1), 1–15.

Jakobus U (2000) Comparison of different techniques for the treatment of lossy dielectric/magnetic bodies within the method of moments formulation. *AEU* **54**, 1–11.

Khayat MA and Wilton DR (2005) Numerical evaluation of singular and near-singular potential integrals. *IEEE Trans. Antennas Propag.* **53**, 3180–3190.

Kolundzija BM and Djordjevic AR (2002) *Electromagnetic Modeling of Composite Metallic and Dielectric Structures*. Artech House.

Ma J, Rokhlin V and Wandzura S (1996) Generalized Gaussian quadrature rules for systems of arbitrary functions. *SIAM J. Numer. Anal.* **33**(3), 971–996.

(ed. Miller EK, Medgyesi-Mitschang L and Newman EH) (1992) *Computational Electromagnetics: Frequency-Domain Method of Moments*. IEEE Press.

Mautz J and Harrington RF (1983) Electromagnetic scattering from a homogeneous material body of revolution. *AEU* **33**, 71–80.

Newman EH (1988) Simple examples of the method of moments in electromagnetics. *IEEE Trans. Educ.* **31**(3), 193–200.

Ozturk AK and Paknys R (2012) Analysis of propagation between rows of conducting cylinders that model solid surfaces using the same surface area rule. *IEEE Trans. Antennas Propag.* **AP-60**, 2602–2606.

Paknys R (1991) The near field of a wire grid model. *IEEE Trans. Antennas Propag.* **AP-39**, 994–999.

Papoulis A (1962) *The Fourier Integral and its Applications*. McGraw-Hill.

Peterson AF, Ray SL and Mittra R (1998) *Computational Methods for Electromagnetics*. IEEE Press.

Rao SM, Wilton DR and Glisson AW (1982) Electromagnetic scattering by surfaces of arbitrary shape. *IEEE Trans. Antennas Propag.* **AP-30**(3), 409–418.

Richmond JH (1965) Digital computer solutions of the rigorous equations for scattering problems. *Proc. IEEE* **53**, 796–804.

Stutzman WL and Thiele GA (2012) *Antenna Theory and Design*. John Wiley & Sons, Inc.

Tsai L (1972) A numerical solution for the near and far fields of an annular ring of magnetic current. *IEEE Trans. Antennas Propag.* **AP-20**(5), 569–576.

Wallenberg RF and Harrington RF (1969) Radiation from apertures in conducting cylinders of arbitrary cross section. *IEEE Trans. Antennas Propag.* **AP-17**(1), 56–62.

Wheless WP and Wurtz LT (1995) Introducing undergraduates to the moment method. *IEEE Trans. Educ.* **38**(4), 193–200.

Problems

6.1 PROGRAM mmtmstrip calculates J_z on a perfectly conducting strip that is illuminated by a TM plane wave at an angle ϕ_i.

    (a) Check the program by calculating J_z on a strip of width 2λ and comparing with Figure 6.4.

    (b) Show that the scattered field can be expressed as

$$E_z^s(\phi^s) = -e^{j\pi/4}\eta\Delta\sqrt{\frac{k}{8\pi}}\sum_{n=1}^{N} a_n e^{jkx_n\cos\phi_s}\frac{e^{-jk\rho}}{\sqrt{\rho}},$$

where x_n are the pulse-basis locations.

(c) Modify the program so that it computes the bistatic echo width where $0 \leq \phi_s \leq 180°$ and $\phi_i = 90°$. Calculate the echo width for a strip of width 2λ. Check your result by comparing with Figure 6.6.

6.2 Repeat Problem 6.1, but this time calculate the monostatic echo width $(\phi = \phi_i = \phi_s)$ where $0 \leq \phi \leq 180°$. Note: The voltage column V in $ZI = V$ changes for every incidence angle, but the L and U factors for $Z = LU$ do not change. So, find the L, U factors only once. Then for other incident angles, recalculate V, but not L, U. Further explanations regarding the LU method are given in Appendix G.

6.3 For the strip in Problem 6.1, the computed surface current J_z is the sum of the lit-side and shadow-side currents. We want to separate them out. This can be done because from J_z and the incident field, we can find the true magnetic field at $y = 0^+$ and $y = 0^-$. From the true magnetic field, we can then find $\mathbf{J}_s$ on both sides of a thick plate.[7] The true surface current on the top surface is

$$\mathbf{J}_s(y = 0^+) = \hat{\mathbf{y}} \times (\mathbf{H}^i - \hat{\mathbf{x}}J_z/2) \qquad (1)$$

and on the bottom surface is

$$\mathbf{J}_s(y = 0^-) = -\hat{\mathbf{y}} \times (\mathbf{H}^i + \hat{\mathbf{x}}J_z/2). \qquad (2)$$

With an incidence angle of $\phi_i = 90°$ and a strip width of 2λ, compute and plot the lit-side and shadow-side surface currents.

6.4 PROGRAM mmtestrip calculates J_x on a perfectly conducting strip that is illuminated by a TE plane wave at an angle ϕ_i.

(a) Using a strip width of 2λ, test the program by calculating J_x and comparing with Figure 6.4.

(b) Show that the scattered field can be expressed as

$$H_z^s(\phi) = e^{j\pi/4}\Delta\sqrt{\frac{k}{8\pi}} \sin\phi \sum_{n=1}^{N} a_n e^{jkx_n \cos\phi} \frac{e^{-jk\rho}}{\sqrt{\rho}}$$

where x_n are the pulse-basis locations.

(c) Modify the program so that it computes the bistatic echo width where $0 \leq \phi_s \leq 180°$ and $\phi_i = 90°$. Calculate the echo width for a strip of width $a = 2\lambda$. Check your result at broadside by comparing with the PO result $\sigma = 2\pi a^2/\lambda$.

6.5 Repeat Problem 6.4, but this time calculate the monostatic echo width $(\phi = \phi_i = \phi_s)$ where $0 \leq \phi \leq 180°$. Apply the considerations regarding the $Z = LU$ factorization, as discussed in Problem 6.2.

6.6 Modify PROGRAM mmtestrip to find the radiation pattern due to a unit strength magnetic line source at $(x, y) = (0, 0)$. This is done by setting the voltage column to $V = -1/(2\Delta)$ at the source location and $V = 0$ elsewhere. Using a strip width of 2λ and $\lambda = 1$ m, plot $20 \log |\eta H_z|$ for $0 \leq \phi \leq 360°$. Check your radiation pattern level at $\phi = 90°$ by comparing it with a magnetic line source of strength $M_0 = 1$ V that is on an infinite ground plane.

---

[7] This can be proven by showing that the total surface current equals the discontinuity of $\mathbf{H}^s$ across the current sheet. It is interesting to note that if we use $\mathbf{J}_s$ from GO, Equation (1) gives us $\mathbf{J}_{s,po}$ on the lit side, and Equation (2) gives $\mathbf{J}_s = 0$ on the shadow side.

6.7 From Section 6.3 the TE scattered electric field on a 2D perfectly conducting strip can be written as

$$E_x^s(x) = \frac{-k\eta}{8} \int_{x'=-a/2}^{a/2} J_x(x')[H_0^{(2)}(k|x-x'|) + H_2^{(2)}(k|x-x'|)]\,dx'$$

or as

$$E_x^s(x) = \frac{-\eta}{4k} \int_{x'=-a/2}^{a/2} J_x(x')\left[\frac{d^2}{dx^2} + k^2\right] H_0^{(2)}(k|x-x'|)\,dx'.$$

Show that these are equivalent. Hint: replace H_0'' with an equivalent expression in terms of H_0 and H_2.

6.8 PROGRAM wire calculates the terminal impedance Z_{in} and current $I(z)$ on a thin wire dipole that has a 1 V generator at its centre.

(a) Use the program to calculate Z_{in}, and plot $|I(z)|$ for a wire that has a radius $a = 0.005\lambda$ and length $\ell = 0.47\lambda$.

(b) Modify the program to calculate the input impedance as a function of length, over $0.1\lambda \le \ell \le 1\lambda$; plot the result.

6.9 Show that the radiation pattern of a thin wire with pulse bases is given by

$$E_\theta(\theta) = \frac{jk\eta\Delta}{4\pi}\sin\theta\frac{e^{-jkr}}{r}\sum_{n=1}^{N} a_n e^{jkz_n\cos\theta}$$

where Δ are the pulse widths, a_n are the pulse amplitudes and z_n are the pulse locations. Use this to calculate and plot the radiation pattern of a dipole antenna that has a radius $a = 0.005\lambda$ and length $\ell = 0.47\lambda$.

6.10 Calculate the backscattering cross section for a thin wire versus the wire length. Assume broadside incidence, $\lambda = 1$ m, a wire radius of 0.005λ and length from 0.1λ to 1.0λ. Solve the problem by making suitable modifications to PROGRAM wire. Check your program by comparing with published results, for example, Richmond (1965).

6.11 Calculate the near-zone scattered field along a z-directed straight thin wire, at $\rho = a$. Assume a broadside incident plane wave of 1 V/m, a wavelength of 1 m, a wire radius of 0.005λ and a wire length of 0.5λ. Solve the problem by making suitable modifications to PROGRAM wire. You will need to do a numerical integration of the current, for each near-field point. Plot both $\text{Re}(E_z^i + E_z^s)$ and $\text{Im}(E_z^i + E_z^s)$ for $-\ell/2 \le z \le \ell/2$. Check your result. At the match points you should see $E_z^i + E_z^s = 0$.

6.12 Calculate the mutual impedance between two z-directed colinear dipole antennas. Assume that each dipole has a wire radius of 0.005λ and length of 0.47λ. One dipole has its terminals at $(0,0,0)$ and the other one is at $(0,0,s_d)$.

Solve the problem by making suitable modifications to PROGRAM wire. Treating the pair of dipoles as a two-port network, first find the currents with a voltage source at Port 1 and a short circuit at Port 2. From the antenna currents you can get the Y matrix. From Y obtain the Z matrix and hence Z_{21}.

Compute and plot R_{21} and X_{21} for $0.5 \le s_d/\lambda \le 3.0$. Check your program by comparing with published results, for example, Stutzman and Thiele (2012, Section 8.7).

6.13 A plane wave $\mathbf{E}^i = \hat{\mathbf{z}}e^{-jkx}$ is incident on a perfectly conducting thin wire that has its endpoints at $(0, 0, -d)$ and $(0, 0, d)$. The wire radius is $a = 0.001\lambda$ and the wire length is $\ell = 2d = 0.1\lambda$. The frequency is 300 MHz. We want to find the scattered field by using the Galerkin MoM with piecewise sinusoidal basis functions; see Newman (1988).

The nth current mode is a piecewise sinusoid. The endpoints are at z_n and z_{n+2}; the centre is at z_{n+1}. It is given by the function

$$F_n(z) = \frac{\sin k(d - |z - z_{n+1}|)}{\sin kd}$$

and the near-zone radiated field of $I(z) = F_n(z)$ is

$$E_z(\rho, z) = G_n(\rho, z) = \frac{30}{\sin kd}\left[-j\frac{e^{-jkr_1}}{r_1} - j\frac{e^{-jkr_2}}{r_2} + j2\cos kd\frac{e^{-jkr_0}}{r_0}\right]$$

where $r_0 = \sqrt{\rho^2 + z^2}$, $r_1 = \sqrt{\rho^2 + (z - d)^2}$ and $r_2 = \sqrt{\rho^2 + (z + d)^2}$. The mutual impedance is generally $Z_{mn} = -\int_m G_n F_m dz$ and for $d \ll \lambda$, self-terms can be approximated by $Z_{mm} \approx 20(kd)^2 + j(120/kd)(1 - \ln(d/a))$.

Using just one basis function, use the MoM to find the wire current and the scattered field at a point $(x, 0, 0)$. Also find the backscatter cross section σ.

6.14 In the previous problem, the excitation is changed to a voltage source at the midpoint of the dipole. The corresponding incident field, assuming a delta-gap generator, is $\mathbf{E}^i = \hat{\mathbf{z}}V_0\delta(z)$ and the voltage column is $V_m = \int_m \mathbf{E}^i \cdot \hat{\mathbf{z}} F_m dz$.
(a) Find the dipole's input impedance.
(b) For a 1 V source, find the far-zone electric field.

6.15 In Section 6.4.2 we formulated the scattering solution for an array of wires along the x axis. It is not that hard to generalize this to wires at arbitrary positions (x_n, y_n). The expressions

$$\rho_n = \sqrt{(x - x_n)^2 + (y - y_n)^2}$$

and

$$\rho_{mn} = \sqrt{(x_m - x_n)^2 + (y_m - y_n)^2}$$

remain true.

Use the MoM to find the echo width for a cylinder of radius b that is modelled by an array of wires at $(x_n, y_n) = (b\cos\phi_n, b\sin\phi_n)$. The wires are at angular positions ϕ_n. The wires all have the same radii a and satisfy the same surface area rule for N wires, that is, $N \times 2\pi a = 2\pi b$.
(a) Using $kb = 9$ and $N = 45$, calculate the backscatter echo width.
(b) Check your answer by comparing to the PO approximate result $\sigma = \pi b$.

6.16 The MoM solution in the previous problem is based on the EFIE. It will fail at the internal resonances of the cylindrical cavity, which occur at $J_n(kb) = 0$.

Investigate this problem by calculating the $\mathbf{Z}$ matrix condition number. This can be found from the optional parameter opt_rcond in SUBROUTINE molerLU; see Appendix G.

In particular, calculate and plot the condition number over the range $1 \leq kb \leq 3$. There will be a problem near $kb = 2.4$, as $J_0(2.4) = 0$.

6.17 An infinite array of z-directed thin wires are at $(x, y) = (ns, 0)$. The wire spacing is s, and n is an integer. The wire radius is a. A TM plane wave is incident at an angle ϕ_i

$$E_z^i(x, y) = E_0 e^{jk(x \cos \phi_i + y \sin \phi_i)}.$$

Using the Poisson summation technique, write a program to calculate the plane wave reflection coefficient Γ. With $\phi_i = 90°$, $a = 0.004\lambda$ and $s = 0.3\lambda$, you should get $|\Gamma| = 0.55$. Using $s = 0.5\lambda$ gives $|\Gamma| = 0.30$. Calculate and plot Γ for $0.1 \leq s/\lambda \leq 0.7$.

7

Finite Element Method

7.1 Introduction

The finite element method (FEM) is a computational method for solving differential and integral equations that arise in engineering. In this chapter the method will be applied to some differential equations, in particular the Laplace, Poisson and Helmholtz wave equations.

The general idea behind the FEM is that it solves for a potential function or a field, which is discretized into N sample values. The sample values, which are unknowns, are then adjusted until some prescribed conditions (based on the field laws) are met.

One approach to find the unknowns involves the direct application of Galerkin's method to the differential equation. Another way is to minimize the energy stored in the system. In this chapter, mostly the latter approach will be used. Although the stored energy approach seems to be less direct, it has the advantage of providing explanations of the concepts of 'essential' and 'natural' boundary conditions.

Energy minimization can be used to solve all sorts of physical problems, both electrical and mechanical. For instance, the shape of a sagging string will adjust itself so that the total amount of stored energy is minimum (subject to the constraint of fixed ends). Similarly, the electric field in a capacitor will adjust itself in such a way so that the stored energy is minimum. In mathematics, the minimization procedure is called a *variational method*. The energy obeys what is known in physics as Hamilton's principle of least action.

Although the idea of minimizing the energy appeals to our common sense, it is not at all obvious that such a method is equivalent to solving a field problem's differential equation with boundary conditions. The relationship between solving a differential equation and the variational method will be described in the section on Galerkin's method.

For a closed region such as the inside of a waveguide or transmission line, the discretized region ends at the outer boundary of the structure. This is a good configuration for the FEM because the computational domain is finite. It turns out that the FEM is also highly useful for field calculations in inhomogeneous regions that are made up of sections of multiple materials, for example, the human body.

For an open structure such as an antenna, the air region surrounding the antenna must be discretized as well. This can lead to a matrix equation with a large number of unknowns. This issue is handled by surrounding the antenna and some of the air region with an absorbing boundary, thus making the computational domain finite in size.

7.2 Laplace's Equation

Many important aspects of the FEM can be learned and appreciated by considering its application to 2D problems. Some essential groundwork will be laid by solving Laplace's equation for a transmission-line

Applied Frequency-Domain Electromagnetics, First Edition. Robert Paknys.
© 2016 John Wiley & Sons, Ltd. Published 2016 by John Wiley & Sons, Ltd.
Companion Website: www.wiley.com/go/paknys9981

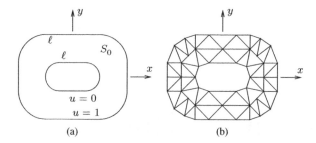

Figure 7.1   (a) Cross-sectional view of a TEM transmission line; region S_0 and bounding surfaces (contours) ℓ. (b) Mesh representation.

problem, as shown in Figure 7.1(a). The inner conductor is at a potential $u = 0$ and the outer conductor is at $u = 1$ V. For the moment, it will be assumed that the region between the conductors is air. We need to find the potential $u(x, y)$ between the conductors. The potential must satisfy Laplace's equation

$$\nabla^2 u = 0 \tag{7.1}$$

and boundary conditions of the Dirichlet or Neumann type:

$$u = \text{const.} \tag{7.2}$$

or

$$\frac{\partial u}{\partial n} = 0. \tag{7.3}$$

The first step is to find a convenient way to express the energy in terms of the unknown potential $u(x, y)$ in the region S_0 between the two conductors. The potential will be represented as a mesh of triangular elements, as shown in Figure 7.1(b).

The second step is develop expressions for the stored energy W in terms of u. The value of u is constrained to known values at boundary points. At all other points, the value of u is unknown and unconstrained. The sample values of $u(x_i, y_i)$ at each vertex i in the mesh will have to be found and adjusted, so that the total energy is minimized.

7.3   Piecewise-Planar Potential

In 2D it is easy to visualize the plot of a function $u(x, y)$ as a surface above the $x - y$ plane. The simplest possible approximation for such a surface is a piecewise-planar approximation, with flat triangles. These triangles are the *finite elements*. Each element is defined by three points (x_i, y_i) in the 2D domain. Such a surface is continuous and has no gaps. Higher-order triangles which are non-planar are also possible. However, in this chapter, the implementations will be limited to the first-order (planar) representation.

If the potential is approximated in a piecewise-planar manner, any one plane will be of the form

$$u(x, y) = a + bx + cy,$$

where a, b and c are constants. The plane will of course fit three potential samples; suppose they are u_1, u_2 and u_3 at (x_1, y_1), (x_2, y_2) and (x_3, y_3). Then,

$$u_1 = u(x_1, y_1) = a + bx_1 + cy_1$$

$$u_2 = u(x_2, y_2) = a + bx_2 + cy_2$$

$$u_3 = u(x_3, y_3) = a + bx_3 + cy_3.$$

This can be written as a matrix equation

$$\begin{bmatrix} u_1 \\ u_2 \\ u_3 \end{bmatrix} = \begin{bmatrix} 1 & x_1 & y_1 \\ 1 & x_2 & y_2 \\ 1 & x_3 & y_3 \end{bmatrix} \begin{bmatrix} a \\ b \\ c \end{bmatrix}.$$

It follows that the sample values give the constants a, b and c:

$$\begin{bmatrix} 1 & x_1 & y_1 \\ 1 & x_2 & y_2 \\ 1 & x_3 & y_3 \end{bmatrix}^{-1} \begin{bmatrix} u_1 \\ u_2 \\ u_3 \end{bmatrix} = \begin{bmatrix} a \\ b \\ c \end{bmatrix}. \tag{7.4}$$

Our goal is to use this plane to represent a triangle having sample values u_i at its three vertices (i.e. the mesh nodes) (x_i, y_i); $i = 1, 2, 3$. An efficient way to do this is to write the plane as

$$u(x, y) = \sum_{i=1}^{3} u_i \alpha_i(x, y). \tag{7.5}$$

The functions $\alpha_i(x, y)$ are *interpolatory*. They have the property that at any given node one of the α_i equals unity and the other two are equal to zero. That is,

$$\alpha_i(x_j, y_j) = \begin{cases} 1; & i = j \\ 0; & i \neq j. \end{cases}$$

With some effort, the system of equations (7.4) can be solved, with the result that

$$\alpha_1(x, y) = \frac{1}{2A} \{(x_2 y_3 - x_3 y_2) + (y_2 - y_3)x + (x_3 - x_2)y\}. \tag{7.6}$$

Cycling through the subscripts modulo-3, that is, $1, 2, 3, 1, 2, \cdots$, gives α_2 and α_3. A is the triangle's area, which can be found from

$$A = \frac{1}{2} |\det A_0|; \quad \text{where } A_0 = \begin{bmatrix} 1 & x_1 & y_1 \\ 1 & x_2 & y_2 \\ 1 & x_3 & y_3 \end{bmatrix}. \tag{7.7}$$

The $\alpha_i(x, y)$ are known by several names. They are called *basis functions*, *expansion functions* or *shape functions*. In any case the planar triangles used in (7.5) are the finite elements. The finite elements will be connected together to form the piecewise-planar representation of $u(x, y)$ over the domain S_0 in Figure 7.1.

7.4 Stored Energy

To solve the field problem we need to find the stored electric energy and minimize it. This can be found by adding up the energies of the individual elements that make up the mesh. If W is the total energy in a collection of M elements and $W^{(e)}$ is the energy in a particular element (e), then

$$W = \sum_{e=1}^{M} W^{(e)}. \tag{7.8}$$

$W^{(e)}$ can be found from the field by using

$$W^{(e)} = \frac{1}{2} \int \mathbf{D} \cdot \mathbf{E} \, dS = \frac{1}{2} \int \epsilon \mathbf{E} \cdot \mathbf{E} \, dS, \tag{7.9}$$

where ϵ is the permittivity at the location of element e. Assuming that ϵ is constant throughout the element and using $\mathbf{E} = -\nabla u$,

$$W^{(e)} = \frac{\epsilon}{2} \int \nabla u \cdot \nabla u \, dS. \tag{7.10}$$

Substituting (7.5) in (7.10) gives

$$W^{(e)} = \frac{1}{2} \sum_{i=1}^{3} \sum_{j=1}^{3} u_i u_j S_{ij}^{(e)}, \tag{7.11}$$

where

$$S_{ij}^{(e)} = \epsilon \int \nabla \alpha_i(x, y) \cdot \nabla \alpha_j(x, y) \, dS. \tag{7.12}$$

In matrix form, (7.11) is

$$W^{(e)} = \frac{1}{2} \mathbf{u}^T \mathbf{S}^{(e)} \mathbf{u}, \tag{7.13}$$

where $\mathbf{u}$ is a column vector of the node potentials u_1, u_2, u_3 and $\mathbf{u}^T$ is its transpose. Using (7.6) for α_i and α_j and (7.12), the elements of the 3×3 matrix $\mathbf{S}^{(e)}$ can be found. They are given by

$$S_{ij}^{(e)} = \frac{\epsilon}{4A} \{ (y_{i+1} - y_{i+2})(y_{j+1} - y_{j+2}) + (x_{i+2} - x_{i+1})(x_{j+2} - x_{j+1}) \}. \tag{7.14}$$

The subscripts i, j follow a modulo-3 order $1, 2, 3, 1, 2, \cdots$, for example, if $i = 3$, then $i + 1 = 1$. The matrix $\mathbf{S}^{(e)}$ is called the *Dirichlet matrix*.

7.5 Connection of Elements

The element mesh represents a continuous function $u(x, y)$. Generally, u has to be differentiable, so continuity is essential. To obtain continuity, the adjacent elements need to be connected. To understand

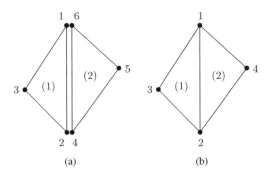

Figure 7.2   (a) Disconnected elements and (b) connected elements.

how this is done, consider a pair of elements as shown in Figure 7.2. Case (a) shows element (1) and element (2); they are disconnected. Element (1) has nodes 1, 2, 3 with potentials u_1, u_2, u_3 and element (2) has nodes 4, 5, 6 with potentials u_4, u_5, u_6. If nodes 1, 6 and 2, 4 get connected then $u_1 = u_6$ and $u_2 = u_4$. This is shown in case (b), where there are now only four nodes. It is clear that the nodes should be renumbered after being connected. If we denote the disconnected potentials as $\boldsymbol{u}_{dis}$ and the connected potentials as $\boldsymbol{u}_{con}$, the relation between them can be described by a connection matrix $\boldsymbol{C}$:

$$
\begin{bmatrix} u_1 \\ u_2 \\ u_3 \\ u_4 \\ u_5 \\ u_6 \end{bmatrix}_{dis} = \begin{bmatrix} 1 & 0 & 0 & 0 \\ 0 & 1 & 0 & 0 \\ 0 & 0 & 1 & 0 \\ 0 & 1 & 0 & 0 \\ 0 & 0 & 0 & 1 \\ 1 & 0 & 0 & 0 \end{bmatrix} \begin{bmatrix} u_1 \\ u_2 \\ u_3 \\ u_4 \end{bmatrix}_{con} \tag{7.15}
$$

or

$$
\boldsymbol{u}_{dis} = \boldsymbol{C}\,\boldsymbol{u}_{con}. \tag{7.16}
$$

The energy is the same, connected or not. For the disconnected case,

$$
W = \frac{1}{2} \begin{bmatrix} u_1 & u_2 & u_3 \end{bmatrix}_{dis} \begin{bmatrix} S_{11}^{(1)} & S_{12}^{(1)} & S_{13}^{(1)} \\ S_{21}^{(1)} & S_{22}^{(1)} & S_{23}^{(1)} \\ S_{31}^{(1)} & S_{32}^{(1)} & S_{33}^{(1)} \end{bmatrix} \begin{bmatrix} u_1 \\ u_2 \\ u_3 \end{bmatrix}_{dis}
$$

$$
+ \frac{1}{2} \begin{bmatrix} u_4 & u_5 & u_6 \end{bmatrix}_{dis} \begin{bmatrix} S_{44}^{(2)} & S_{45}^{(2)} & S_{46}^{(2)} \\ S_{54}^{(2)} & S_{55}^{(2)} & S_{56}^{(2)} \\ S_{64}^{(2)} & S_{65}^{(2)} & S_{66}^{(2)} \end{bmatrix} \begin{bmatrix} u_4 \\ u_5 \\ u_6 \end{bmatrix}_{dis} \tag{7.17}
$$

which can also be written as

$$
W = \frac{1}{2} \boldsymbol{u}_{dis}^T \boldsymbol{S}_{dis}\,\boldsymbol{u}_{dis}, \tag{7.18}
$$

where the disconnected $\boldsymbol{S}$ matrix is defined by

$$
\boldsymbol{S}_{dis} =
\begin{bmatrix}
S_{11}^{(1)} & S_{12}^{(1)} & S_{13}^{(1)} & 0 & 0 & 0 \\
S_{21}^{(1)} & S_{22}^{(1)} & S_{23}^{(1)} & 0 & 0 & 0 \\
S_{31}^{(1)} & S_{32}^{(1)} & S_{33}^{(1)} & 0 & 0 & 0 \\
0 & 0 & 0 & S_{44}^{(2)} & S_{45}^{(2)} & S_{46}^{(2)} \\
0 & 0 & 0 & S_{54}^{(2)} & S_{55}^{(2)} & S_{56}^{(2)} \\
0 & 0 & 0 & S_{64}^{(2)} & S_{65}^{(2)} & S_{66}^{(2)}
\end{bmatrix}
=
\begin{bmatrix}
\boldsymbol{S}^{(1)} & 0 \\
0 & \boldsymbol{S}^{(2)}
\end{bmatrix}. \tag{7.19}
$$

Using (7.16) and (7.18) gives the energy in terms of the connected potentials

$$
W = \frac{1}{2} \boldsymbol{u}_{con}^{T} \boldsymbol{C}^{T} \boldsymbol{S}_{dis} \boldsymbol{C} \boldsymbol{u}_{con}. \tag{7.20}
$$

Therefore, when using connected potentials,

$$
W = \frac{1}{2} \boldsymbol{u}_{con}^{T} \boldsymbol{S} \boldsymbol{u}_{con} \tag{7.21}
$$

where we have defined

$$
\boldsymbol{S} = \boldsymbol{C}^{T} \boldsymbol{S}_{dis} \boldsymbol{C}. \tag{7.22}
$$

Using $\boldsymbol{C}$ in (7.15) with (7.19) and (7.22),

$$
\boldsymbol{S} =
\begin{bmatrix}
S_{11}^{(1)} + S_{66}^{(2)} & S_{12}^{(1)} + S_{64}^{(2)} & S_{13}^{(1)} & S_{65}^{(2)} \\
S_{21}^{(1)} + S_{46}^{(2)} & S_{22}^{(1)} + S_{44}^{(2)} & S_{23}^{(1)} & S_{45}^{(2)} \\
S_{31}^{(1)} & S_{32}^{(1)} & S_{33}^{(1)} & 0 \\
S_{56}^{(2)} & S_{54}^{(2)} & 0 & S_{55}^{(2)}
\end{bmatrix}. \tag{7.23}
$$

If there are N nodes in the connected mesh, $\boldsymbol{S}$ is an $N \times N$ matrix. Since it is related to the entire connected mesh, it is called the *global matrix*. $\boldsymbol{S}^{(e)}$ on the other hand is for one element, so it is a *local matrix*.

Looking at (7.23), Figure 7.3 suggests an easy way of constructing $\boldsymbol{S}$. S_{14} is merely the renamed $S_{65}^{(2)}$. When nodes get connected, their $\boldsymbol{S}$ entries simply get added; for example, $S_{11}^{(1)} + S_{66}^{(2)}$ becomes S_{11}.

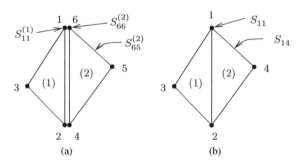

Figure 7.3   Some S_{ij} for (a) disconnected elements and (b) connected elements.

If we follow these rules during the construction of S, the formal matrix multiplication (7.22) becomes unnecessary. In conclusion then, an S constructed in this way along with (7.21) gives the energy.

7.6 Energy Minimization

Consider the two-element mesh in Figure 7.3(b). There are four potentials u_1, u_2, u_3, u_4 at the four nodes. If a node is on a boundary, its value may be fixed, whereas other nodes are free – so their values are unknown. Let us assume that nodes 3 and 4 are fixed and that nodes 1 and 2 are free.

The unknowns u_1, u_2 are now found by minimizing W. Imposing $\partial W / \partial u_1 = 0$ and $\partial W / \partial u_2 = 0$ on (7.21) leads to the equations

$$S_{11}u_1 + S_{12}u_2 + S_{13}u_3 + S_{14}u_4 = 0$$

$$S_{21}u_1 + S_{22}u_2 + S_{23}u_3 + S_{24}u_4 = 0$$

which could also be written as a matrix equation

$$\boldsymbol{Su} = 0. \tag{7.24}$$

To arrive at the above result it is noted from (7.12) that $S_{ij}^{(e)} = S_{ji}^{(e)}$; then by (7.23) we have that $S_{ij} = S_{ji}$. Putting the unknowns on the left and the knowns on the right, we can readily solve the system for u_1 and u_2

$$\begin{bmatrix} S_{11} & S_{12} \\ S_{21} & S_{22} \end{bmatrix} \begin{bmatrix} u_1 \\ u_2 \end{bmatrix} = \begin{bmatrix} -S_{13}u_3 - S_{14}u_4 \\ -S_{23}u_3 - S_{24}u_4 \end{bmatrix}. \tag{7.25}$$

In this simple example we knew how to arrange the equations so that the unknowns u_1, u_2 are on the left and the knowns u_3, u_4 are on the right. However, for a general-purpose program, a better approach to set up the equations is like this

$$\begin{bmatrix} S_{11} & S_{12} & 0 & 0 \\ S_{21} & S_{22} & 0 & 0 \\ 0 & 0 & 1 & 0 \\ 0 & 0 & 0 & 1 \end{bmatrix} \begin{bmatrix} u_1 \\ u_2 \\ u_3 \\ u_4 \end{bmatrix} = \begin{bmatrix} -S_{13}u_3 - S_{14}u_4 \\ -S_{23}u_3 - S_{24}u_4 \\ u_3 \\ u_4 \end{bmatrix}. \tag{7.26}$$

In (7.26), the right side is known. On the left side, u_1 and u_2 are unknown. By pretending that u_3 and u_4 are also unknown (even though they are not), we are able to solve this system for u_1, u_2, u_3, u_4.

The point of this arrangement is not yet apparent. Let us suppose that another person used a different node-numbering scheme, so that u_1, u_3 are known and u_2, u_4 are unknown. Physically, nothing has changed and it is still the same problem. Just the names of the nodes have been changed. The corresponding equations would be

$$\begin{bmatrix} 1 & 0 & 0 & 0 \\ 0 & S_{22} & 0 & S_{24} \\ 0 & 0 & 1 & 0 \\ 0 & S_{42} & 0 & S_{44} \end{bmatrix} \begin{bmatrix} u_1 \\ u_2 \\ u_3 \\ u_4 \end{bmatrix} = \begin{bmatrix} u_1 \\ -S_{21}u_1 - S_{23}u_3 \\ u_3 \\ -S_{41}u_1 - S_{43}u_3 \end{bmatrix}. \tag{7.27}$$

Examining (7.26) and (7.27) we find that the free and fixed u_i can appear in any order, so *the order of the node numbers does not matter*. Therefore, when assigning node numbers in a mesh, we do not have to ensure that all the free nodes come first and all the fixed nodes come second. Organizing the equations this way is useful for a general-purpose program.

Let us now consider an algorithm for setting up the equations in this format. The program should have a list of elements, with each element having three global node numbers, in a connected mesh. For a mesh with N nodes, the matrix is $N \times N$. One element at a time, the 3×3 matrix $S^{(e)}$ of each element is evaluated and, following its global node numbers, is added to the appropriate row/column of the global matrix S. This way, S is built up as each element is encountered. If node i and node j are free, the left side of (7.26) is augmented by $S_{ij} \leftarrow S_{ij} + S_{ij}^{(e)}$. If node i is fixed, then $S_{ii} = 1$ on the left side and u_i fills row i on the right side. If node i is free and node j is fixed, then the right side row i is augmented by a term $-S_{ij}u_j$.

7.7 Natural Boundary Conditions

By using symmetry, the transmission-line problem in Figure 7.1 can be reduced to that in Figure 7.4. The electric field is parallel to the new boundaries which are along the bottom and left edges, so $\partial u/\partial n = 0$ is required there.

At nodes where u is known, the potentials are fixed. At the interior nodes, u is not known, so those potentials are free. At the bottom and left boundaries, it is tempting to enforce $\partial u/\partial n = 0$, as it is the known boundary condition for these edges. However, this will not be done. Rather, these nodes will be left as free and treated as unknowns. The reason for doing this will now be explained.

The energy to be minimized is given, within a constant factor, by

$$W(u) = \frac{1}{2} \int \nabla u \cdot \nabla u \, dS. \tag{7.28}$$

In mathematics, in particular in variational methods, an expression like (7.28) is called a *functional*. $W(u)$ is a function of a function, and it gives a number. This is in contrast to a *function* $f(x)$, which gives a number from another number x.

Let us suppose the exact solution u is perturbed and becomes $u_p = u + \delta v$. The term δ is a number, and v is a function which should be differentiable so that ∇v exists. By definition, we know the exact solution's values on the Dirichlet boundaries, because u is specified there. So, we should assume that $v = 0$ at those points. The perturbed solution in (7.28) gives

$$W(u + \delta v) = \frac{1}{2} \int \nabla(u + \delta v) \cdot \nabla(u + \delta v) \, dS. \tag{7.29}$$

Green's first identity is now applied in 2D, for a surface S_0 with a bounding contour C_0. Along the contour, $d\boldsymbol{\ell} = \hat{\mathbf{n}} d\ell$ and $\hat{\mathbf{n}}$ is the outward normal, perpendicular to C_0. From Appendix B we have

$$\int_{S_0} \nabla f \cdot \nabla g \, dS = -\int_{S_0} g \nabla^2 f \, dS + \oint_{C_0} g \frac{\partial f}{\partial n} d\ell.$$

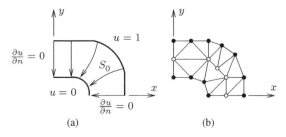

(a)   (b)

Figure 7.4 First quadrant of the transmission line. (a) Electric field lines. (b) Mesh showing fixed nodes • and free nodes ○.

We can use this to rewrite (7.29) as

$$W(u + \delta v) = W(u) + \delta^2 W(v) - \delta \int_{S_0} v \nabla^2 u \, dS + \delta \oint_{C_0} v \frac{\partial u}{\partial n} \, d\ell. \qquad (7.30)$$

Since u is the exact solution, $\nabla^2 u = 0$ and the surface integral is zero. Along the boundary contour, either $v = 0$ at the Dirichlet points, or $\partial u / \partial n = 0$ at the Neumann points, so the contour integral vanishes as well. We are left with

$$W(u + \delta v) = W(u) + \delta^2 W(v). \qquad (7.31)$$

Since $\delta^2 \geq 0$ and $W(v) \geq 0$, Equation (7.31) shows that any deviation from the exact solution u causes the energy to be larger than the exact energy $W(u)$. Therefore, minimizing the energy leads us to the exact solution.

If the potential has an error that is $O(\delta)$, then the error in W is $O(\delta^2)$. Consequently, even when the potential is not highly accurate, the energy calculation can still be quite accurate. This aspect is noteworthy from an engineering point of view, because the energy is often more interesting than the potential.

What would happen if we did not enforce the Neumann part of the boundary condition? The FEM finds the true minimum of W and not an inflection point. Therefore, any terms with an odd power of δ in (7.30) must vanish, so that

$$\oint_{C_0} v \frac{\partial u}{\partial n} \, d\ell = 0.$$

If this is to be true for any arbitrary function v, it implies that minimizing W makes $\partial u / \partial n$ zero along the boundary in some average sense, but does not ensure that $\partial u / \partial n = 0$ at every Neumann point. This is called a *natural boundary condition* because it naturally arises when the energy is minimized. It also shows that energy minimization can be accomplished without necessarily enforcing $\partial u / \partial n = 0$ everywhere along the bottom and left boundaries in Figure 7.4.

Another place where a natural boundary condition occurs is at the interface between two dielectrics. We know that the solution u of an electromagnetic problem must satisfy a differential equation and boundary conditions. When we considered a junction between two dielectrics, the permittivities were accounted for in the element's stored energy (7.9). However, no additional steps were taken to ensure that the boundary conditions for $\mathbf{E}_{tan}$ and D_n are satisfied at the junction. Has something been overlooked? No. It turns out that these conditions are included, as will now be shown.

Figure 7.5(a) shows two connected elements with permittivities ϵ_1 and ϵ_2. Let us denote the potentials as $u^{(1)}$ in element (1) and $u^{(2)}$ in element (2). Just on the left, $\mathbf{E}_{tan}^{(1)} = -\hat{\ell} \, \partial u^{(1)} / \partial \ell$ and just on the right, $\mathbf{E}_{tan}^{(2)} = -\hat{\ell} \, \partial u^{(2)} / \partial \ell$. At the common nodes the potentials are equal, so $u^{(1)} = u^{(2)} = u_1$ at node 1;

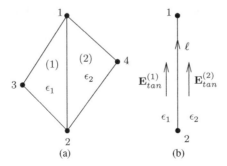

Figure 7.5 (a) Connected elements with permittivities ϵ_1 and ϵ_2. (b) Contour ℓ along the junction.

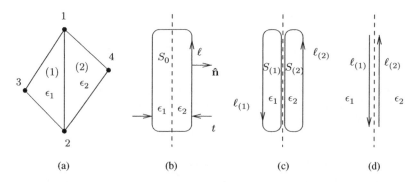

Figure 7.6 (a) Connected elements with permittivities ϵ_1 and ϵ_2. (b) Definition of S_0, ℓ and outward normal for $d\ell = \hat{n}d\ell$. (c) Contour $\ell_{(1)} + \ell_{(2)} = \ell$ bounds $S_{(1)} + S_{(2)} = S_0$. (d) Limit as $t \to 0$.

likewise $u^{(1)} = u^{(2)} = u_2$ at node 2. In between the two nodes, u varies linearly, so the derivative of u along that line is the same in either element. Therefore, $\mathbf{E}_{tan}$ is continuous across that line.

The normal electric flux density D_n is considered next. Figure 7.6(a) shows adjacent mesh elements (1) and (2) with permittivities ϵ_1 and ϵ_2. The stored energy will be affected, as in (7.9), repeated here:

$$W^{(e)} = \frac{1}{2} \int_{S_0} \mathbf{D} \cdot \mathbf{E} \, dS = \frac{1}{2} \int_{S_0} \epsilon \mathbf{E} \cdot \mathbf{E} \, dS. \tag{7.32}$$

As before, we can find the energy associated with a perturbed solution $u_p = u + \delta v$. Integrating over S_0 as shown in Figure 7.6(b), (7.32) becomes

$$W(u + \delta v) = W(u) + \delta^2 W(v) + \delta \int_{S_0} \epsilon \nabla u \cdot \nabla v \, dS. \tag{7.33}$$

Since ϵ is different on the two sides, the integral should be split as in Figure 7.6(c) so that it becomes

$$\epsilon_1 \int_{S_{(1)}} \nabla u \cdot \nabla v \, dS + \epsilon_2 \int_{S_{(2)}} \nabla u \cdot \nabla v \, dS.$$

Applying Green's first identity to each integral and remembering that $\nabla^2 u = 0$, (7.33) becomes

$$W(u + \delta v) = W(u) + \delta^2 W(v) + \delta \left\{ \epsilon_1 \oint_{\ell_{(1)}} v \frac{\partial u}{\partial n} \, d\ell + \epsilon_2 \oint_{\ell_{(2)}} v \frac{\partial u}{\partial n} \, d\ell \right\}. \tag{7.34}$$

The terms with odd powers of δ should vanish, so the bracketed term should be zero.

Figure 7.6(c) shows that the inner parts of the line integrals (the parts along the boundary surface) along $\ell_{(1)}$ and $\ell_{(2)}$ are oppositely directed, so they cancel. If $t \to 0$, the remaining outer parts will appear as in Figure 7.6(d), and $\ell_{(1)} = -\ell_{(2)}$. Denoting $\ell_0 = \ell_{(1)} = -\ell_{(2)}$, the bracketed term in (7.34) becomes

$$\epsilon_1 \int_{\ell_0} v \frac{\partial u}{\partial n} \, d\ell - \epsilon_2 \int_{\ell_0} v \frac{\partial u}{\partial n} \, d\ell = 0.$$

Since $D_n = -\epsilon \partial u / \partial n$, this can be rewritten as

$$\int_{\ell_0} v \left(\mathbf{D}_1 - \mathbf{D}_2 \right) \cdot \hat{n} d\ell = 0. \tag{7.35}$$

This is a natural boundary condition, as it naturally arises from the energy minimization procedure. It states that D_n is continuous at the boundary, in some average sense, with respect to a weighting function v. D_n, however, is not necessarily continuous at every boundary point.

Often, boundary conditions are referred to as either *essential* or else *natural*. Dirichlet boundary conditions are prescribed and are of the essential type. On the other hand, the natural boundary conditions are not specified. They are a consequence that is associated with the properties of the functional.

7.8   Capacitance and Inductance

A TEM transmission line filled with a homogeneous material will have a wave with a propagation velocity that is

$$u = \frac{1}{\sqrt{\mu\epsilon}} = \frac{1}{\sqrt{LC}}, \tag{7.36}$$

where L, C are the inductance and capacitance per unit length. The relationship implies that $LC = \mu\epsilon$. This enables us to find the characteristic impedance from the capacitance alone. It is not necessary to find L. That is,

$$Z_0 = \sqrt{\frac{L}{C}} = \sqrt{\frac{\mu\epsilon/C}{C}} = \frac{1}{uC}. \tag{7.37}$$

We can also find the inductance per unit length, if the capacitance per unit length is known. It is not necessary to solve the magnetic field problem to find L.

Additional considerations are needed if the transmission line is partially filled with a dielectric. For example, consider the shielded microstrip transmission line in Figure 7.7. The rectangular outer shield is at $u = 0$, and the strip conductor is at $u = 1$ V. The field is partially in the air and partially in the dielectric substrate. The propagation velocity in the two regions is necessarily different, so the simple idea described in (7.37) for finding Z_0 cannot be used. Strictly speaking, field continuity requires phase matching at the dielectric–air interface. This cannot be accomplished with a TEM mode. The rigorous solution requires a hybrid mode, which is a combination of TE and TM waves.

A solution with hybrid modes is appealing as it is formally exact. However, it presents some difficulties such as precision problems and spurious modes. These and other issues are beyond the scope of the present discussion but can be found in the references.

Fortunately, in practice a rigorous hybrid-mode solution is not always necessary. With stripline and microstrip lines, it is often a good approximation to assume that the field is quasi-TEM. It is then possible to define an *effective permittivity* ϵ_e that lies somewhere between the air and dielectric permittivity, that is, $\epsilon_0 < \epsilon_e < \epsilon_d$. To find ϵ_e for the transmission line in Figure 7.7, the structure is first solved with air everywhere inside, and a capacitance C_0 is found. The propagation velocity is $u = c$ the speed of light, so

$$c = \frac{1}{\sqrt{LC_0}} = \frac{1}{\sqrt{\mu_0\epsilon_0}}.$$

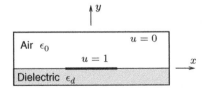

Figure 7.7   Shielded microstrip line.

Next, the problem is solved again, with the actual dielectric–air combination as shown in Figure 7.7. A new capacitance is C_e is obtained. Under the quasi-TEM assumption, the wave will be slower, with a velocity

$$u = \frac{1}{\sqrt{LC_e}} = \frac{1}{\sqrt{\mu_0 \epsilon_e}}.$$

The dielectric does not affect the inductance. Dividing the two results gives the phase velocity

$$u = c\sqrt{\frac{C_0}{C_e}} \tag{7.38}$$

and the characteristic impedance is

$$Z_0 = \sqrt{\frac{L}{C_e}} = \frac{1}{c\sqrt{C_0 C_e}}. \tag{7.39}$$

7.9   Computer Program

PROGRAM fem2d solves Laplace's equation subject to Dirichlet boundary conditions at the fixed nodes. The free nodes obey natural boundary conditions. The permittivity of each element can be specified. The matrix equation is of the form shown in (7.26) or (7.27).

The program uses the Fortran 90 'derived type', which lets us make up variables having any mix of real, integer, logical and character parts (called the 'components'). This is ideal for describing nodes and elements. For instance, by defining the type

```
TYPE node_properties
REAL:: x,y
REAL:: u
LOGICAL:: fixednode
END TYPE node_properties
```

we can declare an array called node:

```
TYPE(node_properties),DIMENSION(100):: node
```

A fixed node #2 at $(0.03, 0.04)$ with a potential $u_2 = 5$ would have the components

```
node(2)%fixednode = .TRUE.
node(2)%x = 0.03
node(2)%y = 0.04
node(2)%u = 5.
```

Similarly, we can define

```
TYPE element_properties
INTEGER, DIMENSION(3):: v
REAL:: er
END TYPE element_properties
```

and declare an array called `element`:

```
TYPE(element_properties),DIMENSION(80):: element
```

An element #8 with vertices at nodes #11, #5 and #6 and a relative permittivity of $\epsilon_r = 9$ would have the components

```
element(8)%v(1)  =  11
element(8)%v(2)  =  5
element(8)%v(3)  =  6
element(8)%er = 9.
```

    The program uses the LU method (see Appendix G) to solve the matrix equations. A commercial FEM code would use more efficient techniques for sparse matrices. However, for learning and development purposes, this code has been written with clarity in mind, at the expense of efficiency. Further details about the implementation are in the code comments.

Example 7.1 (Coaxial Cable Impedance) A circular coaxial cable has an inner radius $a = 0.06$ m, outer radius $b = 0.08$ m and $0 \le \phi \le 360°$. The medium inside is air. Find the characteristic impedance using the FEM and compare it with the exact answer.

Solution: We can take advantage of the circular symmetry by using the mesh in Figure 7.8. Using a portion with $\phi_0 = 15°$ instead of the entire $360°$ reduces the number of unknowns to three. The stored energy in the complete cable is therefore $W \times 360/15$. We can set the potential to $u = 0$ for the nodes at $\rho = a$ and $u = 1$ at $\rho = b$. From symmetry considerations, $\partial u / \partial \phi = 0$ at $\phi = 0$ and $\phi = \phi_0$, so those nodes should be free.

    Using PROGRAM fem2d with the input file coax1.dat gives $W/\epsilon_0 = 0.45646$ for the stored energy in the $15°$ portion or $0.45646 \times 360/15 = 10.955$ for the complete circumference. The stored energy is also given by $W = \frac{1}{2}CV^2$, and since $V = 1$, the capacitance is $C = 2W$. Using (7.37) the characteristic impedance can be found from the capacitance and is

$$Z_0 = \frac{1}{uC} = \frac{1}{(1/\sqrt{\mu_0 \epsilon_0})2W} = \frac{\eta_0}{2W/\epsilon_0} = 17.19 \ \Omega.$$

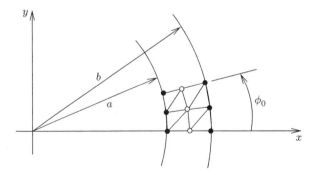

Figure 7.8 Portion of a coaxial cable, with fixed $\bullet$ and free nodes $\circ$ in a mesh, on $0 \le \phi \le \phi_0$.

From (2.127) the exact result is

$$Z_0 = \frac{\eta_0}{2\pi} \ln(b/a) = 17.25 \ \Omega.$$

Due to the variational property of the FEM, the computed stored energy is slightly higher than the true energy. This causes Z_0 to be less than the exact answer. ∎

7.10 Poisson's Equation

To learn how a source is accommodated in an FEM solution, we now consider Poisson's equation

$$\nabla^2 u = g. \tag{7.40}$$

A good example of a physical problem using Poisson's equation is magnetostatic machine design. Structures such as rotors and stators are usually long and can be regarded as z-independent and 2D. A generic configuration is shown in Figure 7.9. The z-directed current is due to a bundle of windings that produces a fairly constant current density in the source region. In the following development, the main objective will be not so much about explaining the theory of magnetostatics, but to show how sources are modelled in the FEM.

For the structure in Figure 7.9, the vector potential and current will have only z components, so in the air region the governing differential equation is

$$\nabla^2 A = -\mu_0 J$$

and $\mu_0 \mathbf{H} = \nabla \times \hat{z}A$ gives the magnetic field. A comparison with (7.40) shows that this is Poisson's equation with $u = A$ and a source $g = -\mu_0 J$. As with Laplace's equation, Dirichlet or Neumann conditions are required on the boundaries of the problem domain. This amounts to specifying particular values of A on the boundaries or else natural boundary conditions. For the case in Figure 7.9, $\mu \gg \mu_0$ and $\mathbf{H}_{tan}$ is continuous at the boundary, so $\mathbf{B}$ has only a normal component. This implies we have a Neumann boundary condition with $\partial A/\partial n = 0$. If $\mathbf{B}$ had only a tangential component, this would require a Dirichlet boundary condition $A = \text{const}$. (See Problem 7.7.)

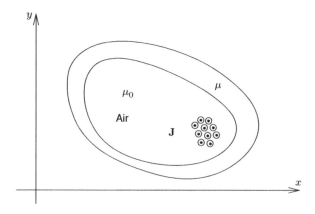

Figure 7.9   An air region and z-directed current density **J**, surrounded by a high-permeability material.

To develop the FEM solution of (7.40), let us begin with one element. The potential u and source g are expanded with piecewise-planar representations:

$$u(x,y) = \sum_{i=1}^{3} u_i \alpha_i(x,y) \tag{7.41}$$

$$g(x,y) = \sum_{i=1}^{3} g_i \alpha_i(x,y). \tag{7.42}$$

The solution u of (7.40) can be found by minimizing the functional

$$F^{(e)}(u) = \frac{1}{2} \int \nabla u \cdot \nabla u \, dS + \int ug \, dS. \tag{7.43}$$

At the boundaries, Dirichlet conditions are prescribed at the fixed nodes, and the natural boundary condition $\partial u/\partial n = 0$ is met at the free nodes.

For the magnetostatic problem, the above functional is related to the energy, but it is not quite the energy.[1] The derivation of this functional takes a few steps and is postponed to Section 7.12.2. It is also discussed in Silvester and Ferrari (1996, Chapter 2).

We can now proceed with the solution. Using the expressions (7.41) and (7.42) in (7.43) gives

$$F^{(e)}(u) = \frac{1}{2} \sum_{i=1}^{3} \sum_{j=1}^{3} u_i u_j S_{ij}^{(e)} + \sum_{i=1}^{3} \sum_{j=1}^{3} u_i g_j T_{ij}^{(e)}, \tag{7.44}$$

where

$$S_{ij}^{(e)} = \int \nabla \alpha_i(x,y) \cdot \nabla \alpha_j(x,y) \, dS \tag{7.45}$$

and

$$T_{ij}^{(e)} = \int \alpha_i(x,y) \alpha_j(x,y) \, dS. \tag{7.46}$$

$S_{ij}^{(e)}$ and $T_{ij}^{(e)}$ are called the *Dirichlet matrix* and the *metric*.

The integral for the metric involves terms with $x, x^2, y, y^2, x - y$ and constant terms. The evaluation is straightforward but tedious. A more elegant approach uses the so-called simplex coordinates (see Silvester and Ferrari 1996, Chapter 4). In any case, the end result turns out to be remarkably simple:

$$T^{(e)} = \frac{A}{12} \begin{bmatrix} 2 & 1 & 1 \\ 1 & 2 & 1 \\ 1 & 1 & 2 \end{bmatrix}, \tag{7.47}$$

where (7.7) gives the triangle's area A.

---

[1] In a 2D magnetostatic problem, the stored energy is $W = \frac{1}{2} \int \mathbf{B} \cdot \mathbf{H} \, dS$. The stored energy can also be written as $W = \frac{1}{2\mu_0} \int \nabla A \cdot \nabla A \, dS$ or as $W = \frac{1}{2} \int AJ \, dS$. The functional (7.43) to be minimized becomes

$$F^{(e)}(A) = \frac{1}{2\mu_0} \int \nabla A \cdot \nabla A \, dS - \int AJ \, dS$$

and from this we can see that $F^{(e)}(A) = -W$. So this is an 'energy-related functional', but it is not quite the energy, because of the minus sign. In any case, we will later set $\partial W/\partial u_i = 0$ so the minus sign does not matter.

If we have multiple elements, their contributions are added. In a disconnected mesh, the functional becomes

$$F(u) = \frac{1}{2} u_{dis}^T S_{dis} u_{dis} + u_{dis}^T T_{dis} g_{dis}. \tag{7.48}$$

In the connected mesh, $u(x, y)$ is continuous. The disconnected mesh potentials can be eliminated by using a connection matrix C so that $u_{dis} = Cu$ and

$$F(u) = \frac{1}{2} u^T S u + u^T C^T T_{dis} g_{dis}, \tag{7.49}$$

where $S = C^T S_{dis} C$. If there are N nodes in the connected mesh, then u is a vector of length N and S is an $N \times N$ matrix.

The next step is to minimize the functional. Suppose we have two elements as in Figure 7.2. As before, it is assumed that u_1, u_2 are free and u_3, u_4 are fixed. Imposing $\partial F/\partial u_1 = 0$ and $\partial F/\partial u_2 = 0$ at the free nodes leads to equations that are very similar to (7.25):

$$\begin{bmatrix} S_{11} & S_{12} \\ S_{21} & S_{22} \end{bmatrix} \begin{bmatrix} u_1 \\ u_2 \end{bmatrix} = \begin{bmatrix} -S_{13}u_3 - S_{14}u_4 \\ -S_{23}u_3 - S_{24}u_4 \end{bmatrix} - \begin{bmatrix} T_{11}^{(1)} & T_{12}^{(1)} & T_{13}^{(1)} & T_{64}^{(2)} & T_{65}^{(2)} & T_{66}^{(2)} \\ T_{21}^{(1)} & T_{22}^{(1)} & T_{23}^{(1)} & T_{44}^{(2)} & T_{45}^{(2)} & T_{46}^{(2)} \end{bmatrix}_{dis} \begin{bmatrix} g_1 \\ g_2 \\ g_3 \\ g_4 \\ g_5 \\ g_6 \end{bmatrix}_{dis}. \tag{7.50}$$

In this result, the forcing function g is given by $g_1, g_2, \cdots, g_6$ at the nodes of the disconnected mesh. (Depending on the source's details, some of the g_i could be zero.)

If we wanted a continuous $g(x, y)$, we could have used $g_{dis} = Cg$ in (7.49) to obtain

$$F(u) = \frac{1}{2} u^T S u + u^T T g, \tag{7.51}$$

where $T = C^T T_{dis} C$. Minimizing this functional by setting $\partial F/\partial u_j = 0$ for the free nodes u_j leads to the matrix equation

$$Su = -Tg. \tag{7.52}$$

This is a generalized version of (7.24) that accounts for sources.

Usually there are no continuity requirements for a forcing function, so this will not be imposed. A practical approach for the magnetostatic problem is to specify g as a constant within each element, providing a staircase approximation. Then we can say that $g_1 = g_2 = g_3 = g^{(1)}$ in element (1) and $g_4 = g_5 = g_6 = g^{(2)}$ in element (2). Doing this and rewriting the $T_{ij}^{(e)}$ in terms of connected node numbers, (7.50) becomes

$$\begin{bmatrix} S_{11} & S_{12} \\ S_{21} & S_{22} \end{bmatrix} \begin{bmatrix} u_1 \\ u_2 \end{bmatrix} = \begin{bmatrix} -S_{13}u_3 - S_{14}u_4 \\ -S_{23}u_3 - S_{24}u_4 \end{bmatrix} - \begin{bmatrix} T_{11}^{(1)} + T_{12}^{(1)} + T_{13}^{(1)} & T_{12}^{(2)} + T_{14}^{(2)} + T_{11}^{(2)} \\ T_{21}^{(1)} + T_{22}^{(1)} + T_{23}^{(1)} & T_{22}^{(2)} + T_{24}^{(2)} + T_{21}^{(2)} \end{bmatrix} \begin{bmatrix} g^{(1)} \\ g^{(2)} \end{bmatrix}. \tag{7.53}$$

As was learned in (7.26), it is expedient to arrange the equations as though there are four unknowns. Applying the idea to (7.53) gives

$$
\begin{bmatrix} S_{11} & S_{12} & 0 & 0 \\ S_{21} & S_{22} & 0 & 0 \\ 0 & 0 & 1 & 0 \\ 0 & 0 & 0 & 1 \end{bmatrix} \begin{bmatrix} u_1 \\ u_2 \\ u_3 \\ u_4 \end{bmatrix} = \begin{bmatrix} -S_{13}u_3 - S_{14}u_4 \\ -S_{23}u_3 - S_{24}u_4 \\ u_3 \\ u_4 \end{bmatrix}
$$

$$
- \begin{bmatrix} T_{11}^{(1)} + T_{12}^{(1)} + T_{13}^{(1)} & T_{12}^{(2)} + T_{14}^{(2)} + T_{11}^{(2)} & 0 & 0 \\ T_{21}^{(1)} + T_{22}^{(1)} + T_{23}^{(1)} & T_{22}^{(2)} + T_{24}^{(2)} + T_{21}^{(2)} & 0 & 0 \\ 0 & 0 & 0 & 0 \\ 0 & 0 & 0 & 0 \end{bmatrix} \begin{bmatrix} g^{(1)} \\ g^{(2)} \\ 0 \\ 0 \end{bmatrix}. \tag{7.54}
$$

The matrix equation (7.54) can be assembled the same way as before, element by element, using the algorithm in Section 7.6. Only one step needs to be added: if element (e) node i is free and node j is fixed or free, the right side row i is augmented by a term $-T_{ij}^{(e)}g^{(e)}$. In summary, except for one minor change, the equation assembly follows the same procedure as Laplace's equation. In a program, the connection matrix and disconnected node numbers are unnecessary.

The result (7.54) highlights an interesting fact about sources and boundary conditions. That is, both the current g and the prescribed potentials u_j on the right-hand side play the same roles, as forcing functions. Even if g were zero, we could still think of the right-hand side u_j as the source of the field. When non-zero boundary values act as the source, it is called *condensation* of the boundary conditions.

7.11   Scalar Wave Equation

Some electromagnetic problems can be solved as a scalar wave equation

$$
\nabla \cdot (p\nabla u) + k^2 qu = g. \tag{7.55}
$$

This is a fairly general form that covers many cases. The functions p and q are arbitrary, so (7.55) can account for inhomogeneous media. A forcing function g can also be included. As special cases, when $p = 1$, $k = 0$ and $g = 0$, it reduces to Laplace's equation, and $p = 1$, $k = 0$ gives Poisson's equation.

The solution of (7.55) can be found by minimizing the functional

$$
F(u) = \frac{1}{2} \int_{S_0} p\nabla u \cdot \nabla u - k^2 qu^2 + 2gu \, dS. \tag{7.56}
$$

The derivation of this functional is described in Section 7.12.2. For the moment, we can just use it and proceed with the solution.

When $u = $ const. is specified at a node, it is a Dirichlet boundary condition. If a node is left free, it is a property of the functional (7.56) that u satisfies the homogeneous Neumann boundary condition $\partial u / \partial n = 0$.

Let us now specialize (7.55) to the case of a homogeneous medium. In this case the wave equation is

$$
\nabla^2 u + k^2 u = g \tag{7.57}
$$

and the functional to be made stationary is

$$F(u) = \frac{1}{2} \int_{S_0} \nabla u \cdot \nabla u - k^2 u^2 + 2gu \, dS. \tag{7.58}$$

The function u on any given element is an interpolatory triangle

$$u(x, y) = \sum_{i=1}^{3} u_i \alpha_i(x, y). \tag{7.59}$$

Similar to (7.21) and (7.49) we can write a functional for the disconnected elements

$$F(u) = \frac{1}{2} u_{dis}^T S_{dis} u_{dis} - \frac{k^2}{2} u_{dis}^T T_{dis} u_{dis} + u_{dis}^T T_{dis} g_{dis}. \tag{7.60}$$

Using the relationships between unconnected and connected quantities

$$u_{dis} = Cu$$

$$S = C^T S_{dis} C$$

$$T = C^T T_{dis} C$$

$$g_{dis} = Cg$$

the functional (7.60) becomes

$$F(u) = \frac{1}{2} u^T S u - \frac{k^2}{2} u^T T u + u^T T g. \tag{7.61}$$

The Dirichlet matrix for each element is

$$S_{ij}^{(e)} = \int \nabla \alpha_i(x, y) \cdot \nabla \alpha_j(x, y) \, dS \tag{7.62}$$

and the metric is

$$T_{ij}^{(e)} = \int \alpha_i(x, y) \alpha_j(x, y) \, dS. \tag{7.63}$$

The next step is to minimize $F(u)$. Taking $\partial F / \partial u_i = 0$ with respect to all free nodes u_i leads to the matrix equation

$$Su - k^2 Tu = -Tg. \tag{7.64}$$

This is essentially a generalized version of (7.52). As a specific example, for the four-node mesh in Figure 7.2 with u_1 and u_2 free and u_3 and u_4 fixed, (7.64) expands out as

$$\begin{bmatrix} S_{11} & S_{12} & S_{13} & S_{14} \\ S_{21} & S_{22} & S_{23} & S_{24} \end{bmatrix} \begin{bmatrix} u_1 \\ u_2 \\ u_3 \\ u_4 \end{bmatrix} - k^2 \begin{bmatrix} T_{11} & T_{12} & T_{13} & T_{14} \\ T_{21} & T_{22} & T_{23} & T_{24} \end{bmatrix} \begin{bmatrix} u_1 \\ u_2 \\ u_3 \\ u_4 \end{bmatrix} = - \begin{bmatrix} T_{11} & T_{12} & T_{13} & T_{14} \\ T_{21} & T_{22} & T_{23} & T_{24} \end{bmatrix} \begin{bmatrix} g_1 \\ g_2 \\ g_3 \\ g_4 \end{bmatrix}. \tag{7.65}$$

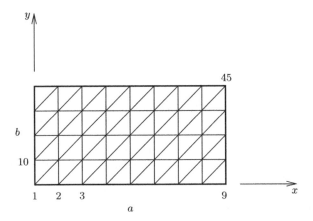

Figure 7.10 Rectangular waveguide with dimensions $a \times b$. There are 64 elements and 45 nodes.

As before, we solve for the unknowns u_1 and u_2. The matrices are built up element by element. The connection matrix C does not have to be stored.

A good problem to illustrate the solution technique is a z-independent waveguide, as shown in Figure 7.10. A rectangular waveguide will be discussed because the exact solution is readily available for comparison purposes. Extension to other cross-sectional shapes is easily done by simply changing the mesh.

Rather than solving for the field due to a particular excitation, we will take a slightly different track and find the waveguide modes and cutoff frequencies. In this case, a forcing function is not necessary. By letting $g = 0$, (7.64) becomes

$$Su = k^2 Tu. \tag{7.66}$$

This is a standard matrix eigenvalue problem, with eigenvalues k^2 and eigenvectors u. If there are N nodes, the solution will give N eigenvalues and N eigenvectors. Each eigenvalue corresponds to a cutoff frequency, and its eigenvector u is N numbers $u_1, u_2, \cdots, u_N$ that specify the shape of the mode. The matrix eigenvalue problem can be solved with SUBROUTINE garbow, which is described in Appendix G.

For TE modes, $u = H_z$. Then, E_x and E_y follow from the curl of H_z as in (3.8)–(3.11). The other field components are zero. Similarly, for TM modes, $u = E_z$, and the same equations give H_x and H_y.

Example 7.2 (Rectangular Waveguide Mode) As an example, let us find the lowest TE cutoff wavenumber and mode for the rectangular waveguide in Figure 7.10. In this case, $u = H_z$ so we are solving

$$\nabla^2 H_z + k^2 H_z = 0.$$

At cutoff, $\beta = 0$ and since $k^2 = k_c^2 + \beta^2$, the cutoff wavenumber is $k = k_c$. As for the boundary conditions, on the metal walls, $\mathbf{E}_{tan} = 0$ is required. Noting that

$$E_x \propto \frac{\partial H_z}{\partial y}$$

$$E_y \propto \frac{\partial H_z}{\partial x}$$

it follows that

$$\frac{\partial H_z}{\partial y} = 0; \quad y = 0, b$$

$$\frac{\partial H_z}{\partial x} = 0; \quad x = 0, a.$$

These are homogeneous Neumann boundary conditions, so the boundary nodes have to be free. The interior nodes are of course free as they are unknowns that must be found. Therefore, *all* the nodes in Figure 7.10 are free.

As a numerical example, suppose that $a = 0.8$ m and $b = 0.4$ m. Using the 45-node mesh in Figure 7.10 and solving a 45×45 matrix eigenvalue equation as in (7.66) gives 45 cutoff wavenumbers. The lowest one is $k_c = 3.95$. The exact result for the TE_{10} mode is $k_c = \pi/a = \pi/0.8 = 3.93$. The agreement is good, as the FEM wavenumber is only 0.51% higher than the exact answer. The corresponding eigenvector has 45 components $u_1, u_2, \cdots, u_{45}$ of which $u_1, u_2, \cdots, u_9$ correspond to the field at nodes along $y = 0$. This is shown in Figure 7.11 and compared with the exact field for the TE_{10} mode

$$H_z = \cos \frac{\pi x}{a} e^{-j\beta z}$$

and here, $\beta = 0$. It is seen that the computed and exact modes are in good agreement. An eigenvector can be scaled by any constant, and in this case it has been normalized to unity at $x = 0$.

In theory the TE_{10} mode does not vary with y, and as expected, it was found that the node values $u_{10}, u_{11}, \cdots, u_{18}$ in the second row and beyond do not differ appreciably from the first row at $y = 0$. In any case, it can be seen that the eigenvector is in good agreement with the theoretical shape of the mode.   ∎

To complete this discussion, we should also consider the TM modes, which satisfy

$$\nabla^2 E_z + k^2 E_z = 0.$$

H_z has been replaced by E_z, but otherwise the differential equations for the TE and TM cases are the same. However, the boundary conditions are different. For the TE case the boundary nodes are free, so

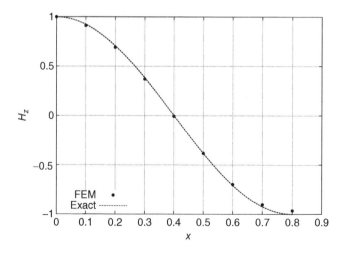

Figure 7.11   Eigenvector and exact H_z for the 45-node mesh. Nodes 1–9 are plotted, which correspond to H_z at $y = 0$.

$\partial H_z / \partial n = 0$ which is the natural (Neumann) boundary condition. For the TM case the boundary nodes are fixed at $u = E_z = 0$ which is the Dirichlet boundary condition. The matrix eigenvalue equation (7.66) should be solved, but only for the *free nodes*. Since the free nodes lie in the interior of the waveguide, the TM case will have slightly fewer unknowns than the TE case for a given mesh.

7.12   Galerkin's Method

There are two ways to develop an FEM solution. We have seen the variational approach, which finds the stationary point of an energy-related functional. Those same equations can be obtained by applying *Galerkin's method*.

An advantage of Galerkin's method is that it deals directly with the differential equation. This is especially helpful in cases where no underlying variational principle can be identified and a functional cannot be found. On the other hand, the variational method is better suited for explaining natural and essential boundary conditions. Galerkin's method will be demonstrated here with an application to Poisson's equation.

We now revisit the element configuration in Figure 7.2, where there are $N = 4$ nodes. u_1, u_2 are free and u_3, u_4 are fixed. Poisson's equation (7.40), repeated here, is

$$\nabla^2 u = g. \tag{7.67}$$

The boundary conditions are of the essential type (Dirichlet) at the fixed nodes

$$u = \text{const.} \tag{7.68}$$

and natural (Neumann) at the free nodes

$$\frac{\partial u}{\partial n} = 0. \tag{7.69}$$

The first step is to multiply both sides of (7.67) by M *testing functions* (also called *weighting functions*) β_i and integrate over the domain S_0 occupied by the elements:

$$\int_{S_0} \beta_i \nabla^2 u \, dS = \int_{S_0} \beta_i g \, dS; \quad i = 1, 2, \cdots, M. \tag{7.70}$$

Here, M is the number of tests, and it will be seen that M should equal the number of unknowns. In this case, $M = 2$, so (7.70) represents two equations. Other properties of the testing functions will be considered shortly.

Next, $u(x, y)$ will be approximated in the usual piecewise-planar manner. This will cause a problem, because (7.70) needs the second derivative of u. Although the piecewise-planar version of u is differentiable, it is not twice differentiable. The way to deal with this is to apply Green's theorem, which is really an integration by parts. Equation (7.70) becomes

$$-\int_{S_0} \nabla \beta_i \cdot \nabla u \, dS + \oint_{C_0} \beta_i \frac{\partial u}{\partial n} \, d\ell = \int_{S_0} \beta_i g \, dS, \tag{7.71}$$

where contour C_0 bounds S_0.

Now, u and g are expanded in terms of the *basis functions* $\alpha_j(x, y)$ as in

$$u(x, y) = \sum_{j=1}^{N} u_j \alpha_j(x, y); \quad g(x, y) = \sum_{j=1}^{N} g_j \alpha_j(x, y)$$

and here, $N = 4$ is the number of nodes in the connected mesh. These are the same interpolatory α_j as in (7.5) that have been used all along.

With these expansions, (7.71) becomes

$$-\sum_{j=1}^{N} u_j \int_{S_0} \nabla \beta_i \cdot \nabla \alpha_j \, dS + \sum_{j=1}^{N} u_j \oint_{C_0} \beta_i \frac{\partial \alpha_j}{\partial n} \, d\ell = \sum_{j=1}^{N} g_j \int_{S_0} \beta_i \alpha_j \, dS.$$

A critical aspect of Galerkin's method is that the basis functions and the testing functions are chosen to be the same. If we let $\beta_i = \alpha_i$, then

$$-\sum_{j=1}^{N} u_j \int_{S_0} \nabla \alpha_i \cdot \nabla \alpha_j \, dS + \sum_{j=1}^{N} u_j \oint_{C_0} \alpha_i \frac{\partial \alpha_j}{\partial n} \, d\ell = \sum_{j=1}^{N} g_j \int_{S_0} \alpha_i \alpha_j \, dS. \tag{7.72}$$

The line integral can be written as

$$\oint_{C_0} \alpha_i \frac{\partial u}{\partial n} \, d\ell.$$

The line integral is zero over the Neumann part of the boundary, where the free nodes 1 and 2 are subject to the natural boundary condition (7.69). The integral is also zero over the Dirichlet part of the boundary, where essential boundary conditions are assigned to the fixed nodes 3 and 4. This occurs because of the interpolatory property of the α_i testing functions. That is, $\alpha_1(x, y) = 0$ at nodes 3 and 4; likewise and $\alpha_2(x, y) = 0$ at nodes 3 and 4. Therefore the line integral vanishes on the entire boundary C_0. Equation (7.72) becomes

$$-\sum_{j=1}^{N} u_j \int_{S_0} \nabla \alpha_i \cdot \nabla \alpha_j \, dS = \sum_{j=1}^{N} g_j \int_{S_0} \alpha_i \alpha_j \, dS. \tag{7.73}$$

The result (7.73) is identical to (7.52) that was obtained by minimizing an energy-related functional. In general, the system of equations obtained from Galerkin's method are equivalent to the variational method.

7.12.1 Discussion

A more formal treatment of the problem in Section 7.12 would employ an *operator L* in a differential equation

$$Lu = g \tag{7.74}$$

and an *inner product*

$$\langle f, g \rangle = \int_{S_0} fg \, dS. \tag{7.75}$$

With these definitions and $L = \nabla^2$, the testing procedure (7.70) takes the form

$$\langle Lu, \beta_i \rangle = \langle g, \beta_i \rangle. \tag{7.76}$$

If u is the exact solution of (7.74), then substituting (7.74) in (7.76) gives

$$\langle g, \beta_i \rangle = \langle g, \beta_i \rangle$$

or, equivalently,

$$\int_{S_0} g\,\beta_i\,dS = \int_{S_0} g\,\beta_i\,dS.$$

Although this is true, it is not useful.

The expression (7.76) becomes useful with an approximate solution; call it $\bar{u}$. Then, $L\bar{u} \approx g$ but $L\bar{u} \neq g$. In this case (7.76) gives

$$\int_{S_0} L\bar{u}\,\beta_i\,dS = \int_{S_0} g\,\beta_i\,dS.$$

The integration can be thought of as an averaging, with respect to some weighting β_i. Hence the β_i are sometimes called weighting functions. The equality means that $L\bar{u}$ equals g, but only in some average sense.

Suppose now that $L = \nabla^2$ and $\bar{u}$ is the piecewise-planar approximation of u. Since $\bar{u}$ is not twice differentiable, an integration by parts should be applied, giving a new L' with only first derivatives. Then $L'\bar{u}$ can be evaluated and (7.76) becomes

$$\langle L'\bar{u}, \beta_i \rangle = \langle g, \beta_i \rangle. \tag{7.77}$$

The procedure (7.77) with $\beta_i = \alpha_i$ along with the approximate $\bar{u}$ leads to the FEM solution. One can say that for all the formulations presented in this chapter, (7.77) *is* the FEM.

The continuity properties of the once-differentiable $\bar{u}$ are weaker than those of the twice-differentiable u, so $\bar{u}$ is called the *weak form* of the solution and u is called the *strong form*. Equation (7.77) is called the *weak form* of the differential equation.

The advantage of a weak form is that the formulation (in this case using planar triangles) is easier to construct than a strong form. Numerical results from the weak form, for all practical purposes, are often just as good as the strong form. For this reason, a lot of emphasis is placed on weak-form solutions.

It should be noted that L could be a differential or integral operator. Therefore, Galerkin's method can be used to solve both differential and integral equations.

The expression (7.77) with $\beta_i = \alpha_i$ is Galerkin's method. However, some other weighting functions $\beta_i \neq \alpha_i$ could have been used. This gives the *method of weighted residuals*. It is possible to define a *residual* r as in

$$Lu - g = r.$$

The residual is ideally zero, but with an approximate solution it will be small. In this method, the weighted residual is set to zero:

$$\langle L'\bar{u}, \beta_i \rangle - \langle g, \beta_i \rangle = \langle r, \beta_i \rangle = 0$$

which is identical to (7.77). Therefore, one can say that Galerkin's method is a special case of the method of weighted residuals, having $\beta_i = \alpha_i$.

Yet another possibility is to implement the *collocation method* (also known as the *point-matching method*). This amounts to making (7.77) hold true at M discrete 'match points'; call them (x_i, y_i). From an implementation point of view, all one has to do is use testing functions β_i that are impulses at the match points (x_i, y_i). The inner product integrals become trivial, and (7.77) becomes

$$L'\bar{u}(x_i, y_i) = g(x_i, y_i); \quad i = 1, 2, \ldots, M. \tag{7.78}$$

By letting M equal the number of unknowns, a square matrix is obtained for the system of equations. The point-matching method is generally less accurate than Galerkin's method and is not as widely used.

7.12.2   *Variational Method*

In broad terms, finding the stationary point of a functional belongs to a branch of mathematical techniques called *variational methods*. We have used this to develop several FEM solutions. This involved minimizing an energy-related functional. It would obviously be useful to have a general-purpose procedure for finding the functional. The following results are stated without proof but can be found in Volakis et al. (1998, Chapter 1), Jin (2002, Chapter 6) and elsewhere. If the functional

$$F(u) = \frac{1}{2}\langle Lu, u\rangle - \langle g, u\rangle \tag{7.79}$$

is rendered stationary, then the resulting u is a solution of the operator equation

$$Lu = g. \tag{7.80}$$

In frequency-domain problems where u could be complex, the functional should be

$$F(u) = \frac{1}{2}\langle Lu, u^*\rangle - \frac{1}{2}\langle g, u^*\rangle - \frac{1}{2}\langle g^*, u\rangle, \tag{7.81}$$

where the '*' denotes the complex conjugate.

The specific case of approximating u as a linear combination $u \approx \sum u_i \alpha_i$ and then making the functional stationary is known as the *Rayleigh–Ritz method*. It leads to the same linear system of equations as Galerkin's method. In the language of variational methods, the $\alpha_i(x, y)$ are called *trial functions* because they represent approximate guesses ('trials') of the true solution. The u_i are called *variational parameters*, which have to be adjusted until the functional is rendered stationary. The resulting *trial solution* is the variational method's approximation of the exact solution.

As an example of using (7.79), we can find the functional for the wave equation (7.55). It is of the form $Lu = g$ where

$$Lu = \nabla \cdot (p\nabla u) + k^2 qu.$$

Applying (7.79) gives

$$F(u) = \frac{1}{2}\int_{S_0} u\nabla \cdot (p\nabla u) + k^2 qu^2 - 2gu \, dS.$$

By integrating the vector identity $f\nabla \cdot \mathbf{A} = \nabla \cdot (f\mathbf{A}) - \mathbf{A} \cdot \nabla f$ over S_0 and applying the divergence theorem to the $\nabla \cdot (f\mathbf{A})$ term, we obtain

$$\int_{S_0} f\nabla \cdot \mathbf{A} \, dS = \oint_{C_0} f\mathbf{A} \cdot \hat{\mathbf{n}} \, d\ell - \int_{S_0} \mathbf{A} \cdot \nabla f \, dS.$$

With $f = u$ and $\mathbf{A} = p\nabla u$, we can use this to rewrite $F(u)$ as

$$F(u) = \frac{1}{2}\int_{C_0} pu\frac{\partial u}{\partial n} \, d\ell - \frac{1}{2}\int_{S_0} p\nabla u \cdot \nabla u - k^2 qu^2 + 2gu \, dS.$$

This is the functional (7.56) used earlier. There is an overall minus sign on F here, but it has no consequence, as we intend to set the derivative of F equal to zero. By letting $k = 0$ and $p = 1$, we obtain the functional (7.43) for Poisson's equation. In both cases, natural and/or essential boundary conditions are used to make the boundary integral over C_0 vanish.

7.13   Vector Wave Equation

In some situations the scalar wave equation is inadequate and the vector version must be solved. The vector wave equation is

$$\nabla \times p\nabla \times \mathbf{u} - k^2 q\mathbf{u} = 0. \tag{7.82}$$

The necessary functional, given here without proof (see Silvester and Ferrari 1996, Chapter 3), is

$$F(\mathbf{u}) = \frac{1}{2}\int_{S_0} \nabla \times \mathbf{u} \cdot p\nabla \times \mathbf{u} - k^2 q\mathbf{u} \cdot \mathbf{u}\, dS. \tag{7.83}$$

The functional has an associated boundary integral, which vanishes if $\mathbf{u}$ obeys the essential (or prescribed) boundary condition

$$\hat{\mathbf{n}} \times \mathbf{u} = \mathbf{u}_{tan}, \tag{7.84}$$

where $\mathbf{u}_{tan}$ is given as a known value. Otherwise, $\mathbf{u}$ meets the natural boundary condition

$$\hat{\mathbf{n}} \cdot \nabla \times \mathbf{u} = 0 \tag{7.85}$$

which is a property of the functional and does not have to be imposed explicitly.

If the formulation is in terms of $\mathbf{u} = \mathbf{E}$, then the essential boundary condition prescribes $\mathbf{E}_{tan}$. The natural boundary condition governs the normal magnetic field, making $H_n = 0$. Likewise, if $\mathbf{u} = \mathbf{H}$, then the essential boundary condition prescribes $\mathbf{H}_{tan}$ and the natural boundary condition is $E_n = 0$. We can choose either an $\mathbf{E}$ or $\mathbf{H}$ as the 'working variable', depending on which boundary conditions are preferable for a given problem.

7.14   Other Element Types

This section provides a summary of other element types and their purposes. The treatment is necessarily short, but readers may consult the references at the end of the chapter for further information.

The triangular finite elements used thus far are a first-order (planar) approximation for the scalar potential u. Higher-order triangles (e.g. quadratic, cubic) are possible and permit the use of larger elements. In 3D, the *tetrahedron* having four faces is used to represent a potential $u(x, y, z)$. Higher-order tetrahedra are also possible.

The first-order triangles are completely determined by the three node values u_i, so they are called *node-based* elements. The tetrahedron with four points is also node based. In either case, they are used to represent a scalar function u. There is another class of *edge-based* elements. These are vector elements and are necessarily more complex, but are better for representing an electromagnetic field. Edge-based elements also admit the possibility of higher-order representations, in both 2D and 3D.

7.14.1   Node-Based Elements

In 2D we have used piecewise-planar node-based elements, with

$$u(x, y) = a + bx + cy.$$

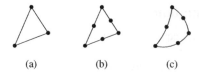

(a)   (b)   (c)

Figure 7.12   (a) First-order element. (b) Second order in potential. (c) Second order in geometry and second order in potential.

Within the domain (x, y) the element edges are along straight lines, so this is also first order (linear) with respect to the geometry. This is shown in Figure 7.12(a).

It is possible to have a higher-order triangle with a quadratic potential

$$u(x, y) = a + bx + cy + dx^2 + ey^2 + fxy.$$

By using a set of equations similar to (7.4), the coefficients $a, b, \cdots, f$ can be obtained by fitting u to six points. This is illustrated in Figure 7.12(b). The domain (x, y) is still meshed in a piecewise-linear fashion, but the range u is now quadratic. With some effort, the representation can be rearranged in the form (Jin 2002, Section 4.7)

$$u(x, y) = \sum_{i=1}^{6} u_i N_i(x, y),$$

where the $N_i(x, y)$ are interpolatory functions having the property

$$N_i(x_j, y_j) = \begin{cases} 1; & i = j \\ 0; & i \neq j. \end{cases}$$

The higher-order representation lets us use a mesh with a smaller number of larger triangles. This is an improvement, but it causes another problem. That is, if a boundary is curved, the straight edges of the triangular mesh might not fit it very well. Therefore it is desirable to have a representation that is of higher order for both u *and* the geometry. Figure 7.12(c) shows an example of an element that is quadratic for both u and the geometry. This is called an *isoparametric element*. The name stems from the fact that both the geometry and the potential have the same degree of approximation.

Besides an isoparametric element, it is possible to have a *subparametric element*, with the geometry represented at a lower order than the potential function. There is also the *superparametric element*, with the geometry represented at a higher order than the potential function. The first-order triangle we have been using is isoparametric. Subparametric elements are common, but superparametric ones are generally not used.

In 3D, a potential $u(x, y, z)$ can be represented in a linear manner by a *tetrahedron*, which has four faces and is defined by four points. This is shown in Figure 7.13(a). These fit together in 3D, without

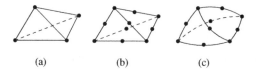

(a)   (b)   (c)

Figure 7.13   (a) First-order tetrahedron. (b) Second-order tetrahedron. (c) Second-order isoparametric tetrahedron.

any gaps. A second-order tetrahedron is specified in terms of 10 points, having 4 at the vertices plus 6 others at the midpoints, as shown in Figure 7.13(b). The second-order isoparametric version is shown in Figure 7.13(c).

In N dimensions we need $N + 1$ or more points to define a shape function. If we use the minimum number $N + 1$, it is called a *simplex element*. A simplex in N dimensions is called an N-simplex. In 1D, the 1-simplex is a straight line. In 2D, the 2-simplex is a triangle. In 3D, the 3-simplex is a tetrahedron.

There are viable alternative elements that are not simplexes. In 2D, there is the quadrilateral, having four points. In 3D we could use a hexahedron, or 'brick element' with six faces and eight points. An advantage of these elements is that compared to simplexes, it is easier to develop the Dirichlet matrix and the metric. A disadvantage is a relative lack of robust and automatic mesh generator software and less geometric flexibility, as compared with simplex elements. The quadrilateral and brick have higher-order and isoparametric versions, the 2D curvilinear quadrilateral and the 3D curvilinear hexahedron (or 'squashed brick').

7.14.2   Spurious Modes

The wave equation is derived from the two Maxwell curl equations. The divergence equations $\nabla \cdot \mathbf{D} = 0$ and $\nabla \cdot \mathbf{B} = 0$ are not needed because they are automatically satisfied (see Section 1.7). However, the wave equation involves second derivatives, so this argument only holds if the wave solution is twice differentiable. The node-based FEM solution is only once differentiable[2] because it is piecewise planar. Therefore, satisfaction of the divergence conditions is not assured. A side effect of not meeting the divergence conditions is that it causes *spurious modes*. These are additional non-physical modes that contaminate the FEM solution for the eigenvalues and eigenvectors. They coexist with the correct modes and could be ignored; however in practice it becomes a nuisance to try and figure out which modes are true and which are spurious. Furthermore, in a node-based mesh, *all* of the field components are continuous. This can cause the normal components to be incorrect – for instance, at an interface between two dielectrics, where E_n has to be discontinuous.

In the node-based FEM for waveguides, the spurious mode problem can be eliminated by using a formulation in terms of the transverse vector fields instead of E_z or H_z. The correct boundary condition for the normal field component is enforced by using a modified functional. The technique can be applied if there is an inhomogeneous ϵ or μ, but not both. Many researchers have used this approach to successfully find the modes for structures such as dielectric waveguides.

Another way to eliminate spurious modes is to use *edge-based* elements. These offer the additional advantage of permitting both inhomogeneous ϵ and μ, simultaneously. However, the importance of edge-based elements goes far beyond solving this specific problem. They form the foundation of a more general and reliable (albeit more complicated) version of the FEM. These elements will now be described.

7.14.3   Edge-Based Elements

In a node-based mesh, all of the field components are continuous. In contrast, edge-based elements enforce continuity of the tangential field, but not the normal field. This makes them highly suitable for representing electric and magnetic fields. As for the normal components, no extra attention is needed because the variational property ensures that they look after themselves through the natural boundary condition, as discussed in Section 7.13.

---

[2] In more mathematical terms, the mesh is C^0 continuous at the element junctions. C^n continuous means a function's nth derivative is continuous, so C^0 means the function is continuous and C^1 means its derivative is continuous. Continuity and spurious modes are further discussed in Jin (2002, Section 5.8.4).

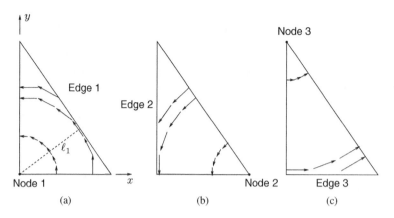

Figure 7.14   Edge-based element, showing (a) $\tau_1(x,y)$, (b) $\tau_2(x,y)$ and (c) $\tau_3(x,y)$.

Suppose we have a z-invariant 2D problem and a triangular element. A transverse vector function **u** (having only x and y components) can be represented with an edge-based element as

$$\mathbf{u}(x,y) = \sum_{i=1}^{3} u_i \tau_i(x,y). \tag{7.86}$$

The unknowns u_i are related to the tangential value of **u** along each element edge. The τ_i are vector shape functions, shown in Figure 7.14. They are interpolatory, with each one having a tangential component equal to unity on one edge and no tangential component on the other two edges. Mathematically expressed, if $\hat{\mathbf{e}}_j$ is a unit vector tangent to edge #j, then

$$\tau_i \cdot \hat{\mathbf{e}}_j = \begin{cases} 1; & i = j \\ 0; & i \neq j. \end{cases}$$

In Figure 7.14(a), if polar coordinates are used, the shape function can be written as $\tau_1(\rho, \phi) = \hat{\phi}\, \rho/\ell_1$; similar expressions exist for τ_2 and τ_3. This is only for illustrative purposes; general-purpose shape functions for 2D triangles, 3D tetrahedrons and many other element types have been worked out and can be found in the references at the end of the chapter.

When edge-based elements are used in a mesh, the tangential field is continuous so that connected triangles having a common edge will have the same u_i (within a $\pm$ sign, depending on how the $\tau_i(x,y)$ are defined).

Since the tangential field is continuous at the junction between elements, $\nabla \times \mathbf{u}$ is finite there (see Problem 7.13). Because of this property, edge-based elements are called *curl-conforming* elements. It turns out that the Dirichlet matrix for edge-based elements involves integrals of the form $\int \tau_i \cdot \tau_j \, dS$ and $\int (\nabla \times \tau_i) \cdot (\nabla \times \tau_j) \, dS$; the curl-conforming property is essential in the latter case.

As an example using edge-based elements, a TE waveguide having a mesh with N_E edges and N_0 nodes could have the magnetic field expressed as

$$\mathbf{H}(x,y) = \sum_{i=1}^{N_E} u_i^e \tau_i(x,y) + \hat{\mathbf{z}} \sum_{i=1}^{N_0} u_i^n \alpha_i(x,y).$$

The first sum uses edge-based shape functions τ_i and the second sum uses node-based scalar shape functions α_i. The coefficients u_i^e and u_i^n are then found from the FEM procedure, via either the variational or Galerkin method. There is no problem with using a node-based representation for H_z, because it is continuous.

7.15   Radiating Structures

The FEM can be used to solve for the fields of antennas and scatterers. The difficulty is that in theory, for any radiating structure the mesh must extend to infinity. In practice, the domain has to be truncated in some way. A few schemes will now be briefly described.

7.15.1   Absorbing Boundary Condition

For illustrative purposes we can consider the simple case in Figure 7.15(a), a perfectly conducting cylinder and a TE field. The cylinder radius is equal to a and there is an absorbing boundary C_b at $\rho = b$. In Figure 7.15(b), it is shown how the contours follow the right-hand rule. The normal $\hat{n}$ points out of S_0, and in the limit the straight segments do not contribute.

The region S_0 on $a \leq \rho \leq b$ is to be meshed with node-based elements, and the working variable is $u = H_z$. A magnetic line source $\mathbf{M}_0 = \hat{z}M_0$ provides the excitation. The wave equation to be solved is

$$\nabla^2 H_z + k^2 H_z = j\omega\epsilon M_0,$$

where $k = \omega\sqrt{\mu\epsilon}$. With $u = H_z$ and $g = j\omega\epsilon M_0$, it is the same as (7.57), repeated here:

$$\nabla^2 u + k^2 u = g. \tag{7.87}$$

The solution will be obtained with the Galerkin method. Being mindful of the new $k^2 u$ term, we can reuse the results from Poisson's equation in Section 7.12. Analogous to (7.71) we have

$$-\int_{S_0} \nabla\beta_i \cdot \nabla u \, dS + \oint_{C_0} \beta_i \frac{\partial u}{\partial n} \, d\ell + \int_{S_0} k^2 \beta_i u \, dS = \int_{S_0} \beta_i g \, dS, \tag{7.88}$$

where S_0 is bounded by the contour $C_0 = C_a + C_b$. Since the contour has two parts, we can write the boundary integral as

$$I_0 = \oint_{C_0} \beta_i \frac{\partial u}{\partial n} \, d\ell = \underbrace{\oint_{C_a} \beta_i \frac{\partial u}{\partial n} \, d\ell}_{I_a} + \underbrace{\oint_{C_b} \beta_i \frac{\partial u}{\partial n} \, d\ell}_{I_b}. \tag{7.89}$$

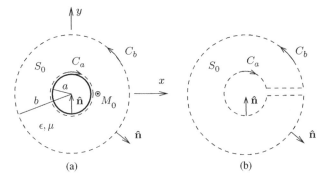

(a)   (b)

Figure 7.15   (a) Metal cylinder with magnetic line source. S_0 is bounded by $C_0 = C_a + C_b$ and an absorbing boundary is at C_b. (b) Relation between S_0, $\hat{n}$, C_a and C_b.

For the reasons discussed in Section 7.12, I_a will be zero. However, no such conditions exist at the outer boundary, so $I_b \neq 0$ and it will have to be found.

The problematic term $\partial u/\partial n$ in I_b needs to be eliminated. If we assume that the boundary at $\rho = b$ is sufficiently far away, the field will behave approximately as $u \propto e^{-jk\rho}/\sqrt{\rho}$. Then, differentiation with respect to ρ gives

$$\frac{\partial u}{\partial \rho} = -u\left(jk + \frac{1}{2\rho}\right). \tag{7.90}$$

This is called the Bayliss–Turkel *absorbing boundary condition* (ABC). It represents the correct termination for an outward-radiating cylindrical wave, far away from the cylinder and any sources. Since the fields are zero beyond $\rho = b$ and there is no reflection, we can imagine the boundary as being absorptive.

Since (7.90) has a derivative with respect to ρ, it is called a first-order ABC. A more accurate second-order version with a $\partial^2 u/\partial \phi^2$ term is also possible (see Volakis et al. 1998, Section 4.4). A distance on the order of two wavelengths might be needed to get an accurate result with the second-order ABC (Peterson and Castillo 1989).

With $\rho = b$ and $d\ell = bd\phi$ in (7.89) and (7.90), the ABC gives

$$I_b = -\int_0^{2\pi} \beta_i u \left(jk + \frac{1}{2b}\right) bd\phi.$$

Next, u is expanded as $\sum u_j \alpha_j$, and Galerkin testing with $\beta_i = \alpha_i$ is used. Equation (7.88) becomes

$$-\sum_{j=1}^{N} u_j \int_{S_0} \nabla \alpha_i \cdot \nabla \alpha_j \, dS - \sum_{j=1}^{N} u_j \int_0^{2\pi} \alpha_i \alpha_j \left(jk + \frac{1}{2b}\right) bd\phi$$

$$+\sum_{j=1}^{N} u_j \int_{S_0} k^2 \alpha_i \alpha_j \, dS = \sum_{j=1}^{N} g_j \int_{S_0} \alpha_i \alpha_j \, dS. \tag{7.91}$$

Equation (7.91) can be solved for the most part with techniques already described in earlier sections. The one new thing we need to examine is the boundary integral, which is a 1D line integral and not a surface integral. Specifically, let us consider the expansion functions α_j that touch the boundary C_b. A portion of the mesh is shown in Figure 7.16. Element (1) shown in case (b) will have $u = u_1\alpha_1 + u_2\alpha_2 + u_3\alpha_3$. However, because of the interpolatory property of the α_j, for all points on the line segment ℓ_{12} joining nodes 1 and 2, we will have $\alpha_3 = 0$. Therefore, $u = u_1\alpha_1 + u_2\alpha_2$ is a 1D linear interpolation of u between nodes 1 and 2, as shown in case (c). Consequently, a knowledge of the two node values is sufficient for evaluation of the boundary integral I_b on that segment. The same approach is then applied to all of the boundary nodes.

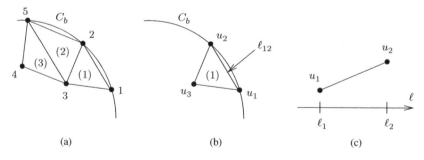

Figure 7.16 (a) Some elements at the boundary C_b. (b) Element (1) with node values u_1, u_2, u_3. (c) Linear interpolation between nodes 1 and 2.

7.15.2 Artificial Absorber

Another way to deal with radiating structures is to enclose them in a kind of anechoic chamber. The chamber walls can be designed to absorb the outgoing waves. Because this is a mathematical absorber, considerable liberties can be taken in choosing the μ, ϵ and σ of the walls. Some very good absorbers can be designed, but since they use 'made-up' materials, they are called artificial absorbers. If the absorber is sufficiently lossy, its backside can be terminated with a perfect conductor, which then makes the computational domain finite in extent.

An absorber that has the same impedance as free space will produce zero reflection at normal incidence. In other words, it should have

$$\eta = \sqrt{\frac{\mu_0\mu_r}{\epsilon_0\epsilon_r}} = 377\,\Omega$$

which implies that $\mu_r = \epsilon_r$. To implement the absorber, it is necessarily finite in thickness and is backed by a perfect electric conductor to terminate the computational domain. To prevent excessive reflection by the metal backing behind the absorber panel, the artificial material should have some loss. Another issue is that this design only works for a normal-incidence plane wave. With a numerical search, useful configurations can be obtained, and it is suggested that $\mu_r = \epsilon_r = 1 - j2.7$ and a material thickness of $0.15\lambda_0$ gives good results (Volakis et al. 1998, Section 4.4).

A more sophisticated absorber design uses anisotropic permittivity and permeability, such that

$$\bar{\bar{\epsilon}}_r = \bar{\bar{\mu}}_r = \begin{bmatrix} c & 0 & 0 \\ 0 & c & 0 \\ 0 & 0 & 1/c \end{bmatrix}, \tag{7.92}$$

where $c = \alpha - j\beta$ is a constant. This configuration makes it possible to have a zero reflection coefficient for all angles of incidence. It is called a *perfectly matched layer* (PML). In practice, a finite thickness and a metal backing are used, so that the computational domain is finite. In this case, the reflection coefficient can be made small, but it will no longer be zero. A typical value is $c = 1 - j1$ and a layer thickness on the order of $\lambda_0/10$. Optimal design criteria are discussed in Sacks et al. (1995) and Kingsland et al. (1996). A consideration for this type of absorber is that the design assumes plane-wave incidence, so placing it too close to the radiating structure can reduce its effectiveness.

Any type of artificial absorber will be meshed, with a resulting increase in the number of unknowns. Therefore, it is desirable that the absorber be as electrically thin as possible. It should be noted that in the above discussions the absorber thickness has been given in terms of λ_0, which is the wavelength outside the absorber and is usually the free-space wavelength.

7.15.3 Boundary Element Method

In the boundary element method (BEM) the outer bounding surface C_b in Figure 7.15 has tangential electric and magnetic fields that are treated as unknowns. These are the *boundary elements*. Via radiation integrals (as in Chapter 4), one can find the field that they produce, outside C_b. Since we are using a Green's function to find the field radiated by the unknown boundary elements, the BEM is a surface integral equation and could just as well be called the method of moments. The region inside C_b, however, is formulated as an FEM mesh. By invoking the continuity of $\mathbf{E}_{tan}$ and $\mathbf{H}_{tan}$ at C_b, one can obtain a system of equations that can be solved for the mesh unknowns inside and the surface unknowns on C_b. This is a 'hybrid' method, because it uses the FEM on the inside and the BEM on the outside. More properly, it is called the BEM/FEM.

The BEM/FEM hybrid combines the best of both types of solutions. A good example application is the computation of fields radiated by a mobile phone in the presence of a human head. Outside, the unbounded and homogeneous region is well suited for an integral-equation formulation. Since the Green's function takes care of the radiation, it is unnecessary to use an ABC or artificial absorber. Inside the head, the FEM is highly suitable for the inhomogeneous materials.

A disadvantage of the BEM/FEM approach is that unlike the pure FEM, the boundary integral contains a Green's function, and the associated matrix for it is not sparse. This makes it more difficult to obtain a rapid solution of the matrix equations. In some cases, the boundary integral can be written as a convolution, and a fast Fourier transform employed to accelerate the solution. The details of the BEM/FEM formulation are omitted here but can be found in the references at the end of the chapter.

7.16 Further Reading

This chapter presented selected topics and applications of the FEM. Readers who want to know more should consult the book by Silvester and Ferrari (1996). Many of the developments and the notation used in this chapter have followed that book. Other helpful books, especially for topics related to radiation, include Volakis et al. (1998), Jin (2002) and Peterson et al. (1998).

The FEM is a projective method, and it rests on the mathematical theories of Hilbert spaces and linear operators. This is important when considering the mathematical requirements for expansion functions, as well as investigating convergence properties and solution accuracy. These concepts are discussed in Silvester and Ferrari (1996) and in greater depth by Dudley (1994) and Jones (1994).

The variational method is a classic technique in mathematical physics. Descriptions from this perspective can be found in the books by Morse and Feshbach (1953, Chapter 3) and Hildebrand (1965, Chapter 2). A good presentation of the variational method with electromagnetic applications is given by Harrington (2001, Chapter 7). Of historical interest, there is the original paper by Galerkin (1915).

The mathematical properties of edge-based elements were thoroughly investigated by Nedelec (1980). Since then, their applications in electromagnetics have steadily grown. It is a highly mathematical paper and not an easy read. Webb (1993) provides a good engineering-oriented review of edge-based elements and their use in electromagnetics.

References

Dudley DG (1994) *Mathematical Foundations of Electromagnetic Theory*. Oxford University Press.

Galerkin BG (1915) Rods and plates: series in some questions of elastic equilibrium of rods and plates. *Vest. Inzh. Tech.* **19**(10), 897–908. (In Russian).

Harrington RF (2001) *Time-Harmonic Electromagnetic Fields*. IEEE Press.

Hildebrand FB (1965) *Methods of Applied Mathematics*. Prentice-Hall.

Jin J (2002) *The Finite Element Method in Electromagnetics*. John Wiley & Sons, Inc.

Jones DS (1994) *Methods in Electromagnetic Wave Propagation*. Oxford University Press.

Kingsland DM, Gong J, Volakis J and Lee JF (1996) Performance of an anisotropic artificial absorber for truncating finite-element meshes. *IEEE Trans. Antennas Propag.* **44**, 975–981.

Morse PM and Feshbach H (1953) *Methods of Theoretical Physics*. McGraw-Hill.

Nedelec JC (1980) Mixed finite elements in R^3. *Numer. Math.* **35**, 315–341.

Peterson AF and Castillo SP (1989) A frequency-domain differential equation formulation for electromagnetic scattering from inhomogeneous cylinders. *IEEE Trans. Antennas Propag.* **AP-37**, 601–607.

Peterson AF, Ray SL and Mittra R (1998) *Computational Methods for Electromagnetics*. IEEE Press.

Sacks ZJ, Kingsland DM, Lee R and Lee JF (1995) A perfectly matched anisotropic absorber for use as an absorbing boundary condition. *IEEE Trans. Antennas Propag.* **43**, 1460–1463.

Silvester PP and Ferrari RL (1996) *Finite Elements for Electrical Engineers*. Cambridge University Press.

Volakis J, Chatterjee A and Kempel LC (1998) *Finite Element Method in Electromagnetics*. John Wiley & Sons, Inc.

Webb JP (1993) Edge elements and what they can do for you. *IEEE Trans. Magn.* **29**(2), 1460–1465.

Problems

7.1   An ideal parallel-plate capacitor is shown in Figure 7.17. Fringing is neglected, so $\partial u/\partial n = 0$ on the sides. The plate width is w and the plate spacing is d. The capacitance per unit length (in z) is $C = \epsilon w/d$. A possible mesh is also shown. Assume that $w = d = 0.02$ m and that the medium is air.

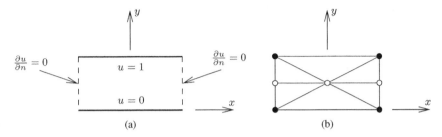

Figure 7.17   (a) Ideal parallel-plate capacitor with no fringing and (b) mesh with fixed • and free ∘ nodes.

    (a) Use the mesh in the figure to solve for the potential u at the nodes. This can be done with `PROGRAM fem2d`.

    (b) Use the program to calculate the stored energy.

    (c) From the theory for an ideal capacitor, find the stored energy.

7.2   Obtain the FEM solution of Problem 7.1 by using a hand calculation.

7.3   A coaxial cable has a centre conductor with a diameter of 5.0 mm and a shield with a diameter of 7.0 mm. The dielectric is polyethylene, having a relative permittivity of $\epsilon_r = 2.25$.

    (a) Find C, L and Z_0 from coaxial cable theory.

    (b) Find C from the FEM. Use symmetry to reduce the size of the problem.

    (c) Examine the convergence of Z_0 to the theoretically exact answer by trying two or three different mesh densities.

7.4   A stripline is shown in Figure 7.18. The centre conductor is at $u = 1$ V and the surrounding shield is at $u = 0$.

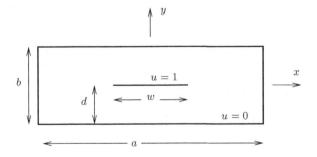

Figure 7.18   Stripline with centre conductor surrounded by ground.

(a) Neglecting all fringing effects and the metal walls at $x = \pm a/2$, consider this as a structure having three parallel plates. What is the potential function and the capacitance per unit length?

(b) Solve this problem using the FEM. Use symmetry to reduce the number of unknowns. Assume that $w = d = 0.02$ m and that $a = 3w$, $b = 2d$. The medium is air. Find C and Z_0. Show a sketch of your mesh and the numerical values of the potentials.

7.5 For the stripline in Figure 7.18, the centre conductor is at $u = 1$ V and the surrounding shield is at $u = 0$. Assume that $w = d = 0.02$ m and that $a = 3w$, $b = 2d$.

(a) Find C, L and Z_0 when the stripline is filled with air.

(b) Find C when the stripline is filled with air for $y > 0$ and has a dielectric with $\epsilon_r = 4$ for $y < 0$.

(c) Using the information from parts (a) and (b) and making the quasi-TEM assumption, obtain the effective dielectric constant, phase velocity and Z_0 for the configuration in part (b).

7.6 Use the stripline in Figure 7.18 to estimate the effect of fringing in a parallel-plate capacitor. Compare the fringing and non-fringing results when $d/w = 1$ and $d/w = 0.1$. Use $a = 3w$ and $b = 2d$. Demonstrate that fringing is less important when d/w is small.

7.7 A 2D magnetostatic problem has a z-directed current density and the vector potential is $\mathbf{A} = \hat{z}A$. At a boundary surface with a tangent coordinate t and normal coordinate n, show that $\mathbf{A}$ and $\mathbf{B}$ are related in the following way:

$$B_n = -\frac{\partial A}{\partial t}; \quad B_t = \frac{\partial A}{\partial n}.$$

Therefore if $\mathbf{B}$ is purely normal to the boundary, the Neumann condition is required for A. If $\mathbf{B}$ is purely tangential, then $A = $ const. is required. You can assume that the boundary is at $y = 0$, the tangent direction is $t = x$ and the normal direction is $n = y$.

7.8 Use the 2D FEM to calculate the cutoff wavenumber for the TE_{10} mode of a rectangular waveguide that is 0.8 m wide and 0.4 m high. Do this with a hand calculation, using two triangles.

7.9 Use the 2D FEM to calculate the cutoff wavenumber k_c for the TE_{10} mode of a rectangular waveguide that is 0.8 m wide and 0.4 m high. Modify PROGRAM fem2d to obtain the S and T matrices. All nodes should be free when calculating S. Find the eigenvalues with SUBROUTINE garbow, described in Appendix G. Perform a convergence study, using 2, 4 and 8 elements. Compare k_c with the exact answer.

7.10 Extend the convergence study of Problem 7.9 to 16, 32 and 64 elements. Calculate k_c and compare it with the exact answer.

7.11 Use the FEM to find the cutoff wavenumber of the TE_{11} mode in a circular waveguide of radius $a = 1$ cm. Compare with the theoretical value of $k_c = 1.841/a$. Ensure that your mesh is good enough to come within 10 % of this value. You can use the mode's symmetry to reduce the number of elements. Show the mesh that was used.

7.12 Solve the differential equation

$$\frac{d^2 f}{dx^2} = 1 - 5x^2; \quad f(0) = f(1) = 0.$$

With basis functions of the form $f_n = x(1 - x^n)$, apply

(a) Point matching at $x = 0.25$, $x = 0.75$.
(b) Galerkin's method.
(c) Make a plot, comparing with the exact solution.

7.13 Show that if the tangential field is continuous at a junction of two elements, the curl is finite (a curl-conforming element). Assume the z-invariant 2D TE case, with two triangular elements that are joined along $x = 0$. Show that a discontinuous tangential field

$$E_y = \begin{cases} 1; & x < 0 \\ 2; & x > 0 \end{cases}$$

gives an infinite curl at $x = 0$, whereas a continuous tangential field

$$E_y = \begin{cases} 1 + x; & x < 0 \\ 1 - x; & x > 0 \end{cases}$$

gives a finite curl at $x = 0$.

8

Uniform Theory of Diffraction

Geometrical optics (GO) provides a physically intuitive model for wave propagation. It allows us to calculate fields in terms of incident, reflected and refracted rays. In 1953 it was a breakthrough when Joseph B. Keller at New York University found a way to extend GO to include diffraction effects. He called it the geometrical theory of diffraction (GTD). Subsequently, this theory was advanced by Robert Kouyoumjian, Prabhakar Pathak and other researchers at Ohio State University, who developed the uniform geometrical theory of diffraction (UTD).

UTD and other ray-based methods require that the object under analysis should be electrically large in terms of the wavelength. In practice this means the size of the object should be at least on the order of a wavelength or more. At the other extreme, these methods become increasingly accurate as the frequency is increased, and there is no upper frequency limit.

The UTD has many applications, for example, the analysis and design of aperture-type antennas, the prediction of radiation characteristics of antennas on complex structures such as satellites, aircraft, and ships, and indoor and outdoor radio propagation modelling.

This chapter begins with a 2D approach for explaining UTD concepts. 2D models can sometimes be used for the realistic modelling of 3D situations, and it also prepares the reader for subsequent developments of the 3D UTD formulations.

UTD wedge diffraction theory is introduced in 2D. It is then used in 3D to find the radiation patterns of slot and monopole antennas on a finite ground plane. GO reflection by convex curved surfaces is treated next. Then we return to the wedge and extensions of the theory that allow for curved wedge faces, curved edges and non-metallic materials. Wedge slope diffraction is presented and provides higher-order terms that can be useful for dipole-like sources. Double and multiple diffraction between multiple edges is given next. When edge-diffracted rays focus to a caustic, the ray solution is singular and this is solved by the method of equivalent edge currents. The diffraction by smooth, convex curved surfaces is described in terms of creeping waves, which literally 'creep' along the curved surface.

8.1 Fermat's Principle

In ray optics, Fermat's principle of least time states that a ray joining two points A and B will follow a path such that the transit time is minimized. A few cases are illustrated in Figure 8.1. Case (a) shows a dielectric half space along with incident, reflected and refracted rays. The incident ray of course follows a straight line from A to B. The reflection point Q_R is located such that the reflected ray transit time from A to B is minimum. The refracted path from A to B_t also gives the minimum time.

Keller extended the work of Fermat, by postulating the existence of new types of rays – diffracted rays, that obey a *generalized* Fermat's principle. These are shown in cases (b) and (c). In case (b) a ray

Applied Frequency-Domain Electromagnetics, First Edition. Robert Paknys.
© 2016 John Wiley & Sons, Ltd. Published 2016 by John Wiley & Sons, Ltd.
Companion Website: www.wiley.com/go/paknys9981

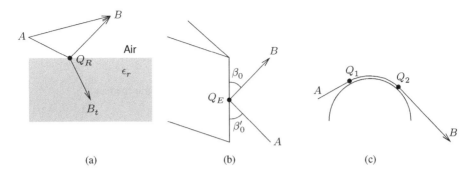

Figure 8.1 Fermat's principle. (a) Direct path, reflection and refraction. (b) Edge-diffracted ray. (c) Creeping wave.

is incident upon the sharp edge of a conducting wedge; a diffracted ray emanates from the edge diffraction point at Q_E. The angles β_0' and β_0 are measured from the edge to the incident and diffracted rays, respectively. Similar to the law of reflection, Keller's law of edge diffraction states that $\beta_0' = \beta_0$, and this minimizes the transit time from A to B. Another diffracted ray type is the surface-diffracted ray, or 'creeping wave', shown in case (c). Here, the ray propagates in free space from A to the attachment point Q_1, propagates as a creeping wave along the curved convex surface, following the geodesic (the shortest path), and then leaves at the launch point Q_2. Both edge-diffracted and surface-diffracted rays serve as corrective contributions that are added to the GO solution to improve the result.

8.2   2D Fields

If the frequency is sufficiently high, wave propagation can be described in terms of rays. In other words, it is required that all distances and obstacle dimensions should be large in terms of the wavelength. To better understand this, consider as an example the cylindrical wave produced by a line source. Strictly speaking, we know that it radiates as a Hankel function $H_0^{(2)}(k\rho)$. However, if $k\rho \gg 1$, the Hankel function can be replaced by its asymptotic approximation

$$H_0^{(2)}(k\rho) \sim \sqrt{\frac{j2}{\pi k}} \frac{e^{-jk\rho}}{\sqrt{\rho}}.$$

Then, from the $e^{-jk\rho}/\sqrt{\rho}$ spatial behaviour and some geometrical considerations, it can be shown that the power flow is along rays emanating from the line source.[1] For this reason, the field is called a *ray optical field*. Close to the source, the field is not ray optical, as it no longer behaves as $e^{-jk\rho}/\sqrt{\rho}$. Since $k\rho \gg 1$ (or equivalently $\lambda \ll 2\pi\rho$) is required, the wavelength has to be small, so this is called a 'high-frequency method'.

8.2.1   Reflection

When $k\rho_0 \gg 1$, an electric/magnetic line source of strength I_0 Amps (or M_0 Volts) produces a field

$$\left.\begin{matrix} E_z^i \\ H_z^i \end{matrix}\right\} = \left\{\begin{matrix} I_0 \, C_e \\ M_0 \, C_m \end{matrix}\right\} \frac{e^{-jk\rho_0}}{\sqrt{\rho_0}},$$

---

[1] The geometrical arguments will be given later in Section 8.4.3, using an astigmatic ray tube.

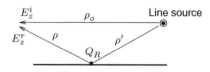

Figure 8.2   Electric line source above a perfect electric conductor.

where ρ_0 is the distance from the source to the field point, $C_e = -\eta\sqrt{jk/8\pi}$ and $C_m = -(1/\eta)\sqrt{jk/8\pi}$.

Suppose that a flat reflecting surface is present, as in Figure 8.2. By considering the source's image, the reflected field can be written as

$$\left.\begin{matrix} E_z^r \\ H_z^r \end{matrix}\right\} = \left\{\begin{matrix} I_0\,C_e \\ M_0\,C_m \end{matrix}\right\} R \frac{e^{-jk(\rho'+\rho)}}{\sqrt{\rho'+\rho}}.$$

This can also be expressed in the format

$$\left.\begin{matrix} E_z^r \\ H_z^r \end{matrix}\right\} = \left\{\begin{matrix} E_z^i(Q_R) \\ H_z^i(Q_R) \end{matrix}\right\} R \sqrt{\frac{\rho'}{\rho'+\rho}}\, e^{-jk\rho}.$$

That is, the reflected field is expressed in terms of the incident field at the reflection point Q_R. ρ' is the distance from the line source to Q_R, and ρ is the distance from Q_R to the field point. R is the reflection coefficient and can be for any material; for the special case of a perfect conductor, $R = \mp 1$ for the E_z and H_z polarizations. This format will be useful later on, to extend the method to reflection by curved surfaces.

8.2.2   Wedge Diffraction

The UTD for a perfectly conducting wedge was developed by Kouyoumjian and Pathak (1974). To gain some familiarity with the method, let us apply GO and UTD to an electric line source in the presence of the metal wedge in Figure 8.3(a). Assuming an electric line source excitation with a TM field, it radiates an incident ray with

$$E_z^i \sim E_0\frac{e^{-jk\rho_0}}{\sqrt{\rho_0}}, \tag{8.1}$$

where E_0 is a constant. By considering the image of the source, the reflected field is

$$E_z^r \sim E_0 R\frac{e^{-jk(\rho'+\rho)}}{\sqrt{\rho'+\rho}}. \tag{8.2}$$

Here, $R = -1$ is the reflection coefficient for the TM case. The TE case can be obtained by replacing E_z with H_z and using $R = +1$.

According to the UTD, the total field is given by GO plus the diffracted part

$$E_z \sim E_z^i U(\pi - \phi + \phi') + E_z^r U(\pi - \phi - \phi') + E_z^d. \tag{8.3}$$

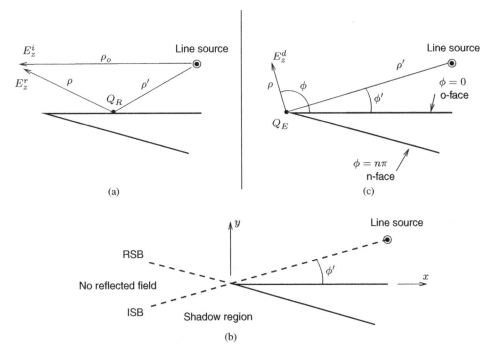

Figure 8.3   (a) GO incident and reflected rays for a line source near a conducting wedge. (b) Incident shadow boundary (ISB) and reflected shadow boundary (RSB). The ISB occurs at $\phi - \phi' = \pi$ and the RSB occurs at $\phi + \phi' = \pi$. (c) Geometry used for the diffraction coefficient.

The reflected field vanishes if the reflection point Q_R wanders off the surface. In the above, unit step functions U are used to turn E_z^i and E_z^r on and off in the appropriate lit and shadow regions of Figure 8.3(b); these are called the *incident shadow boundary* (ISB) and *reflected shadow boundary* (RSB).

The diffracted field E_z^d exists in the entire space outside the wedge, on $0 \le \phi \le n\pi$. Referring to Figure 8.3(c), the diffracted field emanates from the edge diffraction point Q_E as a cylindrical wave. The wave is polarization dependent; for an electric line source (the TM case), the diffracted field is given by

$$E_z^d \sim E_z^i(Q_E)D_s e^{-jk\rho}/\sqrt{\rho} \tag{8.4}$$

and for a magnetic line source (the TE case), it is

$$H_z^d \sim H_z^i(Q_E)D_h e^{-jk\rho}/\sqrt{\rho} \tag{8.5}$$

where $E_z^i(Q_E)$, $H_z^i(Q_E)$ is the field incident on the edge. In $D_{s,h}$ the subscripts s, h refer to the *soft* and *hard* boundary conditions, $E_z = 0$ and $\partial H_z/\partial \phi = 0$. This notation has its historical origins in acoustics, where a soft boundary has pressure $p = 0$ and a hard boundary has velocity $\partial p/\partial n = 0$. For an electromagnetic problem, the TE and TM polarizations correspond to hard and soft. $D_{s,h}$ is derived in Hutchins (1967) and is given by

$$D_{s,h} = \frac{-e^{-j\pi/4}}{2n\sqrt{2\pi k}\sin\beta_0}$$

$$\times \left[\cot\left(\frac{\pi + (\phi - \phi')}{2n}\right) F(kLa^+(\phi - \phi')) \right.$$

$$+ \cot\left(\frac{\pi - (\phi - \phi')}{2n}\right) F(kLa^-(\phi - \phi'))$$

$$\mp \left\{ \cot\left(\frac{\pi + (\phi + \phi')}{2n}\right) F(kLa^+(\phi + \phi')) \right.$$

$$\left. \left. + \cot\left(\frac{\pi - (\phi + \phi')}{2n}\right) F(kLa^-(\phi + \phi')) \right\} \right]. \tag{8.6}$$

The $-$ and $+$ in the $\mp$ term are associated with D_s and D_h, respectively.

The wedge faces are at $\phi = 0$ and $\phi = n\pi$. Hence, they are referred to as the o-face and the n-face. It is noted that although the present discussion is limited to 2D, the above expression for $D_{s,h}$ is also valid in 3D. For the 2D case, we always have the angle $\beta_0 = \pi/2$. The four terms in (8.6) compensate the GO discontinuities associated with the n-face ISB, o-face ISB, n-face RSB and o-face RSB.

The distance parameter L is

$$L = \rho \text{ for plane wave incidence} \tag{8.7a}$$

$$L = \frac{\rho\rho'}{\rho + \rho'} \text{ for cylindrical wave incidence.} \tag{8.7b}$$

The parameter $a^\pm(\beta)$ is related to the angular separation between the field point and the shadow boundaries. It is related to the accounting of included/excluded poles in the contour integral representation of the diffracted field, used in deriving (8.6). The inclusion/exclusion of the poles is in turn directly related to the appearance/disappearance of the incident and reflected fields, as the shadow boundaries are crossed. In general, $a^\pm(\beta)$ is given by

$$a^\pm(\beta) = 2\cos^2\left(\frac{2n\pi N^\pm - \beta}{2}\right),$$

where the $N^\pm$ parameters are integers that most nearly satisfy

$$2\pi n N^+ - \beta = \pi$$

and

$$2\pi n N^- - \beta = -\pi.$$

β represents $\phi \pm \phi'$. For the special case of a half plane ($n = 2$), it greatly simplifies to

$$a^\pm(\beta) = 2\cos^2(\beta/2). \tag{8.8}$$

The transition function F is

$$F(X) = j2\sqrt{X}e^{jX}\int_{\sqrt{X}}^{\infty} e^{-jt^2}\,dt. \tag{8.9}$$

Keller's GTD was able to predict diffraction effects away from the shadow boundaries. However, a problem is that the GTD diffraction coefficient is singular at an ISB or RSB. The UTD solved the singularity problem. Furthermore, the UTD corrects the GO field discontinuities at the shadow boundaries so that the total field is continuous, as it must be. A key distinction between the GTD and UTD is that the UTD contains the transition function $F(kLa)$ which comes from a uniform asymptotic evaluation of the wedge Green's function. By letting $F \to 1$, the UTD reduces to the GTD. In particular, we note that away from the shadow boundaries, kLa is large, so that $F(kLa) \to 1$, and the UTD diffraction coefficient (8.6) reduces to Keller's result

$$D_{s,h} = \frac{e^{-j\pi/4}}{n\sqrt{2\pi k}\sin\beta_0}\sin(\pi/n)$$

$$\times \left[\frac{1}{\cos\pi/n - \cos[(\phi-\phi')/n]} \mp \frac{1}{\cos\pi/n - \cos[(\phi+\phi')/n]}\right]. \tag{8.10}$$

The diffraction coefficient is the leading term of an asymptotic series. Higher-order terms are available (Hutchins 1967) but do not offer any significant improvements in accuracy and are generally not used.

8.2.3 Some Rules for Wedge Diffraction

When applying the UTD, the following rules must be observed:

1. Because the UTD is a high-frequency asymptotic method, the source and field points must not be too close to the edge. This requires the distances ρ' and ρ to be on the order of a wavelength or more.
2. The angles ϕ, ϕ' must always be positive, and they are measured from the o-face. $0 \le \phi' \le n\pi$, and $0 \le \phi \le n\pi$.
3. For the hard case, if $\phi' = 0$ or $\phi' = n\pi$, we have 'grazing incidence'. The surface field has an incident and reflected field component. However, the incident field (not the total) is supposed to be used in the diffraction coefficient calculation. If $H_z^i(Q_E)$ is the field incident on the edge, then

$$H_z^d = H_z^i(Q_E)D_h\frac{e^{-jk\rho}}{\sqrt{\rho}}.$$

On the other hand, if we take $H_z^i(Q_E)$ to represent the sum of the merged incident and reflected rays, we should use

$$H_z^d = \frac{1}{2}H_z^i(Q_E)D_h\frac{e^{-jk\rho}}{\sqrt{\rho}}.$$

4. For the soft case and grazing incidence, $E_z^i(Q_E) = 0$ so there is no diffracted field.

Example 8.1 A perfectly conducting half plane at $y = 0$ and $0 \le x < \infty$ is illuminated by a 1 V/m TM plane wave from above, at an angle $\phi' = 60°$; the wavelength is 1 m. Determine the UTD field at a distance of 3.5 m from the edge.

Solution: The field has three parts

$$E_z = E_z^i + E_z^r + E_z^d.$$

The incident field exists when $\phi - \phi' \le \pi$ and is given by

$$E_z^i = e^{jk(x\cos\phi' + y\sin\phi')}U(\pi - \phi + \phi').$$

The reflected field exists when $\phi + \phi' \le \pi$ and is given by

$$E_z^r = Re^{jk(x\cos\phi' - y\sin\phi')}U(\pi - \phi - \phi'),$$

where $R = -1$ is the reflection coefficient for the TM wave. The unit step functions U turn the incident and reflected fields on and off as needed in the lit and shadow regions. The diffracted field exists for all (ρ, ϕ) with $0 \leq \phi \leq 2\pi$. It is given by

$$E_z^d = E_z^i(0) D_s(\phi, \phi'; L) \frac{e^{-jk\rho}}{\sqrt{\rho}},$$

where $E_z^i(0) = 1$ and $L = \rho$. Using $x = \rho \cos \phi$, $y = \rho \sin \phi$,

$$E_z = e^{jk\rho \cos(\phi - \phi')} U(\pi - \phi + \phi') - e^{jk\rho \cos(\phi + \phi')} U(\pi - \phi - \phi')$$

$$+ D_s(\phi, \phi'; \rho) \frac{e^{-jk\rho}}{\sqrt{\rho}}.$$

Figure 8.4 shows $|E_z|$ when $\rho = 3.5$ m and $\phi' = 60°$. The RSB is at $\phi = 120°$ and the ISB is at $\phi = 240°$. The plot shows that E_z is continuous at the ISB and RSB when the diffracted field is included. The computations were done with PROGRAM utd1.

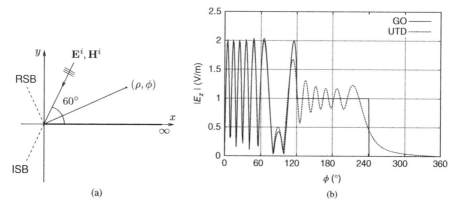

(a)   (b)

Figure 8.4 (a) 1 V/m TM plane wave incident at $\phi' = 60°$ on a perfectly conducting half plane; $\lambda = 1$ m. The RSB is at 120° and the ISB is at 240°. (b) $|E_z|$ at $\rho = 3.5$ m.

In general, for UTD to be accurate, the distances ρ and ρ' should be at least a wavelength or so. However, for the special case of plane wave incidence on half plane, the UTD solution reduces to Sommerfeld's half-plane solution, which is exact. Therefore, the result in the above example is exact.

The ideas in the previous example will now be applied to a strip which has two diffraction points. The TE polarization is used.

Example 8.2 The 2D perfectly conducting strip in Figure 8.5 is at $y = 0$ and the endpoints are at $x = \pm a/2$. It has a magnetic line source at $x = 0$, just above the strip. The GO field for $y > 0$ is given as

$$H_z^{go} = H_0 \frac{e^{-jk\rho}}{\sqrt{\rho}},$$

where H_0 is a constant. Find the UTD far-zone field for $0 \leq \phi \leq 360°$.

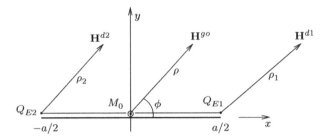

Figure 8.5 A perfectly conducting strip of width a and a magnetic line source.

Solution: The diffraction points Q_{E1}, Q_{E2} are at $x = \pm a/2$. The GO field has two parts, $H_z^{go} = H_z^i + H_z^r$ and, at grazing incidence, $H_z^i = H_z^r$. Consequently, $H_z^i(Q_{E1,2})$ is 1/2 of the total H_z^{go}. The distances from $Q_{E1,2}$ to the far field are $\rho_{1,2} = \rho \mp \frac{a}{2} \cos \phi$. Therefore,

$$
H_z = H_0 \, U(\pi - \phi) \frac{e^{-jk\rho}}{\sqrt{\rho}}
$$

$$
+ \frac{H_0}{2} \frac{e^{-jka/2}}{\sqrt{a/2}} \left(D_{h1} \, e^{-jk(\rho - \frac{a}{2} \cos \phi)} + D_{h2} \, e^{-jk(\rho + \frac{a}{2} \cos \phi)} \right) \frac{1}{\sqrt{\rho}}.
$$

The hard diffraction coefficient is used, with the parameters $\phi_1' = 0$, $\phi_1 + \phi = \pi$, $L = a/2$ in D_{h1} and $\phi_2' = 0$, $\phi_2 = \phi$, $L = a/2$ in D_{h2}. Care must be taken in a computer program to ensure that the angles ϕ_1, ϕ_2 are always positive. ∎

8.2.4   Behaviour Near ISB

It is useful to examine how the diffracted field compensates the discontinuities of GO. This will now be demonstrated for the TM case. The incident field is assumed to be

$$
E_z^i = \frac{e^{-jk\rho_0}}{\sqrt{\rho_0}},
$$

where ρ_0 is the distance from the line source to the field point. As an example consider the ISB and shadowing caused by the o-face. Compensation is provided by the second term from (8.6)

$$
D_s = \frac{-e^{-j\pi/4}}{2n\sqrt{2\pi k}} \cot \left(\frac{\pi - (\phi - \phi')}{2n} \right) F(kLa^-(\phi - \phi')).
$$

The other three terms in D_s are continuous at this ISB and do not need to be considered. At the ISB, $\phi - \phi' = \pi$, so for angles near the ISB it is convenient to let $\phi - \phi' = \pi + \delta$ where δ is small. Just on the lit side of the ISB, $\delta < 0$ and on the shadow side, $\delta > 0$. The cotangent is

$$
\cot \left(\frac{\pi - (\phi - \phi')}{2n} \right) = \cot \left(\frac{-\delta}{2n} \right) \approx \frac{-2n}{\delta}.
$$

With $n = 2$,

$$a = 2\cos^2\left(\frac{\phi - \phi'}{2}\right) = 2\cos^2\left(\frac{\pi + \delta}{2}\right) = 2\sin^2\left(\frac{\delta}{2}\right) \approx 2\left(\frac{\delta}{2}\right)^2$$

from which

$$kLa = k\frac{\rho\rho'}{\rho + \rho'}\frac{\delta^2}{2}.$$

Using the small-argument approximation of the transition function (Kouyoumjian and Pathak 1974)

$$F(X) \approx \sqrt{\pi X}e^{j(X + \pi/4)}$$

leads to

$$D_s = \frac{1}{2}\,\text{sgn}(\delta)\sqrt{\frac{\rho\rho'}{\rho + \rho'}},$$

where $\text{sgn}(\delta) \equiv \delta/|\delta|$ is the signum function; it equals ± 1 according to the sign of δ. At the edge,

$$E_z^i(Q_E) = \frac{e^{-jk\rho'}}{\sqrt{\rho'}}$$

so the diffracted field becomes

$$E_z^d = E_z^i(Q_E)D_s\frac{e^{-jk\rho}}{\sqrt{\rho}} = \frac{1}{2}\,\text{sgn}(\delta)\frac{e^{-jk(\rho + \rho')}}{\sqrt{\rho + \rho'}}.$$

Near the ISB, $\rho_0 \approx \rho + \rho'$ so the incident field on the lit side is

$$E_z^i \approx \frac{e^{-jk(\rho + \rho')}}{\sqrt{\rho + \rho'}}.$$

It is seen that on the lit side, $E_z^d = -E_z^i/2$ and on the shadow side, $E_z^d = E_z^i/2$ which exactly compensates the discontinuity in E_z^i that occurs at the ISB. A similar proof applies for compensation of the reflected field at an RSB.

We can always say that on an ISB, the true field is $1/2$ the GO field; in other words it is down by 6 dB. This figure is very useful when a quick estimate is needed for shadowing by a sharp-edged obstacle.

8.3 Scattering and GTD

In scattering problems, the GTD, not UTD, is used. This is because when both ρ and ρ' approach infinity, the parameter $L = \rho\rho'/(\rho + \rho') \to \infty$ and $F(kLa) \to 1$. The following example will be used to illustrate this point.

Example 8.3 A 2D perfectly conducting strip shown in Figure 8.6 is at $y = 0$ and along $-a/2 \leq x \leq a/2$. It is illuminated by an incident plane wave

$$E_z^i = E_0 e^{jk(x\cos\phi + y\sin\phi)}$$

and $\lambda = 1$ m is assumed. Find the backscatter echo width σ, as a function of the angle ϕ.

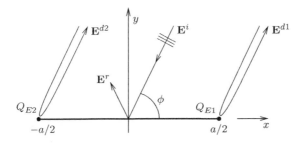

Figure 8.6 TM plane wave incident at an angle ϕ on a perfectly conducting strip of width a, backscatter configuration.

Solution: With plane wave incidence and a far-zone field point, a back-reflected field only exists within a vanishingly small angular sector, centred about $\phi = \pi/2$. Therefore, the reflected field does not contribute to the scattered field, only the diffracted field does. The diffracted field is of the form

$$E_z^d \sim E_z^i(Q_E) D_s \frac{e^{-jkR}}{\sqrt{R}},$$

where R is the distance from an edge to the field point. There are two diffraction points, Q_{E1} and Q_{E2}, at $x = \pm a/2$. For Q_{E1} at $x = a/2$, the incident field at the edge is $E_z^i(Q_{E1}) = E_0 e^{jk(a/2)\cos\phi}$, and the distance from Q_{E1} to the far field is $\rho_1 = \rho - \frac{a}{2}\cos\phi$. For a half plane, $n = 2$. The diffraction angles are $\phi_1 = \phi_1' = \pi - \phi$. Using (8.10), the Keller diffraction coefficient at edge #1 becomes

$$D_{s1} = \frac{-e^{-j\pi/4}}{2\sqrt{2\pi k}}(1 + \sec\phi).$$

Similarly, for Q_{E2} at $x = -a/2$ we have $E_z^i(Q_{E2}) = E_0 e^{-jk(a/2)\cos\phi}$, and the distance from Q_{E2} to the far field, $\rho_2 = \rho + \frac{a}{2}\cos\phi$. Using the diffraction angles $\phi_2 = \phi_2' = \phi$,

$$D_{s2} = \frac{-e^{-j\pi/4}}{2\sqrt{2\pi k}}(1 - \sec\phi).$$

Combining the contributions from both edges gives

$$E_z^d \sim -E_0 \frac{e^{-j\pi/4}}{2\sqrt{2\pi k}} [e^{jk(a/2)\cos\phi}(1 + \sec\phi)e^{jk(a/2)\cos\phi}$$

$$+ e^{-jk(a/2)\cos\phi}(1 - \sec\phi)e^{-jk(a/2)\cos\phi}] \frac{e^{-jk\rho}}{\sqrt{\rho}}.$$

This can be rewritten as

$$E_z^d \sim -E_0 \frac{e^{-j\pi/4}}{\sqrt{2\pi k}} \left[\cos(ka\cos\phi) + jka\frac{\sin(ka\cos\phi)}{ka\cos\phi} \right] \frac{e^{-jk\rho}}{\sqrt{\rho}}.$$

The echo width is

$$\sigma = \lim_{\rho\to\infty} 2\pi\rho \left| \frac{E_z^s}{E_z^i} \right|^2 = \frac{1}{k} \left| \cos(ka\cos\phi) + jka\frac{\sin(ka\cos\phi)}{ka\cos\phi} \right|^2.$$

At broadside, $\phi \to \pi/2$ and

$$\sigma = \frac{1}{k}|1 + jka|^2.$$

At grazing incidence, $\phi \to 0$ and

$$\sigma = \frac{1}{k}.$$

The angle $\phi = \pi/2$ is on the shadow boundary of the reflected field, and as expected, the GTD is singular there. However, the singularities from the two edges cancel, leading to a finite result. At grazing, the echo width is independent of a. ∎

The above example shows that success with GTD in scattering problems requires having a structure with symmetry such that the singular parts of the diffractions cancel.

8.4   3D Fields

The ray fixed coordinate system in Figure 8.7 is used for 3D wedge diffraction. The diffraction point is at Q_E, the source point is at (s', β_0', ϕ') and the field point is at (s, β_0, ϕ). The incident ray is in the direction $\hat{s}'$, and the diffracted ray is along $\hat{s}$. The angles ϕ', ϕ and β_0', β_0 increase in the direction of their corresponding unit vectors. The cross-product relations

$$\hat{s}' \times \hat{\phi}' = \hat{\beta}_0'$$

$$\hat{s} \times \hat{\phi} = \hat{\beta}_0$$

are helpful in determining how the angles are measured.

From the asymptotic analysis of diffraction by a wedge, Keller deduced the *law of edge diffraction*, which states that the incident angle β_0' equals the diffracted angle β_0. Consequently, the diffracted rays form a cone about the edge, at an angle of β_0. This is known as the *Keller cone* and is shown in Figure 8.7(c).

The UTD edge-diffracted field is

$$E_{\beta_0}^d \sim -E_{\beta_0'}^i(Q_E)D_s A(s, s')e^{-jks} \tag{8.11}$$

$$E_\phi^d \sim -E_{\phi'}^i(Q_E)D_h A(s, s')e^{-jks}. \tag{8.12}$$

This is sometimes written as a matrix equation

$$\begin{bmatrix} E_{\beta_0}^d \\ E_\phi^d \end{bmatrix} \sim - \begin{bmatrix} D_s & 0 \\ 0 & D_h \end{bmatrix} \begin{bmatrix} E_{\beta_0'}^i(Q_E) \\ E_{\phi'}^i(Q_E) \end{bmatrix} A(s, s')e^{-jks}.$$

Another form uses the more compact dyadic notation[2]

$$\mathbf{E}^d(s) \sim \mathbf{E}^i(Q_E) \cdot \overline{\mathbf{D}}A(s, s')e^{-jks},$$

---

[2] A dyad is used to transform the magnitude and direction of a vector. It is written as two vectors, side by side. For instance, if $\overline{\mathbf{D}} = 3\hat{x}\hat{y}$ and $\mathbf{A} = 5\hat{y}$, then $\overline{\mathbf{D}} \cdot \mathbf{A} = 3\hat{x}\hat{y} \cdot 5\hat{y} = 15\hat{x}(\hat{y} \cdot \hat{y}) = 15\hat{x}$. The dyad can be interpreted as a matrix operator

$$\overline{\mathbf{D}} \cdot \mathbf{A} = \begin{bmatrix} 0 & 3 \\ 0 & 0 \end{bmatrix} \begin{bmatrix} 0 \\ 5 \end{bmatrix} = \begin{bmatrix} 15 \\ 0 \end{bmatrix}.$$

To distinguish dyads from vectors, they are denoted in boldface with a bar above.

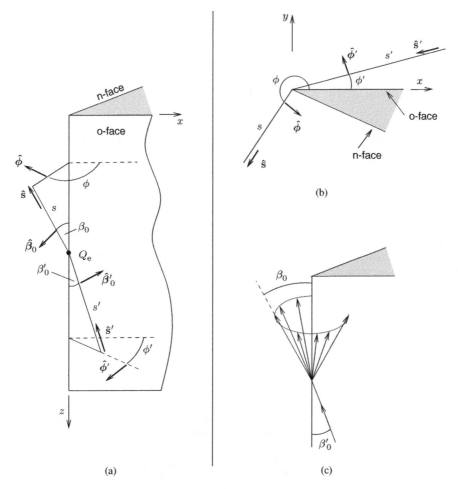

Figure 8.7 (a) 3D edge-fixed coordinates, (b) cross-sectional view. The source point is at (s', ϕ', β_0') and the field point is at (s, ϕ, β_0). (c) Keller cone of edge-diffracted rays; $\beta_0' = \beta_0$.

where the dyadic diffraction coefficient is

$$\overline{\mathbf{D}} = -\hat{\beta}_0'\hat{\beta}_0 D_s - \hat{\phi}'\hat{\phi}D_h.$$

The spread factor is

$$A(s, s') = \frac{1}{\sqrt{s}} \text{ for plane wave incidence} \qquad (8.13a)$$

$$A(s, s') = \sqrt{\frac{s'}{s(s + s')}} \text{ for spherical wave incidence.} \qquad (8.13b)$$

The distance parameter is

$$L = s\sin^2\beta_0 \text{ for plane wave incidence} \qquad (8.14a)$$

$$L = \frac{ss'}{s + s'}\sin^2\beta_0 \text{ for spherical wave incidence.} \qquad (8.14b)$$

Since the incident, reflected and diffracted fields behave locally as plane waves, they obey the relation

$$\eta\mathbf{H} = \hat{\mathbf{s}} \times \mathbf{E},$$

where $\eta = \sqrt{\mu/\epsilon}$ and $\hat{\mathbf{s}}$ is a unit vector in the direction of the ray path. It is noted that the diffraction coefficient (8.6) is valid in 2D or 3D, and the only difference between the two cases is that $\sin\beta_0 = 1$ in 2D.

8.4.1   Slot Antenna on a Finite Ground Plane

In this section we will apply the UTD to find the radiation pattern in the $\phi = 0$ plane for a small x-polarized slot on a finite rectangular ground plane. The configuration is shown in Figure 8.8. The plate has edge lengths a and b. According to the generalized Fermat principle, the shortest and longest diffracted ray paths define the locations of the two diffraction points which are Q_{E1} and Q_{E2} at $x = \pm a/2$.

The x-polarized slot can be modelled as a y-directed magnetic dipole $\mathbf{p}_m = \hat{\mathbf{y}}p_m$. The GO field is defined as the slot's field on an infinite ground. In the $\phi = 0$ plane this is

$$E_\theta^{go} = E_0\frac{e^{-jkr}}{r},$$

where E_0 is a constant that is related to the strength of $\mathbf{p}_m$. The field E_θ^{go} represents the merged incident and reflected fields due to $\mathbf{p}_m$ and its image, so this is a case of grazing incidence.

The edge diffractions shown in Figure 8.8 can be calculated using (8.12) with (8.13b), (8.14b) for spherical wave incidence. The form of the far-field edge diffractions (using $s \gg s'$) is

$$E_{\psi_0} \sim -E_{\psi_0'}^i(Q_E)\, D_h(\psi, \psi')\, \frac{\sqrt{s'}}{s}\, e^{-jks}.$$

To avoid confusion with the spherical coordinate angle ϕ, the diffraction angles are denoted here as ψ, ψ', measured from the o-face. The necessary parameters follow from the inspection of Figure 8.8 and are

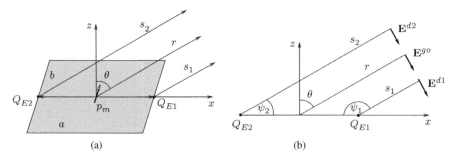

(a)   (b)

Figure 8.8   (a) Perfectly conducting rectangular plate and magnetic current p_m. (b) GO and diffracted fields.

Table 8.1   Edge diffraction parameters, rectangular plate.

	$\mathbf{E}^i(Q_E)$	$\hat{\psi}'$	$\hat{\psi}$	ψ'	ψ	s'	s	L
Edge 1	$-\hat{\mathbf{z}}(E_0/a)\,e^{-jka/2}$	$\hat{\mathbf{z}}$	$\hat{\boldsymbol{\theta}}$	0	$\pi/2+\theta$	$a/2$	$r-(a/2)\sin\theta$	$a/2$
Edge 2	$\hat{\mathbf{z}}(E_0/a)\,e^{-jka/2}$	$\hat{\mathbf{z}}$	$-\hat{\boldsymbol{\theta}}$	0	$\pi/2-\theta$	$a/2$	$r+(a/2)\sin\theta$	$a/2$

summarized in Table 8.1. Note that $E^i(Q_E) = \frac{1}{2}E^{go}(Q_E)$ is used, because of grazing incidence. The field is then

$$E_\theta \sim E_\theta^{go} + E_\theta^{d1} + E_\theta^{d2}$$

and of course $E_\theta^{go} = 0$ below the ground plane. For $0 \le \theta \le \pi$ this becomes

$$E_\theta \sim E_0 U(\pi/2 - \theta)\,\frac{e^{-jkr}}{r}$$

$$+ \frac{E_0}{\sqrt{2a}}\,e^{-jka/2}\left(D_h(\psi_1)e^{jk(a/2)\sin\theta} + D_h(\psi_2)e^{-jk(a/2)\sin\theta}\right)\frac{e^{-jkr}}{r}. \qquad (8.15)$$

The reader might be wondering why there are no diffraction points on the other two edges and that is a good question. There could be a Q_{E3} on the edge along $y = b/2$ and Q_{E4} on the edge along $y = -b/2$. Considering Q_{E3}, the incident ray has $\psi' = 0$, $\hat{\psi}' = \hat{\mathbf{z}}$; the diffracted ray has $\psi = \pi/2$, $\hat{\psi} = \hat{\mathbf{y}}$. For Q_{E4}, the incident ray has $\psi' = 0$, $\hat{\psi}' = \hat{\mathbf{z}}$; the diffracted ray has $\psi = \pi/2$, $\hat{\psi} = -\hat{\mathbf{y}}$. Since the $\hat{\psi}$ for Q_{E3} and Q_{E4} are oppositely directed, their contributions to the diffracted field cancel and they can be ignored.

In the $y - z$ plane, Q_{E3} is at $(x, y, z) = (0, b/2, 0)$ and Q_{E4} is at $(x, y, z) = (0, -b/2, 0)$. The dipole p_m has a null in those directions, so $\mathbf{E}^i(Q_{E3})$ and $\mathbf{E}^i(Q_{E4})$ are zero and once again, there is no net contribution from those edges. In the general case outside the xz and $y - z$ planes, Q_{E3} and Q_{E4} can contribute to the diffracted field.

We saw that edge diffraction compensates for the appearance/disappearance of incident fields at an ISB and the appearance/disappearance of reflected fields at an RSB. It is also possible to experience the appearance/disappearance of an edge-diffracted field, and this will now be discussed.

In the present problem, the diffraction points Q_{E3} and Q_{E4} may not even exist. Remember that Keller's law of edge diffraction has to be satisfied. For instance, as θ ranges from 0 to 90°, Q_{E3} will move from $(x, y, z) = (0, b/2, 0)$ to some other point $(x_0, b/2, 0)$, where $0 \le x_0 \le a/2$. At some intermediate value of θ, the point Q_{E3} will wander off the edge at $x = a/2$. At this point the diffraction abruptly disappears. The disappearing diffracted field will cause a discontinuity in the total field. We can call this a 'diffraction shadow boundary' (DSB) in analogy with the ISB and RSB discussed before. The disappearance of an edge diffraction is compensated by *corner diffraction*. Corner diffracted rays emanate as spherical waves from the four corners of the plate. In the xz plane, Q_{E1} and Q_{E2} are in the middle of the edge at $y = 0$ and are in no danger of disappearing. Therefore, corner diffraction effects can be safely neglected. In other planes, corner diffraction may make a significant contribution to the total field. Seeing discontinuities in edge-diffracted fields tells us that corner diffraction is needed.

An empirical corner diffraction coefficient was developed by Sikta et al. (1983); it can also be found in Pathak (1988). The accuracy is sufficient for several cases of practical interest. A uniform asymptotic formula based on the two edges forming the corner was later obtained by Hill and Pathak (1991). A rigorous but more difficult to use technique based on an exact solution for the vertex has been obtained by Ozturk et al. (2011).

8.4.2   Monopole Antenna on a Finite Ground Plane

In this section we will apply the UTD to find the radiation pattern in the $\phi = 0$ plane for a short monopole on a finite rectangular ground plane. This can be done with only minor modifications of the slot result in the previous section.

The GO field is defined as the monopole field on an infinite ground and is given by

$$E_\theta^{go}(r,\theta) = E_0 \sin\theta \; \frac{e^{-jkr}}{r},$$

where E_0 is a constant. The field E_θ^{go} represents the merged incident and reflected fields due to the source and its image, so this is a case of grazing incidence.

The main difference from a slot is that the monopole produces an electric field with an even symmetry. Therefore,

$$\mathbf{E}^i(Q_{E1}) = \mathbf{E}^i(Q_{E2}) = \frac{1}{2}\mathbf{E}^{go}(a/2, \pi/2) = -\frac{1}{2}\hat{\mathbf{z}}\, E_0 \; \frac{e^{-jka/2}}{a/2}.$$

Aside from the sign change on $\mathbf{E}^i(Q_{E2})$, the solution is the same as for the slot. Following the same procedure as before, the field for $\phi = 0, 0 \le \theta \le \pi$ is found to be

$$E_\theta \sim E_0 U(\pi/2 - \theta)\; \sin\theta \; \frac{e^{-jkr}}{r}$$
$$+ \frac{E_0}{\sqrt{2a}}\, e^{-jka/2}\left(D_h(\psi_1)e^{jk(a/2)\sin\theta} - D_h(\psi_2)e^{-jk(a/2)\sin\theta}\right)\frac{e^{-jkr}}{r}. \qquad (8.16)$$

The UTD radiation patterns for the slot and monopole are plotted and compared in Figure 8.9. The plate width is $a = 6\lambda$. The b dimension does not affect the pattern. The patterns are normalized so that $E^{go} = 0$ dB at 90°. The slot antenna radiates strongly in the $\theta = 0$ direction whereas the monopole has a null. The slot illuminates the two edges with an odd-symmetric electric field whereas for the monopole it is even symmetric. Therefore, the edge diffractions for the two cases have different phase relationships. As a consequence, the slot produces a maximum at 180° whereas the monopole has a null.

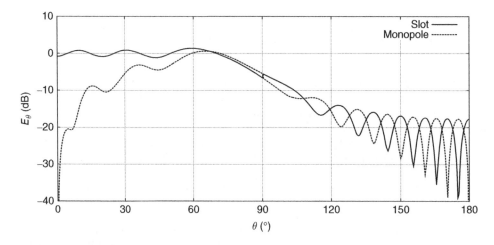

Figure 8.9   UTD radiation patterns for a slot and a monopole on a rectangular plate, of width $a = 6\lambda$.

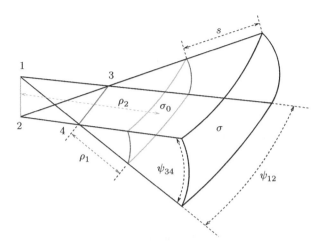

Figure 8.10 Astigmatic ray tube. Caustic 1–2 is associated with ρ_2 and ψ_{12}; caustic 3–4 is associated with ρ_1 and ψ_{34}. (Colour version at www.wiley.com/go/paknys9981.)

There is a shadow boundary at 90°, and the total field there is exactly 6 dB below the GO field level, for the reasons discussed in Section 8.2.4. This is a general characteristic of any antenna pattern that experiences shadowing by an edge. The minor discontinuity at 90° occurs because the edge diffraction from Q_{E2} is not continuous there. This effect diminishes with increasing plate size. The discontinuity could be corrected by adding double diffraction, as described in Section 8.9.

8.4.3   Astigmatic Fields

When finding the reflection by curved surfaces and diffraction by edges, the resulting rays do not necessarily emanate from a point focus (a spherical wave) or a line focus (a cylindrical wave). Rather, there is a more complex situation involving *astigmatism*. This is described by an *astigmatic ray tube*, as shown in Figure 8.10. It has two focal lines, or *caustics*, denoted as 1–2 and 3–4. The propagating wavefront is shown at two instances in time and is denoted as σ_0 and σ. Plane, cylindrical and spherical waves are all special cases of an astigmatic ray tube.

According to GO, (a) the electric and magnetic fields are parallel to the wavefront and (b) the power flow is perpendicular to the wavefront. Therefore, power is conserved within the ray tube. As a result, the power crossing the areas σ_0 and σ has to be the same. From the geometry, the wavefront areas are

$$\sigma_0 = \psi_{12}\rho_2\psi_{34}\rho_1; \quad \sigma = \psi_{12}(\rho_2 + s)\psi_{34}(\rho_1 + s)$$

If the total power crossing σ_0 or σ is P, then

$$\frac{\text{power density at } \sigma}{\text{power density at } \sigma_0} = \frac{P/\sigma}{P/\sigma_0} = \frac{\sigma_0}{\sigma} = \frac{\rho_1\rho_2}{(\rho_1 + s)(\rho_2 + s)}$$

from which

$$\frac{\text{amplitude at } \sigma}{\text{amplitude at } \sigma_0} = \sqrt{\frac{\rho_1\rho_2}{(\rho_1 + s)(\rho_2 + s)}} \equiv A(s).$$

$A(s)$ is associated with the geometrical spreading out of the rays, so it is called the *spread factor*. In addition, as the wave propagates from σ_0 to σ, a phase e^{-jks} is acquired. Therefore, the electric fields at σ_0 and σ are related by

$$\mathbf{E}(\sigma) = \mathbf{E}(\sigma_0)\sqrt{\frac{\rho_1\rho_2}{(\rho_1+s)(\rho_2+s)}}e^{-jks}. \tag{8.17}$$

The GO field behaves locally as a plane wave so that the magnetic field can be found from the relation $\mathbf{H} = \hat{s} \times \mathbf{E}/\eta$.

The special case of 2D can be obtained by letting $\rho_2 \to \infty$. Equation (8.17) becomes

$$\mathbf{E}(\sigma) = \mathbf{E}(\sigma_0)\sqrt{\frac{\rho_1}{\rho_1+s}}e^{-jks} \tag{8.18}$$

which is a cylindrical wave with its focus at the caustic 3–4 in Figure 8.10.

The ray tube description (8.17) assumed that ρ_1 and ρ_2 are positive numbers and that $s \geq 0$; this leads to diverging rays. We can interchange the positions of σ_0 and σ (while keeping $s = 0$ at σ_0). This describes converging rays, as might occur with a lens or a concave reflector. In this case, some of the terms inside the square root of $A(s)$ can be negative. When evaluating $A(s)$, it is important to remember this property: upon propagation through a caustic, the field picks up an extra phase shift of $+\pi/2$, that is, $\sqrt{-1} = e^{+j\pi/2}$ (Kouyoumjian and Pathak 1974).

The astigmatic ray tube is used to describe incident, reflected, refracted and diffracted rays, in the most general ray optical terms. Even for the special case of a spherical incident wave, it is a necessary concept because an edge-diffracted wave is astigmatic. A reflected wave is spherical, but only when the reflecting surface is flat; otherwise it is astigmatic. The following sections will describe astigmatic reflected and diffracted fields in more detail.

8.4.4 Reflection

Both an incident and reflected field can be associated with an astigmatic ray tube. Before tackling the general case, we will examine a simpler case with a point source and a flat reflecting surface, as shown in Figure 8.11(a). Since the source and image are spherical waves, there is no astigmatism.

s' is the distance from the source to the reflection point Q_R, and s is the distance from Q_R to the field point. This is a special case of the astigmatic tube (8.17) having $\rho_1 = \rho_2 = s'$. The reflected field's

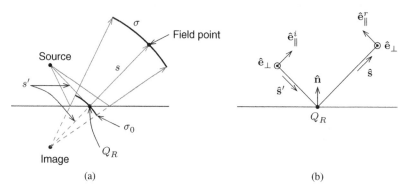

Figure 8.11 Point source and reflected field. (a) Reflected rays showing ray tube cross sections σ_0 and σ. (b) Unit vectors used in the dyadic reflection coefficient $\overline{\overline{R}}$.

ray tube reference surface σ_0 is chosen to be at Q_R, so that $\mathbf{E}(\sigma_0) = \mathbf{E}(Q_R)$. Therefore, the reflected field is

$$\mathbf{E}^r(s) = \mathbf{E}(Q_R) \frac{s'}{s' + s} e^{-jks}. \qquad (8.19)$$

$\mathbf{E}(Q_R)$ is the reflected field at $s = 0$ and it can be expressed in terms of the incident field at the reflection point, $\mathbf{E}^i(Q_R)$. This is best accomplished by using a dyadic reflection coefficient

$$\overline{\mathbf{R}} = \hat{\mathbf{e}}_\perp \hat{\mathbf{e}}_\perp \Gamma_\perp - \hat{\mathbf{e}}_\parallel^i \hat{\mathbf{e}}_\parallel^r \Gamma_\parallel \qquad (8.20)$$

so that $\mathbf{E}(Q_R) = \mathbf{E}^i(Q_R) \cdot \overline{\mathbf{R}}$. Here, $\Gamma_\perp$ and $\Gamma_\parallel$ are the Fresnel reflection coefficients. For a perfect conductor, $\Gamma_\perp = \Gamma_\parallel = -1$. More generally, they are given by (2.37) and (2.49). The unit vectors in (8.20) are shown in Figure 8.11(b). The plane of incidence contains $\hat{\mathbf{s}}'$, $\hat{\mathbf{e}}_\parallel^i$ and $\hat{\mathbf{n}}$ such that $\hat{\mathbf{s}}' \times \hat{\mathbf{e}}_\parallel^i = \hat{\mathbf{e}}_\perp$. Likewise, the plane of reflection contains $\hat{\mathbf{s}}$, $\hat{\mathbf{e}}_\parallel^r$ and $\hat{\mathbf{n}}$ such that $\hat{\mathbf{s}} \times \hat{\mathbf{e}}_\parallel^r = \hat{\mathbf{e}}_\perp$. The incident and reflection planes are in the same plane, so they have the same $\hat{\mathbf{e}}_\perp$. For a parallel polarized incident wave, $\mathbf{E}^i(Q_R)$ is in the direction $\hat{\mathbf{e}}_\parallel^i$; likewise for a perpendicular polarized incident wave, $\mathbf{E}^i(Q_R)$ is in the direction $\hat{\mathbf{e}}_\perp$.

Using the dyadic reflection coefficient, the reflected field at a distance s from Q_R can then be written in terms of $\mathbf{E}^i(Q_R)$ as

$$\mathbf{E}^r(s) = \mathbf{E}^i(Q_R) \cdot \overline{\mathbf{R}} \frac{s'}{s' + s} e^{-jks}. \qquad (8.21)$$

If the reflecting surface is curved or the source is astigmatic, the reflected field will be astigmatic. This more general case will be considered in Section 8.5.

8.4.5   Edge Diffraction

The most general form of the edge-diffracted field can account for a straight or curved edge. In either case, the diffracted field is expressed in terms of the astigmatic ray tube, as shown in Figure 8.12. The first caustic 3–4 is positioned on the edge, at the diffraction point Q_E. The second caustic 1–2 is at a distance ρ_c from the edge. The ray tube is used to relate the incident field $\mathbf{E}^i(Q_E)$ to the diffracted field $\mathbf{E}^d(s)$. Analogous to reflection, the edge-diffracted field can be expressed as

$$\mathbf{E}^d(s) = \mathbf{E}^i(Q_E) \cdot \overline{\mathbf{D}} \sqrt{\frac{\rho_c}{s(\rho_c + s)}} e^{-jks}. \qquad (8.22)$$

This is a special case of the ray tube, whereby $\mathbf{E}^i(Q_E) \cdot \overline{\mathbf{D}} = \mathbf{E}(\sigma_0)$ and $\mathbf{E}^d(s) = \mathbf{E}(\sigma)$. $\rho_2 = \rho_c$ is the distance from Q_E to the second caustic, and $\rho_1 = 0$. The dyad $\overline{\mathbf{D}}$ is the diffraction coefficient, and s is the distance from Q_E to the field point. The incident and diffraction directions, as discussed earlier, must obey Keller's law of edge diffraction. The reference surface σ_0 can be at any arbitrary position, so it is possible to have σ_0 at the edge. To ensure that $\mathbf{E}^d(s)$ is finite, $\overline{\mathbf{D}}$ is defined in such a way that

$$\lim_{\rho_1 \to 0} \sqrt{\rho_1}\, \overline{\mathbf{D}}$$

remains finite.

In 2D, the second caustic is at $\rho_c \to \infty$ so that the diffracted field spread factor becomes

$$A(\rho_c, s) = \sqrt{\frac{\rho_c}{s(s + \rho_c)}} = \frac{1}{\sqrt{s}}. \qquad (8.23)$$

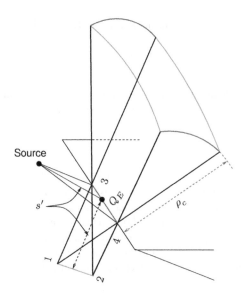

Figure 8.12 Edge-diffracted ray tube. Caustic 3–4 is on the diffracting edge. Both the source and the caustic 1–2 are at the same distance s' from Q_E. (Colour version at www.wiley.com/go/paknys9981.)

For a point source in 3D and a straight edge, $\rho_c = s'$ so that

$$A(\rho_c, s) = \sqrt{\frac{\rho_c}{s(s + \rho_c)}} = \sqrt{\frac{s'}{s(s + s')}}. \tag{8.24}$$

8.4.6   Curved Edge

If the edge of the wedge is curved, the location of the second caustic is given by

$$\frac{1}{\rho_c} = \frac{1}{\rho_e^i} + \frac{\hat{n}_e \cdot (\hat{s} - \hat{s}')}{a_e \sin^2 \beta_o}. \tag{8.25}$$

The spread factor follows from (8.24) using ρ_c from (8.25). In the above, ρ_e^i is the incident wavefront radius of curvature in the plane containing the incident ray direction $\hat{s}'$ and edge tangent $\hat{e}$. a_e is the edge radius of curvature and $\hat{n}_e$ is normal to the edge curvature. $\hat{n}_e$ points away from the radius of curvature. Equivalently, the radius of curvature's centre point and the vectors $\hat{n}_e$, $\hat{e}$ are coplanar. a_e is always positive and does not depend on the curve being convex or concave.

8.4.7   Monopole on a Disc

In Section 8.4.2 we solved the UTD problem of a short monopole on a rectangular plate. With only minor modifications we can find the fields for a monopole on a circular disc, as shown in Figure 8.13.

The disc has a radius a and an electric monopole of strength p_e is at the centre. The GO field is defined as the monopole field on an infinite ground and is given by

$$E_\theta^{go}(r, \theta) = E_0 \sin \theta \, \frac{e^{-jkr}}{r},$$

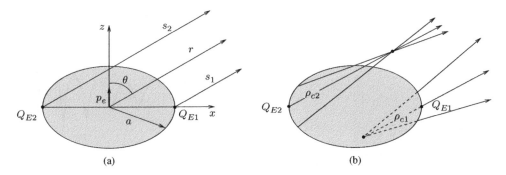

Figure 8.13   (a) Perfectly conducting circular disc and monopole p_e. (b) Diffracted ray caustics.

where E_0 is a constant that is related to p_e. The field E_θ^{go} represents the merged incident and reflected fields due to the source and its image, so this is a case of grazing incidence.

The only things different from the rectangular plate are the caustic distances and their effect on the spread factors. Using (8.25) we find that

$$\rho_{c1} = \frac{a}{\sin\theta}; \quad \rho_{c2} = \frac{-a}{\sin\theta}.$$

In the far field $s \gg \rho_c$ and since spread factors are amplitude terms, we can let $s \approx r$. With this in mind, the spread factor becomes

$$A(s) = \sqrt{\frac{\rho_c}{s(\rho_c + s)}} \approx \frac{\sqrt{\rho_c}}{r}.$$

An important point regarding edge 2 is that the caustic distance ρ_{c2} is negative. The square root should be evaluated as

$$\sqrt{\rho_{c2}} = \sqrt{\frac{-a}{\sin\theta}} = e^{+j\pi/2}\sqrt{\frac{a}{\sin\theta}}.$$

This is because the rays diffracting from the concave edge at Q_{E2} form a caustic, which is crossed as the diffracted wave propagates from Q_{E2} to the far field. Generally, propagation through a caustic causes a phase shift of $+\pi/2$; see Section 8.4.3 for further discussion.

Following these considerations, the spread factors for edges 1 and 2 become

$$A_1 = \frac{\sqrt{a/\sin\theta}}{r}; \quad A_2 = e^{j\pi/2}\frac{\sqrt{a/\sin\theta}}{r}.$$

The L parameters are $L_{1,2} = a$ and the rest of the procedure is similar to the rectangular plate in Section 8.4.2. The resulting field is

$$E_\theta \sim E_0 U(\pi/2 - \theta)\,\sin\theta\,\frac{e^{-jkr}}{r}$$

$$+ \frac{E_0}{2\sqrt{a}\sin\theta}\,e^{-jka}\left(D_h(\psi_1)e^{jka\sin\theta} - e^{j\pi/2}\,D_h(\psi_2)e^{-jka\sin\theta}\right)\frac{e^{-jkr}}{r}. \quad (8.26)$$

Notice that when $\theta = 0°$ or $180°$ the caustic distances ρ_{c1} and ρ_{c2} become infinite. Referring to Figure 8.13(b) it can be seen that the edge-diffracted rays become parallel to the z axis, forming a caustic.

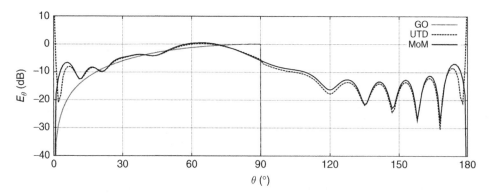

Figure 8.14   Radiation pattern for a short monopole on a circular disc, radius $a = 3\lambda$. Comparison of GO, UTD and MoM.

Therefore, it is no longer possible to find the field from rays. Another way to recognize there is a problem is by noticing that all the edge-diffracted rays are the same length, so it becomes impossible to apply the generalized Fermat's principle and identify the shortest and longest diffracted ray paths. Since the diffracted field emanates from a continuum of points along the entire edge, it has to be found from a line integral around the edge, using *equivalent edge currents*. These are described in Section 8.10.

The radiation pattern computed from (8.26) is shown in Figure 8.14. The disc radius is $a = 3\lambda$ and the patterns are normalized so that $E^{go} = 0$ dB at $90°$. The UTD is in good agreement with the MoM result for most angles. UTD is also very useful because it provides a way to physically interpret the radiation pattern in terms of rays. The two edge diffractions form the pattern below the disc. The diffraction from Q_{E1} compensates the GO discontinuity at the $90°$ shadow boundary. On the topside, the edge diffractions act to improve the overall result.

The UTD solution is seen to be singular at the diffracted ray caustics $\theta = 0°$ and $\theta = 180°$, as discussed above. The correct answer for the diffracted field at the caustic is in fact zero; this can be seen by considering the circular symmetry of the diffracted rays or from the formally exact MoM solution. It can also be shown using an equivalent edge current solution.

The UTD shows a small discontinuity at $90°$ because the diffraction from Q_{E2} is discontinuous. However, this can be solved with double diffraction, which is described in Section 8.9.

As a practical note, the monopole on a disc is an important benchmark. It is used as a test standard for qualifying the performance of aircraft antennas – which are required to provide a pattern coverage that is equal to or better than a monopole on a disc. The book by Weiner et al. (1987) is entirely devoted to this type of antenna.

8.5   Curved Surface Reflection

8.5.1   2D Reflection

Earlier, we examined the reflection by a planar surface. The ray tube concept will now be used to extend GO reflection to curved surfaces. A line source above a curved surface is shown in Figure 8.15. Because of the surface curvature, the image of the source is shifted. From (8.18) it follows that the reflected field is

$$\left.\begin{matrix} E^r_z \\ H^r_z \end{matrix}\right\} = \left\{\begin{matrix} E^i_z(Q_R) \\ H^i_z(Q_R) \end{matrix}\right\} R\sqrt{\frac{\rho_c}{\rho_c + \rho}}e^{-jk\rho}. \tag{8.27}$$

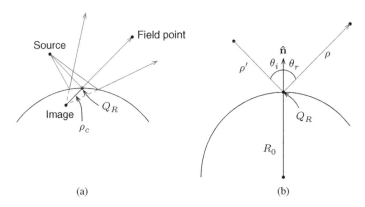

Figure 8.15 Line source and curved surface reflected field. (a) The reflected ray caustic distance ρ_c is measured between the image and Q_R. (b) The reflecting surface radius of curvature at Q_R is R_0; the law of reflection requires $\theta_i = \theta_r$.

The quantity ρ_c in the spread factor is the caustic distance for the reflected field. It depends on the curvatures of the incident wavefront and the reflecting surface at Q_R. For the special case of 2D, it can be shown that (see Problem 8.22)

$$\frac{1}{\rho_c} = \frac{1}{\rho'} + \frac{2}{R_0 \cos \theta_i}. \tag{8.28}$$

ρ' is the distance from the source to the reflection point, R_0 is the reflecting surface's radius of curvature (positive, for a convex surface) and θ_i is the angle between the incident ray and the surface normal at Q_R.

8.5.2 3D Reflection

In the most general case, the reflected field can be astigmatic. This will happen if the reflecting surface is curved. It can also happen with a flat reflecting surface if the incident wavefront is astigmatic. We shall denote the caustic distances as ρ_1^i and ρ_2^i for the incident field and ρ_1^r and ρ_2^r for the reflected field. The astigmatic tube expression in (8.17) provides the spread factor and phase. Since GO makes a local plane wave assumption, we can take a hint from (8.21) and account for the reflection coefficient. Therefore, the reflected field is

$$\mathbf{E}^r(s) = \mathbf{E}_i(Q_R) \cdot \overline{\mathbf{R}} \sqrt{\frac{\rho_1^r \rho_2^r}{(\rho_1^r + s)(\rho_2^r + s)}} \, e^{-jks}. \tag{8.29}$$

Similar to the 2D case (with only one caustic) in Figure 8.15, the 3D reflected field emanates from two caustics that are at ρ_1^r and ρ_2^r, behind the reflecting surface. For the special case of a spherical incident wave and a flat reflecting surface, $\rho_1^r = \rho_2^r = s'$, where s' is the distance from the source to Q_R. Then, Equation (8.29) reduces to (8.21).

Finding the astigmatic reflected field caustic positions ρ_1^r and ρ_2^r involves some differential geometry. The reflecting surface has two principal radii of curvature R_1 and R_2. These are found by considering a plane that cuts the surface and contains the normal $\hat{n}$ at Q_R. The intersection of the plane with the surface is a curve. As the plane is rotated about $\hat{n}$, there will be orientations that either maximize or minimize the curvature. These extrema define the principal radii of curvature. The two planes are perpendicular whenever $R_1 \neq R_2$. For an astigmatic incident field, the general expressions for R_1 and R_2

are in Kouyoumjian and Pathak (1974). For the case of spherical incidence, they simplify to

$$\frac{1}{\rho_{1,2}^r} = \frac{1}{s'} + \frac{1}{\cos\theta_i}\left[\frac{\sin^2\theta_2}{R_1} + \frac{\sin^2\theta_1}{R_2}\right]$$

$$\pm\sqrt{\frac{1}{\cos^2\theta_i}\left[\frac{\sin^2\theta_2}{R_1} + \frac{\sin^2\theta_1}{R_2}\right]^2 - \frac{4}{R_1R_2}}. \tag{8.30}$$

Here, $\theta_{1,2}$ are the angles between the incident ray direction $\hat{s}'$ and $\hat{U}_{1,2}$. The unit vectors $\hat{U}_{1,2}$ are tangent to the surface at Q_R and are associated with the principal radii of curvature R_1, R_2.

Example 8.4 A z-polarized plane wave $E_z^i = E_0 e^{-jkx}$ illuminates a circular cylinder of radius a that is centred at $(x,y) = (0,0)$. Find the scattered field at a point on the negative x axis at a distance $\rho_s = -x$. Find the echo width in the backscatter direction.

Solution: The reflection point Q_R is at $(x,y) = (-a,0)$. At the reflection point, the radius of curvature is $R_0 = a$, the reflection coefficient is $R = -1$, and the incident field is $E_z^i(Q_R) = E_0 e^{jka}$. Because of plane wave incidence, $\rho' \to \infty$. Therefore, the caustic distance is $\rho_c = a/2$. For backscatter, the field point is at $x = -\rho_s$ (x is negative) so the distance ρ between Q_R and the far-field point is at $\rho = \rho_s - a$. The reflected field becomes

$$E_z^r(\rho_s) = E_0 e^{jka}(-1)\sqrt{\frac{a/2}{a/2 + (\rho_s - a)}}\, e^{-jk(\rho_s - a)}.$$

Since $\rho_s \gg a$ the spread factor can be simplified. Using the echo width definition

$$\sigma = \lim_{\rho_s \to \infty} 2\pi\rho_s \frac{|E_z^r|^2}{|E_z^i|^2}$$

leads to $\sigma = \pi a$. ∎

Example 8.5 A z-polarized plane wave $E_z^i = E_0 e^{-jkx}$ illuminates a sphere of radius a that is centred at the origin. Find the scattered field at a point on the negative x axis at a distance $r_s = -x$. Find the radar cross section in the backscatter direction.

Solution: The reflection point Q_R is at $(x,y,z) = (-a,0,0)$. The incident field at Q_R is $\mathbf{E}^i(Q_R) = \hat{z}E_0 e^{jka}$ and $\hat{e}_\parallel^i = -\hat{e}_\parallel^r = \hat{z}$ so that $\overline{\mathbf{R}} = -\hat{z}\hat{z}$ and

$$\mathbf{E}^i(Q_R) \cdot \overline{\mathbf{R}} = \hat{z}E_0 e^{jka} \cdot (-\hat{z}\hat{z}) = -\hat{z}E_0 e^{jka}.$$

The sphere's principal radii of curvature are both equal to a, and for plane wave incidence, $\rho' \to \infty$. From (8.30) it can be shown that the caustic distances for the reflected field are $\rho_1^r = \rho_2^r = a/2$. For backscatter, the field point is at $x = -r_s$ (x is negative) so the distance s between Q_R and the field point is $s = r_s - a$. The reflected field becomes

$$\mathbf{E}^r(r_s) = -\hat{z}E_0 e^{jka}\frac{a/2}{a/2 + (r_s - a)}e^{-jk(r_s - a)}.$$

Since $r_s \gg a$ the spread factor can be simplified. Using the RCS definition

$$\sigma = \lim_{r_s \to \infty} 4\pi r_s^2 \frac{|\mathbf{E}^r|^2}{|\mathbf{E}^i|^2}$$

leads to $\sigma = \pi a^2$. ∎

8.6   Curved Wedge Face

If a wedge face is curved, it affects the reflection. It is possible to generalize the wedge diffraction coefficient (8.6) by enforcing field continuity at the shadow boundaries. There are four types of shadow boundaries: the n-face and o-face ISBs and n-face and o-face RSBs. At each shadow boundary, one of the four terms in (8.6) compensates a particular discontinuity, while the other three terms remain continuous there. It is useful to identify the responsible terms by replacing the distance parameters L in (8.6) by L^{in}, L^{io}, L^{rn} and L^{ro} so that

$$
\begin{aligned}
D_{s,h} = {} & \frac{-e^{-j\pi/4}}{2n\sqrt{2\pi k}\sin\beta_0} \\
& \times \left[\cot\left(\frac{\pi + (\phi - \phi')}{2n}\right) F(kL^{in}a^+(\phi - \phi')) \right. \\
& + \cot\left(\frac{\pi - (\phi - \phi')}{2n}\right) F(kL^{io}a^-(\phi - \phi')) \\
& \mp \left\{ \cot\left(\frac{\pi + (\phi + \phi')}{2n}\right) F(kL^{rn}a^+(\phi + \phi')) \right. \\
& \left. \left. + \cot\left(\frac{\pi - (\phi + \phi')}{2n}\right) F(kL^{ro}a^-(\phi + \phi')) \right\} \right].
\end{aligned}
\tag{8.31}
$$

For generality, we will assume an astigmatic incident field. Enforcing continuity at an ISB leads to

$$
L^i = \frac{s(\rho_e^i + s)\rho_1^i \rho_2^i}{\rho_e^i(\rho_1^i + s)(\rho_2^i + s)}\sin^2\beta_o,
\tag{8.32}
$$

where $\rho_{1,2}^i$ are the principal radii of curvature of the incident wavefront and ρ_e^i is as before, in (8.25). The face curvatures do not affect the incident field shadowing, so we can simply let $L^{in} = L^{io} = L^i$. Enforcing continuity at the RSBs leads to

$$
L^r = \frac{s(\rho_e^r + s)\rho_1^r \rho_2^r}{\rho_e^r(\rho_1^r + s)(\rho_2^r + s)}\sin^2\beta_o
\tag{8.33}
$$

where $\rho_{1,2}^r$ are the principal radii of curvature of the reflected wavefront and ρ_e^r is the reflected wavefront radius of curvature in the plane containing the reflected ray $\hat{s}$ direction and edge tangent $\hat{e}$. In this case, L^{ro} and L^{rn} depend on the face curvatures. Similar to (8.25), we have

$$
\frac{1}{\rho_e^r} = \frac{1}{\rho_e^i} - \frac{2(\hat{n}_e \cdot \hat{n})(\hat{s}' \cdot \hat{n})}{a_e \sin^2\beta_o}.
\tag{8.34}
$$

At the diffraction point, $\hat{n}$ is normal to the wedge face and differs for the o-face and n-face.

8.7   Non-Metallic Wedge

In the previous section it was seen that each term in the diffraction coefficient compensates a specific discontinuity. This can be extended to a non-metallic wedge, having o-face and n-face reflection coefficients Γ^o, Γ^n and o-face and n-face transmission coefficients T^o, T^n. A heuristic modification can be made, enforcing continuity of the fields at the shadow boundaries, with the result that

$$
D = \frac{-e^{-j\pi/4}}{2n\sqrt{2\pi k}\sin\beta_0}
$$

$$
\times \left[(1 - T^n)\cot\left(\frac{\pi + (\phi - \phi')}{2n}\right) F(kL^{in}a^+(\phi - \phi')) \right.
$$

$$
+ (1 - T^o)\cot\left(\frac{\pi - (\phi - \phi')}{2n}\right) F(kL^{io}a^-(\phi - \phi'))
$$

$$
\Gamma^n \cot\left(\frac{\pi + (\phi + \phi')}{2n}\right) F(kL^{rn}a^+(\phi + \phi'))
$$

$$
\left. + \Gamma^o \cot\left(\frac{\pi - (\phi + \phi')}{2n}\right) F(kL^{ro}a^-(\phi + \phi')) \right]. \tag{8.35}
$$

The reflection and transmission coefficients are evaluated at their respective shadow boundaries. For a perfectly conducting wedge, $T^n = T^o = 0$ and $\Gamma^n = \Gamma^o = \mp 1$ so it reduces to (8.31).

This approach gives good results in many practical situations; however, it does not account for surface wave diffraction, which could be present on a dielectric or dielectric-coated metallic surface. The method is described in Burnside and Burgener (1983), which treated the diffraction by a thin dielectric slab.

8.8 Slope Diffraction

The edge diffraction expressions (8.11), (8.12) apply when the incident field at the edge is constant or slowly varying with respect to ϕ'. If the incident field at the edge has a non-zero slope, that is, it is changing with respect to ϕ', then *slope diffraction* may be needed. If the incident field is non-zero and the slope of the field is non-zero, then both types of diffractions will be present.

As an example of an incident field with a slope, Figure 8.16 shows a dipole illuminating a wedge. The importance of slope diffraction becomes more apparent if the dipole is aligned so that $\mathbf{E}^i(Q_E) = 0$. In this case, the diffracted field is zero, but the slope diffracted field is non-zero.

For a source at (ρ', ϕ') and field point at (ρ, ϕ), the slope diffracted field is given by

$$
E_{\beta_0}^d \sim -e_s A(s, s') e^{-jks} \tag{8.36}
$$

$$
E_{\phi}^d \sim -e_h A(s, s') e^{-jks}, \tag{8.37}
$$

where

$$
e_{s,h} = \frac{1}{jk\sin\beta_0} \frac{\partial D_{s,h}}{\partial\phi'} \frac{\partial E}{\partial n}. \tag{8.38}
$$

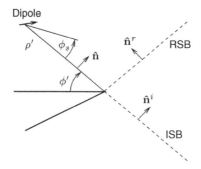

Figure 8.16 Parameters for slope diffraction by a dipole.

$\partial D/\partial\phi'$ is the slope diffraction coefficient, and $\partial E^i/\partial n$ is the derivative of the incident field evaluated at Q_E:

$$\frac{\partial E^i}{\partial n} = \frac{1}{\rho'}\frac{\partial E^i}{\partial\phi_s}. \tag{8.39}$$

A general form that allows for curved faces can be obtained by differentiating $D_{s,h}$ in (8.31), with the result that

$$\frac{\partial D_{s,h}}{\partial\phi'}\frac{\partial E}{\partial n} = \frac{-e^{-j\pi/4}}{4n^2\sqrt{2\pi k}\sin^2\beta_0}$$

$$\times\left[\csc^2\left(\frac{\pi+(\phi-\phi')}{2n}\right)F_s(kL^{in}a^+(\phi-\phi'))\frac{\partial E^{in}}{\partial n^i}\right.$$

$$-\csc^2\left(\frac{\pi-(\phi-\phi')}{2n}\right)F_s(kL^{io}a^-(\phi-\phi'))\frac{\partial E^{io}}{\partial n^i}$$

$$\pm\left\{\csc^2\left(\frac{\pi+(\phi+\phi')}{2n}\right)F_s(kL^{rn}a^+(\phi+\phi'))\frac{\partial E^{rn}}{\partial n^r}\right.$$

$$\left.\left.-\csc^2\left(\frac{\pi-(\phi+\phi')}{2n}\right)F_s(kL^{ro}a^-(\phi+\phi'))\frac{\partial E^{ro}}{\partial n^r}\right\}\right]. \tag{8.40}$$

$\hat{\mathbf{n}}^i$ points to the lit side, and $\hat{\mathbf{n}}^r$ points to the reflection side. The slope transition function F_s is

$$F_s(X) = j2X\,(1-F(X)). \tag{8.41}$$

In 2D we can use

$$E^d \sim \frac{1}{jk}\frac{\partial D_s}{\partial\phi'}\frac{\partial E^i}{\partial n}\frac{e^{-jk\rho}}{\sqrt{\rho}} \tag{8.42}$$

$$H^d \sim \frac{1}{jk}\frac{\partial D_h}{\partial\phi'}\frac{\partial H^i}{\partial n}\frac{e^{-jk\rho}}{\sqrt{\rho}}. \tag{8.43}$$

8.9   Double Diffraction

An edge-diffracted ray can encounter another edge and get diffracted a second time. Suppose we have a strip as in Figure 8.17. The illumination is assumed to be TE. It can be a plane wave or line source; this is not important. What matters is that $H_z^d(\phi)$ is discontinuous, as $H_z^d(0) \neq H_z^d(360°)$. This occurs because the diffraction by Q_{E1} is based on a half plane that extends to $x = +\infty$. The discontinuity can be eliminated by introducing H_z^{ddo}, which compensates for shadowing of H_z^d by the o-face, and H_z^{ddn}, which compensates for shadowing of H_z^d by the n-face. The total will then be continuous at $\phi = 0$.

Edge diffraction requires an incident field that is ray optical. Therefore, with double diffraction the second edge must lie outside the transition region of the first edge – in other words, away from any shadow boundaries of the first edge. This limitation is encountered, for example, when calculating the backscatter of a strip or plate at grazing incidence. The trailing edge happens to be right on the shadow boundary of the leading edge, so the first-order diffraction illuminating the trailing edge is not ray optical. The remedy for this problem is to expand the non-ray optical field from the first edge diffraction as a plane wave spectrum (Tiberio and Kouyoumjian 1984). Then, the spectral components can be treated one by one, as plane waves that are incident on the second edge. Adding up the contributions gives the diffraction by the second edge. An alternate approximate technique for multiple diffraction is also possible (Bach Andersen 1997), using slope diffraction.

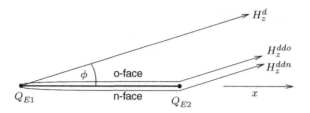

Figure 8.17   TE double diffraction by a strip.

If the polarization were TM, then the grazing rays on the metal plate would be zero, and there would be no double diffraction. Therefore, for this polarization the double diffraction is zero. On the other hand, the slope of the field incident on Q_{E2} would not be zero, so a slope diffraction contribution could be found.

It is possible to add third and even higher-order diffraction. For some geometries, it is possible to combine *all* of the diffractions up to infinity, in closed form. This technique is called the *self-consistent GTD* (Wang 1976) and it gives very good results when it can be applied. Its limitation is that it uses GTD, so it can only be used away from shadow boundaries.

Because of the ray optical problem in multiple diffraction, general-purpose computer programs do not consider diffractions beyond second order.

8.10   GTD Equivalent Edge Currents

For some geometries the edge-diffracted rays form a caustic. A way to recognize there is a problem is by considering the generalized Fermat's principle, which requires us to identify the shortest and longest ray paths. This becomes impossible if all the edge-diffracted rays have the same length.

As an example, the radiation pattern of a monopole on a circular disc generally involves two edge diffraction points. However, in the normal direction we no longer have two points but a continuum of edge contributions. Similar problems occur with the on-axis radiation pattern of a reflector antenna with a circular rim and the broadside backscatter from a flat plate of any shape.

The continuum of edge contributions can be added up with an integration, using a GTD equivalent edge current (GTDEEC). These are generally useful whenever an edge diffraction caustic is encountered. GTDEECs are fictitious line currents that are assumed to flow along the edge ℓ of a body. The idea is shown in Figure 8.18. In case (a) an incident plane wave diffracts off a wedge and produces a field at a point (r, θ, ϕ). It is postulated that in case (b) an electric line current $I(z)$ in free space will produce the same field as case (a). Requiring the two fields to be equal then defines $I(z)$.

The incident field is a plane wave in the direction (θ', ϕ')

$$\mathbf{E}^i = \hat{\boldsymbol{\beta}}_0' E_0 \, e^{-jk(x \sin \theta' \cos \phi' + y \sin \theta' \sin \phi' - z \cos \theta')}.$$

The GTD diffracted field is

$$E_{\beta_0}^d = -E_{\beta_0'}^i(Q_E) \, D_s^k \, \frac{e^{-jks}}{\sqrt{s}}$$

or

$$E_{\beta_0}^d = -E_0 e^{jkz_e \cos \beta_0'} \, D_s^k \, \frac{e^{-jks}}{\sqrt{s}}. \tag{8.44}$$

The proposed equivalent current is

$$I(z) = I_0 e^{jkz \cos \beta_0'}.$$

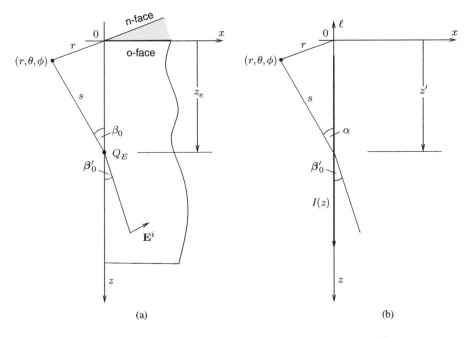

Figure 8.18   (a) Edge diffraction and (b) equivalent current $I(z)$.

The vector potential for this current is

$$A(z) = \frac{\mu}{4\pi} I_0 \int_{-\infty}^{\infty} e^{jkz' \cos \beta_0'} \frac{e^{-jks}}{s} \, dz',$$

where

$$s = \sqrt{x^2 + y^2 + (z - z')^2} = \sqrt{\rho^2 + (z - z')^2}.$$

The phase of the integrand is

$$\psi(z') = -ks + kz' \cos \beta_0'$$

and it is stationary when

$$\psi'(z') = k \frac{z - z'}{s} + k \cos \beta_0' = 0$$

so that $\alpha = \beta_0'$. Therefore, the stationary point occurs at Q_E. Denoting $z' = z_e$ and using $\psi''(z_e) = -k \sin^2 \beta_0' / s$ in the stationary phase formula from Section 5.5 gives

$$A_z = \sqrt{\frac{2\pi s}{k |\sin^2 \beta_0'|}} \frac{\mu I_0}{4\pi} e^{j(kz_e \cos \beta_0' - \pi/4)} \frac{e^{-jks}}{s}$$

$$= \frac{\mu I_0}{\sqrt{8\pi k} |\sin \beta_0'|} e^{j(kz_e \cos \beta_0' - \pi/4)} \frac{e^{-jks}}{\sqrt{s}}.$$

Since $E_{\beta_0} = -j\omega \sin \alpha A_z$, the line source field is

$$E_{\beta_0} = -\eta I_0 \sqrt{\frac{k}{8\pi}} \, e^{j(kz_e \cos \beta_0' + \pi/4)} \frac{e^{-jks}}{\sqrt{s}}. \tag{8.45}$$

The wedge-diffracted field (8.44) must equal the line source field (8.45) so that[3]

$$E_{\beta_0}^d = \underbrace{-E_0 e^{jkz_e \cos \beta_0'} D_s^k \frac{e^{-jks}}{\sqrt{s}}}_{E_{\beta_0'}^i(Q_E)} = -\eta \sqrt{\frac{k}{8\pi}} e^{j\pi/4} \underbrace{I_0 e^{jkz_e \cos \beta_0'}}_{I(z_e)} \frac{e^{-jks}}{\sqrt{s}}. \tag{8.46}$$

Therefore,

$$I(z_e) = \frac{1}{\eta} \sqrt{\frac{8\pi}{k}} e^{-j\pi/4} E_{\beta_0'}^i(Q_E) D_s^k.$$

To reach the final form of the equivalent current, we change to a variable $\ell = -z$. Then, $\mathbf{E}^i$ projected onto the edge is in the $\hat{\boldsymbol{\ell}}$ direction. With this notation we obtain $E_{\beta_0'}^i(Q_E) = \hat{\boldsymbol{\ell}} \cdot \mathbf{E}^i(Q_E)/\sin \beta_0'$.[4]

When the equivalent currents are used in a line integral, there may or may not be a stationary point, so we should drop the Q_E notation and simply use the coordinate ℓ along the edge. The other polarization is associated with a magnetic source and can be obtained from duality. The final form is then (Pathak 1988)

$$I(\ell) = -\frac{1}{\eta} e^{-j\pi/4} \sqrt{\frac{8\pi}{k}} \frac{\hat{\boldsymbol{\ell}} \cdot \mathbf{E}^i(\ell)}{\sin \beta_0'} D_s^k(\beta_0, \phi, \phi') \tag{8.47}$$

$$M(\ell) = -\eta e^{-j\pi/4} \sqrt{\frac{8\pi}{k}} \frac{\hat{\boldsymbol{\ell}} \cdot \mathbf{H}^i(\ell)}{\sin \beta_0'} D_h^k(\beta_0, \phi, \phi'), \tag{8.48}$$

where the Keller diffraction coefficients $D_{s,h}^k$ are given in (8.10).

The GTDEECs are non-physical, as their value depends on the angle of observation. Also, nothing forces $M(\ell)$ to be zero on a conductor, and in general, both $I(\ell)$ and $M(\ell)$ will be present. Nevertheless, they are useful for solving the caustic problem. Once they are known, a radiation line integral along the edge ℓ gives the edge-diffracted wave. From (4.31) and (4.32),

$$\mathbf{E}^d = \frac{jk}{4\pi} \int_\ell (\eta \hat{\mathbf{R}} \times \hat{\mathbf{R}} \times \hat{\boldsymbol{\ell}}' I(\ell') + \hat{\mathbf{R}} \times \hat{\boldsymbol{\ell}}' M(\ell')) \frac{e^{-jkR}}{R} \, d\ell' \tag{8.49}$$

and

$$\mathbf{H}^d = \frac{-jk}{4\pi} \int_\ell \left(\hat{\mathbf{R}} \times \hat{\boldsymbol{\ell}}' I(\ell') - \frac{1}{\eta} \hat{\mathbf{R}} \times \hat{\mathbf{R}} \times \hat{\boldsymbol{\ell}}' M(\ell') \right) \frac{e^{-jkR}}{R} \, d\ell', \tag{8.50}$$

where $\mathbf{R} = \mathbf{r} - \mathbf{r}'$; note that the plane wave relationship $\eta \mathbf{H}^d = \hat{\mathbf{R}} \times \mathbf{E}^d$ applies. In the far field, $\hat{\mathbf{R}} \approx \hat{\mathbf{r}}$ and $R \approx r - \hat{\mathbf{r}} \cdot \mathbf{r}'$.

Outside of any caustic regions and on the Keller cone, the radiation integrals (8.49), (8.50) possess a stationary phase point, and an asymptotic evaluation recovers (8.22). So, the edge diffraction formula (8.22) with $\overline{\mathbf{D}} = \overline{\mathbf{D}}^k$ can be thought of as a special case of the GTDEEC approach. If there is no caustic, it is preferable to use the simpler (8.22) rather than (8.49), (8.50).

---

[3] The equality in (8.46) is possible because D_s^k does not depend on distance. The equality cannot hold with the UTD diffraction coefficient D_s, because it is range dependent in the transition regions.

[4] The equivalent current was developed with the assumption that we are on the Keller cone, so we could just as well have used $\sin \beta_0$ instead of $\sin \beta_0'$. However, the latter choice is preferred, because $D_{s,h}^k$ contains a $1/\sin \beta_0$ term. Consequently, the overall behaviour of the equivalent current is as $1/(\sin \beta_0' \sin \beta_0)$ which satisfies reciprocity. This is a desirable heuristic extension that accommodates field points off the Keller cone, whereby $\beta_0' \neq \beta_0$.

8.11   Surface Ray Diffraction

When a ray is incident on a curved surface, it launches a *creeping wave*. This type of wave was first described by G. N. Watson in 1918, who was investigating radio wave propagation along the Earth's surface. In the 1940s and 1950s, Fock in the Soviet Union, Pekeris in the United States and others generalized creeping wave theory for surfaces with variable curvature and impedance boundaries. In 1956 Keller introduced the concept of surface-diffracted rays which paved the way for a GTD description. Some 20 years later, Pathak and co-workers developed UTD surface ray expressions for perfectly conducting surfaces that are valid across shadow boundaries.

The UTD provides a ray optical description of curved surface diffraction in terms of direct, reflected and surface-diffracted rays, as shown in Figure 8.19(a). Besides direct and reflected rays, there are creeping waves. These become 'attached' at Q_1 and Q_3. Then they propagate along the surface, decaying exponentially in amplitude as they shed energy tangentially, in the forward direction. The rays then leave the surface at the 'launch' points Q_2 and Q_4. Upon leaving the surface, they radiate as cylindrical waves in 2D or as astigmatic ray tubes in 3D.

Creeping waves can encircle the surface multiple times, as shown in Figure 8.19(b). On perfect conductors the decay rate is large, so the two rays in Figure 8.19(a) are usually sufficient and multiple encirclements, which occur in both the clockwise and counterclockwise directions, can be ignored. If the metal has a dielectric coating or if it is a reactive surface impedance, the decay rate can be greatly reduced. In these cases multiple encirclements can become important.

To illustrate the application of UTD for curved surfaces, the following sections will examine the diffraction by a perfectly conducting circular cylinder. There are three separate cases: scattering, radiation and coupling. The first case is the configuration in Figure 8.19(a) and is called the *scattering problem*. The source and field points are away from the cylindrical surface, and the diffraction coefficients are in terms of the Pekeris functions $p(\xi)$ and $q(\xi)$. The scattering problem requires that the source and field points be at least a wavelength or so away from the surface – otherwise the asymptotic theory on which it is based becomes inaccurate.

The second case is called the *radiation problem*. It has the source on the surface and the field point away from the surface. The diffraction coefficients are in terms of the Fock radiation functions $\tilde{g}(\xi)$ and $g(\xi)$. This formulation is used for surface-mounted antennas.

The third case is called the *coupling problem*. It has the source and field points on the surface. The diffraction coefficients are in terms of the Fock coupling functions $u(\xi)$ and $v(\xi)$. This formulation is used to find the mutual coupling between surface-mounted antennas.

8.11.1   Scattering

The UTD scattering formulation (Pathak et al. 1980) includes incident, reflected and creeping waves. The illumination can be an electric or magnetic line source, and the geometry is given in Figure 8.20.

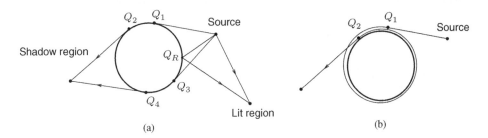

Figure 8.19   Diffraction by a circular cylinder. (a) Incident and reflected rays in lit region; creeping waves in shadow region. (b) Multiple encirclement by a creeping wave.

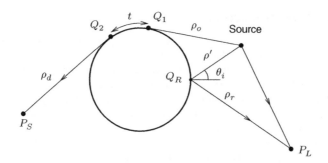

Figure 8.20 Scattering by a circular cylinder. Incident and reflected rays reach P_L in the lit region; creeping wave reaches P_S in the shadow region.

In the uniform theory it is necessary that the incident plus reflected rays smoothly transition to a creeping wave field in the shadow region. The ordinary GO reflection coefficient cannot provide this behaviour; instead a different uniform reflection coefficient is required. Otherwise, the 2D reflected field is of the same form as in (8.27) and is given by

$$
\left.\begin{array}{c} E_z^r(P_L) \\ H_z^r(P_L) \end{array}\right\} = \left\{\begin{array}{c} E_z^i(Q_R)\,\mathcal{R}_s \\ H_z^i(Q_R)\,\mathcal{R}_h \end{array}\right\} \sqrt{\frac{\rho_c}{\rho_c + \rho_r}}\, e^{-jk\rho_r},
\tag{8.51}
$$

where ρ_c is the caustic distance (8.28) and ρ_r is the distance from Q_R to the field point. The uniform reflection coefficient is

$$
\mathcal{R}_{s,h} = -\sqrt{\frac{-4}{\xi^L}}\, e^{-j(\xi^L)^3/12} \left[\frac{e^{-j\pi/4}}{2\xi^L\sqrt{\pi}}(1 - F(X^L)) + \hat{P}_{s,h}(\xi^L) \right].
\tag{8.52}
$$

Here, ξ^L is the lit-region Fock parameter

$$
\xi^L = -2m(Q_R)\cos\theta_i
\tag{8.53}
$$

and $m(Q_R)$ is the curvature parameter

$$
m(Q_R) = \left(\frac{k\rho_g(Q_R)}{2}\right)^{1/3}.
\tag{8.54}
$$

The quantity $\rho_g(Q_R)$ denotes the reflecting surface's local radius of curvature at Q_R. For a circular cylinder of radius a, we have that $\rho_g(Q_R) = a$ and $m = (ka/2)^{1/3}$. We also have X^L given by

$$
X^L = 2kL\cos^2\theta_i; \quad L = \frac{\rho'\rho_r}{\rho' + \rho_r}.
\tag{8.55}
$$

It is noted that if the cylinder is large and θ_i is not near $90°$, then $\xi^L \ll 0$ is large and negative. Using large-argument approximations for the functions in $\mathcal{R}_{s,h}$, it can be shown that the reflection coefficient approaches $\mathcal{R}_{s,h} \to \mp 1$.

In the shadow region, the 2D diffracted field is

$$
\left.\begin{array}{c} E_z^d(P_S) \\ H_z^d(P_S) \end{array}\right\} = \left\{\begin{array}{c} E_z^i(Q_1)\,\mathcal{D}_s \\ H_z^i(Q_1)\,\mathcal{D}_h \end{array}\right\} \frac{e^{-jk\rho_d}}{\sqrt{\rho_d}}.
\tag{8.56}
$$

The surface ray diffraction coefficient is

$$
\mathcal{D}_{s,h} = -\sqrt{m(Q_1)m(Q_2)}\sqrt{\frac{2}{k}}\left[\frac{e^{-j\pi/4}}{2\xi\sqrt{\pi}}(1-F(X^d)) + \hat{P}_{s,h}(\xi)\right]e^{-jkt}. \tag{8.57}
$$

The Fock parameter is

$$
\xi = \int_{Q_1}^{Q_2}\frac{m(t)}{\rho_g(t)}\,dt \tag{8.58}
$$

the curvature parameter is

$$
m(t) = \left(\frac{k\rho_g(t)}{2}\right)^{1/3} \tag{8.59}
$$

and X^d, L are given by

$$
X^d = \frac{kL\xi^2}{2m(Q_1)m(Q_2)}; \quad L = \frac{\rho_o\rho_d}{\rho_o + \rho_d}. \tag{8.60}
$$

On the surface ray path (the geodesic) from Q_1 to Q_2, the local radius of curvature is denoted as $\rho_g(t)$, where the parameter t is used to denote that ρ_g may vary along the arc length. On an elliptic cylinder, ρ_g will vary along the geodesic. For a circular cylinder of radius a, we have that $\rho_g = a$ is a constant. Therefore, if $\phi_{1,2}$ are the angular positions of Q_1 and Q_2, then $t = a(\phi_2 - \phi_1)$ is the arc length between the two points, $\xi = mt/a$ and $m = (ka/2)^{1/3}$.

Both $\mathcal{R}_{s,h}$ and $\mathcal{D}_{s,h}$ use the transition function $F(X)$ given by (8.9) and the Pekeris caret functions $\hat{P}_{s,h}(\xi)$ which are

$$
\hat{P}_s(\xi) = \frac{e^{-j\pi/4}}{\sqrt{\pi}}\int_{-\infty}^{\infty}\frac{V(t)}{W_2(t)}e^{-j\xi t}\,dt; \quad \hat{P}_h(\xi) = \frac{e^{-j\pi/4}}{\sqrt{\pi}}\int_{-\infty}^{\infty}\frac{V'(t)}{W_2'(t)}e^{-j\xi t}\,dt. \tag{8.61}
$$

In the above, $V(t)$ and $W_2(t)$ are Fock-type Airy functions that are defined in terms of Miller-type Airy functions $\mathrm{Ai}(t)$, $\mathrm{Bi}(t)$ as

$$
j2V(t) = W_1(t) - W_2(t); \quad W_{1,2} = \sqrt{\pi}(\mathrm{Bi}(t) \pm j\mathrm{Ai}(t)).
$$

The Airy functions arise because Hankel functions that appear in the cylinder scattering solution have been replaced by their Watson approximation (C.17). The Pekeris caret functions can also be expressed as

$$
\hat{P}_{s,h}(\xi) = \left\{\begin{matrix}p^*(\xi)\\q^*(\xi)\end{matrix}\right\}e^{-j\pi/4} - \frac{e^{-j\pi/4}}{2\xi\sqrt{\pi}}, \tag{8.62}
$$

where $p(\xi)$ and $q(\xi)$ are the soft and hard Pekeris functions. These can be calculated using FUNCTION pfun(x) and FUNCTION qfun(x) and are plotted in Figure 8.21.

8.11.2   Radiation

The UTD radiation formulation (Pathak et al. 1981) is used when the source is on the surface. In the lit region there is a direct ray, and in the shadow region there are creeping waves. Creeping waves can contribute in the lit region as well. The geometry is given in Figure 8.22.

There are three types of sources that can radiate on a perfect electric conductor and they are shown in Figure 8.22. The magnetic line source (b) and radial electric line dipole (d) produce a TE field that is easily described in terms of H_z. The magnetic line dipole (c) has a TM field in terms of E_z.

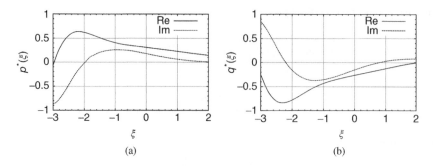

Figure 8.21   Pekeris functions $p^*(\xi)$ and $q^*(\xi)$.

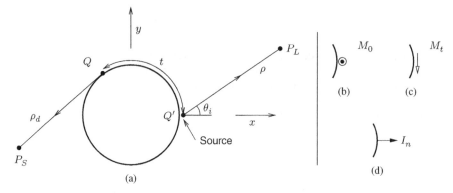

Figure 8.22   (a) Radiation by a source on a circular cylinder. Direct ray reaches P_L in the lit region; creeping wave reaches P_S in the shadow region. Source types: (b) magnetic line source M_0, (c) magnetic line dipole M_t and (d) electric line dipole I_n.

The magnetic sources will be discussed first. Referring to Figure 8.22, the source is at Q'. At a field point P_L in the lit region,

$$\left.\begin{array}{r} E_z(P_L) \\ H_z(P_L) \end{array}\right\} = -\sqrt{\frac{k}{8\pi}}\,e^{j\pi/4}\left\{\begin{array}{l} M_t\,S^\ell(\xi^\ell) \\ \frac{M_0}{\eta}\,H^\ell(\xi^\ell) \end{array}\right\}\frac{e^{-jk\rho}}{\sqrt{\rho}}. \tag{8.63}$$

In the TE case a magnetic line source $\mathbf{M} = \hat{\mathbf{z}}M_0$ carries its current in the $+z$ direction. In the TM case a magnetic line dipole $\mathbf{M} = \hat{\mathbf{t}}M_t$ is oriented such that $\hat{\mathbf{t}} \times \hat{\mathbf{n}} = \hat{\mathbf{z}}$ where $\hat{\mathbf{n}}$ is the surface normal at Q'. Here, ξ^ℓ is the lit-region Fock parameter

$$\xi^\ell = -m(Q')\cos\theta_i \tag{8.64}$$

and $m(Q')$ is the curvature parameter

$$m(Q') = \left(\frac{k\rho_g(Q')}{2}\right)^{1/3}. \tag{8.65}$$

The quantity $\rho_g(Q')$ denotes the surface's local radius of curvature at Q'. For a circular cylinder of radius a, we have that $\rho_g(Q') = a$ and $m = (ka/2)^{1/3}$.

It is noted that in the deep lit region for a large cylinder, $\xi^\ell \ll 0$ is large and negative. Using large-argument approximations, it can be shown that

$$S^\ell(\xi^\ell) \approx 2\cos\theta_i; H^\ell(\xi^\ell) \approx 2.$$

Using these approximations in (8.63), one finds that E_z and H_z reduce to that of a magnetic line dipole or magnetic line source on a flat ground plane. The factor of 2 is the doubling by image theory, and the $\cos\theta_i$ factor is the element pattern of the magnetic line dipole for the TM case. Therefore, one can see that $S^\ell(\xi^\ell)$ and $H^\ell(\xi^\ell)$ can be thought of as modifying factors that correct for the presence of a curved ground plane.

Another important point is that in the $\theta_i = 90°$ direction, a magnetic dipole M_t on a flat ground plane would not radiate at all; the element pattern is zero. On the other hand, the same source on a curved ground *will* radiate.

In the shadow region the diffracted field is

$$\left.\begin{matrix} E_z(P_S) \\ H_z(P_S) \end{matrix}\right\} = -\sqrt{\frac{k}{8\pi}}\, e^{j\pi/4} \left\{\begin{matrix} M_t\, S(\xi) \\ \frac{M_0}{\eta}\, H(\xi) \end{matrix}\right\} e^{-jkt} \left(\frac{\rho_g(Q)}{\rho_g(Q')}\right)^{1/6} \frac{e^{-jk\rho_d}}{\sqrt{\rho_d}}. \tag{8.66}$$

The Fock parameter is

$$\xi = \int_{Q'}^{Q} \frac{m(t)}{\rho_g(t)}\, dt \tag{8.67}$$

and the curvature parameter is

$$m(t) = \left(\frac{k\rho_g(t)}{2}\right)^{1/3}. \tag{8.68}$$

For a circular cylinder of radius a, $\rho_g = a$ is a constant. If Q' is at $\phi' = 0$ and Q is at ϕ, then $t = a\phi$ is the arc length between the two points, $\xi = mt/a$ and $m = (ka/2)^{1/3}$. The term $(\rho_g(Q)/\rho_g(Q'))^{1/6} = 1$.

In (8.63), the lit-region functions $S^\ell(\xi^\ell)$ and $H^\ell(\xi^\ell)$ are

$$S^\ell(\xi^\ell) = \frac{-j}{m(Q')}\, e^{-j(\xi^\ell)^3/3}\, \tilde{g}(\xi^\ell); \quad H^\ell(\xi^\ell) = e^{-j(\xi^\ell)^3/3}\, g(\xi^\ell) \tag{8.69}$$

and in (8.66) the shadow-region functions $S(\xi)$ and $H(\xi)$ are

$$S(\xi) = \frac{-j}{m(Q')}\, \tilde{g}(\xi); \quad H(\xi) = g(\xi). \tag{8.70}$$

$\tilde{g}(\xi)$ and $g(\xi)$ are called the soft and hard Fock radiation functions. They are given by

$$\tilde{g}(\xi) = \frac{1}{\sqrt{\pi}} \int_{C_0} \frac{e^{-j\xi t}}{W_2(t)}\, dt; \quad g(\xi) = \frac{1}{\sqrt{\pi}} \int_{C_0} \frac{e^{-j\xi t}}{W_2'(t)}\, dt, \tag{8.71}$$

where the contour C_0 starts at $\infty \exp(-j2\pi/3)$ and ends at ∞ in the complex t-plane. Similar to the scattering problem, the Fock-type Airy function W_2 comes about because a Hankel function that appears in the cylinder radiated-field solution was replaced by its Watson approximation.

For computational purposes it is convenient to define $\tilde{G}(\xi)$ and $G(\xi)$ such that

$$\tilde{G}(\xi) = \begin{cases} e^{-j\xi^3/3}\tilde{g}(\xi); \ \xi < 0 \\ \tilde{g}(\xi); \ \xi \geq 0 \end{cases} \tag{8.72}$$

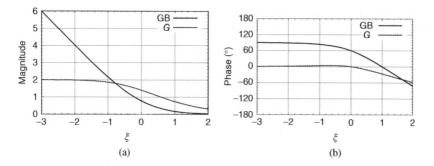

Figure 8.23 Fock radiation functions $\tilde{G}(\xi)$ and $G(\xi)$. (a) Magnitude and (b) phase.

$$G(\xi) = \begin{cases} e^{-j\xi^3/3}g(\xi); \ \xi < 0 \\ g(\xi); \ \xi \geq 0. \end{cases} \tag{8.73}$$

$\tilde{G}(\xi)$ and $G(\xi)$ can be calculated using FUNCTION GB(x) and FUNCTION G(x); they are plotted in Figure 8.23.

8.11.2.1 Radial Electric Dipole

The lit-region field for the radial dipole in Figure 8.22(d) is

$$H_z(P_L) = \sqrt{\frac{k}{8\pi}}\, e^{j\pi/4}\, I_n\, H^\ell(\xi^\ell)\sin\theta_i\, \frac{e^{-jk\rho}}{\sqrt{\rho}}. \tag{8.74}$$

In the shadow region the diffracted field is

$$H_z(P_S) = \sqrt{\frac{k}{8\pi}}\, e^{j\pi/4}\, I_n\, H(\xi)\,\mathrm{sgn}(y)\, e^{-jkt}\left(\frac{\rho_g(Q)}{\rho_g(Q')}\right)^{1/6} \frac{e^{-jk\rho_d}}{\sqrt{\rho_d}} \tag{8.75}$$

and for a circular cylinder $(\rho_g(Q)/\rho_g(Q'))^{1/6} = 1$.

Referring to Figure 8.22(a) and (d), the lit region H_z for this source is odd symmetric with respect to $\theta_i = 0$. In (8.74), this gets accounted for by the $\sin\theta_i$ term, with $-90° \leq \theta_i \leq 90°$. In the shadow region, the $\mathrm{sgn}(y)$ term in (8.75) takes care of the odd symmetry requirement.

Equation (8.74) can be deduced by recognizing the TE field is similar to the magnetic line source excitation in Equation (8.63) and then including the $\sin\theta_i$ element pattern of the radial electric dipole. Equation (8.75) is similar to (8.66) and the final form follows from requiring (8.75) to equal the lit-region expression (8.74) at the shadow boundary.

8.11.3 Coupling

The UTD coupling formulation (Pathak and Wang 1981) is used when the source and field points are on the surface. This is useful for calculating the coupling between two surface-mounted antennas. The geometrical definitions for the counterclockwise ray are given in Figure 8.24. The source is at Q' where the local tangent, normal and binormal vectors obey the relation $\hat{\mathbf{t}}' \times \hat{\mathbf{n}}' = \hat{\mathbf{b}}'$; at the field point Q, we have $\hat{\mathbf{t}} \times \hat{\mathbf{n}} = \hat{\mathbf{b}}$. For the 2D case in Figure 8.24, $\hat{\mathbf{b}}' = \hat{\mathbf{b}} = -\hat{\mathbf{z}}$.

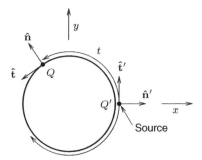

Figure 8.24 Coupling problem, source on a circular cylinder. Creeping wave reaches point Q in the shadow region.

The source is assumed to be a $\hat{b}'$-directed magnetic line source of strength M_0 or a $\hat{t}'$-directed magnetic line dipole of strength M_t. This can be written as

$$\mathbf{M} = \hat{b}' M_0 + \hat{t}' M_t. \tag{8.76}$$

According to the UTD, the magnetic field at Q is

$$\mathbf{H}(Q) = -\sqrt{\frac{k}{8\pi}} \frac{e^{j\pi/4}}{\eta} \mathbf{M} \cdot (\hat{b}'\hat{b} F_s + \hat{t}'\hat{t} G_s) e^{-jkt}, \tag{8.77}$$

where

$$F_s = \sqrt{\frac{jk}{2}} \left\{ m(Q') \left(\frac{\rho_g(Q)}{\rho_g(Q')} \right)^{1/6} \right\}^{-1} \xi^{-1/2} e^{-j\pi/4} v(\xi) \tag{8.78}$$

and

$$G_s = -\sqrt{\frac{jk}{2}} \left\{ m(Q') \left(\frac{\rho_g(Q)}{\rho_g(Q')} \right)^{1/6} \right\}^{-3} \xi^{-3/2} e^{-j3\pi/4} u(\xi). \tag{8.79}$$

For a circular cylinder of radius a, $\rho_g = a$ is a constant. If Q' is at $\phi' = 0$ and Q is at ϕ, then $t = a\phi$ is the arc length between the two points, $\xi = mt/a$ and $m = (ka/2)^{1/3}$. The term $(\rho_g(Q)/\rho_g(Q'))^{1/6} = 1$.

The coupling functions G_s and F_s use the soft and hard Fock coupling functions $u(\xi)$ and $v(\xi)$:

$$u(\xi) = \frac{\xi^{3/2} e^{j3\pi/4}}{\sqrt{\pi}} \int_{C_0} \frac{W_2'(t)}{W_2(t)} e^{-j\xi t} dt; \quad v(\xi) = \frac{\xi^{1/2} e^{j\pi/4}}{\sqrt{\pi}} \int_{C_0} \frac{W_2(t)}{W_2'(t)} e^{-j\xi t} dt, \tag{8.80}$$

where the contour C_0 starts at $\infty \exp(-j2\pi/3)$ and ends at ∞ in the complex t-plane. $u(\xi)$ and $v(\xi)$ can be calculated using FUNCTION FU(x) and FUNCTION FV(x); they are plotted in Figure 8.25.

On a very large cylinder, if $t = a\phi$ remains fixed while a increases, it becomes 'flat' and we obtain

$$\xi = m\phi = (ka/2)^{1/3}(t/a) = (k/2)^{1/3} t/a^{2/3} \to 0.$$

The Fock functions become $u(0) = 1$ and $v(0) = 1$. Equation (8.77) reduces to $\mathbf{H}$ for a source on a flat ground plane. Therefore, $u(\xi)$ and $v(\xi)$ can be thought of as correction terms that account for a ground plane's curvature.

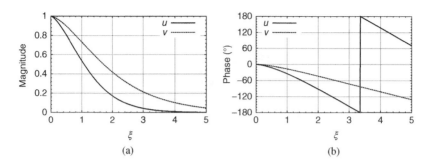

Figure 8.25   Fock coupling functions $u(\xi)$ and $v(\xi)$.

8.11.4   2D and 3D Radiation

The preceding sections described the UTD formulations for 2D scattering, radiation and coupling. 2D conveys most, but not all, of the key concepts. In this section, the 2D and 3D radiation problem will be used to present concepts that are unique to 3D. A full discussion of 3D for scattering, coupling and radiation is beyond the scope of this chapter; however, the material can be found in McNamara et al. (1990); Pathak (1988) and other references at the end of the chapter.

Figure 8.26 shows the radiation problem in both 2D and 3D. Let us begin with 2D. Using (8.63) for the lit region,

$$H_z(P_L) = -\sqrt{\frac{k}{8\pi}}\,e^{j\pi/4}\,\frac{M_0}{\eta}\,H^\ell(\xi^\ell)\,\frac{e^{-jk\rho}}{\sqrt{\rho}}$$ (8.81)

and using (8.66) for the shadow region,

$$H_z(P_S) = -\sqrt{\frac{k}{8\pi}}\,e^{j\pi/4}\,\frac{M_0}{\eta}\,H(\xi)\,e^{-jkt}\,\frac{e^{-jk\rho_d}}{\sqrt{\rho_d}},$$ (8.82)

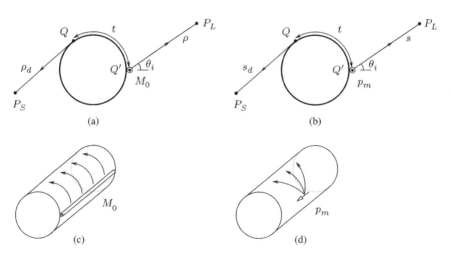

Figure 8.26   (a, c) Magnetic line source M_0 and 2D radiation; (b, d) magnetic dipole p_m and 3D radiation. The cylinders are z-directed and the radius is a.

where $\xi^\ell = -m\cos\theta_i$ and $\xi = mt/a$. In these expressions we can recognize the large-argument form of the Hankel function (C.13). With this in mind, they can be rewritten as

$$H_z(P_L) = \left[\frac{-kM_0}{4\eta} H_0^{(2)}(k\rho)\right] H^\ell(\xi^\ell) \tag{8.83}$$

and

$$H_z(P_S) = \left[\frac{-kM_0}{4\eta} H_0^{(2)}(k\rho_d)\right] H(\xi) e^{-jkt}. \tag{8.84}$$

The terms in square brackets represent free-space radiation from Q' in the lit region and Q in the shadow region.

In 3D, assuming the magnetic dipole p_m and field point are in the $x - y$ plane, the UTD expressions (Pathak 1988) can be obtained and turn out to be

$$H_z(P_L) = \left[\frac{-jkp_m}{4\pi\eta} \frac{e^{-jks}}{s}\right] H^\ell(\xi^\ell) \tag{8.85}$$

and

$$H_z(P_S) = \left[\frac{-jkp_m}{4\pi\eta} \frac{e^{-jks_d}}{\sqrt{s_d(t+s_d)}}\right] H(\xi) e^{-jkt}. \tag{8.86}$$

It is customary to use the distances ρ, ρ_d in 2D and s, s_d in 3D. In the far zone, $s_d \gg t$ and (8.86) becomes

$$H_z(P_S) = \left[\frac{-jkp_m}{4\pi\eta} \frac{e^{-jks_d}}{s_d}\right] H(\xi) e^{-jkt}. \tag{8.87}$$

The square bracket terms in (8.85) and (8.87) represent free-space propagation, so conceptually these expressions are of exactly the same form as in 2D. In the shadow region, if $s_d \gg t$ is not satisfied, then the 2D and 3D formulas have a fundamental difference. A physical description of why this is so is given in Figure 8.26(c) and (d). In 2D, the surface rays from the line source do not spread, whereas in 3D there is a spreading away from the point source. This effect appears in the spread factor $1/\sqrt{s_d(t+s_d)}$ in Equation (8.86).[5] The spread factor does not depend on what kind of point source is used and applies just as well to an axial or azimuthal magnetic dipole or an electric monopole on a circular cylinder.

In summary, the radiation patterns for 2D and for the $x - y$ plane in 3D are the same. Therefore, if only the principal plane pattern is needed, a 2D model is sufficient.

---

[5] The reader is referred to Pathak et al. (1981) and Pathak (1988, Table 3). In short, the 3D spread factor has two parts. First, there is surface ray spreading along the curved surface, from Q' to Q. Second, there is astigmatic spreading, away from the launch point Q.

To find the surface ray spreading from Q' to Q on a circular cylinder, the caustic distance for the surface ray tube is $\rho^d = t$, and $\sqrt{d\psi_0/d\psi} = 1$. Using $\rho^d d\psi = d\eta(Q)$ leads to the surface ray spread factor

$$\sqrt{\frac{d\psi_0}{d\eta(Q)}} = \frac{1}{\sqrt{t}}.$$

The spreading beyond point Q is accounted for by an astigmatic tube having

$$\sqrt{\frac{\rho^d}{s_d(\rho^d + s_d)}} = \sqrt{\frac{t}{s_d(t + s_d)}}.$$

Combining these two spread factors gives

$$\frac{1}{\sqrt{s_d(t + s_d)}}.$$

8.12   Further Reading

Hansen (1981) contains a collection of early key papers on GTD/UTD and a historical perspective. McNamara et al. (1990) present a practical and detailed introduction to the UTD and its applications. The books by James (1986) and Borovikov and Kinber (1994) are written at a more advanced level; the latter has a Russian perspective and it contains material that historically had been less accessible in the West. The book chapter by Pathak (1988) contains a good summary of the theoretical formulations as well as practical techniques. Pathak (1992) describes UTD for the analysis of horn, reflector and other aperture-type antennas.

The UTD is highly suitable for modelling antennas on complex platforms such as aircraft, ships and satellites. A significant part of the challenge is implementing the ray tracing. These applications are discussed by Burnside and Marhefka (1988), Marhefka and Burnside (1992), Pathak (1992) and Saez de Adana et al. (2011).

Besides the UTD there other high-frequency asymptotic methods that have not been covered in this chapter, for example, the spectral theory of diffraction STD, the uniform asymptotic theory UAT and the boundary layer method. UTD and these other techniques are reviewed in a treatise by Bouche et al. (1997).

References

Bach Andersen J (1997) UTD multiple-edge transition zone diffraction. *IEEE Trans. Antennas Propag.* **AP-45**(7), 1093–1097.

Borovikov VA and Kinber B (1994) *Geometrical Theory of Diffraction*. IEE.

Bouche D, Molinet F and Mittra R (1997) *Asymptotic Methods in Electromagnetics*. Springer-Verlag.

Burnside WD and Burgener KW (1983) High frequency scattering by a thin dielectric slab. *IEEE Trans. Antennas Propag.* **AP-31**(1), 104–110.

Burnside WD and Marhefka RJ (1988) Antennas on aircraft, ships, or any large, complex environment. In *Antenna Handbook, Theory Application and Design* (ed. Lo YT and Lee SW) Van Nostrand Reinhold pp. 20.1–20.100.

(ed. Hansen RC) (1981) *Geometric Theory of Diffraction*. IEEE Press.

Hill KC and Pathak PH (1991) A UTD solution for the EM diffraction by a corner in a plane angular sector. *IEEE AP-S/URSI International Symposium, London Ontario, Canada*.

Hutchins DL (1967) *Asymptotic Series Describing the Diffraction of a Plane Wave by a Two-dimensional Wedge of Arbitrary Angle*. PhD thesis. Ohio State University.

James GL (1986) *Geometrical Theory of Diffraction for Electromagnetic Waves*. Peter Peregrinus.

Kouyoumjian RG and Pathak PH (1974) A uniform geometrical theory of diffraction for an edge in a perfectly conducting surface. *Proc. IEEE* **62**, 1448–1461.

Marhefka RJ and Burnside WD (1992) Antennas on complex platforms. *Proc. IEEE* **80**, 204–208.

McNamara DA, Pistorius CWI and Malherbe JAG (1990) *Introduction to the Uniform Geometrical Theory of Diffraction*. Artech House.

Ozturk AK, Paknys R and Trueman CW (2011) Vertex diffracted edge waves on a perfectly conducting plane angular sector. *IEEE Trans. Antennas Propag.* **AP-59**(3), 888–897.

Pathak PH (1988) Techniques for high frequency problems. In *Antenna Handbook, Theory Application and Design* (ed. Lo YT and Lee SW) Van Nostrand Reinhold pp. 4.1–4.117.

Pathak PH (1992) High frequency techniques for antenna analysis. *Proc. IEEE* **80**, 44–65.

Pathak PH, Burnside WD and Marhefka RJ (1980) A uniform UTD analysis of the diffraction of electromagnetic waves by a smooth convex surface. *IEEE Trans. Antennas Propag.* **AP-28**, 609–622.

Pathak PH and Wang N (1981) Ray analysis of mutual coupling between antennas on a convex surface. *IEEE Trans. Antennas Propag.* **AP-29**, 911–922.

Pathak PH, Wang N, Burnside WD and Kouyoumjian RG (1981) A uniform GTD solution for the radiation from sources on a smooth convex surface. *IEEE Trans. Antennas Propag.* **AP-29**, 609–621.

Saez de Adana F, Gutiérrez O, González I, Cátedra MF and Lozano L (2011) *Practical Applications of Asymptotic Techniques in Electromagnetics*. Artech House.

Sikta FA, Burnside WD, Chu TT and Peters L (1983) First-order equivalent current and corner diffraction scattering from flat plate structures. *IEEE Trans. Antennas Propag.* **AP-31**(4), 584–589.

Tiberio R and Kouyoumjian RG (1984) Calculation of the high-frequency diffraction by two nearby edges illuminated at grazing incidence. *IEEE Trans. Antennas Propag.* **AP-32**(11), 1186–1196.

Wang N (1976) Self-consistent GTD formulation for conducting cylinders with arbitrary convex cross section. *IEEE Trans. Antennas Propag.* **AP-24**(4), 463–468.

Weiner MM, Cruze SP, Li CC and Wilson WJ (1987) *Monopole Elements on Circular Ground Planes*. Artech House.

Problems

8.1 PROGRAM utd1 in Example 8.1 implements the TM UTD solution for a plane wave incident on a half plane.

  (a) Test the program by reproducing the result in the example.
  (b) Change the field point distance to $\rho = 2\lambda$; calculate and plot $|E_z(\phi)|$.
  (c) With $\rho = 2\lambda$ calculate and plot Re $E_z(\phi)$ and Im $E_z(\phi)$ near the ISB. Show that the diffracted field equals $-\frac{1}{2}E_z^{go}$ on the lit side, and $+\frac{1}{2}E_z^{go}$ on the shadow side, thus compensating the discontinuity in E_z^{go} at the incident shadow boundary $\phi = 240°$.

8.2 Show that the four-term UTD wedge diffraction coefficient $D_{s,h}$ in (8.6) reduces to Keller's result (8.10) away from any shadow or reflection boundaries (i.e. when $F(kLa) \to 1$).

8.3 With $n = 2$, show that the UTD diffraction coefficient (8.6) can be simplified to

$$D_{s,h} = \frac{-e^{-j\pi/4}}{2\sqrt{2\pi k}} \left\{ \frac{F(kLa(\phi - \phi'))}{\cos((\phi - \phi')/2)} \mp \frac{F(kLa(\phi + \phi'))}{\cos((\phi + \phi')/2)} \right\}$$

This is known as *Sommerfeld's half-plane solution*.

8.4 A half plane is illuminated by a plane wave at $\phi' = 45°$. Accurately sketch the transition regions in the $x - y$ plane (no program). Assume that the transition region boundary is defined by the condition $kLa = 2\pi$, making $F(kLa) \approx 1$. In the transition region, the diffracted field is not ray optical. That is, it no longer behaves as $e^{-jk\rho}/\sqrt{\rho}$. This can be important when considering multiple diffractions by two or more edges.

8.5 An electric line source at $(x', y') = (0, 1)$ m produces an incident field $E_z^i = E_0 e^{-jk\ell}/\sqrt{\ell}$ where ℓ is the distance from the source to the field point (x, y). A perfectly conducting strip is at $y = 0$ and has endpoints at $x = 0$ and $x = 2$ m. Compute and plot the incident plus reflected GO field magnitude $|E_z^i + E_z^r|$ at $y = 2$ m, for -2 m $\leq x \leq 2$ m. Assume that $E_0 = 1$ and $\lambda = 1$ m.

8.6 A magnetic line source on a 2D strip is described in Example 8.2. Write a program to calculate the radiation pattern. This can be done by modifying PROGRAM utd1. Assume that $H_0 = 1$ and $\lambda = 1$ m.

  (a) With $a = 2$ m, calculate and plot the radiation pattern $|H_z|$ in dB for $0 \leq \phi \leq 360°$.
  (b) Repeat, for $a = 10$ m.
  (c) Check your result. Near $\phi = 90°$ the field should be dominated by GO, so that $20 \log (H_z) \approx 0$ dB.

8.7 A plane wave $E_z^i = E_0\, e^{jk(x\cos\phi' + y\sin\phi')}$ is incident on a 2D perfectly conducting strip that is at $y = 0$, with endpoints at $x = \pm a/2$. The field point (ρ, ϕ) is in the near field. It is assumed that $\rho \geq a/2$ and that the incident wave is restricted to $0 \leq \phi' \leq 90°$.

Write a program to calculate $|E_z|$ for $0 \leq \phi \leq 360°$. The two edge-diffracted fields contribute for all ϕ. There will be RSBs in the range $0° \leq \phi \leq 180°$ and ISBs in the range $180° \leq \phi \leq 360°$.

You can program it by extending PROGRAM utd1 as needed. Compute the near zone $|E_z|$ for an incidence angle of $\phi' = 30°$, a strip of width $a = 2$ m and the field point at $\rho = 2$ m. Assume $E_0 = 1$ V/m and $\lambda = 1$ m.

8.8 A 2D perfectly conducting strip is at $y = 0$ and has endpoints at $x = \pm a/2$. It is illuminated by an incident plane wave $H_z^i = H_0 e^{jk(x\cos\phi + y\sin\phi)}$.

(a) Find the GTD backscatter echo width σ, as a function of the angle ϕ. The formulation is similar to Example 8.3.
(b) Plot the echo width for $0 \leq \phi \leq 90°$ with $a = 2$ m and 5 m. Assume $\lambda = 1$ m.
(c) Compare with the PO solution of Example 4.10. Which of these solutions is correct at wide angles and for what reasons?

8.9 For the GTD solution in Problem 8.8 add the second-order GTD contributions. Assume broadside incidence $\phi = 90°$. There will be four second-order mechanisms – two on the top surface and two on the bottom surface of the strip, as shown in Figure 8.27. Show that the echo width is

$$\sigma = \frac{1}{k}\left|(jka - 1) + 4\sqrt{\frac{2}{\pi ka}}\, e^{-j(ka + \pi/4)}\right|^2.$$

The $(jka - 1)$ term is from first-order effects and the second term is due to multiple diffraction.

Figure 8.27   Second-order TE diffractions on a strip.

8.10 Derive the expression (8.16) for the UTD radiation pattern in the $\phi = 0$ plane for a short monopole at the centre of a finite rectangular plate. The plate is oriented as in Figure 8.8 and the monopole GO field is

$$E^{go} = E_0 \sin\theta\, \frac{e^{-jkr}}{r}.$$

Compute and plot the radiation pattern when $a = 1\lambda, 3\lambda$ and 6λ.

8.11 Find the UTD radiation pattern in the $\phi = 0$ plane for an x-polarized TE_{10} rectangular waveguide aperture in a finite rectangular plate. The plate is oriented as in Figure 8.8. The waveguide aperture is centred at the origin and has dimensions $a_0 \times b_0$ (and $b_0 < a_0$). On an infinite ground plane, the waveguide radiation pattern is of the form

$$E_\theta^{go}(r, \theta) = E_0\, f(\theta)\, \frac{e^{-jkr}}{r}.$$

The UTD solution is similar to the case of slot excitation in Section 8.4.1 except that the $\mathbf{E}^i(Q_E)$ terms are affected by the radiation pattern of the slot.

(a) Find the waveguide radiation pattern $f(\theta)$ for an infinite ground plane.
(b) Find the radiation pattern for the waveguide and finite rectangular plate.
(c) Compute and plot the radiation pattern for a plate width of $a = 6\lambda$ and a WR90 waveguide aperture having $a_0 = 0.9''$ and $b_0 = 0.4''$. The frequency is 10 GHz.

8.12 Find the edge-diffracted field for a short monopole on a disc, as in Figure 8.13. Assume $\theta = 0$. Use the GTDEECs $I(\phi')$ and $M(\phi')$. Evaluate the radiation integrals, which will be line integrals of the form $\int(\cdot)\,ad\phi'$. You should find that they are zero. Since $E^{go} = 0$ the total field is zero on the z axis.

8.13 Find the GTDEECs $I(\phi')$ and $M(\phi')$ for a short monopole on a disc, as in Figure 8.13. Do this for arbitrary θ. The solution should be of the form

$$E_\theta \sim E_\theta^{go} + E_\theta^d,$$

where E_θ^d is due to the GTDEECs.

Find an expression for the radiation integral and evaluate it using numerical integration. With $a = 3\lambda$ compare your result with the radiation patterns shown in Figure 8.14. Your solution should correct for the caustic problem near $\theta = 0°$ and $180°$.

8.14 Find the UTD radiation pattern in the $\phi = 0$ plane for a y-directed magnetic dipole $\mathbf{p}_m = \hat{\mathbf{y}}p_m$ on a disc of radius a. The disc is oriented as in Figure 8.13 and the dipole radiates a GO field

$$E_\theta^{go} = E_0 \frac{e^{-jkr}}{r}.$$

Compute and plot the radiation pattern for $0 \le \theta \le 180°$ when $a = 3\lambda$.

8.15 Develop the GTD radiation pattern solution for the parallel plate waveguide in Figure 8.28. The aperture is in the $x = 0$ plane and extends over $-a/2 \le y \le a/2$. The waveguide field is $E_y = E_0 e^{-jkx}$ and is TEM with respect to x. The radiation pattern comes from the two diffracted

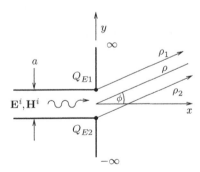

Figure 8.28   Parallel plate waveguide of width a in an infinite ground plane.

rays at Q_{E1} and Q_{E2}; in the far zone there is no GO contribution from the aperture. These are $90°$ wedges so the diffraction coefficients have $n = 3/2$. Use a formulation in terms of H_z. Plot the pattern for $-90° \leq \phi \leq 90°$ when $a = 2\lambda$.

8.16 Find the radiation pattern for a TEM parallel plate waveguide in Figure 8.28 by using aperture integration. The waveguide field is $E_y = E_0 e^{-jkx}$. Using the radiation integral (4.108), find H_z from the aperture $\mathbf{M}_s$ (and do not use $\mathbf{J}_s$). Plot the pattern for $-90° \leq \phi \leq 90°$ when $a = 2\lambda$. You can use this to check the GTD solution in Problem 8.15.

8.17 The E-plane and H-plane patterns of a pyramidal horn can be obtained from 2D UTD. The horn is shown in Figure 8.29. In this problem the E-plane pattern will be found.

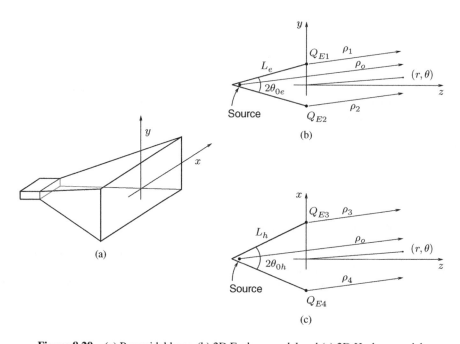

Figure 8.29 (a) Pyramidal horn. (b) 2D E-plane model and (c) 2D H-plane model.

The E-plane horn is modelled by a pair of metal plates with lengths L_e; the throat angle is $2\theta_{0e}$. An x-directed magnetic line source at the throat produces a GO field

$$H_x^{go} = H_0 \frac{e^{-jk\rho_0}}{\sqrt{\rho_0}},$$

where H_0 is a constant. The GO field illuminates the diffraction points Q_{E1} and Q_{E2}. In the far zone, the GO and diffracted rays are all parallel. The radiation pattern takes the form

$$H_x = H_x^{go} + H_x^i(Q_{E1})D_{h1}\frac{e^{-jk\rho_1}}{\sqrt{\rho}} + H_x^i(Q_{E2})D_{h2}\frac{e^{-jk\rho_2}}{\sqrt{\rho}}.$$

Find the UTD radiation pattern solution. Plot the pattern for $0 \le \theta \le 360°$ when $2\theta_{0e} = 23°$ and $L_e = 15\lambda$.

8.18 For the horn in Problem 8.17, use Keller's GTD to find the H_x radiation pattern at $\theta = 0°$ and $\theta = 180°$. Assuming that $2\theta_{0e} = 23°$ and $L_e = 15\lambda$, use this to calculate the antenna's front-to-back ratio in dB. This is a hand calculation; a program is not needed.

8.19 The main beam of an E-plane horn will be maximum when the rays from the throat and the edge diffractions are in phase. For the horn in Problem 8.18, use Keller's GTD to find the H_x radiation pattern at $\theta = 0°$. Find the relation between $2\theta_{0e}$ and L_e to maximize $|H_x|$. If $L_e = 15\lambda$, what value of $2\theta_{0e}$ is needed?

8.20 Find the radiation pattern for a 2D E-plane horn in Figure 8.29(b) by using aperture integration. The aperture is at $z = 0$. The aperture field is $H_x = H_0 e^{-jkR_0}/\sqrt{R_0}$, where H_0 is a constant and R_0 is the distance from the line source to a point in the aperture.

Using radiation integrals of the form (4.108) and (4.109), find H_x from the aperture $\mathbf{M}_s$ and $\mathbf{J}_s$. You will have to evaluate the radiation integrals numerically. Plot the pattern for $0 \le \theta \le 360°$ when $2\theta_{0e} = 23°$ and $L_e = 15\lambda$. You can compare this with the UTD solution in Problem 8.17.

8.21 Find the directivity of the E-plane horn in Problem 8.17. The 2D directivity is $D = 2\pi\rho S_{max}/P_t$ where S_{max} is the maximum power density. Find S_{max} at $\theta = 0°$ from Keller's GTD. The input power to the horn P_t can be found by integrating the Poynting vector of the GO field over the antenna aperture at $\rho_0 = L_e$, $-\theta_{0e} \le \theta \le \theta_{0e}$. Evaluate D when $2\theta_{0e} = 23°$ and $L_e = 15\lambda$.

8.22 Using the 3D expression (8.30) for the reflected field caustic positions $\rho_{1,2}^r$, derive the 2D result (8.28) for ρ_c. Do this by using the general expression with $\rho_1^i = \rho'$, $\rho_2^i \to \infty$ and $R_2 \to \infty$.

8.23 A plane wave $E_z^i = E_0 e^{-jkx}$ is incident on a perfectly conducting 2D cylinder of radius a. Use GO to find the far-zone reflected field, magnitude and phase. Your result should be in the form $E_z^s(\phi) = f(\phi) e^{-jk\rho}/\sqrt{\rho}$.

8.24 A plane wave $E_z^i = E_0 e^{-jkx}$ is incident on a perfectly conducting 2D cylinder of radius a. Use GO to find the near-zone reflected field. Compute and plot the magnitude of the total field, $|E_z^i + E_z^r|$. Use a cylinder radius of $a = 1\lambda$ and a field point at $\rho = 2\lambda$. Although the field point is at (ρ, ϕ), make the plot versus the incident angle θ_i, as defined at the reflection point.

Hint: The distance from the reflection point to the field point at (ρ, ϕ) can be expressed in terms of a, ρ, θ_i and no other quantities.

8.25 An electric line source at $(x, y) = (0, 0)$ has a field $E_z^i = E_0 e^{-jk\rho}/\sqrt{\rho}$. A perfectly conducting semicircular reflector as in Figure 8.30(a) has a radius a and it extends over $90° \le \phi \le 270°$.

(a) Find the GO reflected field at a point (ρ, ϕ) when $\rho > a$. Note that the rays pass through a caustic, so the spread factor must pick up a phase term of $e^{+j\pi/2}$.

(b) Consider what happens as $\rho \to \infty$. What field components are only present in the near field?

8.26 Consider a perfectly conducting semicircular cylinder of radius a as shown in Figure 8.30(b). The right side is a cylinder and the left side at $x = 0$ is a flat strip of width $2a$. The incident field is a plane wave $H_z^i = H_0 e^{jk(x\cos\phi + y\sin\phi)}$.

For $0° < \phi \le 90°$, the scattered field comes from the reflection Q_R from the curved face plus an edge diffraction Q_{E1}. The other edge Q_{E2} is shadowed. For $90° < \phi \le 180°$, the scattered field comes from Q_{E1} and Q_{E2}.

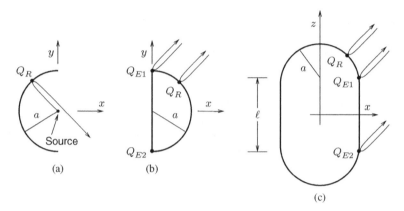

Figure 8.30 Some geometries with curved surface reflection. (a) 2D concave reflector. (b) 2D semicircular cylinder. (c) 3D cylinder and hemispherical endcaps.

Each edge is a 90° wedge so the wedge angle parameter is $n = 3/2$. Also one wedge face is curved; that face will require a modified L parameter.

Calculate and plot the monostatic echo width $\sigma(\phi)$ for $0° \le \phi \le 180°$, when $a = 2$ m and $\lambda = 1$ m. Check your answer; we know from GO that at $\phi = 0°$ a cylinder has $\sigma = \pi a$, and from PO at $\phi = 180°$ a strip of width $2a$ has $\sigma = k(2a)^2$.

8.27 A cylinder with hemispherical endcaps is shown in Figure 8.30(c). The cylinder length (excluding the endcaps) is ℓ. The cylinder and the endcap radii are a. The incident field is a plane wave $E_\theta^i = E_0 e^{jk(x\sin\theta + z\cos\theta)}$. A reflection point exists on the endcap hemisphere; the total field follows from the endcap reflection plus two edge diffractions.

The junction between the cylinder and hemisphere is smooth; this is a wedge with a wedge angle of 180°, so the wedge angle parameter is $n = 1$. Each wedge has curved faces, which affects the L parameters, and a curved edge, which affects the spread factor.

  (a) Find the monostatic RCS $\sigma(\theta)$ for $0 \le \theta \le 90°$. Plot the RCS in dBsm (dB with a reference level of 1 m²) using $\ell = 10$ m, $a = 1$ m and $\lambda = 1$ m.
  (b) Evaluate the limit as $\theta \to 90°$.
  (c) Why does the solution fail as $\theta \to 0°$?

8.28 Find the 2D UTD diffracted field for the thin, flat metal/dielectric junction in Figure 8.31. The o-face $x > 0$ is a PEC sheet, and the n-face $x < 0$ is a thin dielectric slab. This is a flat wedge, with the wedge angle parameter $n = 1$. The illumination is assumed to be a magnetic line source of strength M_0 at (x', y').

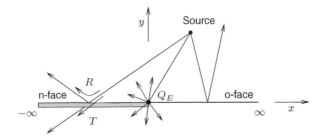

Figure 8.31 Diffraction at the junction of a PEC half plane and a thin dielectric slab.

The solution must use the four-term diffraction coefficient, which includes both the o- and n-faces. You will have to consider five phenomena: (i) the direct ray, (ii) reflection from the metal, (iii) reflection from the dielectric, (iv) transmission through the dielectric and (v) edge diffraction. You can assume that the dielectric slab reflection and transmission coefficients are known, and given as $R(\theta_i)$ and $T(\theta_i)$ where θ_i is measured between the surface normal $\hat{n} = \hat{y}$ and the incident ray. Show that the diffracted field

$$H_z^d = H^i(Q_E)[(1-T)D_1 + D_2 + RD_3 + D_4]e^{-jk\rho}/\sqrt{\rho}$$

provides the necessary compensations at the ISBs and RSBs. We associate D_1 with the n-face ISB, D_2 with the o-face ISB, D_3 with the n-face RSB and D_4 with the o-face RSB.

8.29 A plane wave $E_\theta^i = E_0 e^{jk(x\sin\theta_i + z\cos\theta_i)}$ is incident on a circular metal disc of radius a and lying in the $x - y$ plane.

(a) Using the GTD, find an expression for the bistatic radar cross section $\sigma(\theta_s)$ in the $\phi = 0$ plane.

(b) For $\theta_i = 30°$ and $a = 3$ m, $\lambda = 1$ m, calculate $\sigma(\theta_s)$ for $0° \leq \theta_s \leq 360°$.

8.30 The H-plane pattern of a pyramidal horn can be modelled in 2D as in Figure 8.29(c), using an 'apparent source' at the throat of two plates. The throat angle is $2\theta_{0h}$ and the plate length is L_h. The GO field for $-\theta_{0h} \leq \theta \leq \theta_{0h}$ at a distance ρ_0 from the throat is given by

$$\mathbf{E}^{go} = \hat{y}E_0 \frac{e^{-jk\rho_0}}{\sqrt{\rho_0}} \cos\left(\frac{\pi\tan\theta}{2\tan\theta_{0h}}\right).$$

The field diffracted by Q_{E3} is

$$E^{d3} = \frac{1}{2jk}\frac{-1}{L_h}\frac{-\pi}{\sin 2\theta_{0h}}\frac{e^{-jkL_h}}{\sqrt{L_h}}\frac{\partial D_s}{\partial\phi'}\bigg|_{(Q_{E3})}\frac{e^{-jk\rho_0}}{\sqrt{\rho_0}}e^{jkL_h\cos(\theta-\theta_{0h})}$$

and a similar expression exists for the other edge Q_{E4}.

There is a 1/2 factor on the slope diffraction coefficient because of grazing incidence. Also, $\partial E^i/\partial n$ is a derivative with respect to the direction of $\hat{n}$, so that when $\phi' = 0$, we have $\partial E^i/\partial n = (1/\rho')\partial E^i/\partial\phi'$. Using

$$\frac{\partial}{\partial\theta}\cos\left(\frac{\pi\tan\theta}{2\tan\theta_{0h}}\right) = \frac{-\pi}{\sin 2\theta_{0h}}$$

leads to the final result.

Calculate the E_y radiation pattern in dB, using $2\theta_{0h} = 27°$ and $L_h = 15\lambda$.

8.31 Develop the GTD radiation pattern solution for the parallel plate waveguide in Figure 8.28. The aperture is in the $x = 0$ plane and extends over $-a/2 \leq y \leq a/2$. The waveguide field is the dominant mode $E_z = E_0 \cos(\pi y/a) \, e^{-j\beta x}$ where $\beta = \sqrt{k^2 - (\pi/a)^2}$ which is TM (i.e. TM with respect to z or TE with respect to x).

The radiation pattern comes from the two diffracted rays at Q_{E1} and Q_{E2}; in the far zone there is no GO contribution from the aperture. These are $90°$ wedges so the diffraction coefficients have $n = 3/2$. Use a formulation in terms of E_z.

You will need to use slope diffraction, as the field incident on the edges is zero. At Q_{E1} we have that $\partial E^i/\partial n = -\partial E^i/\partial y = \pi/a$, and a similar result applies to Q_{E2}.

(a) Plot the pattern for $-90° \leq \phi \leq 90°$ when $a = 2\lambda$.

(b) You should find that the solution is singular as $\phi \to 0$. Is this physically reasonable?

8.32 The TM parallel plate waveguide field for the previous problem can be decomposed into two plane waves, that is,

$$E_z = E_0 \frac{e^{j\pi y/a} + e^{-j\pi y/a}}{2} \, e^{-j\beta x},$$

where $\beta = \sqrt{k^2 - (\pi/a)^2}$. Does the GTD solution using incident plane waves give the same result as the slope diffraction approach? Plot the pattern when $a = 2\lambda$.

8.33 Develop the aperture integration radiation pattern solution for the TM parallel plate waveguide in Figure 8.28. The aperture is in the $x = 0$ plane and extends over $-a/2 \leq y \leq a/2$. The waveguide field is $E_z = E_0 \cos(\pi y/a) \, e^{-j\beta x}$ where $\beta = \sqrt{k^2 - (\pi/a)^2}$.

Using the radiation integral (4.110), find E_z from the aperture $\mathbf{M}_s$ (and do not use $\mathbf{J}_s$). Plot the pattern for $-90° \leq \phi \leq 90°$ when $a = 2\lambda$. You can compare this with the two previous GTD solutions.

8.34 PROGRAM utd1 in Example 8.1 implements the TM UTD solution for a plane wave incident on a half plane. Take the normal derivative of E_z on the half-plane surface; note that

$$\frac{\partial}{\partial y} = \frac{1}{\rho} \frac{\partial}{\partial \phi}.$$

From this we can find H_x and J_z on the surface. This will involve the slope diffraction coefficient. Modify the program so that it can calculate J_z on the top and bottom surfaces. Use it to obtain J_z on the top and bottom faces of the half plane, for a plane wave incident at $\phi' = 90°$. Compare your result with Figure 5.6.

8.35 A metal cylinder with a radius a has a plane wave incident $E_z^i = E_0 e^{-jkx}$.

(a) Using the GO current, obtain a stationary phase solution for the scattered field $E_z^s(\phi)$ and the 2D bistatic echo width $\sigma(\phi)$. Consider the contributions of the integration endpoints at $\phi' = \pm \pi/2$ and whether or not a stationary phase point exists.

(b) Check your result by comparing with the exact solution, using $a = 3$ m and $\lambda = 1$ m.

8.36 A z-directed magnetic line source of strength $M_0 = 1$ Volt illuminates a perfectly conducting cylinder of radius a. The geometry is shown in Figure 8.32(a). The source is at a height $h = 1$ m above the cylinder and at $\phi = 0$; the wavelength is $\lambda = 1$ m. The cylinder size is $ka = 10$.

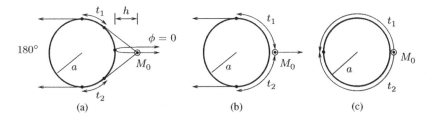

Figure 8.32 UTD rays for a circular cylinder. (a) Scattering, (b) radiation and (c) coupling.

(a) Use the incident and reflected rays to calculate $|H_z|$ at $\phi = 0$.
(b) Use the two diffracted rays to calculate $|H_z|$ at $\phi = 180°$. Assume far-field conditions, so that the two rays are parallel.
(c) Find the front/back ratio $|H_z(0)/H_z(180°)|$ in dB.
(d) Repeat the calculations for $ka = 5$ and 20.

8.37 A z-directed magnetic line source of strength $M_0 = 1$ Volt illuminates a perfectly conducting cylinder of radius a. The geometry is shown in Figure 8.32(a). The source is at a height $h = 1$ m above the cylinder and at $\phi = 0$; the wavelength is $\lambda = 1$ m. The cylinder size is $ka = 10$.

(a) At what angle is $t_1 = 0$?
(b) Calculate the far-zone UTD radiation pattern in dB for $0 \le \phi \le 180°$.

8.38 A z-directed magnetic line source of strength $M_0 = 1$ Volt is on the surface of a perfectly conducting cylinder of radius a. The geometry is shown in Figure 8.32(b). The source is at $\phi = 0$; the wavelength is $\lambda = 1$ m. The cylinder size is $ka = 10$.

(a) Use the direct ray to calculate $|H_z|$ at $\phi = 0$.
(b) Use the two diffracted rays to calculate $|H_z|$ at $\phi = 180°$. Assume far-field conditions, so that the two rays are parallel.
(c) Find the front/back ratio $|H_z(0)/H_z(180°)|$ in dB.
(d) Repeat the calculations for $ka = 5$ and 20.

8.39 A z-directed magnetic line source of strength $M_0 = 1$ Volt is on the surface of a perfectly conducting cylinder of radius a. The geometry is shown in Figure 8.32(b). The source is at $\phi = 0$; the wavelength is $\lambda = 1$ m. The cylinder size is $ka = 10$. Calculate the far-zone UTD radiation pattern in dB for $0 \le \phi \le 180°$.

(a) For $0 \le \phi \le 90°$ use the direct and t_2 rays.
(b) For $90° \le \phi \le 180°$ use the t_1 and t_2 rays.
(c) For $0 \le \phi \le 90°$ use the direct and t_2 rays, and add a t_1 ray that takes a long counterclockwise path. You should find that its contribution is weak.

8.40 A z-directed magnetic line source of strength $M_0 = 1$ Volt is on the surface of a perfectly conducting cylinder of radius a. The field point is on the surface and the geometry is shown in Figure 8.32(c). The source is at $\phi = 0$; the wavelength is $\lambda = 1$ m. The cylinder size is $ka = 10$.

(a) Use the two diffracted rays to calculate $|H_z|$ at $\phi = 180°$.
(b) Calculate $|H_z|$ for $10° \le \phi \le 180°$.

(c) Using only the t_1 ray (the shorter one of the two), calculate $|H_z|$ for $10° \leq \phi \leq 180°$. Plot it versus t_1 where $t_1 = a\phi$, and compare it with the same source on a flat ground plane. The flat ground plane should have a stronger field.

8.41 Find a 2D coupling formulation for the radial dipole on a cylinder in Figure 8.33. This involves the TE (hard) polarization. Case (a) shows a z-directed magnetic line source $M^a(Q')$ and its field $E^a(Q)$. From the local plane wave behaviour, $E^a(Q) = -\eta H_z(Q)$. Now, assume that an electric line dipole of strength $I^b(Q)$ is at Q. From reciprocity obtain $H^b(Q')$.

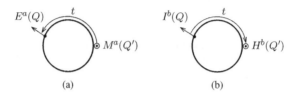

                            (a)   (b)

Figure 8.33   Coupling formulation for an electric dipole. (a) Source M^a produces E^a. (b) Source I^b produces H^b.

9

Physical Theory of Diffraction

Physical optics uses geometrical optics to estimate the surface current. This current is then used in a radiation integral to find the scattered field. PO often works very well for electrically large and smooth bodies. However, in spite of its great success in many applications, the use of an approximate current can lead to inherent limitations of its accuracy. The physical theory of diffraction (PTD) is an important high-frequency technique that addresses this problem. The PTD provides additional terms – and when these are added to a PO solution, it improves the accuracy.

The PTD was invented by Pyotr Ufimtsev in the Soviet Union, in the early 1960s. The original technical report of Ufimtsev (1962) was essentially unknown in the West until an English translation appeared in 1971. Subsequently, the PTD became a widely used technique for the analysis and design of antennas, radar cross-section analysis, and stealth design.

It is interesting to note that Ufimtsev's work to overcome the deficiencies of physical optics took place at the same time that Keller in the United States was finding a way to overcome the shortcomings of geometrical optics, using geometrical theory of diffraction (GTD). Although the PTD and GTD are different methods, they do have some interesting interrelationships, and this will be discussed later on.

The PTD theory will first be developed in 2D. Its use will be demonstrated by calculating the radiation pattern of a reflector antenna. It will then be generalized to 3D and applied to the scattering by a circular disc.

9.1 PO and an Edge

The PO approximation and its possible shortcomings can be understood by considering a perfectly conducting half plane, illuminated by a plane wave. The exact solution was discussed in Chapter 5 and is revisited now. Figure 9.1 shows the exact surface current induced by a broadside-incident plane wave

$$\mathbf{E}^i = \hat{\mathbf{x}} E_0 e^{jky}$$

having $E_0 = 1$ V/m and $\lambda = 1$ m. It can be seen that, as one would expect, the shadow-side surface current is relatively weak – but it is not zero. Also, the current is continuous, as the flow 'wraps around' the edge at $x = 0$.

For this same case, GO predicts a surface current

$$\mathbf{J}_s^{go} \approx 2\hat{\mathbf{n}} \times \mathbf{H}^i = \hat{\mathbf{x}} \, \frac{2E_0}{\eta} = \hat{\mathbf{x}} \, 0.053 \text{ A/m}$$

on the lit side, and $\mathbf{J}_s^{go} \approx 0$ on the shadow side.

Applied Frequency-Domain Electromagnetics, First Edition. Robert Paknys.
© 2016 John Wiley & Sons, Ltd. Published 2016 by John Wiley & Sons, Ltd.
Companion Website: www.wiley.com/go/paknys9981

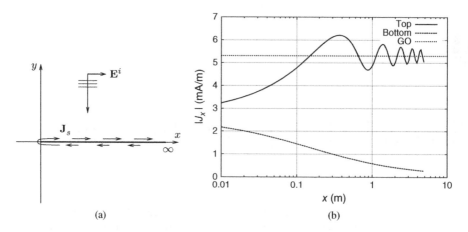

Figure 9.1 (a) TE plane wave incident on a perfectly conducting half plane. (b) Exact surface current and GO approximation; $E_0 = 1$ V/m and $\lambda = 1$ m.

The GO current is exact for an infinite ground plane, but for any other shape, it will be an approximation. It often works reasonably well for large smooth bodies and away from sharp edges. Figure 9.1 shows that a few wavelengths away from the edge the GO current is a good approximation that passes through the average of the exact result. However, towards the edge it becomes increasingly inaccurate. The geometry of a half plane is only slightly more complicated than an infinite ground plane, yet even this simple case is challenging the accuracy. Note that PO uses currents obtained from GO, so we will always use the terms *PO integral*, *PO approximation* and *GO current*. We will avoid the term 'PO current' that sometimes appears in the literature.

In Ufimtsev's terminology, the GO current is called the *uniform* part, and a remainder, defined as the exact current minus the uniform part, is called the *nonuniform* part. Therefore, the nonuniform part serves as a correction to the uniform part. From Figure 9.1, we can see that the nonuniform part (i.e. the deviation from the uniform part) is somewhat concentrated towards the edge, so its radiation is called a *fringe wave*.

We are usually more interested in the fields than the currents. We could use a radiation integral with the nonuniform current to find the fringe wave. However we will not do this. We will find the fringe wave via indirect considerations – so an explicit radiation integral with the nonuniform current will not even be needed. This correction *field*, the fringe wave, will be in the form of an edge contribution, similar to GTD.[1]

9.2 Asymptotic Evaluation

This section will evaluate the PO radiation integral for a magnetic line source illuminating a perfectly conducting half plane, shown in Figure 9.2. The asymptotic evaluation of the integral will be done with the stationary phase method. This will enable us to find the fringe wave, which is the PTD correction of PO. The incident field is

$$H_z^i = H_0 \frac{e^{-jkR_0}}{\sqrt{R_0}}. \tag{9.1}$$

---

[1] The nonuniform current is not confined to the edge but is somewhat spread out over the surface, as implied by Figure 9.1. It is possible to obtain the fringe wave field by direct integration of the nonuniform surface current. We will not cover this approach, but interested readers can consult Ufimtsev (2014, Chapter 7).

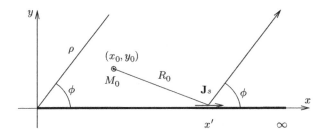

Figure 9.2   Magnetic line source M_0 and surface current $\mathbf{J}_s$ on a half plane.

The distance from the source at (x_0, y_0) to $(x', 0)$ is

$$R_0 = \sqrt{(x_0 - x')^2 + y_0^2} \tag{9.2}$$

and the constant H_0 is related to the source strength M_0 by

$$H_0 = -\frac{e^{j\pi/4}}{\eta}\sqrt{\frac{k}{8\pi}}\, M_0. \tag{9.3}$$

The GO current $\mathbf{J}_s = 2\hat{n} \times \mathbf{H}^i$ becomes

$$J_x = 2H_0 \frac{e^{-jkR_0}}{\sqrt{R_0}}. \tag{9.4}$$

The scattered field which is radiated by the current $J_x(x')$ is found via the radiation integral (4.109), which in this case becomes

$$H_z^s = e^{j\pi/4}\sqrt{\frac{k}{8\pi}}\frac{e^{-jk\rho}}{\sqrt{\rho}} \int_0^\infty \hat{z} \cdot (\hat{x} \times \hat{\rho}) e^{jk\hat{\rho}\cdot\rho'} J_x(x')\, dx'. \tag{9.5}$$

The current is at $\rho' = \hat{x}x'$, and the field point is at (ρ, ϕ) in the direction $\hat{\rho} = \hat{x}\cos\phi + \hat{y}\sin\phi$. Using the current (9.4) in the radiation integral (9.5) gives

$$H_z^s = 2H_0 e^{j\pi/4}\sqrt{\frac{k}{8\pi}}\sin\phi\frac{e^{-jk\rho}}{\sqrt{\rho}} \int_0^\infty \frac{e^{-jkR_0}}{\sqrt{R_0}} e^{jkx'\cos\phi}\, dx'. \tag{9.6}$$

The next step is to evaluate the integral in (9.6), which is

$$I(k) = \int_0^\infty \frac{e^{-jkR_0}}{\sqrt{R_0}} e^{jkx'\cos\phi}\, dx'. \tag{9.7}$$

With the high-frequency assumption, k will be large so we can evaluate $I(k)$ asymptotically by the method of stationary phase. Before doing this, however, it is helpful to look at the behaviour of the integrand, which is shown in Figure 9.3. In this example, the source is at $(x_0, y_0) = (2, 1)$ m and the wavenumber is $k = 4\pi$ rad/m. The far-field point is at $\phi = 45°$. From Figure 9.3 it can be seen that over most of the integration, the rapid oscillations will cause the cancellation of almost equal positive

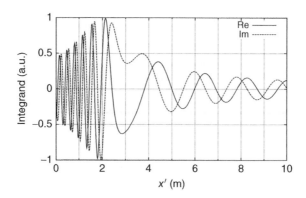

Figure 9.3   Integrand of (9.7) in arbitrary units, with $k = 4\pi$ rad/m, the source at $(x_0, y_0) = (2, 1)$ m, and the field point at $\phi = 45°$.

and negative areas, so there is no net contribution. The only parts that do not cancel occur around the stationary phase point at $x' = 3$ m and at the endpoint $x' = 0$.

We will now apply the stationary phase method to evaluate (9.7). We can use the formula from Section 5.5. With one endpoint at $x = 0$ and the other at $x = \infty$, the formula becomes

$$I(k) = \int_0^\infty f(x)e^{jk\theta(x)}\, dx \sim I^{sp}(k) + I^{ep}(k) + O(k^{-3/2}), \tag{9.8}$$

where the stationary phase point x_s is defined by $\theta'(x_s) = 0$ and contributes

$$I^{sp}(k) = f(x_s)\sqrt{\frac{2\pi}{k|\theta''(x_s)|}}\, e^{j[k\theta(x_s)+(\pi/4)\,\mathrm{sgn}\,\theta''(x_s)]}. \tag{9.9}$$

The endpoint contribution is

$$I^{ep}(k) = -\frac{1}{k}\frac{f(0)}{\theta'(0)}\, e^{j[k\theta(0)-(\pi/2)]}. \tag{9.10}$$

Note that (9.8)–(9.10) are only valid if the stationary point is not too close to the endpoint. For present purposes, this will not be an obstacle.

In (9.7) we can identify

$$f(x') = \frac{1}{\sqrt{(x'-x_0)^2 + y_0^2}} \tag{9.11}$$

and

$$\theta(x') = x'\cos\phi - \sqrt{(x'-x_0)^2 + y_0^2}. \tag{9.12}$$

First we will find $I^{sp}(k)$. Knowing $f(x')$ and $\theta(x')$ we can use (9.9) to obtain

$$I^{sp}(k) \sim \frac{1}{|\sin\phi|}\sqrt{\frac{2\pi}{k}}\, e^{j[k(x_0\cos\phi - y_0\sin\phi)-\pi/4]}. \tag{9.13}$$

From (9.6) and (9.13), we can obtain the stationary phase part of H_z^s. Noting that $\sin\phi/|\sin\phi| = \mathrm{sgn}\,y$, it becomes

$$H_z^{sp} \sim H_0\,\mathrm{sgn}\,y\frac{e^{-jk\rho}}{\sqrt{\rho}}\, e^{jk(x_0\cos\phi - y_0\sin\phi)}. \tag{9.14}$$

The result (9.14) is the far-field radiation by a magnetic line source at the image point $(x_0, -y_0)$. It is *exactly* the same result we would get from the direct application of geometrical optics for a line source above ground. This leads to a most important fact: *Evaluating a PO integral by the method of stationary phase is equivalent to using geometrical optics.* Another important point related to this is that solving

$$\theta'(x) = \cos\phi - \frac{x - x_0}{\sqrt{(x - x_0)^2 + y_0^2}} = 0$$

gives the stationary point

$$x_s = x_0 + y_0 \cot\phi$$

which has the following geometrical interpretation: *The stationary phase point x_s occurs at exactly the same place as the point of specular reflection that is associated with geometrical optics.*

For some combinations of source position (x_0, y_0) and field point angle ϕ, there will be no reflection point and consequently no stationary phase point, so that $H_z^{sp} \sim 0$. If the field point is slightly above/below the half plane, H_z^i and H_z^{sp} will either add (on the lit side) or cancel (on the shadow side) in accordance with the sgn y term.

Now we will find $I^{ep}(k)$. Knowing $f(x')$ and $\theta(x')$, we can use (9.10) to obtain

$$I^{ep}(k) \sim -\frac{1}{k} \frac{e^{j(k\sqrt{x_0^2+y_0^2}-\pi/2)}}{\sqrt{x_0^2+y_0^2}\cos\phi + x_0}(x_0^2 + y_0^2)^{1/4} \tag{9.15}$$

and with (9.6),

$$H_z^{ep} \sim 2H_0 e^{j\pi/4}\sqrt{\frac{k}{8\pi}}\sin\phi\frac{e^{-jk\rho}}{\sqrt{\rho}}\frac{-1}{k}\frac{e^{j(k\sqrt{x_0^2+y_0^2}-\pi/2)}}{\sqrt{x_0^2+y_0^2}\cos\phi + x_0}(x_0^2 + y_0^2)^{1/4}. \tag{9.16}$$

The endpoint contribution can be expressed in GTD-style coordinates. If the line source is at (ρ', ϕ'), then $x_0 = \rho'\cos\phi'$ and $\rho' = \sqrt{x_0^2 + y_0^2}$. Equation (9.16) becomes

$$H_z^{ep} \sim H_0 \underbrace{\frac{e^{-jk\rho'}}{\sqrt{\rho'}}}_{H_z^i(Q_E)}\underbrace{\frac{-e^{-j\pi/4}}{\sqrt{2\pi k}}\frac{\sin\phi}{\cos\phi + \cos\phi'}}_{D_h^{po}}\frac{e^{-jk\rho}}{\sqrt{\rho}}, \tag{9.17}$$

where, as usual, $H_z^i(Q_E)$ denotes the incident field at the edge-diffraction point Q_E. We have also defined a *PO diffraction coefficient* D_h^{po}. With some effort, it can be rewritten as

$$D_h^{po} = \frac{-e^{-j\pi/4}}{\sqrt{2\pi k}}\frac{\sin(\phi/2)\cos(\phi/2)}{\cos((\phi-\phi')/2)\cos((\phi+\phi')/2)}. \tag{9.18}$$

With the asymptotic evaluation complete, the incident (9.1), stationary phase (9.14) and endpoint (9.17) contributions can now be combined to obtain the total field

$$H_z = H_z^i + H_z^{sp} + H_z^{ep}. \tag{9.19}$$

9.2.1 PO Endpoint Correction

Equation (9.19) hints at the close relationship between PTD and GTD. There is however a critical difference. The PO diffraction coefficient is not quite the same as Keller's GTD diffraction coefficient, which for the half plane is

$$D_h^k = \frac{-e^{-j\pi/4}}{\sqrt{2\pi k}} \frac{\cos(\phi/2)\cos(\phi'/2)}{\cos((\phi-\phi')/2)\cos((\phi+\phi')/2)}. \tag{9.20}$$

Since PO is based on an approximate current, the endpoint contribution will be incorrect. The GTD, on the other hand, does not use an approximate current. Therefore the PO can be improved by subtracting out the PO endpoint contribution and replacing it with the GTD version. In other words, the PO can be corrected by adding the fringe wave

$$H_z^u \sim H_z^i(Q_E) D_h^u \frac{e^{-jk\rho}}{\sqrt{\rho}}, \tag{9.21}$$

where D_h^u is the Keller GTD diffraction coefficient minus the erroneous PO endpoint

$$D_h^u = D_h^k - D_h^{po}. \tag{9.22}$$

A remarkable thing happens in (9.22). One of the problems with GTD is that D_h^k is singular on shadow boundaries. So is the PO version D_h^{po}. However, in the difference (9.22), the singularities cancel, and D_h^u is finite everywhere.

In summary, then, the PTD procedure uses

$$H_z \sim H_z^i + H_z^{po} + H_z^u. \tag{9.23}$$

The scattered field is calculated with ordinary PO, which uses the GO current. In practice this is almost always a numerical integration. Then, the fringe wave (9.21) is added as a correction.

9.2.2 Relationship Between PTD and GTD

The PTD formulation (9.23) is closely related to both GTD and UTD. By making an asymptotic evaluation of the PO integral (i.e. using (9.8)), we recover GTD. From this viewpoint, GTD can be thought of as a special case of the more general PTD. It is noted that the confluence of the stationary (reflection) point x_s and the edge occurs when the field angle ϕ is on a shadow boundary. Equation (9.8) is not valid in this case, and it is also where GTD fails. A more general *uniform asymptotic* version of (9.8) can overcome this limitation. Using a uniform asymptotic evaluation of the radiation integral recovers the UTD.

9.2.3 General Formulas

The fringe wave was derived for a magnetic line source. However, the ideas hold just as well for an electric line source. Therefore,

$$H_z^u \sim H_z^i(Q_E) D_h^u \frac{e^{-jk\rho}}{\sqrt{\rho}}; \quad \text{TE case} \tag{9.24}$$

$$E_z^u \sim E_z^i(Q_E) D_s^u \frac{e^{-jk\rho}}{\sqrt{\rho}}; \quad \text{TM case.} \tag{9.25}$$

For the special case of a half plane, the relevant diffraction coefficients are

$$D_h^k = \frac{-e^{-j\pi/4}}{\sqrt{2\pi k}} \frac{\cos(\phi/2)\cos(\phi'/2)}{\cos((\phi-\phi')/2)\cos((\phi+\phi')/2)} \tag{9.26}$$

$$D_s^k = \frac{-e^{-j\pi/4}}{\sqrt{2\pi k}} \frac{-\sin(\phi/2)\sin(\phi'/2)}{\cos((\phi-\phi')/2)\cos((\phi+\phi')/2)} \tag{9.27}$$

$$D_h^{po} = \frac{-e^{-j\pi/4}}{\sqrt{2\pi k}} \frac{\sin(\phi/2)\cos(\phi/2)}{\cos((\phi-\phi')/2)\cos((\phi+\phi')/2)} \tag{9.28}$$

$$D_s^{po} = \frac{-e^{-j\pi/4}}{\sqrt{2\pi k}} \frac{-\sin(\phi'/2)\cos(\phi'/2)}{\cos((\phi-\phi')/2)\cos((\phi+\phi')/2)}. \tag{9.29}$$

The PTD diffraction coefficients are

$$D_h^u = D_h^k - D_h^{po} = \frac{-e^{-j\pi/4}}{\sqrt{2\pi k}} \cos(\phi/2)\frac{\cos(\phi'/2) - \sin(\phi/2)}{\cos((\phi-\phi')/2)\cos((\phi+\phi')/2)} \tag{9.30}$$

$$D_s^u = D_s^k - D_s^{po} = \frac{-e^{-j\pi/4}}{\sqrt{2\pi k}} \sin(\phi'/2)\frac{-\sin(\phi/2) + \cos(\phi'/2)}{\cos((\phi-\phi')/2)\cos((\phi+\phi')/2)}. \tag{9.31}$$

In 3D, similar expressions exist involving a dyadic diffraction coefficient. Analogous to the GTD edge-diffraction expression in Equation (8.22), we have

$$\mathbf{E}^u(s) = \mathbf{E}^i(Q_E) \cdot \overline{\mathbf{D}}^u \sqrt{\frac{\rho_c}{s(\rho_c + s)}} e^{-jks}. \tag{9.32}$$

The PTD diffraction coefficient is

$$\overline{\mathbf{D}}^u = \overline{\mathbf{D}}^k - \overline{\mathbf{D}}^{po}$$

or

$$\overline{\mathbf{D}}^u = -\hat{\boldsymbol{\beta}}_0'\hat{\boldsymbol{\beta}}_0(D_s^k - D_s^{po}) - \hat{\boldsymbol{\phi}}'\hat{\boldsymbol{\phi}}(D_h^k - D_h^{po}).$$

The edge is not limited to a half plane, and it can also be a wedge with an exterior angle of $n\pi$. The GTD diffraction coefficients $D_{s,h}^k$ can be found in Equation (8.10), and the PO diffraction coefficients (Pathak 1988) are

$$D_{s,h}^{po} = \frac{-e^{-j\pi/4}}{2\sqrt{2\pi k}} \frac{1}{\sin\beta_0} \left[\tan\left(\frac{\phi-\phi'}{2}\right) \mp \tan\left(\frac{\phi+\phi'}{2}\right)\right] ; \text{ o-face is lit}$$

$$= \frac{-e^{-j\pi/4}}{2\sqrt{2\pi k}} \frac{1}{\sin\beta_0} \left[\mp\tan\left(\frac{\phi+\phi'}{2}\right) \mp \tan\left(\frac{2n\pi - (\phi+\phi')}{2}\right)\right] ; \text{ both faces are lit}$$

$$= \frac{-e^{-j\pi/4}}{2\sqrt{2\pi k}} \frac{1}{\sin\beta_0} \left[-\tan\left(\frac{\phi-\phi'}{2}\right) \mp \tan\left(\frac{2n\pi - (\phi+\phi')}{2}\right)\right] ; \text{ n-face is lit} \tag{9.33}$$

For a 2D half plane, $n = 2$ and $\sin\beta_0 = 1$ so that (9.33) reduces to (9.28) and (9.29).

9.3   Reflector Antenna

A 2D reflector antenna will be used to demonstrate the application of PO and PTD. The parabolic cylinder in Figure 9.4 is illuminated by a magnetic line source of strength M_0, located at the focal point $(x, y) = (f, 0)$. The reflector surface is defined by

$$x = \frac{y^2}{4f} \tag{9.34}$$

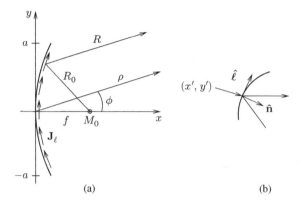

Figure 9.4 (a) 2D parabolic reflector antenna and magnetic line source of strength M_0; the surface current is $\mathbf{J}_\ell$. (b) Unit normal $\hat{\mathbf{n}}$ and tangent $\hat{\boldsymbol{\ell}}$ at the surface point (x', y').

and the diameter is $D = 2a$. The incident field at (x', y') on the antenna surface is

$$H_z^i = H_0 \frac{e^{-jkR_0}}{\sqrt{R_0}}. \tag{9.35}$$

The distance from the source at $(f, 0)$ to (x', y') on the parabolic surface is

$$R_0 = x' + f. \tag{9.36}$$

The constant H_0 is related to the source strength M_0 by

$$H_0 = -\frac{e^{j\pi/4}}{\eta} \sqrt{\frac{k}{8\pi}}\, M_0. \tag{9.37}$$

9.3.1   PO Part

The scattered field that is radiated by the current $\mathbf{J}_\ell$ on the reflector surface C_0 is found via the radiation integral (4.109), repeated here

$$H_z^s = e^{j\pi/4} \sqrt{\frac{k}{8\pi}} \frac{e^{-jk\rho}}{\sqrt{\rho}} \int_{C_0} \hat{\mathbf{z}} \cdot (\hat{\boldsymbol{\ell}} \times \hat{\boldsymbol{\rho}}) e^{jk\hat{\boldsymbol{\rho}}\cdot\boldsymbol{\rho}'}\, J_\ell(\boldsymbol{\rho}')\, d\ell'. \tag{9.38}$$

To find the reflector's surface normal and tangent vectors, the surface equation (9.34) can be rewritten as

$$g(x, y) = x - \frac{y^2}{4f} = 0.$$

The normal to any surface $g = \text{const.}$ follows from the gradient, so

$$\hat{\mathbf{n}} = \pm \frac{\nabla g}{|\nabla g|} = \frac{\hat{\mathbf{x}} - \hat{\mathbf{y}}(y/2f)}{\sqrt{1 + (y/2f)^2}}$$

where the $\pm$ sign was chosen as $+$ to give a right-pointing normal. The tangent vector is

$$\hat{\ell} = \hat{z} \times \hat{n} = \frac{\hat{x}(y/2f) + \hat{y}}{\sqrt{1 + (y/2f)^2}}.$$

The GO current is $\mathbf{J}_s = 2\hat{n} \times \mathbf{H}^i$ and the ℓ component is $J_\ell = \hat{\ell} \cdot \mathbf{J}_s$. Using the integration variables y' and $x' = y'^2/4f$ along with (9.35) and (9.36) for H_z^i leads to

$$J_\ell = -2H_0 \frac{e^{-jk[(y'^2/4f) + f]}}{\sqrt{(y'^2/4f) + f}}. \qquad (9.39)$$

The field point is in the direction $\hat{\rho} = \hat{x}\cos\phi + \hat{y}\sin\phi$, and $J_\ell(x', y')$ is at $\rho' = \hat{x}x' + \hat{y}y' = \hat{x}(y'^2/4f) + \hat{y}y'$. Since $d\ell = \sqrt{(dx)^2 + (dy)^2}$, we have $d\ell' = \sqrt{(y'/2f)^2 + 1}\, dy'$. Using the radiation integral (9.38) with the GO current (9.39) gives

$$H_z^s = \frac{jkM_0}{4\pi\eta}\frac{e^{-jk\rho}}{\sqrt{\rho}} \int_{-a}^{a} [(y'/2f)\sin\phi - \cos\phi]\, e^{jk[(y'^2/4f)\cos\phi + y'\sin\phi]}$$

$$\cdot \frac{e^{-jk[(y'^2/4f) + f]}}{\sqrt{(y'^2/4f) + f}}\, dy'. \qquad (9.40)$$

This integral has to be done numerically. The integrand is oscillatory, so care is required to ensure an adequate sampling of at least 10 points per wavelength.

The distance from the line source to the far field is $\rho - f\cos\phi$. The phase of the incident field should be adjusted accordingly so that (9.35) becomes

$$H_z^i = H_0 e^{jkf\cos\phi}\frac{e^{-jk\rho}}{\sqrt{\rho}}. \qquad (9.41)$$

Finally, the results (9.40) and (9.41) are added to obtain the radiation pattern

$$H_z = H_z^i + H_z^s. \qquad (9.42)$$

It should be remembered that $\mathbf{J}_\ell$ is a surface-equivalent current acting in free space, so the incident field in (9.42) radiates everywhere; there is no 'blockage' by the reflector. On the right side of the reflector, the incident and scattered fields tend to add, forming the antenna's main beam, whereas on the left side, they tend to cancel, producing the relatively weak shadow-region field.

9.3.2 PTD Part

The ray geometry for the PTD part is shown in Figure 9.5. The line source at $x = f$ illuminates the edges at Q_{E1} and Q_{E2}. The two diffractions add up to produce

$$H_z^u = H_z^i(Q_{E1})\, D_h^u(\psi_1, \psi_1')\frac{e^{-jk\rho_1}}{\sqrt{\rho_1}} + H_z^i(Q_{E2})\, D_h^u(\psi_2, \psi_2')\frac{e^{-jk\rho_2}}{\sqrt{\rho_2}}.$$

The edges are at $(x, y) = (a^2/4f, \pm a)$. The far-field distances are

$$\rho_1 = \rho - [(a^2/4f)\cos\phi + a\sin\phi]$$

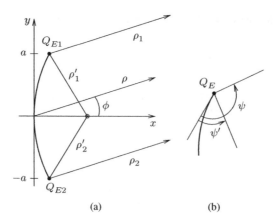

(a)   (b)

Figure 9.5 (a) 2D parabolic reflector, showing edge diffraction points Q_{E1} and Q_{E2}. (b) Local geometry showing tangent plane and diffraction angles.

and

$$\rho_2 = \rho - [(a^2/4f)\cos\phi - a\sin\phi].$$

The distances ρ_1' and ρ_2' can be found the same way as R_0 in the PO solution, so that

$$\rho_1' = \rho_2' = a^2/4f + f.$$

Therefore

$$H_z^u = H_0 \frac{e^{-jk(a^2/4f+f)}}{\sqrt{a^2/4f+f}} \frac{e^{-jk\rho}}{\sqrt{\rho}}$$

$$\cdot \left\{ D_h^u(\psi_1, \psi_1') e^{jk[(a^2/4f)\cos\phi + a\sin\phi]} + D_h^u(\psi_2, \psi_2') e^{jk[(a^2/4f)\cos\phi - a\sin\phi]} \right\}. \tag{9.43}$$

The diffraction coefficients are found from (9.30) with ϕ, ϕ' replaced by ψ, ψ'. These angles can be found from the local tangent vector $\hat{\ell}$ at the edges. Similar to (9.42), the contributions (9.40), (9.41) and (9.43) are added to obtain the radiation pattern

$$H_z = H_z^i + H_z^s + H_z^u. \tag{9.44}$$

When $\phi > 90°$ the diffraction emanating from Q_{E2} will be shadowed by the reflector, so it should be set to zero. Q_{E2} also cannot contribute when $\psi_2 > 360°$, which involves the region in between the tangent plane at Q_{E2} and the back side of the reflector. Nevertheless, the Q_{E2} contribution will reappear at some angle $90° < \phi < 180°$. Similar remarks apply for Q_{E1}, and these shadowing effects must be included in any computer program.

Figure 9.6 shows the normalized radiation pattern of the reflector, calculated from PO, PTD and a formally exact MoM benchmark. The reflector diameter is $D = 2a = 10\lambda$, and $f/D = 0.3$. The results in Figure 9.6 show that PO is already very good for calculating the main beam, so PTD is not necessary there. At the backside, adding PTD causes a big improvement and leads to excellent agreement with the formally exact MoM. The reason for this improvement is that the backside field is weak, and it comes from the almost equal and opposite H_z^i and H_z^s cancelling out. Therefore even a small correction to H_z^s can greatly affect the cancellation and the accuracy of the total field. In this example, the shadowing of Q_{E2} by the reflector has no visible effect at $90°$ and causes a minor discontinuity at $130°$.

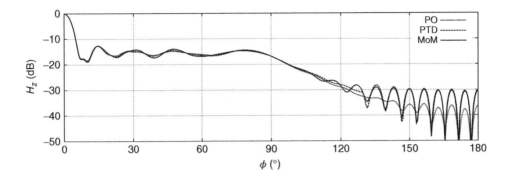

Figure 9.6 Comparison of PO, PTD and MoM radiation patterns for the parabolic reflector. $f/D = 0.3$ and $D = 10\lambda$.

9.4 RCS of a Disc

In this section the PTD will be used to compute the radar cross section of a perfectly conducting circular disc. This will show how the PTD is used in a 3D problem. The disc is shown in Figure 9.7. The disc radius is a, and the backscatter case is assumed, with the incident and backscatter directions in the $\phi = 0$ plane. The incident field is a θ-polarized plane wave

$$\mathbf{E}^i = \hat{\boldsymbol{\theta}}\, E_0\, e^{jk(x\sin\theta + z\cos\theta)}. \tag{9.45}$$

The location of the diffraction points at Q_{E1} and Q_{E2} are in accordance with the generalized Fermat's principle, appearing along the shortest and longest diffracted ray paths. The diffraction points will be relevant to the PTD formulation.

9.4.1 PO Part

The GO surface current follows from $\mathbf{J}_s = 2\hat{n} \times \mathbf{H}^i$ and (9.45) so that

$$\mathbf{J}_s = \hat{\mathbf{x}}\frac{E_0}{\eta}e^{jkx'\sin\theta} = \hat{\mathbf{x}}\frac{2E_0}{\eta}e^{jk\rho'\cos\phi'\sin\theta}.$$

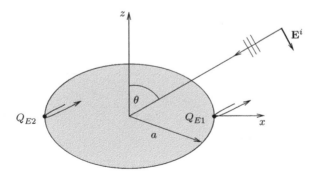

Figure 9.7 Perfectly conducting circular disc and θ-polarized incident plane wave.

The vector potential integral is carried out over the disc surface

$$A_x = \frac{\mu}{4\pi} \frac{e^{-jkr}}{r} \int_0^a \int_0^{2\pi} J_{sx}(\rho',\phi') e^{jk\hat{\mathbf{r}}\cdot\mathbf{r}'} \rho' d\phi' d\rho',$$

where

$$\hat{\mathbf{r}}\cdot\mathbf{r}' = (\hat{\mathbf{x}}\sin\theta + \hat{\mathbf{z}}\cos\theta)\cdot(\hat{\mathbf{x}}\rho'\cos\phi' + \hat{\mathbf{y}}\rho'\sin\phi') = \rho'\sin\theta\cos\phi'.$$

The integrals can be done in terms of Bessel functions by noticing that

$$\int_0^{2\pi} e^{ju\cos\phi'} d\phi' = 2\pi J_0(u)$$

and

$$\int x J_0(x)\, dx = x J_1(x).$$

The scattered field comes from the transverse part of $\mathbf{A}$

$$E_\theta^s = -j\omega A_x \cos\theta.$$

The PO scattered field is therefore

$$E_\theta^{po} = -jka^2 E_0 \cos\theta \frac{J_1(2ka\sin\theta)}{2ka\sin\theta} \frac{e^{-jkr}}{r}. \tag{9.46}$$

9.4.2   PTD Part

The PTD edge diffractions shown in Figure 9.8 can be calculated using (9.32). Aside from using $\overline{\mathbf{D}}^u$ instead of $\overline{\mathbf{D}}$, the procedure is the same as for GTD. The form of the far-field solution (and using $s \gg \rho_c$) is

$$E_\theta^u \sim E_\theta^i(Q_E)\, D_h^u(\psi)\, \frac{\sqrt{\rho_c}}{s}\, e^{-jks}.$$

The necessary parameters mostly follow from Figure 9.8 and are summarized in Table 9.1. Equation (8.25) gives the caustic distances ρ_c.

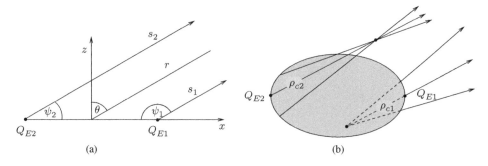

Figure 9.8   Circular disc, (a) edge-diffracted rays and (b) edge-diffracted ray caustics.

Table 9.1 Edge diffraction parameters.

	$E_\theta^i(Q_E)$	ψ	s	ρ_c
Edge 1	$E_0 e^{jka\sin\theta}$	$\pi/2 + \theta$	$r - a\sin\theta$	$a/2\sin\theta$
Edge 2	$E_0 e^{-jka\sin\theta}$	$\pi/2 - \theta$	$r + a\sin\theta$	$-a/2\sin\theta$

An important point regarding edge 2 is that the square root for the spread factor should be evaluated as

$$\sqrt{\rho_{c2}} = \sqrt{\frac{-a}{2\sin\theta}} = e^{+j\pi/2}\sqrt{\frac{a}{2\sin\theta}}.$$

This is because the rays diffracting from the concave edge at Q_{E2} form a caustic, which is crossed as the diffracted wave propagates from Q_{E2} to the far field. Generally, propagation through a caustic causes a phase shift of $+\pi/2$; see Section 8.4.3 for further discussion.

The diffraction coefficient (9.30) with $\phi = \phi'$ for backscatter and $\phi \to \psi$ is

$$D_h^u(\psi) = \frac{-e^{-j\pi/4}}{\sqrt{2\pi k}}\cos(\psi/2)\frac{\cos(\psi/2) - \sin(\psi/2)}{\cos\psi} = \frac{-e^{-j\pi/4}}{2\sqrt{2\pi k}}\frac{\cos\psi + 1 - \sin\psi}{\cos\psi}.$$

Adding up the two edge contributions, we can obtain

$$E_\theta^u \sim \frac{E_0}{2}\sqrt{\frac{a}{\pi k\sin\theta}}\left(\cos t + j\frac{\cos\theta - 1}{\sin\theta}\sin t\right)\frac{e^{-jkr}}{r}, \tag{9.47}$$

where $t = 2ka\sin\theta - \pi/4$.

The total scattered field can now be calculated from the sum of the PO part (9.46) and PTD part (9.47)

$$E_\theta^s = E_\theta^{po} + E_\theta^u \tag{9.48}$$

and E_θ^s with (1.91) gives the radar cross section σ.

The RCS is shown in Figure 9.9 and is compared with a formally exact MoM solution. The disc radius is $a = 3\lambda$ and the frequency is 10 GHz. The result shows that adding the PTD part moves the RCS in the right direction. However, the scattered field should be zero at grazing incidence, $\theta \to 90°$ and it is not. This shortcoming occurs because PTD approximates the disc as half planes at Q_{E1} and Q_{E2}. Since the half planes are semi-infinite, they support nonphysical currents outside the boundary of the disc. A modified procedure that terminates these currents (see Ufimtsev 2014, Chapter 6) can be used to improve the solution. The GTD counterpart of this issue is multiple diffraction.

Another point to mention is that near broadside ($\theta = 0$), the first term of the PTD part (9.47) behaves as $1/\sqrt{\sin\theta}$ so it is singular (the second term becomes zero). In this example, E^u could be accurately calculated for an angle as small as $1°$. However, E^u is truly singular when $\theta = 0$ so it cannot be used there. Since PO is already very good near broadside, the PTD part could simply be left out for small angles.

The reason for the PTD singularity can be understood by looking at Figure 9.8(b). As $\theta \to 0$ the caustics ρ_{c1} and ρ_{c2} move to infinity so the edge-diffracted rays become parallel. The contributions no longer come from just two points Q_{E1} and Q_{E2}, but the entire rim. Therefore, the broadside direction is a caustic of edge-diffracted rays, and ray theory cannot be used.

Another way to recognize that there is a problem is by considering the generalized Fermat's principle, which requires us to identify the shortest and longest ray paths. This becomes impossible at broadside, where all the edge-diffracted rays have the same length.

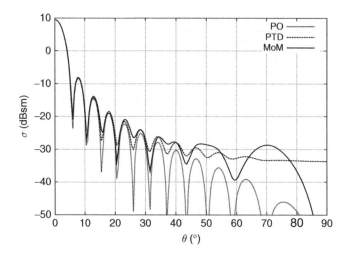

Figure 9.9 Circular disc, monostatic radar cross section σ, and comparison of PO, PTD and MoM. $0\,\mathrm{dBsm} = 1\ \mathrm{m}^2$, disc radius $a = 3\lambda$ and $f = 10$ GHz.

9.5 PTD Equivalent Edge Currents

We saw that the backscatter from a circular disk involves two diffraction points. At broadside, we no longer have two points but a continuum of edge contributions. These contributions can be added up with an integration, using a PTD equivalent edge current (PTDEEC). These are generally useful whenever an edge-diffraction caustic is encountered.

PTDEECs are fictitious line currents that are assumed to flow along the edge ℓ of a body. They are like the GTD equivalent edge currents (GTDEECs) described in Section 8.10 and follow the same edge-fixed coordinates. PTDEECs correct for fringe wave caustics and improve PO in the same way that GTDEECs correct for edge-diffracted ray caustics and improve GO. The only difference is that $D_{s,h}^{k}$ is replaced by $D_{s,h}^{u}$. Adopting the expressions from Section 8.10 leads to

$$I^{u}(\ell) = -\frac{1}{\eta} e^{-j\pi/4} \sqrt{\frac{8\pi}{k}} \frac{\hat{\boldsymbol{\ell}} \cdot \mathbf{E}^{i}(\ell)}{\sin \beta_0'} D_s^{u}(\beta_0, \phi, \phi') \tag{9.49}$$

and

$$M^{u}(\ell) = -\eta e^{-j\pi/4} \sqrt{\frac{8\pi}{k}} \frac{\hat{\boldsymbol{\ell}} \cdot \mathbf{H}^{i}(\ell)}{\sin \beta_0'} D_h^{u}(\beta_0, \phi, \phi'). \tag{9.50}$$

The PTDEECs, like GTDEECs, are nonphysical, as their value depends on the angle of observation. Also, nothing forces $M^{u}(\ell)$ to be zero on a conductor, and in general, both $I^{u}(\ell)$ and $M^{u}(\ell)$ will be present. Nevertheless, they are useful for solving the caustic problem. PTDEECs use $D_{s,h}^{u}$, so unlike GTDEECs, they remain finite on shadow boundaries.

Once the PTDEECs are known, a radiation line integral along the edge ℓ gives the fringe wave. From (4.31) and (4.32),

$$\mathbf{E}^{u} = \frac{jk}{4\pi} \int_{\ell} (\eta \hat{\mathbf{R}} \times \hat{\mathbf{R}} \times \hat{\boldsymbol{\ell}}' I^{u}(\ell') + \hat{\mathbf{R}} \times \hat{\boldsymbol{\ell}}' M^{u}(\ell')) \frac{e^{-jkR}}{R}\, d\ell' \tag{9.51}$$

and

$$\mathbf{H}^u = \frac{-jk}{4\pi} \int_\ell \left(\hat{\mathbf{R}} \times \hat{\boldsymbol{\ell}}' I^u(\ell') - \frac{1}{\eta} \hat{\mathbf{R}} \times \hat{\mathbf{R}} \times \hat{\boldsymbol{\ell}}' M^u(\ell') \right) \frac{e^{-jkR}}{R} \, d\ell', \tag{9.52}$$

where $\mathbf{R} = \mathbf{r} - \mathbf{r}'$; note that the plane wave relationship $\eta \mathbf{H}^u = \hat{\mathbf{R}} \times \mathbf{E}^u$ applies. In the far field, $\hat{\mathbf{R}} \approx \hat{\mathbf{r}}$ and $R \approx r - \hat{\mathbf{r}} \cdot \mathbf{r}'$.

Suppose that for the disc in the previous section, the radiation integral is evaluated by stationary phase. Off broadside this recovers the two-point solution, with stationary phase points at Q_{E1} and Q_{E2}. As $\theta \to 0$ the stationary points disappear but the integral still gives a meaningful $\mathbf{E}^u$. Therefore the two-point method, and more generally (9.32), can be thought of as a special case of the PTDEEC approach. If there is no caustic, then it is preferable to use the simpler (9.32) rather than (9.51) and (9.52).

9.6 Further Reading

The PTD was first described in a technical report by Ufimtsev (1962). His recent book (Ufimtsev 2014) reviews the fundamentals as well as the current state of the art. Knott (2004, Chapter 5) discusses GTD, PTD and other high-frequency methods for radar scattering. The book chapter by Pathak (1988) reviews the PTD, GTD and UTD and also describes their interrelations.

References

Knott EF, Shaeffer JF and Tuley MT (2004) *Radar Cross Section*. SciTech Publishing, Raleigh NC.

Pathak PH (1988) Techniques for high frequency problems. In *Antenna Handbook, Theory Application and Design* (ed. Lo YT and Lee SW) Van Nostrand Reinhold, pp. 4.1–4.117.

Ufimtsev PY (1962) Metod krayevykh voln v fizicheskoy teorii difraktskii (method of edge waves in the physical theory of diffraction). Technical report, Sovetskoe Radio, Moscow, USSR. English translation in DTIC Report AD733203, Defense Technical Information Center, Alexandria VA.

Ufimtsev PY (2014) *Fundamentals of the Physical Theory of Diffraction*. John Wiley & Sons, Inc.

Problems

9.1 A 2D parabolic reflector antenna has a diameter D and a focal length f. The illumination is a magnetic line source.

(a) In the radiation integral (9.40), show that $\sqrt{(y'^2/4f) + f} \approx \sqrt{f}$ in the denominator, is a good approximation when $(D/4f)^2 \ll 1$.

(b) Using this approximation, evaluate the radiation integral in closed form, at $\phi = 0$.

(c) Find the source strength M_0 so that $H_z^s = 1$ at $\phi = 0$ (the far-field factor is suppressed).

9.2 Compute and plot the radiation pattern of a 2D reflector having a diameter of 10λ and $f/D = 0.3$. The illumination is a magnetic line source. Use the PO equations (9.40) and (9.41) with numerical integration using Simpson's rule. To obtain three figures of accuracy for $|H_z|$ (linear, not in dB), how many slices are required at $\phi = 0$ and at $\phi = 90°$?

9.3 Compute and plot the radiation pattern of a 2D reflector having a diameter of 10λ and $f/D = 0.3$. The illumination is a magnetic line source. Use the PTD equations (9.40), (9.41) and (9.43).

9.4 Develop the PO formulation for a 2D reflector having a diameter D and focal length f. The illumination is an electric line source of strength I_0. Use numerical integration with Simpson's rule to compute and plot the radiation pattern when $D = 10\lambda$ and $f/D = 0.3$.

9.5 Develop the PTD formulation for a 2D reflector having a diameter D and focal length f. The illumination is an electric line source of strength I_0. Compute and plot the PO and PTD radiation patterns when $D = 10\lambda$ and $f/D = 0.3$.

9.6 For a half plane illuminated by an electric line source of strength I_0, obtain D_s^k from the general case (8.10), derive the PO endpoint contribution, D_s^{po}, and obtain D_s^u in (9.31).

9.7 A 2D perfectly conducting strip is at $y = 0$ and along $-a/2 \leq x \leq a/2$. It is illuminated by an incident plane wave

$$E_z^i = E_0 e^{jk(x\cos\phi + y\sin\phi)}.$$

Find the PO and PTD scattered field. Use this to show that at $\phi = \pi/2$ the echo width is

$$\sigma = \frac{1}{k}|jka + 1|^2.$$

The first term comes from PO and the second one comes from the fringe wave.

9.8 A 2D perfectly conducting strip is at $y = 0$ and along $-a/2 \leq x \leq a/2$. It is illuminated by an incident plane wave

$$H_z^i = H_0 e^{jk(x\cos\phi + y\sin\phi)}.$$

Find the PO and PTD scattered field. Use this to show that at $\phi = \pi/2$ the echo width is

$$\sigma = \frac{1}{k}|jka - 1|^2.$$

The first term comes from PO and the second one comes from the fringe wave.

9.9 For Problem 9.7, plot the echo width for $0 \leq \phi \leq 90°$ with $a = 2\lambda$ and 10λ.

9.10 For Problem 9.8, plot the echo width for $0 \leq \phi \leq 90°$ with $a = 2\lambda$ and 10λ.

9.11 We can consider the PTD strip scattered field in Problem 9.8 as having three parts: the PO, Keller and PO endpoint correction terms:

$$H_z^s \sim H_z^{po} + H_z^k - H_z^{ep}.$$

Evaluate the endpoint contribution and prove that for this problem $H_z^{po} = H_z^{ep}$. Therefore the PTD and GTD solutions for strip scattering are identical.

9.12 A plane wave in the $\phi = 0$ plane

$$\mathbf{E}^i = \hat{\boldsymbol{\theta}}\, E_0 e^{jk(x\sin\theta + z\cos\theta)}$$

is incident on a perfectly conducting disc of radius a as in Figure 9.7. We want to find the backscattered field at $\theta = 0$ due to the fringe wave.

(a) Find the expressions for the PTDEECs, $I^u(\phi)$ and $M^u(\phi)$.

(b) Evaluate the radiation integral along the rim of the disc and obtain the fringe wave $\mathbf{E}^u$. You should find out that the result is zero. Therefore, at $\theta = 0$, there is a first-order contribution from the PO part, but not from the fringe wave.

9.13 In Problem 9.12 PTD looks after the first-order effects. The second-order effects can be added in by using GTD. Do this for backscatter at $\theta = 0$. The relevant rays for the top surface are illustrated in Figure 9.10. There is a second set of rays on the bottom surface.

    (a) Use GTD to find the diffracted field from Q_{E1} that is incident at the *second* diffraction point Q_{E2}. Note that there is a phase shift associated with the rays passing through a caustic at the centre of the disc, which affects the spread factor.

    (b) Using the field incident at Q_{E2}, find the GTD equivalent currents $I(\phi)$ and $M(\phi)$. You should find that $I(\phi) = 0$ whereas $M(\phi)$ is the same for the top-surface and bottom-surface rays.

    (c) Evaluate a line integral of $M(\phi')$ along the edge of the disc to obtain the backscattered field. Add this to the first-order part and show that the radar cross section is

$$\frac{\sigma}{\sigma_{po}} = \left| 1 + \frac{e^{-j(2ka+\pi/4)}}{\sqrt{\pi}(ka)^{3/2}} \right|^2.$$

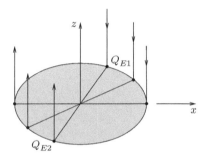

Figure 9.10 Double-diffraction backscatter from a perfectly conducting circular disc, broadside incidence.

9.14 A ϕ-polarized plane wave in the $\phi = 0$ plane

$$\mathbf{E}^i = \hat{\phi}\, E_0 e^{jk(x\sin\theta + z\cos\theta)}$$

is incident on a perfectly conducting disc of radius a, similar to Figure 9.7.

    (a) Find the PO and PTD backscattered field assuming $\theta \neq 0$.

    (b) Compute and plot the RCS $\sigma(\theta)$ at 10 GHz when $a = 3\lambda$.

9.15 Use PO to find the radiation pattern of a vertical monopole on a disc of radius a. The disc is in the $x - y$ plane and the monopole has a strength of p_e. Its magnetic field on the lit side of the disc is

$$H_\phi(\rho) = \frac{p_e k^2}{2\pi} e^{-jk\rho} \left(\frac{j}{k\rho} + \frac{1}{(k\rho)^2} \right).$$

The radiation integral will be of the form

$$\int_{\rho'=\delta}^{a} \int_{\phi'=0}^{2\pi} \hat{\rho}'(\phi') H_{\phi'}(\rho') e^{jk\hat{\mathbf{r}}\cdot\boldsymbol{\rho}'} \rho' d\phi' d\rho'$$

and the surface integral on $0 \leq \rho' \leq \delta$ is negligible when δ is small.

(a) Find the radiation integral for $E_\theta(\theta)$ in the $\phi = 0$ plane.
(b) Show that when $k\rho' \ll 1$, the integral over $0 \le \rho' \le \delta$ is negligible.
(c) Find the far-zone pattern E_θ when $a = 3$ m and $\lambda = 1$ m. Use numerical integration and plot the result for $0° \le \theta \le 180°$.
(d) Compare your answer with the UTD result in Figure 8.14.

9.16 For the monopole on a disc in the previous problem, add the PTD part using two diffraction points, and recalculate the radiation pattern.

10

Scalar and Dyadic Green's Functions

A Green's function is the impulse response of a linear system. The concept was introduced by British mathematical physicist George Green in the 1830s. Since then it has been applied to many engineering problems.

In linear system theory, the response to an input $x(t)$ is obtained by a convolution $y(t) = x(t) \otimes h(t)$, and if the input is $x(t) = \delta(t)$, then the output is $y(t) = h(t)$; hence $h(t)$ is the impulse response, or Green's function, for the system.

In frequency-domain electromagnetics, an impulse function is used to represent a point source in space, as $\delta(|\mathbf{r} - \mathbf{r}'|)$. The Green's function is the field due to a point source. The response to an arbitrary source distribution can then be found from the point source response and a convolution.

A Green's function can be used to express the vector potentials $\mathbf{A}, \mathbf{F}$ or fields $\mathbf{E}, \mathbf{H}$ in terms of their sources. In 2D, the field formulations can be in terms of a scalar component E_z or H_z, so a scalar Green's function is sufficient. In 3D, the vector directions of the source and the field are different and related to each other in a complicated way. It is therefore necessary to use a dyadic Green's function, which can be thought of as a 3×3 matrix, having nine scalar Green's functions.

Green's function concepts are first described for $\mathbf{A}$ in 3D and for fields in 2D. Then, the dyadic concept is presented, for 3D fields. Using Green's second identity, one obtains field expressions in terms of surface integrals. These turn out to be exactly the same expressions that were obtained from the surface equivalence principle in Chapter 4.

Green's functions are used in the method of moments. There is a singularity at the source point – that is, when $\mathbf{r} = \mathbf{r}'$. The singularity as it appears in a volume integral equation is discussed. This is applied to find the scattering by a dielectric rod.

10.1 Impulse Response

A solution of the wave equation

$$(\nabla^2 + k^2)G(\mathbf{r}, \mathbf{r}') = -\delta(|\mathbf{r} - \mathbf{r}'|) \tag{10.1}$$

Applied Frequency-Domain Electromagnetics, First Edition. Robert Paknys.
© 2016 John Wiley & Sons, Ltd. Published 2016 by John Wiley & Sons, Ltd.
Companion Website: www.wiley.com/go/paknys9981

is the *free-space Green's function*

$$G(\mathbf{r}, \mathbf{r}') = \frac{e^{-jkR}}{4\pi R} \tag{10.2}$$

where $\mathbf{R} = \mathbf{r} - \mathbf{r}'$ and $R = |\mathbf{r} - \mathbf{r}'|$. As a rule, $\mathbf{R}$ is always a vector from the source point to the field point, and we will follow the convention that the first argument of G is the field point and the second argument is the source point. In (10.1) and (10.2), the variable $\mathbf{r}$ denotes the field point position and the constant $\mathbf{r}'$ gives the impulse source position.

The Green's function can be used to obtain the solution of the wave equation

$$(\nabla^2 + k^2)A(\mathbf{r}) = F(\mathbf{r}), \tag{10.3}$$

where F is a forcing function. The solution is

$$A(\mathbf{r}) = \int_V F(\mathbf{r}') \, G(\mathbf{r}, \mathbf{r}') \, dx'dy'dz'. \tag{10.4}$$

The domain of integration V covers the region where the forcing function F is nonzero.

From linear systems theory, we know that the output $y(t)$ can be found from the input $x(t)$ and the system impulse response $h(t)$ via a convolution

$$y(t) = \int_{-\infty}^{\infty} x(\lambda)h(t-\lambda) \, d\lambda. \tag{10.5}$$

Noting the similarity of (10.4) and (10.5), we see that (10.4) is a convolution of the forcing function F and impulse response G.

In the following discussion, several manipulations with $\mathbf{r}$ and $\mathbf{r}'$ will be developed. Depending on how G is to be used, either $\mathbf{r}$ or $\mathbf{r}'$ might be the variable, and either the field point or the source point might be the variable. For instance, by interchanging the primed and unprimed quantities in (10.1), it becomes

$$(\nabla'^2 + k^2)G(\mathbf{r}', \mathbf{r}) = -\delta(|\mathbf{r}' - \mathbf{r}|) \tag{10.6}$$

so that $\mathbf{r}'$ is a variable field point and $\mathbf{r}$ is fixed. From (10.2) it is clear that G has the reciprocal property

$$G(\mathbf{r}, \mathbf{r}') = G(\mathbf{r}', \mathbf{r}). \tag{10.7}$$

Since the delta function is even and G is reciprocal, comparing (10.1) and (10.6) implies that

$$\nabla^2 G(\mathbf{r}, \mathbf{r}') = \nabla'^2 G(\mathbf{r}', \mathbf{r}). \tag{10.8}$$

It is noted that although G is reciprocal, its derivative is not

$$\nabla G(\mathbf{r}, \mathbf{r}') = -\nabla G(\mathbf{r}', \mathbf{r}) \tag{10.9}$$

so it is quite important to pay attention to the order of $\mathbf{r}$ and $\mathbf{r}'$. Some other useful properties of ∇G are

$$\nabla G(\mathbf{r}, \mathbf{r}') = -\nabla' G(\mathbf{r}, \mathbf{r}') \tag{10.10}$$

$$\nabla^2 G(\mathbf{r}, \mathbf{r}') = \nabla'^2 G(\mathbf{r}, \mathbf{r}'). \tag{10.11}$$

Using (10.11) in (10.1) gives

$$(\nabla'^2 + k^2)G(\mathbf{r}, \mathbf{r}') = -\delta(|\mathbf{r} - \mathbf{r}'|) \tag{10.12}$$

so that $\mathbf{r}'$ is a variable source point. Interchanging primed and unprimed coordinates gives

$$(\nabla^2 + k^2)G(\mathbf{r}',\mathbf{r}) = -\delta(|\mathbf{r}'-\mathbf{r}|) \tag{10.13}$$

so that $\mathbf{r}$ is a variable source point.[1]

10.2 Green's Function for A

A rectangular component $A = A_x, A_y, A_z$ of the vector potential $\mathbf{A}(\mathbf{r})$ satisfies the scalar equation

$$(\nabla^2 + k^2)A(\mathbf{r}) = -\mu J(\mathbf{r}). \tag{10.14}$$

The equation

$$(\nabla^2 + k^2)G(\mathbf{r}',\mathbf{r}) = -\delta(|\mathbf{r}'-\mathbf{r}|) \tag{10.15}$$

has a Green's function

$$G(\mathbf{r}',\mathbf{r}) = \frac{e^{-jkR}}{4\pi R}, \tag{10.16}$$

where $R = |\mathbf{r}'-\mathbf{r}|$. A solution of (10.14) can be found by multiplying (10.14) by G and (10.15) by A, subtracting them, and integrating with respect to the source coordinate $\mathbf{r}$ over a volume V enclosing the current source J to obtain

$$\int_V A(\mathbf{r})\nabla^2 G(\mathbf{r}',\mathbf{r}) - G(\mathbf{r}',\mathbf{r})\nabla^2 A(\mathbf{r})\,dV = \int_V -\delta(|\mathbf{r}'-\mathbf{r}|)A(\mathbf{r}) + \mu J(\mathbf{r})G(\mathbf{r}',\mathbf{r})\,dV. \tag{10.17}$$

Next, we apply Green's second identity (B.43) to the source coordinates

$$\int_V f\nabla^2 g - g\nabla^2 f\,dV = \oint_S f\frac{\partial g}{\partial n} - g\frac{\partial f}{\partial n}\,dS \tag{10.18}$$

to convert the left-hand side of (10.17) into a surface integral. The volume V is bounded by a closed surface S. The normal derivatives are taken in the direction of the outward normal to the surface. V can be a sphere, so that $\partial/\partial n = \partial/\partial r$. The sifting property of the delta function is applied to the right-hand side of (10.17) so that it becomes

$$\oint_S A(\mathbf{r})\frac{\partial G(\mathbf{r}',\mathbf{r})}{\partial r} - G(\mathbf{r}',\mathbf{r})\frac{\partial A(\mathbf{r})}{\partial r}\,dS = -A(\mathbf{r}') + \mu\int_V J(\mathbf{r})G(\mathbf{r}',\mathbf{r})\,dV. \tag{10.19}$$

We now invoke the Sommerfeld radiation condition (see Section 4.5)

$$\lim_{r\to\infty} r\left(\frac{\partial A(r)}{\partial r} + jkA(r)\right) = 0 \tag{10.20}$$

---

[1] $\delta(|\mathbf{r}-\mathbf{r}'|)$ is equivalent to $\delta(x-x')\delta(y-y')\delta(z-z')$. In cylindrical coordinates, it can be written as $\delta(\rho-\rho')\delta(\phi-\phi')\delta(z-z')/\rho$, and in spherical, it is $\delta(r-r')\delta(\theta-\theta')\delta(\phi-\phi')/r^2\sin\theta$. In any case,

$$\int_V \delta(|\mathbf{r}-\mathbf{r}'|)dV = 1$$

when $\mathbf{r}'$ is in V.

which is a requirement that the outward-travelling waves decay at least as fast as $1/r$. If both $A \propto 1/r$ and $G \propto 1/r$, then with $dS = r^2 \sin \theta d\theta d\phi$, the surface integral in (10.19) will vanish when the sphere radius $r \to \infty$. Under these conditions, (10.19) becomes

$$A(\mathbf{r}') = \mu \int_V J(\mathbf{r})G(\mathbf{r}',\mathbf{r}) \, dV. \tag{10.21}$$

When deriving (10.21), it was convenient to use the source point $\mathbf{r}$ as the working variable and regard $\mathbf{r}'$ as a constant. However, when applying (10.21), it is most common to treat $\mathbf{r}'$ as the variable of integration and $\mathbf{r}$ as the field point. Interchanging the primed and unprimed quantities in (10.21) gives an equivalent representation for $A(\mathbf{r})$

$$A(\mathbf{r}) = \mu \int_V J(\mathbf{r}')G(\mathbf{r},\mathbf{r}') \, dV' \tag{10.22}$$

which is recognizable as the vector potential solution for the components A_x, A_y, A_z. The integral in (10.22) is a convolution, and $A(\mathbf{r}) = \mu G(\mathbf{r}, 0)$ is the impulse response to a current $J(\mathbf{r}') = \delta(|\mathbf{r}'|)$.

A similar development can be carried out for the 2D case, where the wave equation is

$$(\nabla^2 + k^2)G(\boldsymbol{\rho}', \boldsymbol{\rho}) = -\delta(|\boldsymbol{\rho}' - \boldsymbol{\rho}|). \tag{10.23}$$

The radiation condition in 2D requires outward-travelling waves that decay at least as fast as $1/\sqrt{\rho}$. The Green's function that meets this condition is

$$G(\boldsymbol{\rho}, \boldsymbol{\rho}') = \frac{1}{j4} H_0^{(2)}(k|\boldsymbol{\rho} - \boldsymbol{\rho}'|). \tag{10.24}$$

10.3   2D Field Solutions Using Green's Functions

Figure 10.1 shows a surface area S bounded by a contour ℓ. If ψ and ϕ are arbitrary functions that are twice differentiable, then Green's second identity (B.46) (with modifications for 2D) states that

$$\int_S \psi \nabla^2 \phi - \phi \nabla^2 \psi \, dS = \oint_\ell (\phi \nabla \psi - \psi \nabla \phi) \cdot \hat{\mathbf{n}} \, d\ell, \tag{10.25}$$

where the line integral is taken in the direction of increasing ℓ. We are interested in solving the wave equation in 2D

$$(\nabla^2 + k^2)u(\boldsymbol{\rho}) = f(\boldsymbol{\rho}) \tag{10.26}$$

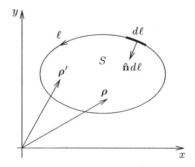

Figure 10.1   2D surface S and bounding curve ℓ in the $x - y$ plane.

by using a Green's function that satisfies

$$(\nabla^2 + k^2)G(\boldsymbol{\rho}', \boldsymbol{\rho}) = -\delta(|\boldsymbol{\rho}' - \boldsymbol{\rho}|). \tag{10.27}$$

The problem is assumed to be in the $x - y$ plane so that $\boldsymbol{\rho} = \hat{\mathbf{x}}x + \hat{\mathbf{y}}y$ and $\boldsymbol{\rho}' = \hat{\mathbf{x}}x' + \hat{\mathbf{y}}y'$.

In (10.25) we let $\psi = G$ and $\phi = u$ and use (10.26), (10.27) to replace $\nabla^2 u$ and $\nabla^2 G$ so that the left-hand side of (10.25) becomes

$$\int_S G\nabla^2 u - u\nabla^2 G \, dS = \int_S G[f - k^2 u] - u[-\delta(|\boldsymbol{\rho}' - \boldsymbol{\rho}|) - k^2 G] \, dS$$

$$= \int_S G(\boldsymbol{\rho}', \boldsymbol{\rho})f(\boldsymbol{\rho}) \, dS + \begin{cases} u(\boldsymbol{\rho}'); \ \boldsymbol{\rho}' \text{ inside } S \\ 0; \ \boldsymbol{\rho}' \text{ outside } S. \end{cases}$$

Noting that $\hat{\mathbf{n}} \cdot \nabla u = \partial u/\partial n$ and $\hat{\mathbf{n}} \cdot \nabla G = \partial G/\partial n$, this becomes

$$\left. \begin{array}{l} \boldsymbol{\rho}' \text{ inside } S; \ u(\boldsymbol{\rho}') \\ \boldsymbol{\rho}' \text{ outside } S; \ 0 \end{array} \right\} = -\int_S f(\boldsymbol{\rho})G(\boldsymbol{\rho}', \boldsymbol{\rho}) \, dS$$

$$+ \oint_\ell u(\boldsymbol{\rho})\frac{\partial G(\boldsymbol{\rho}', \boldsymbol{\rho})}{\partial n} - G(\boldsymbol{\rho}', \boldsymbol{\rho})\frac{\partial u(\boldsymbol{\rho})}{\partial n} \, d\ell. \tag{10.28}$$

The integral over S represents the incident field due to the source $f(\boldsymbol{\rho})$. The integral over ℓ represents the scattered field in terms of the equivalent sources u and $\partial u/\partial n$ on the boundary ℓ.

When the observer is within S, the two integrals add up to give the total field $u(\boldsymbol{\rho}')$. On the other hand, when the observer is outside S, then the left-hand side of (10.28) is zero. For this reason, (10.28) is sometimes called the *null field integral equation*.

When deriving (10.28), it was convenient to use $\boldsymbol{\rho}$ as the working variable. However, when applying (10.28) it is most common to use $\boldsymbol{\rho}$ as the field point. By interchanging the primed and unprimed quantities, we obtain

$$\left. \begin{array}{l} \boldsymbol{\rho} \text{ inside } S; \ u(\boldsymbol{\rho}) \\ \boldsymbol{\rho} \text{ outside } S; \ 0 \end{array} \right\} = -\int_S f(\boldsymbol{\rho}')G(\boldsymbol{\rho}, \boldsymbol{\rho}') \, dS'$$

$$+ \oint_\ell u(\boldsymbol{\rho}')\frac{\partial G(\boldsymbol{\rho}, \boldsymbol{\rho}')}{\partial n'} - G(\boldsymbol{\rho}, \boldsymbol{\rho}')\frac{\partial u(\boldsymbol{\rho}')}{\partial n'} \, d\ell'. \tag{10.29}$$

Here, n' remains the inward-pointing normal direction.

10.3.1 2D TM Fields

In this case, the electric field has only a z component and the magnetic field has only x and y components, that is, $\mathbf{E} = \hat{\mathbf{z}}E_z$ and $\mathbf{H} = \hat{\mathbf{t}}H_t$. Here, $\hat{\mathbf{n}}$ points into S as shown in Figure 10.1 and $\hat{\mathbf{t}}$ represents a unit tangent vector that satisfies the relation

$$\hat{\mathbf{t}} \times \hat{\mathbf{n}} = \hat{\mathbf{z}}$$

at any point along the boundary contour ℓ. A z-directed current distribution in S will generate an electric field that must satisfy

$$(\nabla^2 + k^2)E_z(\boldsymbol{\rho}) = j\omega\mu J_z(\boldsymbol{\rho}). \tag{10.30}$$

Comparing this with (10.26), we see that $u(\boldsymbol{\rho}) = E_z(\boldsymbol{\rho})$ and $f(\boldsymbol{\rho}) = j\omega\mu J_z(\boldsymbol{\rho})$. Equation (10.28) becomes

$$\left.\begin{array}{l} \boldsymbol{\rho}' \text{ inside } S; \; E_z(\boldsymbol{\rho}') \\ \boldsymbol{\rho}' \text{ outside } S; \; 0 \end{array}\right\} = -j\omega\mu \int_S J_z(\boldsymbol{\rho})G(\boldsymbol{\rho}',\boldsymbol{\rho})\, dS$$

$$+ \oint_\ell E_z(\boldsymbol{\rho})\frac{\partial G(\boldsymbol{\rho}',\boldsymbol{\rho})}{\partial n} - G(\boldsymbol{\rho}',\boldsymbol{\rho})\frac{\partial E_z(\boldsymbol{\rho})}{\partial n}\, d\ell. \tag{10.31}$$

The first integral represents the impressed or incident field. The second integral represents the scattered field; the quantities E_z and $\partial E_z/\partial n$ are related to surface magnetic and electric currents on contour ℓ. The second and third integrals in (10.31) can be written in terms of a $\hat{\mathbf{t}}$-directed boundary magnetic current $\hat{\mathbf{t}}M_{\ell t}$ and a $\hat{\mathbf{z}}$-directed electric current $\hat{\mathbf{z}}J_{\ell z}$ on contour ℓ so that

$$\left.\begin{array}{l} \boldsymbol{\rho}' \text{ inside } S; \; E_z(\boldsymbol{\rho}') \\ \boldsymbol{\rho}' \text{ outside } S; \; 0 \end{array}\right\} = -j\omega\mu \int_S J_z(\boldsymbol{\rho})G(\boldsymbol{\rho}',\boldsymbol{\rho})\, dS$$

$$- \oint_\ell M_{\ell t}(\boldsymbol{\rho})\frac{\partial G(\boldsymbol{\rho}',\boldsymbol{\rho})}{\partial n}\, d\ell - j\omega\mu \oint_\ell G(\boldsymbol{\rho}',\boldsymbol{\rho})J_{\ell z}(\boldsymbol{\rho})\, d\ell. \tag{10.32}$$

10.3.2   2D TE Fields

In this case, $\mathbf{H} = \hat{\mathbf{z}}H_z$ and $\mathbf{E} = \hat{\mathbf{t}}E_t$. As before, $\hat{\mathbf{n}}$ points into S and $\hat{\mathbf{t}}$ represents a unit tangent vector. Applying duality to the TM case leads to

$$(\nabla^2 + k^2)H_z(\boldsymbol{\rho}) = j\omega\epsilon M_z(\boldsymbol{\rho}) \tag{10.33}$$

and

$$\left.\begin{array}{l} \boldsymbol{\rho}' \text{ inside } S; \; H_z(\boldsymbol{\rho}') \\ \boldsymbol{\rho}' \text{ outside } S; \; 0 \end{array}\right\} = -j\omega\epsilon \int_S M_z(\boldsymbol{\rho})G(\boldsymbol{\rho}',\boldsymbol{\rho})\, dS$$

$$+ \oint_\ell J_{\ell t}(\boldsymbol{\rho})\frac{\partial G(\boldsymbol{\rho}',\boldsymbol{\rho})}{\partial n}\, d\ell - j\omega\epsilon \oint_\ell G(\boldsymbol{\rho}',\boldsymbol{\rho})M_{\ell z}(\boldsymbol{\rho})\, d\ell. \tag{10.34}$$

10.3.3   Free-Space Interpretation

The only requirement on G is that it satisfies the Helmholtz equation (10.27). For the moment, let us assume that the free space Green's function is chosen. This is equivalent to assuming that the currents radiate in free space. G_f is given by

$$G_f(\boldsymbol{\rho}',\boldsymbol{\rho}) = \frac{1}{j4}H_0^{(2)}(kR) \tag{10.35}$$

with

$$R = \sqrt{(x'-x)^2 + (y'-y)^2}. \tag{10.36}$$

The normal derivative of (10.35) will be needed. To find it, we first consider an x-directed magnetic current $M_{\ell t} = M_{\ell x}$ at (x,y), as shown in Figure 10.2. In (10.36) it is important to remember that the

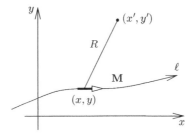

Figure 10.2   An x-directed magnetic current $M_{\ell x}$ at (x, y) with the field point at (x', y').

source point is the variable. The order of ρ and ρ' does not affect G, but it does affect its normal derivative because $\partial R/\partial y = -\partial R/\partial y'$. From (10.35) and (10.36), it follows that

$$\frac{\partial G_f}{\partial y} = -\frac{jk}{4}\hat{\mathbf{z}} \cdot (\hat{\mathbf{x}} \times \hat{\mathbf{R}}) H_1^{(2)}(kR). \tag{10.37}$$

The result in (10.37) can be generalized for an arbitrarily oriented transverse magnetic current $M_{\ell t}$ by replacing $x \to t$ and $y \to n$ so that

$$\frac{\partial G_f}{\partial n} = -\frac{jk}{4}\hat{\mathbf{z}} \cdot (\hat{\boldsymbol{\ell}} \times \hat{\mathbf{R}}) H_1^{(2)}(kR). \tag{10.38}$$

Using (10.35) and (10.38) in (10.32), the TM case gives us

$$\left.\begin{matrix} \rho' \text{ inside } S; \ E_z(\rho') \\ \rho' \text{ outside } S; \ 0 \end{matrix}\right\} = -\frac{\omega\mu}{4}\int_S J_z(\rho)\, H_0^{(2)}(kR)\, dS$$

$$+ \frac{jk}{4}\oint_\ell M_{\ell t}(\rho)\, \hat{\mathbf{z}} \cdot (\hat{\boldsymbol{\ell}} \times \hat{\mathbf{R}})\, H_1^{(2)}(kR)\, d\ell - \frac{\omega\mu}{4}\oint_\ell J_{\ell z}(\rho)\, H_0^{(2)}(kR)\, d\ell. \tag{10.39}$$

The first term can be interpreted as the incident field. Inside S, the second plus third terms produce the scattered field, whereas outside S, they produce the negative incident field, making the total equal to zero.

It is interesting to note that the above result (10.39) is telling us the same thing as the surface equivalence principle. That is, the currents on ℓ act together with J_z to produce the original field $\mathbf{E}$, $\mathbf{H}$ inside S and zero fields outside S. The surface currents on ℓ are related to the field discontinuity across the boundary.

10.3.4   Special Green's Functions

Returning to (10.28), we note that it can be simplified by imposing (a) $u = 0$ and $G = 0$ on ℓ (Dirichlet boundary conditions) or (b) $\partial u/\partial n = 0$ and $\partial G/\partial n = 0$ on ℓ (Neumann boundary conditions). Then the surface integrals in (10.28) are zero and the total field $u(\rho')$ at any point ρ' in S can be found from

$$u(\rho') = -\int_S f(\rho)G(\rho', \rho)\, dS. \tag{10.40}$$

We can see that a Green's function satisfying certain boundary conditions is desirable because the boundary integral along ℓ would no longer be required.

As an example, let us consider the implication of choosing a Green's function $G = G_p$ that satisfies the Helmholtz equation (10.27) and meets the Dirichlet boundary condition $G_p = 0$ along the plane $y = 0$. It is assumed that the region of interest S is the upper half space $y \geq 0$. The source point is at (x, y) and the field point is at (x', y'). Image theory can be used to obtain

$$G_p = \frac{1}{j4} H_0^{(2)}(kR_1) - \frac{1}{j4} H_0^{(2)}(kR_2),$$ (10.41)

where

$$R_1 = \sqrt{(x' - x)^2 + (y' - y)^2}$$ (10.42)

$$R_2 = \sqrt{(x' - x)^2 + (y' + y)^2}.$$ (10.43)

It is seen that as $y \to 0$ $R_1 \to R_2$ and $G_p \to 0$ so that the Dirichlet condition is met.

Also of interest is the normal derivative when $y \to 0$. Differentiation of (10.41) with respect to y and letting $R_1 = R_2 = R$ leads to

$$\frac{\partial G_p}{\partial y} = -\frac{jk}{2} \hat{z} \cdot (\hat{x} \times \hat{R}) H_1^{(2)}(kR).$$ (10.44)

As expected from image theory, (10.44) is double the free-space result (10.37). Since $G_p = 0$ on the boundary, (10.32) becomes

$$E_z(\rho') = -j\omega\mu \int_S J_z(\rho) G_p(kR) \, dS - \int_{-\infty}^{\infty} M_{\ell x}(x) \frac{\partial G_p(kR)}{\partial y} \, dx.$$ (10.45)

This equation gives E_z for any current distribution J_z above a ground plane. The first integral provides the incident plus reflected field. On the conducting plane at $y = 0$, $M_{\ell x} = -E_z = 0$ so the second integral is zero.

To summarize, the advantage of using a special Green's function is that only an integration over the known (impressed) sources is needed to find the total field. In this section, it was not too difficult to find G_p for a ground plane; however a general disadvantage of using a special Green's function is that for a complex-shaped body, it may be difficult or even impossible to find.

10.4   3D Dyadic Green's Functions

It was seen that in 2D TM problems, an electric current J_z is related to an electric field E_z through an integral

$$E_z(\rho) = -j\omega\mu_0 \int_S G(\rho, \rho') J_z(\rho') \, dS'.$$

In contrast, a z-directed electric dipole in 3D has an electric field with E_r and E_θ components. Generally, 3D problems are more complex because the current and the field can point in different directions. To accommodate such situations, it is useful to extend the idea of a vector (having three elements) to a *dyadic* (having nine elements). A vector is written in terms of unit vectors $\hat{x}, \hat{y}, \hat{z}$, whereas dyadics are written in terms of pairs of unit vectors, for example, $\hat{x}\hat{z}$, $\hat{y}\hat{y}$ and so forth.

Dyadics will be denoted in boldface, with an extra bar. A simple example of a dyadic is

$$\bar{C} = 12\hat{x}\hat{y} + 8\hat{x}\hat{x}.$$

A dyadic can be dotted with a vector. For example, if $\mathbf{A} = \hat{\mathbf{x}}$, then

$$\mathbf{A} \cdot \bar{\mathbf{C}} = \hat{\mathbf{x}} \cdot (12\hat{\mathbf{x}}\hat{\mathbf{y}} + 8\hat{\mathbf{x}}\hat{\mathbf{x}}) = 12\hat{\mathbf{x}} \cdot \hat{\mathbf{x}}\hat{\mathbf{y}} + 8\hat{\mathbf{x}} \cdot \hat{\mathbf{x}}\hat{\mathbf{x}} = 12\hat{\mathbf{y}} + 8\hat{\mathbf{x}}$$

whereas

$$\bar{\mathbf{C}} \cdot \mathbf{A} = (12\hat{\mathbf{x}}\hat{\mathbf{y}} + 8\hat{\mathbf{x}}\hat{\mathbf{x}}) \cdot \hat{\mathbf{x}} = 12\hat{\mathbf{x}}\hat{\mathbf{y}} \cdot \hat{\mathbf{x}} + 8\hat{\mathbf{x}}\hat{\mathbf{x}} \cdot \hat{\mathbf{x}} = 8\hat{\mathbf{x}}.$$

It is seen that $\mathbf{A} \cdot \bar{\mathbf{C}} \neq \bar{\mathbf{C}} \cdot \mathbf{A}$ and in general the dot product of a vector with a dyadic is not commutative. Two dyadics can be dotted, and the result is a dyadic. It is possible to interpret dyadic manipulations in terms of matrix algebra. For example, if

$$\bar{\mathbf{C}} = C_{xx}\hat{\mathbf{x}}\hat{\mathbf{x}} + C_{xy}\hat{\mathbf{x}}\hat{\mathbf{y}} + C_{yx}\hat{\mathbf{y}}\hat{\mathbf{x}} + C_{yy}\hat{\mathbf{y}}\hat{\mathbf{y}}$$

$$\mathbf{B} = B_x\hat{\mathbf{x}} + B_y\hat{\mathbf{y}}$$

$$\mathbf{A} = A_x\hat{\mathbf{x}} + A_y\hat{\mathbf{y}}$$

then $\bar{\mathbf{C}} \cdot \mathbf{B} = \mathbf{A}$ can be interpreted as

$$\begin{bmatrix} C_{xx} & C_{xy} \\ C_{yx} & C_{yy} \end{bmatrix} \begin{bmatrix} B_x \\ B_y \end{bmatrix} = \begin{bmatrix} A_x \\ A_y \end{bmatrix}.$$

A vector dotted with a vector gives a scalar; however a vector dotted with a dyadic gives a vector. This will be useful in developing a dyadic Green's function $\bar{\mathbf{G}}(\mathbf{r}, \mathbf{r}')$ that relates a current vector $\mathbf{J}$ to its field vector $\mathbf{E}$ via an integral

$$\mathbf{E}(\mathbf{r}) = -j\omega\mu_0 \int_V \bar{\mathbf{G}}(\mathbf{r}, \mathbf{r}') \cdot \mathbf{J}(\mathbf{r}') \, dV'.$$

10.5 Some Dyadic Identities

Some useful dyadic identities can be established by expanding them out in matrix notation. If $\mathbf{A}, \mathbf{B}$ are vectors, $\bar{\mathbf{C}}$ is a dyadic and $\bar{\mathbf{C}}_t$ is its transpose, it can be shown that

$$\mathbf{A} \cdot \bar{\mathbf{C}} = \bar{\mathbf{C}}_t \cdot \mathbf{A}$$

$$\mathbf{A} \times \bar{\mathbf{C}} = (\bar{\mathbf{C}}_t \times \mathbf{A})_t.$$

It is possible to extend previously known vector identities to include dyadics. Consider the representation

$$\bar{\mathbf{C}} = \mathbf{C}_x\hat{\mathbf{x}} + \mathbf{C}_y\hat{\mathbf{y}} + \mathbf{C}_z\hat{\mathbf{z}}.$$

If the dyadic is the *last* term in an expression involving vectors and a dyadic, it can be treated as though only vectors were involved. The presence of the dyadic term means that we now have three vector equations instead of one. For example,

$$\mathbf{A} \cdot (\mathbf{B} \times \mathbf{C}) = (\mathbf{A} \times \mathbf{B}) \cdot \mathbf{C}$$

extends to

$$\mathbf{A} \cdot (\mathbf{B} \times \bar{\mathbf{C}}) = (\mathbf{A} \times \mathbf{B}) \cdot \bar{\mathbf{C}}.$$

On the other hand, the vector identity

$$\mathbf{A} \cdot \mathbf{B} = \mathbf{B} \cdot \mathbf{A}$$

is not extendable, as

$$\mathbf{A} \cdot \bar{\mathbf{C}} \neq \bar{\mathbf{C}} \cdot \mathbf{A}.$$

If $\bar{\mathbf{C}}$ is not the last term in an expression, we can often manipulate it into that position. Hence, many vector identities can be extended to include dyadics. This leads to the following useful identities

$$\nabla \cdot (\nabla \times \bar{\mathbf{C}}) = 0$$

$$\nabla \times (\nabla \times \bar{\mathbf{C}}) = \nabla(\nabla \cdot \bar{\mathbf{C}}) - \nabla^2 \bar{\mathbf{C}}$$

$$\nabla \cdot (\mathbf{A} \times \nabla \times \bar{\mathbf{C}}) = (\nabla \times \mathbf{A}) \cdot (\nabla \times \bar{\mathbf{C}}) - \mathbf{A} \cdot (\nabla \times \nabla \times \bar{\mathbf{C}})$$

$$\nabla \cdot [(\nabla \times \mathbf{A}) \times \bar{\mathbf{C}}] = (\nabla \times \nabla \times \mathbf{A}) \cdot \bar{\mathbf{C}} - (\nabla \times \mathbf{A}) \cdot (\nabla \times \bar{\mathbf{C}}).$$

If ψ is a scalar function,

$$\nabla \cdot (\psi \bar{\mathbf{C}}) = (\nabla \psi) \cdot \bar{\mathbf{C}} + \psi \nabla \cdot \bar{\mathbf{C}}$$

$$\nabla \times (\psi \bar{\mathbf{C}}) = (\nabla \psi) \nabla \times \bar{\mathbf{C}} + \psi \nabla \times \bar{\mathbf{C}}.$$

The unit dyad $\bar{\mathbf{I}}$ is defined by

$$\bar{\mathbf{I}} = \hat{\mathbf{x}}\hat{\mathbf{x}} + \hat{\mathbf{y}}\hat{\mathbf{y}} + \hat{\mathbf{z}}\hat{\mathbf{z}}$$

and has the properties

$$\mathbf{A} \cdot \bar{\mathbf{I}} = \bar{\mathbf{I}} \cdot \mathbf{A} = \mathbf{A}$$

$$\bar{\mathbf{C}} \cdot \bar{\mathbf{I}} = \bar{\mathbf{C}}.$$

10.6   Solution Using a Dyadic Green's Function

The inhomogeneous vector Helmholtz wave equation can be solved by using a dyadic Green's function. In a homogeneous medium characterized by $k = \omega\sqrt{\mu\epsilon}$, the vector Helmholtz equation is given by

$$\nabla \times \nabla \times \mathbf{E}(\mathbf{r}) - k^2\mathbf{E}(\mathbf{r}) = -j\omega\mu\mathbf{J}(\mathbf{r}). \tag{10.46}$$

The dyadic Green's function satisfies the differential equation

$$\nabla \times \nabla \times \bar{\mathbf{G}}(\mathbf{r}, \mathbf{r}') - k^2\bar{\mathbf{G}}(\mathbf{r}, \mathbf{r}') = \bar{\mathbf{I}}\delta(|\mathbf{r} - \mathbf{r}'|). \tag{10.47}$$

Let us post-dot multiply (10.46) by $\cdot\bar{\mathbf{G}}$ and pre-dot multiply (10.47) by $\mathbf{E}\cdot$, subtract the two equations, and integrate over the volume V. This gives us

$$\left.\begin{array}{r} \mathbf{r}' \text{ inside } V; \; \mathbf{E}(\mathbf{r}') \\ \mathbf{r}' \text{ outside } V; \; 0 \end{array}\right\} = -j\omega\mu \int_V \mathbf{J}(\mathbf{r}) \cdot \bar{\mathbf{G}}(\mathbf{r}, \mathbf{r}') \, dV$$

$$+ \int_V \mathbf{E}(\mathbf{r}) \cdot \nabla \times \nabla \times \bar{\mathbf{G}}(\mathbf{r}, \mathbf{r}') \, dV - \int_V \nabla \times \nabla \times \mathbf{E}(\mathbf{r}) \cdot \bar{\mathbf{G}}(\mathbf{r}, \mathbf{r}') \, dV. \tag{10.48}$$

The second and third integrals can be changed to surface integrals by using Green's second identity for a vector and dyadic (B.46). They become

$$\int_S \hat{\mathbf{n}} \cdot [(\nabla \times \mathbf{E}(\mathbf{r})) \times \bar{\mathbf{G}}(\mathbf{r}, \mathbf{r}')] \, dS + \int_S \hat{\mathbf{n}} \cdot [\mathbf{E}(\mathbf{r}) \times (\nabla \times \bar{\mathbf{G}}(\mathbf{r}, \mathbf{r}'))] \, dS. \tag{10.49}$$

The vector identity $\mathbf{A} \cdot (\mathbf{B} \times \mathbf{C}) = \mathbf{C} \cdot (\mathbf{A} \times \mathbf{B})$ allows us to obtain $\mathbf{A} \cdot (\mathbf{B} \times \bar{\mathbf{C}}) = (\mathbf{A} \times \mathbf{B}) \cdot \bar{\mathbf{C}} = \bar{\mathbf{C}}_t \cdot (\mathbf{A} \times \mathbf{B})$, which can be used to simplify some of the terms in (10.49)

$$\hat{\mathbf{n}} \cdot (\nabla \times \mathbf{E}(\mathbf{r})) \times \bar{\mathbf{G}}(\mathbf{r}, \mathbf{r}') = (\hat{\mathbf{n}} \times (\nabla \times \mathbf{E}(\mathbf{r}))) \cdot \bar{\mathbf{G}}(\mathbf{r}, \mathbf{r}') = -j\omega\mu \mathbf{Js}(\mathbf{r}) \cdot \bar{\mathbf{G}}(\mathbf{r}, \mathbf{r}') \tag{10.50}$$

$$\hat{\mathbf{n}} \cdot (\mathbf{E}(\mathbf{r}) \times (\nabla \times \bar{\mathbf{G}}(\mathbf{r}, \mathbf{r}'))) = (\hat{\mathbf{n}} \times \mathbf{E}(\mathbf{r})) \cdot (\nabla \times \bar{\mathbf{G}}(\mathbf{r}, \mathbf{r}')) = -\mathbf{M_s}(\mathbf{r}) \cdot (\nabla \times \bar{\mathbf{G}}(\mathbf{r}, \mathbf{r}')). \tag{10.51}$$

Equation (10.48) can be rewritten using (10.49), (10.50) and (10.51) with the result that

$$\left. \begin{array}{l} \mathbf{r}' \text{ inside } V; \ \mathbf{E}(\mathbf{r}') \\ \mathbf{r}' \text{ outside } V; \ 0 \end{array} \right\} = -j\omega\mu \int_V \mathbf{J}(\mathbf{r}) \cdot \bar{\mathbf{G}}(\mathbf{r}, \mathbf{r}') \, dV$$

$$- j\omega\mu \int_S \mathbf{J_s}(\mathbf{r}) \cdot \bar{\mathbf{G}}(\mathbf{r}, \mathbf{r}') \, dS - \int_S \mathbf{M_s}(\mathbf{r}) \cdot (\nabla \times \bar{\mathbf{G}}(\mathbf{r}, \mathbf{r}')) \, dS. \tag{10.52}$$

In the above equation, the field point is at $\mathbf{r}' = $ constant, and $\mathbf{r}$ is the variable of integration over the domain containing $\mathbf{J}(\mathbf{r})$, $\mathbf{J_s}(\mathbf{r})$ and $\mathbf{M_s}(\mathbf{r})$. If desired, the roles of $\mathbf{r}$ and $\mathbf{r}'$ can be interchanged to obtain

$$\left. \begin{array}{l} \mathbf{r} \text{ inside } V; \ \mathbf{E}(\mathbf{r}) \\ \mathbf{r} \text{ outside } V; \ 0 \end{array} \right\} = -j\omega\mu \int_V \mathbf{J}(\mathbf{r}') \cdot \bar{\mathbf{G}}(\mathbf{r}', \mathbf{r}) \, dV'$$

$$- j\omega\mu \int_S \mathbf{J_s}(\mathbf{r}') \cdot \bar{\mathbf{G}}(\mathbf{r}', \mathbf{r}) \, dS' - \int_S \mathbf{M_s}(\mathbf{r}') \cdot (\nabla' \times \bar{\mathbf{G}}(\mathbf{r}', \mathbf{r})) \, dS'. \tag{10.53}$$

10.7 Symmetry Property of $\bar{\mathbf{G}}$

In general, reciprocity must hold for fields expressed in terms of the dyadic Green's function. Suppose that a current element $\mathbf{J}_1(\mathbf{r}_1)$ at $\mathbf{r}_1$ produces a field $\mathbf{E}_1(\mathbf{r}_2)$ at $\mathbf{r}_2$, and a current element $\mathbf{J}_2(\mathbf{r}_2)$ at $\mathbf{r}_2$ produces a field $\mathbf{E}_2(\mathbf{r}_1)$ at $\mathbf{r}_1$. The Lorentz reciprocity theorem requires that

$$\mathbf{E}_1(\mathbf{r}_2) \cdot \mathbf{J}_2(\mathbf{r}_2) = \mathbf{E}_2(\mathbf{r}_1) \cdot \mathbf{J}_1(\mathbf{r}_1)$$

from which
$$(\mathbf{J}_1(\mathbf{r}_1) \cdot \bar{\mathbf{G}}(\mathbf{r}_1, \mathbf{r}_2)) \cdot \mathbf{J}_2(\mathbf{r}_2) = (\mathbf{J}_2(\mathbf{r}_2) \cdot \bar{\mathbf{G}}(\mathbf{r}_2, \mathbf{r}_1)) \cdot \mathbf{J}_1(\mathbf{r}_1).$$

This can be interpreted as

$$\mathbf{J}_1(\mathbf{r}_1) \cdot \underbrace{(\bar{\mathbf{G}}(\mathbf{r}_1, \mathbf{r}_2) \cdot \mathbf{J}_2(\mathbf{r}_2))}_{E_2(\mathbf{r}_1)} = \mathbf{J}_2(\mathbf{r}_2) \cdot \underbrace{(\bar{\mathbf{G}}(\mathbf{r}_2, \mathbf{r}_1) \cdot \mathbf{J}_1(\mathbf{r}_1))}_{E_1(\mathbf{r}_2)}$$

which implies that we must have

$$\mathbf{J}_2(\mathbf{r}_2) \cdot \bar{\mathbf{G}}(\mathbf{r}_2, \mathbf{r}_1) = \bar{\mathbf{G}}(\mathbf{r}_1, \mathbf{r}_2) \cdot \mathbf{J}_2(\mathbf{r}_2).$$

Since $\mathbf{J}_2$ is arbitrary, this implies

$$\bar{\mathbf{G}}(\mathbf{r}_2, \mathbf{r}_1) = \bar{\mathbf{G}}_t(\mathbf{r}_1, \mathbf{r}_2) \tag{10.54}$$

which is the *symmetry property* of the dyadic Green's function. This is useful when $\mathbf{r}$ and $\mathbf{r}'$ need to be interchanged. For instance,

$$\bar{\mathbf{G}}_t(\mathbf{r}', \mathbf{r}) = \bar{\mathbf{G}}(\mathbf{r}, \mathbf{r}') \tag{10.55}$$

$$\bar{\mathbf{G}}_t(\mathbf{r}, \mathbf{r}') = \bar{\mathbf{G}}(\mathbf{r}', \mathbf{r}). \tag{10.56}$$

This can be used to show that

$$\mathbf{J}(\mathbf{r}') \cdot \bar{\mathbf{G}}(\mathbf{r}', \mathbf{r}) = \bar{\mathbf{G}}_t(\mathbf{r}', \mathbf{r}) \cdot \mathbf{J}(\mathbf{r}') = \bar{\mathbf{G}}(\mathbf{r}, \mathbf{r}') \cdot \mathbf{J}(\mathbf{r}'). \tag{10.57}$$

It can be shown[2] that the symmetry property also applies to the curl of the dyadic Green's function in the following way:

$$\nabla \times \bar{\mathbf{G}}(\mathbf{r}, \mathbf{r}') = (\nabla' \times \bar{\mathbf{G}}(\mathbf{r}', \mathbf{r}))_t. \tag{10.58}$$

From (10.55), (10.56) and (10.58), it can be seen that interchanging primed and unprimed quantities is equivalent to transposing the dyadic. It is noted that in rectangular coordinates, the operators ∇ and ∇' represent

$$\nabla = \hat{\mathbf{x}}\frac{\partial}{\partial x} + \hat{\mathbf{y}}\frac{\partial}{\partial y} + \hat{\mathbf{z}}\frac{\partial}{\partial z}$$

$$\nabla' = \hat{\mathbf{x}}\frac{\partial}{\partial x'} + \hat{\mathbf{y}}\frac{\partial}{\partial y'} + \hat{\mathbf{z}}\frac{\partial}{\partial z'}.$$

The symmetry relations are sometimes used to obtain a slightly different-looking version of (10.53) involving $\bar{\mathbf{G}}(\mathbf{r}, \mathbf{r}')$ instead of $\bar{\mathbf{G}}(\mathbf{r}', \mathbf{r})$.

10.8   Interpretation of the Radiation Integrals

Let us reexamine (10.53) for the case when the boundary surface S recedes to infinity. The surface integrals vanish and only the first term remains. This gives us $\mathbf{E}$ due to an electric current $\mathbf{J}$. Applying duality, the $\mathbf{H}$ field at a point in V due to a magnetic current $\mathbf{M}$ is

$$\mathbf{H}(\mathbf{r}) = -j\omega\epsilon \int_V \mathbf{M}(\mathbf{r}') \cdot \bar{\mathbf{G}}(\mathbf{r}', \mathbf{r}) \, dV'. \tag{10.59}$$

By taking the curl, we obtain the electric field

$$\mathbf{E}(\mathbf{r}) = -\int_V \nabla \times \mathbf{M}(\mathbf{r}') \cdot \bar{\mathbf{G}}(\mathbf{r}', \mathbf{r}) \, dV'. \tag{10.60}$$

---

[2] See Tai (1993, Section 4.5) for the JH reciprocity theorem.

One would expect this expression to resemble the third term in (10.53) since it too is a field contribution due to a magnetic current. To establish this correspondence, we exploit the symmetry properties of $\bar{\mathbf{G}}$ and its curl to obtain

$$\nabla \times \{\mathbf{M}(\mathbf{r}') \cdot \bar{\mathbf{G}}(\mathbf{r}', \mathbf{r})\} = \nabla \times \{\bar{\mathbf{G}}_t(\mathbf{r}', \mathbf{r}) \cdot \mathbf{M}(\mathbf{r}')\}$$

$$= \nabla \times \bar{\mathbf{G}}(\mathbf{r}, \mathbf{r}') \cdot \mathbf{M}(\mathbf{r}') = \mathbf{M}(\mathbf{r}') \cdot (\nabla \times \bar{\mathbf{G}}(\mathbf{r}, \mathbf{r}'))_t$$

$$= \mathbf{M}(\mathbf{r}') \cdot \nabla' \times \bar{\mathbf{G}}(\mathbf{r}', \mathbf{r})$$

so that (10.60) becomes

$$\mathbf{E}(\mathbf{r}) = -\int_V \mathbf{M}(\mathbf{r}') \cdot \nabla' \times \bar{\mathbf{G}}(\mathbf{r}', \mathbf{r}) \, dV'. \tag{10.61}$$

Replacing $\mathbf{M}(\mathbf{r}')dV'$ with $\mathbf{M}_s(\mathbf{r}')dS'$ completes the proof that this is of the same form as the third term in (10.53).

10.9 Free Space Dyadic Green's Function

The vector potential $\mathbf{A}$ can be used to obtain the electric field from $\mathbf{J}$. That is,

$$\mathbf{E} = -j\omega\mathbf{A} - j(\omega/k^2)\nabla\nabla \cdot \mathbf{A}, \tag{10.62}$$

where

$$\mathbf{A}(\mathbf{r}) = \mu \int_V \mathbf{J}(\mathbf{r}') \cdot \bar{\mathbf{G}}(\mathbf{r}', \mathbf{r}) \, dV' \tag{10.63}$$

and in free space $\bar{\mathbf{G}}(\mathbf{r}', \mathbf{r})$ is given by

$$\bar{\mathbf{G}}(\mathbf{r}', \mathbf{r}) = \bar{\mathbf{I}}\frac{e^{-jkR}}{4\pi R}; \quad R = |\mathbf{r}' - \mathbf{r}|. \tag{10.64}$$

This Green's function is a solution to

$$(\nabla'^2 + k^2)\bar{\mathbf{G}}(\mathbf{r}', \mathbf{r}) = -\bar{\mathbf{I}}\,\delta(|\mathbf{r}' - \mathbf{r}|) \tag{10.65}$$

and satisfies the radiation condition.

Equation (10.64) is the dyadic Green's function for $\mathbf{A}$. That is, it enables us to find $\mathbf{A}$ from $\mathbf{J}$. We will now find another Green's function $\bar{\mathbf{G}}_e$, which gives $\mathbf{E}$ in terms of $\mathbf{J}$. Using (10.62) and (10.63), we obtain

$$\mathbf{E}(\mathbf{r}) = -j\omega\mu \int_V \mathbf{J}(\mathbf{r}') \, g(\mathbf{r}', \mathbf{r}) \, dV' - \frac{j\omega\mu}{k^2} \int_V \nabla\nabla \cdot \mathbf{J}(\mathbf{r}') \, g(\mathbf{r}', \mathbf{r}) \, dV', \tag{10.66}$$

where

$$g(\mathbf{r}', \mathbf{r}) = \frac{e^{-jkR}}{4\pi R}.$$

Since ∇ only operates on unprimed coordinates, $\nabla \cdot \mathbf{J}(\mathbf{r}')g = (\nabla g) \cdot \mathbf{J}(\mathbf{r}')$. Furthermore, $\nabla\nabla g$ is a symmetric dyad (see Problem 10.5) so that we may write

$$\nabla\nabla \cdot \mathbf{J}(\mathbf{r}')g(\mathbf{r}',\mathbf{r}) = \nabla\nabla g(\mathbf{r}',\mathbf{r}) \cdot \mathbf{J}(\mathbf{r}')$$
$$= \mathbf{J}(\mathbf{r}') \cdot (\nabla\nabla g(\mathbf{r}',\mathbf{r}))_t = \mathbf{J}(\mathbf{r}') \cdot \nabla\nabla g(\mathbf{r}',\mathbf{r})$$

and then (10.66) becomes

$$\mathbf{E}(\mathbf{r}) = -j\omega\mu \int_V \mathbf{J}(\mathbf{r}') \cdot \left(\bar{\mathbf{I}} + \frac{\nabla\nabla}{k^2} \right) g(\mathbf{r}',\mathbf{r})\, dV'. \tag{10.67}$$

In the above result, we can identify

$$\bar{\mathbf{G}}_e(\mathbf{r}',\mathbf{r}) = \left(\bar{\mathbf{I}} + \frac{\nabla\nabla}{k^2} \right) \frac{e^{-jkR}}{4\pi R}. \tag{10.68}$$

In free space, $\epsilon = \epsilon_0$, $\mu = \mu_0$, and $k = k_0 = \omega\sqrt{\mu_0\epsilon_0}$, so we can write (10.67) as

$$\mathbf{E}(\mathbf{r}) = -j\omega\mu_0 \int_V \mathbf{J}(\mathbf{r}') \cdot \bar{\mathbf{G}}_e(\mathbf{r}',\mathbf{r})\, dV'. \tag{10.69}$$

$\bar{\mathbf{G}}_e(\mathbf{r}',\mathbf{r})$ gives us the electric field due to an arbitrary current $\mathbf{J}$ in free space, and it is known as the *free space electric dyadic Green's function*.

10.10   Dyadic Green's Function Singularity

In principle, (10.69) can be used to find the electric field at $\mathbf{r}$, for an arbitrary current distribution $\mathbf{J}(\mathbf{r}')$ as shown in Figure 10.3. This configuration arises when using a volume polarization current $\mathbf{J} = j\omega(\epsilon - \epsilon_0)\mathbf{E}$ to represent a dielectric body with a permittivity ϵ. The volume current acts in free space and radiates the scattered field everywhere, both inside and outside the body.

The field point $\mathbf{r}$ in Figure 10.3 can be outside or inside the source region V as shown in cases (a) and (b). In case (b), a problem arises, as $R = |\mathbf{r} - \mathbf{r}'| \to 0$ occurs during the integration. The $\nabla\nabla g$ term in $\bar{\mathbf{G}}_e(\mathbf{r}',\mathbf{r})$ is singular to $O(1/R^3)$, whereas dV' is proportional to R^2, so the integral does not converge. This difficulty is caused by the interchange of integration and differentiation when using (10.62) and (10.63) and to obtain (10.69). The singularity must be avoided when carrying out the integration.

When $\mathbf{r} \approx \mathbf{r}'$, the singularity can be avoided by excluding a small spherical volume V_0 from the integration as shown in Figure 10.3(c). Equation (10.69) becomes

$$\mathbf{E}(\mathbf{r}) = -j\omega\mu_0 \int_{V-V_0} \mathbf{J}(\mathbf{r}') \cdot \bar{\mathbf{G}}_e(\mathbf{r}',\mathbf{r})\, dV' - j\omega\mu_0 \int_{V_0} \mathbf{J}(\mathbf{r}') \cdot \bar{\mathbf{G}}_e(\mathbf{r}',\mathbf{r})\, dV'. \tag{10.70}$$

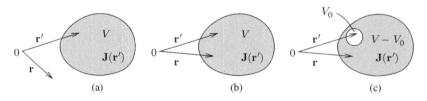

Figure 10.3   Field point $\mathbf{r}$ is (a) outside and (b) inside the source region V. (c) Exclusion of V_0 from integration.

The first integral over $V - V_0$ is nonsingular and can be evaluated by any convenient means, usually a numerical integration. The second integral gives $\mathbf{E}$ due to $\mathbf{J}$ within V_0. It will be shown that a uniform electric current $\mathbf{J}$ within a tiny spherical region V_0 in free space generates the following field at its centre:

$$\mathbf{E}(\mathbf{r}) = \frac{j}{3\omega\epsilon_0}\mathbf{J}(\mathbf{r}). \tag{10.71}$$

The contribution to $\mathbf{E}$ from volume V_0 depends on the fact that the excluded volume is spherical in shape; the result (10.71) was obtained by J. Van Bladel in 1961.

Replacing the second integral in (10.70) by (10.71) gives

$$\mathbf{E}(\mathbf{r}) = -j\omega\mu_0\int_{V-V_0}\mathbf{J}(\mathbf{r}')\cdot\bar{\mathbf{G}}_e(\mathbf{r}',\mathbf{r})\,dV' + \frac{j}{3\omega\epsilon_0}\mathbf{J}(\mathbf{r}). \tag{10.72}$$

Equation (10.71) does not depend on the radius of the sphere. However, the result is based on a quasistatic analysis, so V_0 should be small in terms of the dielectric wavelength. It is important to notice that as the volume V_0 shrinks to zero, its field contribution does *not* vanish.

It is possible to develop expressions similar to (10.71) for other exclusion shapes (Yaghjian, 1980). For example, one could exclude a small cube, which would be convenient if carrying out a numerical integration in rectangular coordinates. Another useful shape is a long and thin cylinder – a 'needle'. In this case, the contribution from V_0 vanishes, and a correction term is not necessary.

Another way to address the singularity problem is to use an alternative representation of $\bar{\mathbf{G}}_e$. There is an eigenfunction series expansion that can be used instead of (10.68). The expansion is in terms of spherical vector wave functions. The series is term by term integrable and does not suffer from the singularity problem (Yaghjian, 1980).

10.10.1 Derivation of Equation (10.71)

Without a loss of generality, we can assume that $\mathbf{J}$ is z directed. Then $\mathbf{J} = \hat{\mathbf{z}}J_0$ is constant within a small spherical volume V_0 of radius a that is centred at the origin. The electric field $\mathbf{E} = -j\omega\mathbf{A} - \nabla V$ has two parts. First, the 'J part' of E_z is found from

$$A_z = \frac{\mu_0}{4\pi}\int_{V_0}J_0\frac{e^{-jkR}}{R}\,dV'.$$

At the centre of the sphere, $r = 0$ and $r' = R$. Since $a \ll \lambda$,

$$E_z^J(0,0,0) = -j\omega\mu_0 J_0\frac{a^2}{2}.$$

To find the 'rho part', the charges have to be known. Since $\mathbf{J}$ is constant within the sphere, $\nabla\cdot\mathbf{J} = 0$ and there is no charge there. Because $\mathbf{J} \neq 0$ inside and $\mathbf{J} = 0$ outside the sphere, there will be ρ_s on the surface. The discontinuity of J_r is related to ρ_s. From the continuity equation, $J_r = j\omega\rho_s$ and using the conversion $J_r = J_z\cos\theta$, the surface charge density is

$$\rho_s = \frac{J_0\cos\theta}{j\omega}.$$

To find $\mathbf{E}$, the quasistatic approximation can be used; this means taking the static solution and multiplying it by $e^{j\omega t}$ (and the $e^{j\omega t}$ factor is assumed and suppressed). Therefore

$$\mathbf{E} = \int_{S_0}\frac{\rho_s\hat{\mathbf{R}}}{4\pi\epsilon_0 R^2}\,dS'$$

from which

$$E_z(0,0,0) = \int_{S_0} \frac{\rho_s \,\hat{\mathbf{z}} \cdot \hat{\mathbf{R}}}{4\pi\epsilon_0 a^2} \, dS'.$$

Using $\hat{\mathbf{z}} \cdot \hat{\mathbf{R}} = -\cos\theta'$ and carrying out the surface integral gives

$$E_z^\rho(0,0,0) = j\frac{J_0}{3\omega\epsilon_0}.$$

Combining the two contributions to E_z gives

$$E_z(0,0,0) = E_z^J + E_z^\rho = -j\omega\mu_0 J_0 \frac{a^2}{2} + j\frac{J_0}{3\omega\epsilon_0}.$$

When a is small, the first term can be ignored.

10.11 Dielectric Rod

The importance of properly treating the Green's function singularity will now be demonstrated by examining the MoM solution for the scattering by a slender dielectric rod. In this case, the situation of $|\mathbf{r} - \mathbf{r}'| \to 0$ occurs. The resulting singularity requires due consideration. The formulation follows Richmond (1965).

The rod's radius is $a \ll \lambda_1$ and it is oriented along $-L/2 \le z \le L/2$. Inside the dielectric, $k_1 = k_0\sqrt{\epsilon_r} = 2\pi/\lambda_1$, and in the air, $k_0 = 2\pi/\lambda_0$. The incident field is a plane wave

$$E_z^i = E_0 e^{jk_0 x}. \tag{10.73}$$

The field inside the rod can be represented by N Fourier modes

$$E_z(\rho, z) = \sum_{n=0}^{N-1} A_n \cos\frac{2n\pi z}{L} J_0(k_0\rho\sqrt{\epsilon_r - (n\lambda/L)^2}). \tag{10.74}$$

In the above, we have used $k_\rho^2 + k_z^2 = k_1^2$ where $k_z = 2n\pi/L$ and $k_\rho = k_0\sqrt{\epsilon_r - (n\lambda/L)^2}$.

The rod can be replaced by a volume equivalent current in free space

$$J_z(\rho, z) = j\omega\epsilon_0(\epsilon_r - 1)E_z(\rho, z). \tag{10.75}$$

The z component of the scattered field at any point on the z axis is

$$E_z^s(0, z) = \frac{-j}{\omega\epsilon_0} \int_{z'=-L/2}^{L/2} \int_{\phi'=0}^{2\pi} \int_{\rho'=0}^{a} J_z(\rho', z') \, F(\rho', z', z)\rho' d\rho' d\phi' dz', \tag{10.76}$$

where

$$F(\rho', z', z) = \frac{e^{-jk_0 R}}{4\pi R^5} \left[(1 + jk_0 R)(2R^2 - 3\rho'^2) + (k_0\rho' R)^2\right] \tag{10.77}$$

and $R = \sqrt{\rho'^2 + (z - z')^2}$. Equation (10.76) can be rewritten as

$$E_z^s(0, z) = 2\pi(\epsilon_r - 1) \int_{z'=-L/2}^{L/2} \int_{\rho'=0}^{a} E_z(\rho', z') \, F(\rho', z', z)\rho' d\rho' dz'. \tag{10.78}$$

Employing point matching on the z axis at $z = z_m$ requires that

$$E_z^i(0, z_m) = E_z(0, z_m) - E_z^s(0, z_m) \tag{10.79}$$

or

$$E_0 = \sum_{n=0}^{N-1} A_n \cos \frac{2n\pi z_m}{L} - 2\pi(\epsilon_r - 1) \int_{z'=-L/2}^{L/2} \int_{\rho'=0}^{a}$$

$$\sum_{n=0}^{N-1} A_n \cos \frac{2n\pi z'}{L} J_0(k_0\rho'\sqrt{\epsilon_r - (n\lambda/L)^2}) \, F(\rho', z', z_m) \, \rho' d\rho' dz'. \tag{10.80}$$

One finds that modes of the form $\cos(n\pi z'/L)$ with odd n do not contribute and can be omitted. This is because they are associated with E_ρ, which should be zero at the rod ends; see Problem 10.12.

For a short rod and broadside incidence, the $n = 0$ term is sufficient (whereas oblique incidence would require more terms). Using one match point at $z = 0$, (10.80) becomes

$$E_0 = A_0 - A_0 \, 2\pi(\epsilon_r - 1) \int_{z'=-L/2}^{L/2} \int_{\rho'=0}^{a} J_0(k_1\rho') \, F(\rho', z', 0)\rho' d\rho' dz'. \tag{10.81}$$

Performing a double numerical integration gives A_0 and the field inside the rod.

10.11.1   Numerical Considerations

The integrand has terms on the order of $1/R$, $1/R^2$ and $1/R^3$, so the integration involves a singularity at $\rho' = 0$ that must be avoided. This is accomplished by replacing the lower integration limit with $\rho' = \delta$, where $\delta \ll a$ is a very small number. The approximation is justified by noting that excluding $\rho' < \delta$ is equivalent to replacing the solid rod by a rod with its centre drilled out – a 'needle-shaped' detour. Since $\mathbf{E}_{tan}$ is continuous, E_z on the z axis for the solid and drilled rods are about the same; the E_z around the centre of the air cavity is hardly affected. More rigorously, it can be shown that the needle detour has a contribution of $O(\delta^2)$. In theory the error can be made to vanish. In practice, the error can be made arbitrarily small, but numerical integration becomes difficult for very small values of δ. The needle-shaped detour was used by Richmond (1965).

Another way to avoid the singularity is to use a small spherical detour, as described in Section 10.10. Unlike the needle detour, the spherical contribution is stronger and it does not vanish, so it must be accounted for. For a uniform current $\mathbf{J}$ within a small spherical region that is centred at the origin, its field at the centre of the sphere is

$$\mathbf{E}(0) = \frac{j}{3\omega\epsilon_0}\mathbf{J}(0). \tag{10.82}$$

To calculate $E_z^s(0)$ with a spherical detour, a numerical integration is carried out for points $R > \delta$. The remaining contribution due to the volume current within $R \leq \delta$ is found by using (10.75) in (10.82). Adding the two contributions, (10.81) becomes

$$E_0 = A_0 - A_0 \, 2\pi(\epsilon_r - 1) \int_{V-V_0} J_0(k_1\rho') \, F(\rho', z', 0)\rho' d\rho' dz' + \frac{1}{3}(\epsilon_r - 1)E_0, \tag{10.83}$$

where $V - V_0$ is the volume of the rod minus the spherical detour. Since E_z is constant within V_0, the exact size of the detour does not matter, as long as it is much smaller than the wavelength in the dielectric. A detour with $\delta = a$ is a satisfactory choice for the thin rod.

10.12   Further Reading

Scalar and dyadic Green's functions are thoroughly covered in Collin (1991, Chapter 2). The book by Tai (1993) is entirely devoted to these topics. Other helpful books include Van Bladel (2007, Chapter 7) and Rothwell and Cloud (2008, Chapter 5).

References

Collin RE (1991) *Field Theory of Guided Waves*. Oxford University Press.
Richmond JH (1965) Digital computer solutions of the rigorous equations for scattering problems. *Proc. IEEE* **53**, 796–804.
Rothwell EJ and Cloud MJ (2008) *Electromagnetics*. CRC Press.
Tai CT (1993) *Dyadic Green's Functions in Electromagnetic Theory*. IEEE Press.
Van Bladel J (2007) *Electromagnetic Fields*. John Wiley & Sons, Inc.
Yaghjian AD (1980) Electric dyadic Green's functions in source regions. *Proc. IEEE* **68**, 248–263.

Problems

10.1   Derive the expression (10.32) for the 2D TM fields. First, show that on ℓ,

$$\frac{\partial E_z}{\partial n} = -j\omega\mu J_{\ell z}.$$

Do this by taking $\hat{\mathbf{n}} \times$ of Faraday's law, that is, $\hat{\mathbf{n}} \times (\nabla \times \mathbf{E}) = -j\omega\mu(\hat{\mathbf{n}} \times \mathbf{H})$. We can identify the right-hand side as the boundary current. The left-hand side can be expanded and simplified by using the vector identity $\mathbf{A} \times (\mathbf{B} \times \mathbf{C}) = (\mathbf{A} \cdot \mathbf{C})\mathbf{B} - (\mathbf{A} \cdot \mathbf{B})\mathbf{C}$ with ∇ treated as a vector. It is useful to note that

$$\hat{\mathbf{n}} \cdot \nabla f = \frac{\partial f}{\partial n}$$

which suggests the following operator:

$$(\hat{\mathbf{n}} \cdot \nabla) = \frac{\partial}{\partial n}.$$

Second, show that on ℓ,

$$E_z = -M_{\ell t}\frac{\partial G}{\partial n}.$$

We need to find the electric field due to a boundary magnetic dipole $\hat{\mathbf{t}}M_{\ell t}$. To obtain this, use the known field of an electric line source. Suppose that $\boldsymbol{\rho}_a$ is a point within S and $\boldsymbol{\rho}_b$ is a point on the boundary curve ℓ. It is known that an electric line current I_{z1} at $\boldsymbol{\rho}_a$ will produce a field $E_z = -j\omega\mu I_{z1}G$ at $\boldsymbol{\rho}_b$. Use this result, together with the reciprocity theorem, to obtain

$$E_z = -\nabla \times (\hat{\mathbf{z}}G) \cdot \hat{\mathbf{t}}M_{\ell t}.$$

From $\mathbf{A} \cdot (\mathbf{B} \times \mathbf{C}) = \mathbf{B} \cdot (\mathbf{C} \times \mathbf{A})$, along with $\nabla \cdot (\phi\mathbf{A}) = \mathbf{A} \cdot \nabla\phi + \phi\nabla \cdot \mathbf{A}$ and the fact that $\nabla M_{\ell t} \perp \hat{\mathbf{n}}$ to simplify the expression, obtain the final result for E_z.

10.2   Analogous to Equation (10.39), use duality to find an expression for H_z in the TE case.

10.3 In the following, answer true or false:
  (a) $\mathbf{A} \cdot (\mathbf{B} \times \bar{\mathbf{C}}) = (\mathbf{A} \times \mathbf{B}) \cdot \bar{\mathbf{C}}$
  (b) $\mathbf{A} \cdot (\mathbf{B} \times \bar{\mathbf{C}}) = \bar{\mathbf{C}}_t \cdot (\mathbf{A} \times \mathbf{B})$
  (c) $\mathbf{A} \cdot (\mathbf{B} \times \bar{\mathbf{C}}) = \bar{\mathbf{C}} \cdot (\mathbf{A} \times \mathbf{B})$.

10.4 Using symmetry properties, obtain a slightly different-looking version of Equation (10.53) that involves $\bar{\mathbf{G}}(\mathbf{r}, \mathbf{r}')$ and its curl.

10.5 Verify that $\nabla \nabla g(\mathbf{r}', \mathbf{r})$ in Equation (10.67) is a symmetric dyad by expanding it in rectangular coordinates.

10.6 Analogous to Equation (10.69), find a magnetic dyadic Green's function $\bar{\mathbf{G}}_m(\mathbf{r}', \mathbf{r})$ that gives $\mathbf{H}$ due to an electric current $\mathbf{J}$.

10.7 A slender dielectric rod with radius $a \ll \lambda_1$ is oriented along the z axis, on $-L/2 \leq z \leq L/2$. The relative permittivity is ϵ_r and λ_1 is the wavelength in the dielectric. The incident field is $\mathbf{E}^i = \hat{z} E_0 e^{-jk_0 x}$. We wish to find the backscatter RCS.
  Use a double numerical integration (see Appendix G) to evaluate (10.81) and obtain the constant A_0. The singularity at $R = 0$ is avoided by changing the limits of integration to $\delta \leq \rho' \leq a$ where $\delta \ll a$. The solution will correspond to that for a dielectric rod, with a small hole drilled along its axis.
  To improve the numerical integration, dz' should be replaced by $d\theta'$ where $z' = \rho' \tan \theta'$. Assume that $f = 9.53\,\text{GHz}, a = 0.0625$ inches, $\epsilon_r = 2.54$ and $L = 0.5, 0.6, 0.7, \ldots, 1.0$ inches. Tabulate A_0 for these cases.

10.8 Find the scattered field E_z^s from the dielectric rod's polarization current. Use the far-zone approximation for A_z

$$A_z = \frac{\mu_0 e^{-jk_0 r}}{4\pi r} \int_{-L/2}^{L/2} \int_0^{2\pi} \int_0^a e^{jk_0 \hat{\mathbf{r}} \cdot \mathbf{r}'} J_z(\mathbf{r}') \, \rho' d\rho' d\phi' dz'.$$

For the ϕ' integration, use the identity

$$J_0(x) = \frac{1}{2\pi} \int_0^{2\pi} e^{jx \cos \phi} d\phi.$$

You should end up with a result in terms of

$$\int_0^a J_0(k_0 \rho) J_0(k_1 \rho) \rho d\rho = \frac{\rho}{k_0^2 - k_1^2} [k_0 J_0(k_1 \rho) J_1(k_0 \rho) - k_1 J_0(k_0 \rho) J_1(k_1 \rho)]_{\rho=0}^a.$$

10.9 Compute the backscatter RCS of the dielectric rod, for $L = 0.5, 0.6, 0.7, \ldots, 1.0$ inches. Use your numerical data for A_0 the expression for E_z^s from the previous two problems. Compare your results with Richmond (1965, Figure 5).

10.10 For the dielectric rod described in Problem 10.7, use a double numerical integration to evaluate (10.80) with a needle-shaped detour and N unknowns. Using $N = 20$, calculate the constants A_n and find the RCS. Compare your results with the $N = 1$ case in Problem 10.9.

10.11 For the dielectric rod described in Problem 10.7, use a double numerical integration to evaluate (10.83) and obtain the constant A_0. In this case, a spherical detour is used. Use this to find the RCS. Assume that $L = 0.8$ inches with all other parameters unchanged.

10.12 The cosine modes used to expand E_z in a thin dielectric rod can be obtained from a vector potential

$$A_z = J_0(k_t \rho) \cos(2n\pi z/L)$$

where $k_t^2 + (2n\pi/L)^2 = k_1^2$. Show that these modes have $E_\phi = 0$ and that having $\partial E_z/\partial z = 0$ at $z = \pm L/2$ implies that $E_\rho = 0$ at the ends.

11

Green's Functions Construction I

When a forcing function or *source term* is present, the wave equation is inhomogeneous, and the forced response is called the Green's function. In this chapter a method for finding 1D Green's functions will be described. The 1D solutions will then used to construct 2D and 3D solutions via a generalized separation of variables (GSV). The resulting 2D and 3D Green's functions are in terms of contour integrals in the complex plane.

The contour integrals possess singularities. In any kind of open boundary problem, the integrand will contain a complex-valued square root function, which has a branch cut; with a closed boundary, poles are present. Evaluation of integrals containing these types of singularities will be illustrated by finding the Green's functions for a waveguide and for a wedge.

11.1 Sturm–Liouville Problem

To begin, the *Sturm–Liouville Problem* (SLP) is a class of differential equations with boundary conditions, having a solution in terms of orthogonal eigenfunctions. The 1D wave equation is a special case of an SLP. The inhomogeneous SLP is of the form

$$[L + \lambda w(x)]\phi(x) = f(x), \tag{11.1}$$

where L is the *self-adjoint operator*

$$L(\cdot) = \frac{d}{dx}\left[p(x)\frac{d}{dx}\right](\cdot) - q(x)(\cdot). \tag{11.2}$$

$p(x)$ and $q(x)$ are assumed to be real and piecewise continuous on the interval $a \leq x \leq b$. As this is a boundary value problem, the boundary conditions are imposed at $x = a$ and $x = b$. The function $w(x)$ is known as the *weight function* and λ is a parameter. In general, it is possible to put any second-order differential operator

$$D = P(x)\frac{d^2}{dx^2} + Q(x)\frac{d}{dx} + R(x)$$

in the form (11.2).

Applied Frequency-Domain Electromagnetics, First Edition. Robert Paknys.
© 2016 John Wiley & Sons, Ltd. Published 2016 by John Wiley & Sons, Ltd.
Companion Website: www.wiley.com/go/paknys9981

11.2   Green's Second Identity

Let us suppose that we have functions $u(x)$ and $v(x)$ that are twice differentiable. Then,

$$vLu = v\left[\frac{d}{dx}\left[p(x)\frac{du}{dx}\right] - q(x)u(x)\right] = v(pu')' - quv.$$

But $(vpu')' = v(pu')' + v'pu'$ so that $vLu = (vpu')' - v'pu' - quv$. Similarly, we can show that $uLv = (upv')' - u'pv' - qvu$. By subtracting these two expressions and integrating from a to b, we obtain

$$\int_a^b vLu - uLv \; dx = p(u'v - uv')|_{x=a}^b. \qquad (11.3)$$

This useful expression is Green's second identity. It establishes a relation between an integration involving $u(x)$ and $v(x)$ on the interval $a \le x \le b$, and their values at the boundary points a, b.

11.3   Hermitian Property

The operator L is said to be Hermitian if

$$\int_a^b uLv \; dx = \int_a^b vLu \; dx.$$

By considering Green's second identity, we can see that this happens if

$$p(u'v - uv')|_{x=a}^b = 0.$$

This can be expanded out, using determinant notation

$$-p(b)\begin{vmatrix} u(b) & v(b) \\ u'(b) & v'(b) \end{vmatrix} + p(a)\begin{vmatrix} u(a) & v(a) \\ u'(a) & v'(a) \end{vmatrix} = 0.$$

Since the boundary conditions at $x = a$ are assumed to be independent of those at $x = b$, each term must be zero:

$$p(b)\begin{vmatrix} u(b) & v(b) \\ u'(b) & v'(b) \end{vmatrix} = 0 \quad \text{and} \quad p(a)\begin{vmatrix} u(a) & v(a) \\ u'(a) & v'(a) \end{vmatrix} = 0. \qquad (11.4)$$

There are six ways to satisfy the conditions (11.4) at the boundaries $x = a$, $x = b$:

1. $p(a) = 0$, $p(b) = 0$ with u, v bounded at $x = a$, $x = b$.
2. Periodic boundary conditions, where u, v, u', v' and p repeat their values and have the same values at $x = a$, $x = b$.
3. Impedance boundary conditions: where $p(a) \ne 0$ and $u(a)v'(a) - u'(a)v(a) = 0$. This can be expressed as $u(a)v'(a) = u'(a)v(a)$ or

$$\frac{u(a)}{u'(a)} = \frac{v(a)}{v'(a)} = \text{const.} = \gamma_a \qquad (11.5)$$

and a similar expression at the other endpoint $x = b$. In physical problems, γ_a is related to an impedance or admittance at the boundary.

4. Dirichlet boundary conditions: $u(a) = 0$, $v(a) = 0$ and $u(b) = 0$, $v(b) = 0$. This is a special case of the impedance boundary condition when $\gamma_a = 0$.
5. Neumann boundary conditions: $u'(a) = 0$, $v'(a) = 0$ and $u'(b) = 0$, $v'(b) = 0$. This is a special case of the impedance boundary condition when $\gamma_a \to \infty$.
6. Sommerfeld radiation condition: Equation (4.73) states that waves of the form $e^{\pm jkx}$ are admissible. Suppose that $u(x) = c_1 e^{-jkx}$ and $v(x) = c_2 e^{-jkx}$ where c_1, c_2 are constants. Then $u(a)v'(a) - u'(a)v(a) = 0$ and $u(b)v'(b) - u'(b)v(b) = 0$. It also holds true for an e^{jkx} wave.

In passing, we note that it is possible to satisfy (11.4) with combinations of the cases above, for example, Dirichlet conditions at one end and Neumann conditions at the other end. In other words, one could have $u(a) = 0$ and $u'(b) = 0$; $v(a) = 0$ and $v'(b) = 0$.

11.4 Particular Solution

We seek a particular solution to (11.1) where $w(x)$ and $f(x)$ are assumed to be piecewise continuous and λ is an arbitrary parameter.

Linear differential equations with constant coefficients can be solved with the method of undetermined coefficients. The particular solution of (11.1) requires a more involved technique, which will now be shown.

Let us define a 1D Green's function G which satisfies

$$[L' + \lambda w(x')]G(x', x) = -\delta(x' - x), \tag{11.6}$$

where L' operates on x'. For now, we can avoid looking at the boundary conditions. Multiplying (11.1) by $G(x', x)$ and (11.6) by $\phi(x')$ we obtain

$$G(x', x)L'\phi(x') + \lambda w(x')\phi'(x')G(x', x) = f(x')G(x', x)$$

$$\phi(x')L'G(x', x) + \lambda w(x')\phi'(x')G(x', x) = -\phi(x')\delta(x' - x).$$

By subtracting them, integrating from a to b, and using Green's second identity, we obtain

$$\phi(x) = -\int_a^b f(x')G(x', x)\,dx' + p(x')\left[G(x', x)\frac{d\phi(x')}{dx'} - \phi(x')\frac{dG(x', x)}{dx'}\right]_{x'=a}^b. \tag{11.7}$$

This represents a formal solution to (11.1), assuming that $G(x', x)$ is known. The function $\phi(x)$ is hence in terms of the forcing function $f(x)$ and the values of $\phi(x')$ and its derivatives at the boundaries. If we arrange that $\phi(x')$ and $G(x', x)$ meet the *same* boundary conditions, then the last term in (11.7) vanishes, leaving simply

$$\phi(x) = -\int_a^b f(x')G(x', x)\,dx'. \tag{11.8}$$

11.5 Properties of the Green's Function

From (11.6) we see that $L'G$ or G must have a singularity at $x = x'$. In this development, let us assume that G is continuous at $x = x'$. By integrating over a small interval about x', (11.6) becomes

$$\lim_{\epsilon \to 0}\left[\int_{x'-\epsilon}^{x'+\epsilon}\frac{d}{dx}\left(p\frac{dG}{dx}\right)dx - \int_{x'-\epsilon}^{x'+\epsilon}(q - \lambda w)G\,dx\right] = -1.$$

The second integral vanishes in the limit, with the result that

$$p \left. \frac{dG}{dx} \right|_{x=x'-0}^{x'+0} = -1$$

hence $p\, dG/dx$ has a discontinuity of -1 at $x = x'$. If we would have assumed that G is discontinuous at $x = x'$, we would find that $p\, dG/dx$ does not have the proper singularity to satisfy the delta function. Hence, our original assumption that G is continuous at $x = x'$ is correct. From similar reasoning we find that when $x \neq x'$, G and $p\, dG/dx$ are continuous.

We next examine the symmetry property of G. Let x, x' and x'' be three points in the interval (a, b). The points x and x' are fixed; they can be interpreted as source points. x'' is a variable and can be interpreted as a field point. The differential equations

$$[L'' + \lambda w(x'')]G(x'', x) = -\delta(x'' - x)$$
$$[L'' + \lambda w(x'')]G(x'', x') = -\delta(x'' - x')$$

are satisfied by the Green's functions $G(x'', x)$ and $G(x'', x')$ so that

$$\int_a^b G(x'', x')L''G(x'', x) - G(x'', x)L''G(x'', x')\, dx''$$

$$= -\int_a^b G(x'', x')\delta(x'' - x)\, dx'' + \int_a^b G(x'', x)\delta(x'' - x)\, dx'' = 0 \qquad (11.9)$$

This result equals zero because of the Hermitian property of L with respect to the boundary conditions (11.4), which both $G(x'', x')$ and $G(x'', x)$ must satisfy. Performing the integration gives us

$$G(x, x') = G(x', x)$$

which is the symmetry property of the Green's function. The physical interpretation is that the field at x due to a source at x' is the same as the field at x' due to a source at x. Hence, reciprocity is satisfied.

11.6   UT Method

The *UT method* for constructing a 1D Green's function is now described. To begin, the Green's function should have the following properties:

1. For $x \neq x'$, G satisfies
$$[L + \lambda w(x)]\, G(x, x') = 0.$$

2. For $x = x'$, $G(x, x')$ is continuous and $dG(x, x')/dx$ is discontinuous by $-1/p(x')$.
3. For all points in the interval (a, b), $G(x, x') = G(x', x)$.

These properties will now be used to construct $G(x, x')$. The Green's function is defined as the solution to (11.6), subject to the boundary conditions

$$\gamma_1 G(x, x') + \gamma_2 \left. \frac{dG(x, x')}{dx} = 0 \right|_{x=a, b} \qquad (11.10)$$

or the conditions (11.4). It is proposed to construct the solution for G in two intervals: $a \leq x \leq x'$ and $x' \leq x \leq b$. Continuity conditions at $x = x'$ are used to find the constant C. We let

$$G(x, x') = \frac{U(x)T(x')}{C} \quad \text{if } x \leq x'$$

$$= \frac{U(x')T(x)}{C} \quad \text{if } x \geq x', \tag{11.11}$$

where C is a constant and U, T are two independent solutions of

$$[L + \lambda w(x)] \begin{Bmatrix} U(x) \\ T(x) \end{Bmatrix} = 0.$$

The function $U(x)$ satisfies the boundary condition at $x = a$, and $T(x)$ satisfies the boundary condition at $x = b$, so that

$$\gamma_1 U(a) + \gamma_2 \left. \frac{dU}{dx} \right|_{x=a} = 0$$

$$\gamma_1 T(b) + \gamma_2 \left. \frac{dT}{dx} \right|_{x=b} = 0. \tag{11.12}$$

The form of (11.10) ensures that the three required properties for G are satisfied, and the boundary conditions (11.11) ensure that (11.9) is satisfied. The constant C can be found from the jump property,

$$\frac{U(x')T'(x') - U'(x')T(x')}{C} = \frac{-1}{p(x')}$$

which leads to $C = W(T, U)$ where $W(T, U)$ is the *conjunct* of T, U and is defined by

$$W(T, U) = p(x')[T(x')U'(x') - T'(x')U(x')]$$

$$= p(x')w(T, U)$$

and $w(T, U)$ is the Wronskian of T, U. It is useful to note that the conjunct is independent of x'. This can be shown by noting that $dW/dx' = 0$. Moreover, $W(T, U) \neq 0$ because T, U are assumed to be two independent solutions. Hence, (11.10) becomes

$$G(x, x') = \frac{U(x)T(x')}{W(T, U)} \quad \text{if } x \leq x'$$

$$= \frac{U(x')T(x)}{W(T, U)} \quad \text{if } x \geq x'.$$

This can be compactly expressed as

$$G(x, x') = \frac{U(x_<)T(x_>)}{W(T, U)}, \tag{11.13}$$

where $x_<$ is the smaller of x and x', and $x_>$ is the larger of x and x'.

Example 11.1 (Neumann Boundary) Use the UT method to solve

$$\left(\frac{d^2}{dx^2} + k^2\right) G(x, x') = -\delta(x - x'); \quad 0 \leq x \leq b$$

subject to the Neumann boundary conditions

$$\left.\frac{dG}{dx}\right|_{x=0} = 0; \quad \left.\frac{dG}{dx}\right|_{x=b} = 0.$$

Solution: This is an SLP with $p(x) = 1$, $q(x) = 0$, $w(x) = 1$ and $\lambda = k^2$. $U(x)$ and $T(x)$ satisfy

$$\left(\frac{d^2}{dx^2} + k^2\right) \left\{\begin{matrix} U(x) \\ T(x) \end{matrix}\right\} = 0.$$

Since $U'(0) = 0$ then $U(x) = \cos kx$. Since $T'(b) = 0$ then $T(x) = \cos k(b - x)$. The conjunct is $W(T, U) = -k \cos k(b - x') \sin kx' = -k \cos kx' \sin k(b - x') = -k \sin kb$ from which

$$G(x, x') = \frac{-\cos kx_< \cos k(b - x_>)}{k \sin kb}. \qquad\blacksquare$$

Example 11.2 (Radiation Condition) Use the UT method to solve

$$\left(\frac{d^2}{dx^2} + k^2\right) G(x, x') = -\delta(x - x'); \quad -\infty < x < \infty$$

subject to the Sommerfeld radiation condition

$$\frac{dG}{dx} - jkG = 0; \quad x \to -\infty$$

$$\frac{dG}{dx} + jkG = 0; \quad x \to \infty$$

Solution: These are the 1D radiation conditions for $e^{j\omega t}$ time dependence. G represents a wave travelling in the $+x$ direction as $x \to \infty$ and in the $-x$ direction as $x \to -\infty$.

This is an SLP with $p(x) = 1$, $q(x) = 0$, $w(x) = 1$ and $\lambda = k^2$. $U(x)$ and $T(x)$ satisfy

$$\left(\frac{d^2}{dx^2} + k^2\right) \left\{\begin{matrix} U(x) \\ T(x) \end{matrix}\right\} = 0.$$

$U(x) = e^{jkx}$ and $T(x) = e^{-jkx}$ satisfy the radiation condition at $x = \mp\infty$ respectively. The conjunct is $W(T, U) = jk(e^{-jkx'}e^{jkx'} + e^{-jkx'}e^{jkx'}) = j2k$ from which

$$G(x, x') = \frac{e^{-jk|x-x'|}}{j2k}. \qquad\blacksquare$$

Example 11.3 (Legendre Equation) Use the UT method to solve Legendre's equation

$$\left[\frac{1}{\sin\theta}\left(\frac{d}{d\theta}\sin\theta\frac{d}{d\theta}\right) + \nu(\nu + 1)\right] G(\theta, \theta') = -\frac{\delta(\theta - \theta')}{\sin\theta'}; \quad 0 \leq \theta \leq \theta_0$$

for $\nu \neq$ integer. The boundary conditions are

$$G(\theta, \theta') \text{ is finite at } \theta = 0$$

and

$$\left. \frac{dG}{d\theta} \right|_{\theta=\theta_0} = 0.$$

Solution: This is an SLP with $p(\theta) = \sin\theta$, $q(\theta) = 0$, $w(\theta) = \sin\theta$ and $\lambda = \nu(\nu+1)$. $U(\theta)$ and $T(\theta)$ satisfy

$$\left[\left(\frac{d}{d\theta} \sin\theta \frac{d}{d\theta} \right) + \nu(\nu+1)\sin\theta \right] \left\{ \begin{matrix} U(\theta) \\ T(\theta) \end{matrix} \right\} = 0.$$

The solutions are $P_\nu(\cos\theta)$ and $P_\nu(-\cos\theta)$. $U(\theta) = P_\nu(\cos\theta)$ is finite at $\theta = 0$. At the other boundary, $T(\theta) = C_1 P_\nu(\cos\theta) + C_2 P_\nu(-\cos\theta)$. Enforcing

$$T'(\theta) = C_1 P_\nu'(\cos\theta)(-\sin\theta) + C_2 P_\nu'(-\cos\theta)(\sin\theta) = 0$$

at $\theta = \theta_0$ (the prime denotes differentiation with respect to $\cos\theta$) gives $C_1 = P_\nu'(-\cos\theta_0)$, $C_2 = -P_\nu'(\cos\theta_0)$ and a solution

$$T(\theta) = P_\nu'(-\cos\theta_0)P_\nu(\cos\theta) - P_\nu'(\cos\theta_0)P_\nu(-\cos\theta).$$

It can be shown that the conjunct is $W(T, U) = (2/\pi)\sin\nu\pi P_\nu'(\cos\theta_0)$ and so

$$G(\theta, \theta') = \frac{P_\nu(\cos\theta_<)[P_\nu'(-\cos\theta_0)P_\nu(\cos\theta_>) - P_\nu'(\cos\theta_0)P_\nu(-\cos\theta_>)]}{(2/\pi)\sin\nu\pi P_\nu'(\cos\theta_0)}. \qquad \blacksquare$$

11.6.1   *Independent Solutions of the SLP*

The SLP

$$[L + \lambda w(x)]\, y(x) = 0$$

has two linearly independent solutions $y_1(x)$ and $y_2(x)$. If $y_1(x)$ is known, then the other solution $y_2(x)$ can be found.

From the properties of Green's functions, we know that

$$y_2'(x)y_1(x) - y_2(x)y_1'(x) = C/p(x),$$

where C is a constant. For a Green's function we saw that C is equal to the conjunct. Since we are dealing with a homogeneous differential equation, C can be any constant. Dividing by $y_1^2(x)$ we obtain

$$\frac{y_2'(x)y_1(x) - y_2(x)y_1'(x)}{y_1^2(x)} = \frac{C}{p(x)y_1^2(x)}$$

which can be rewritten as

$$\left(\frac{y_2(x)}{y_1(x)} \right)' = \frac{C}{p(x)y_1^2(x)}.$$

Integration then gives us the second solution in terms of the first one

$$y_2(x) = C y_1(x) \int \frac{dx}{p(x)y_1^2(x)}.$$

11.7   Discrete and Continuous Spectra

The Green's function satisfies the differential equation

$$(L + \lambda w(x))G(x, x') = -\delta(x - x'); \quad a \le x \le b \tag{11.14}$$

and the boundary conditions

$$\gamma_1 G(x, x') + \gamma_2 \frac{dG(x, x')}{dx} = 0|_{x=a,b}. \tag{11.15}$$

When $x \ne x'$, G is a solution to a homogeneous differential equation that meets the given boundary conditions. Consider the related SLP

$$[L + \lambda_n w(x)]\phi_n(x) = 0; \quad a \le x \le b \tag{11.16}$$

with boundary conditions at a, b. It is known that the eigenvalues λ_n are real and that the eigenfunctions $\phi_n(x)$ are orthogonal. These functions can be used to represent an arbitrary function $f(x)$ on $a \le x \le b$ as a Fourier series:

$$f(x) = \sum_n c_n \phi_n(x). \tag{11.17}$$

Using an inner product that is defined by

$$\langle u, v \rangle = \int_a^b u(x)v^*(x)w(x)dx \tag{11.18}$$

we can use orthogonality to obtain the c_n. If we also assume that the eigenfunctions are *orthonormal*, then

$$\langle \phi_n, \phi_m \rangle = \delta_{mn} \tag{11.19}$$

and

$$c_n = \int_a^b f(x')\phi_n^*(x')w(x')dx'. \tag{11.20}$$

Here, δ_{mn} is the 'Kronecker delta', that is, $\delta_{mn} = 0$ when $m \ne n$ and $\delta_{mn} = 1$ when $m = n$.

Let us assume that $G(x, x')$ can be represented as a Fourier series:

$$G(x, x') = \sum_k g_k \phi_k(x). \tag{11.21}$$

We need to find the coefficients g_k. This is done by using the three equations:

$$LG\ (x, x') = -\delta(x - x') - \lambda w(x)G(x, x') \tag{11.22a}$$

$$L\phi_n^*(x) = -\lambda_n w(x)\phi_n^*(x) \tag{11.22b}$$

$$\int_a^b \phi_n(x)\phi_m^*(x)w(x)dx = \delta_{mn}. \tag{11.22c}$$

From this we can obtain

$$\int_a^b \phi_n^*(x)LG(x, x') - G(x, x')L\phi_n^*(x)dx = -\lambda g_n + \lambda_n g_n - \phi_n^*(x') = 0. \tag{11.23}$$

The integral is zero because we have assumed that $\phi_n^*(x)$ and $G(x, x')$ satisfy the same boundary conditions, and we have used the Hermitian property of L. It is also noted that the λ_n for the SLP are real and that $w(x)$ is real. From (11.23) we see that

$$g_n = \frac{\phi_n^*(x')}{\lambda_n - \lambda}$$

from which we obtain

$$G(x, x') = \sum_n \frac{\phi_n^*(x')\phi_n(x)}{\lambda_n - \lambda}. \tag{11.24}$$

This is an alternative to the UT form of G discussed earlier.

11.7.1 Complete Set of Eigenfunctions

Completeness is an important concept, because it indicates whether or not an eigenfunction expansion of a function is possible. As a simple example, suppose we have a set of functions $\{1, \sin(n\pi x/L), \cos(n\pi x/L)\}$. It is well known that these can be used in a trigonometric Fourier series to represent any square-integrable function on $(-L, L)$. So, they form what is called a *complete set*. On the other hand, the set of functions $\{\sin(n\pi x/L)\}$ on $(-L, L)$ cannot represent an even function. Therefore, they are an incomplete set.

For simple problems having trigonometric functions as the eigenfunctions, the properties of completeness and orthogonality are so well understood that further discussion is unnecessary. On the other hand, when less familiar higher transcendental functions[1] are involved, a systematic method for determining completeness is useful. Completeness is in fact an advanced mathematics topic that is well beyond the scope of this book. For our purposes here, it suffices to know that the eigenfunctions of the SLP form a complete set (Arfken and Weber, 2005, Chapter 10).

A systematic way to obtain a complete set of eigenfunctions $\{\phi_n(x)\}$ will now be described. From the residue theorem (see Appendix E) we have that

$$\oint_{ccw} \frac{f(z)}{z - z_0} \, dz = j2\pi f(z_0),$$

where ccw is a counterclockwise closed contour of any shape enclosing z_0 in the complex z plane. Since (11.24) has simple poles at $\lambda = \lambda_n$, we have the relation

$$\frac{1}{j2\pi} \oint_{ccw} G(x, x'; \lambda) \, d\lambda = -\sum_n \phi_n^*(x')\phi_n(x). \tag{11.25}$$

Suppose that we found G by some other method such as the UT method. From (11.25) we would then be able to find $\sum \phi_n^*(x')\phi_n(x)$ from which we could then identify the complete set of eigenfunctions $\{\phi_n(x)\}$. Equation (11.25) is called the *completeness relation*.

In general a Green's function may have not only pole singularities, but branch cuts as well. In the previous example only simple poles are present, though (11.25) can be readily generalized to include other types of singularities. The analytic properties of poles and branch cuts are discussed in Appendix E.

---

[1] Transcendental functions are those having an infinite series, for example, $\sin x$. 'Higher transcendental functions' refers to other more exotic special functions, for example, Bessel functions (Erdelyi et al. 1953).

A formal procedure for finding the complete set of eigenfunctions is to (i) construct G by using the UT method, (ii) examine its singularities in the form of residues, branch points and branch cuts and (iii) sum the residues and branch cut contributions to obtain

$$\sum_n \phi_n^*(x')\phi_n(x) + \int \phi^*(x';\beta)\phi(x;\beta)d\beta.$$

This set of $\phi_n(x)$ plus $\phi(x;\beta)$ is the complete set of functions required to represent an arbitrary piecewise continuous function.

11.7.2 Another Representation of $\delta(x-x')$

An alternate representation of $\delta(x-x')$ will now be obtained. This will be useful later on for the development of Green's functions in 2D and 3D.

From (11.17) to (11.20) we saw that an arbitrary piecewise continuous function $f(x)$ can be represented as

$$f(x) = \int_a^b \sum_n \phi_n^*(x')\phi_n(x)w(x')f(x')dx', \tag{11.26}$$

where the order of summation and integration has been interchanged to facilitate the discussion. It is known that

$$f(x) = \int_a^b \delta(x-x')f(x')dx'. \tag{11.27}$$

Comparing (11.26) and (11.27) suggests that

$$\delta(x-x') = \sum_n \phi_n^*(x')\phi_n(x)w(x') \tag{11.28}$$

which shows us that the Dirac delta function may be represented by using a complete set of eigenfunctions. Equation (11.28) suggests that (11.25) can be rewritten more generally as

$$\frac{1}{j2\pi}\oint_{ccw} G(x,x';\lambda)\,d\lambda = -\sum_n \phi_n^*(x')\phi_n(x) = -\frac{\delta(x-x')}{w(x')}. \tag{11.29}$$

This gives us a contour integral representation of the delta function. The contour in the λ plane must enclose all the singularities of the Green's function. If G has branch cut singularities, then there is also a continuous spectrum of eigenvalues, and an integral must be added to the sum in (11.29).

11.7.3 A Discrete Spectrum of Eigenfunctions

As an example, we will now use these concepts to find the complete set of orthonormal eigenfunctions that satisfy the differential equation

$$\left[\frac{d^2}{dx^2} + \lambda_n\right]\phi_n(x) = 0; \quad 0 \le x \le b \tag{11.30}$$

and boundary conditions

$$\phi_n(0) = 0, \phi_n(b) = 0. \tag{11.31}$$

We will use the completeness relation

$$\frac{1}{j2\pi} \oint_{ccw} G(x, x'; \lambda) \, d\lambda = -\sum_n \phi_n^*(x')\phi_n(x) = -\frac{\delta(x - x')}{w(x')}, \tag{11.32}$$

where $w(x) = 1$ and the integration contour encloses all the singularities of $G(x, x'; \lambda)$ in the complex λ plane in a counterclockwise sense.

The required Green's function satisfies

$$\left[\frac{d^2}{dx^2} + \lambda\right] G(x, x'; \lambda) = -\delta(x - x'); \ 0 \leq x \leq b \tag{11.33}$$

$$G(0, x') = 0, G(b, x') = 0. \tag{11.34}$$

Using the UT method, G is found to be

$$G(x, x'; \lambda) = \frac{\sin \sqrt{\lambda} x_< \sin \sqrt{\lambda}(b - x_>)}{\sqrt{\lambda} \sin \sqrt{\lambda} b}. \tag{11.35}$$

The next step is to find all the singularities of G. It might first appear that there is a branch cut associated with $\sqrt{\lambda}$. This is actually not the case, as (11.35) is not affected by the sign choice for $\pm\sqrt{\lambda}$. Therefore, G has only poles. It is also noted that

$$\lim_{\lambda \to 0} \frac{\sin \sqrt{\lambda} x_< \sin \sqrt{\lambda}(b - x_>)}{\sqrt{\lambda} \sin \sqrt{\lambda} b} = \frac{x_<(b - x_>)}{b}$$

so we see that $\lambda = 0$ is not a pole. The only singularities in G are simple poles, which are at

$$\lambda_n = \left(\frac{n\pi}{b}\right)^2; \quad n = 1, 2, 3, \ldots . \tag{11.36}$$

Because the integrand has only simple poles, the integral can be evaluated with the residue theorem (see Appendix E) so that

$$\oint_{ccw} \frac{p(z)}{q(z)} \, dz = j2\pi \sum_p \frac{p(z)}{q'(z)}\bigg|_{z=z_p},$$

where $q(z_p) = 0$ at the pole z_p and $p(z)$ is analytic at z_p. To apply this to (11.35), we identify

$$p(\lambda) = \frac{\sin \sqrt{\lambda} x_< \sin \sqrt{\lambda}(b - x_>)}{\sqrt{\lambda}}$$

$$q(\lambda) = \sin \sqrt{\lambda} b$$

so that (11.32) becomes

$$\frac{1}{j2\pi} \oint \frac{\sin \sqrt{\lambda} x_< \sin \sqrt{\lambda}(b - x_>)}{\sqrt{\lambda} \sin \sqrt{\lambda} b} \, d\lambda = \sum_{n=1}^{\infty} \frac{2 \sin \sqrt{\lambda_n} x_< \sin \sqrt{\lambda_n}(b - x_>)}{b \cos \sqrt{\lambda_n} b}. \tag{11.37}$$

This is not in the symmetric form (11.32) that would allow us to identify $\{\phi_n(x)\}$, so we need to rearrange (11.37). Using

$$\sin\sqrt{\lambda_n}(b-x_>) = \underbrace{\sin\sqrt{\lambda_n}b\cos\sqrt{\lambda_n}x_>}_{=0} - \cos\sqrt{\lambda_n}b\sin\sqrt{\lambda_n}x_>$$

we can rewrite (11.37) as

$$\frac{1}{j2\pi}\oint_{ccw} G(x,x';\lambda)\,d\lambda = -\sum_{n=1}^{\infty}\frac{2}{b}\sin\sqrt{\lambda_n}x_<\sin\sqrt{\lambda_n}x_>. \tag{11.38}$$

Comparing (11.38) and (11.32) shows us that the complete orthonormal set of eigenfunctions for this problem is

$$\phi_n(x) = \sqrt{\frac{2}{b}}\sin\sqrt{\lambda_n}x\ ;\ n=1,2,3,\ldots \tag{11.39}$$

and the corresponding eigenvalues are

$$\lambda_n = \left(\frac{n\pi}{b}\right)^2;\ n=1,2,3,\ldots. \tag{11.40}$$

These results can be readily used to obtain the alternative representation of G

$$\begin{aligned}G(x,x';\lambda) &= \sum_n \frac{\phi_n^*(x')\phi_n(x)}{\lambda_n - \lambda}\\ &= \frac{2}{b}\sum_{n=1}^{\infty}\frac{\sin(n\pi x'/b)\sin(n\pi x/b)}{(n\pi/b)^2 - \lambda}.\end{aligned} \tag{11.41}$$

11.7.4   A Continuous Spectrum of Eigenfunctions

We wish to find the complete set of eigenfunctions which satisfy the differential equation

$$\left[\frac{d^2}{dx^2} + \lambda\right]\phi(x;\beta) = 0;\quad -\infty < x < \infty \tag{11.42}$$

and the radiation condition for $e^{j\omega t}$ time dependence as $|x| \to \infty$. Since the boundaries are at infinity, the appropriate completeness relation is given by

$$\frac{1}{j2\pi}\oint_{ccw} G(x,x';\lambda)\,d\lambda = -\int \phi^*(x';\beta)\phi(x;\beta)d\beta = -\frac{\delta(x-x')}{w(x')}. \tag{11.43}$$

The Green's function associated with (11.43) must satisfy the radiation condition. From Example 11.2 with $k^2 = \lambda$, it is

$$G(x,x';\lambda) = \frac{e^{-j\sqrt{\lambda}|x-x'|}}{j2\sqrt{\lambda}}. \tag{11.44}$$

Substituting (11.44) into (11.43) gives

$$\frac{1}{j2\pi}\oint_{ccw} \frac{e^{-j\sqrt{\lambda}|x-x'|}}{j2\sqrt{\lambda}}\,d\lambda = -\delta(x-x'). \tag{11.45}$$

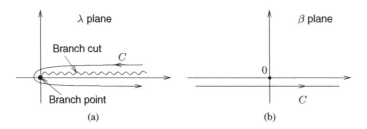

Figure 11.1 (a) Branch cut in the complex λ plane and counterclockwise integration path C. (b) Integration path C in the complex β plane.

The Green's function (11.44) has a branch point singularity at $\lambda = 0$. To ensure that the function $\sqrt{\lambda}$ is single valued with respect to λ, we need to define a branch cut. If the condition $\text{Im}(\sqrt{\lambda}) < 0$ is imposed, convergence of the integral (11.45) will be assured. Hence, the phase of $\sqrt{\lambda}$ will be chosen in accordance with the following rule:

$$-\pi < \arg \sqrt{\lambda} \leq 0$$

or equivalently $-2\pi < \arg \lambda \leq 0$.

This rule is shown in Figure 11.1(a).

With $\sqrt{\lambda} = \beta$, $d\beta = d\lambda/2\sqrt{\lambda}$, (11.45) becomes

$$\frac{-1}{2\pi} \int_{-\infty}^{\infty} e^{-j\beta|x-x'|} \, d\beta = -\delta(x - x'). \tag{11.46}$$

The integration path in the complex β plane is shown in Figure 11.1(b).

It is noted that (11.46) is not sensitive to the sign in the exponent. This can be seen by changing the dummy variable of integration $\beta \to -\beta$. Hence, $|x - x'|$ can be replaced by $(x - x')$ so that

$$\frac{-1}{2\pi} \int_{-\infty}^{\infty} e^{-j\beta x} e^{j\beta x'} d\beta = -\delta(x - x')$$

from which we may identify the terms

$$\phi^*(x'; \beta) = \frac{1}{\sqrt{2\pi}} e^{j\beta x'}, \quad \phi(x; \beta) = \frac{1}{\sqrt{2\pi}} e^{-j\beta x} \tag{11.47}$$

as the complete spectrum of continuous eigenfunctions. Since this is a complete orthonormal set, we can use it to construct an alternative representation for G. By considering the discrete version (11.24) of G, the continuous version, by analogy, is

$$G(x, x'; \lambda) = \int_{-\infty}^{\infty} \frac{\phi^*(x'; \beta)\phi(x; \beta)}{\beta^2 - \lambda} \, d\beta \tag{11.48}$$

from which we obtain

$$G(x, x'; \lambda) = \frac{1}{2\pi} \int_{-\infty}^{\infty} \frac{e^{-j\beta(x-x')}}{\beta^2 - \lambda} \, d\beta. \tag{11.49}$$

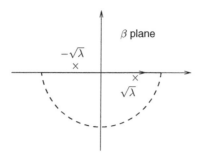

Figure 11.2 β plane showing the poles of (11.49) at $\beta = \pm\sqrt{\lambda}$ and the closed integration path.

We can 'close the loop' with these concepts by evaluating (11.49) to recover the original representation of G in (11.44). Equation (11.49) can be evaluated by residues at the two poles $\beta = \pm\sqrt{\lambda}$. If we assume a slightly lossy medium, then Im $\sqrt{\lambda} < 0$ and the poles will be offset from the real β axis as in Figure 11.2.

If we assume $x > x'$, then the integration path can be closed in the lower half β plane, and the integral along this semicircle will vanish. Evaluation of the residue at the pole $\beta = \sqrt{\lambda}$ gives us

$$G = (-j2\pi)\frac{1}{2\pi}\frac{e^{-j\beta(x-x')}}{\frac{d}{d\beta}(\beta^2-\lambda)}\bigg|_{\beta=\sqrt{\lambda}}$$

$$= \frac{e^{-j\sqrt{\lambda}(x-x')}}{j2\sqrt{\lambda}}$$

which is identical to the original Green's function (11.44).

If $x < x'$ is assumed, a similar procedure applies, except that the β plane contour is now closed in the upper half plane instead of the lower half plane.

11.8   Generalized Separation of Variables

The method of separation of variables allows us to express the solution of a 2D or 3D partial differential equation as a product of 1D solutions. However, the method only works for a homogeneous PDE. It is possible to generalize the procedure to the inhomogeneous case, with a delta function as the forcing function. Then, 1D Green's functions can be used to construct 2D and 3D Green's functions. We will call this procedure the *generalized separation of variables*, or GSV. A description of the GSV now follows.

To solve the 3D homogeneous wave equation,

$$\left[\frac{\partial^2}{\partial x^2} + \frac{\partial^2}{\partial y^2} + \frac{\partial^2}{\partial z^2} + k^2\right]\phi(x,y,z) = 0 \tag{11.50}$$

the method of separation of variables assumes that the 3D solution can be expressed as a product of 1D solutions

$$\phi(x,y,z) = X(x)Y(y)Z(z). \tag{11.51}$$

Substitution of (11.51) in (11.50) leads to

$$X''YZ + XY''Z + XYZ'' + k^2XYZ = 0$$

and dividing by XYZ gives us

$$\underbrace{\frac{X''(x)}{X(x)}}_{-\lambda_x} + \underbrace{\frac{Y''(y)}{Y(y)}}_{-\lambda_y} + \underbrace{\frac{Z''(z)}{Z(z)}}_{-\lambda_z} + k^2 = 0,$$

where λ_x, λ_y and λ_z are constants that are known as 'separation constants'. This equation can be rewritten as

$$\frac{d^2 X}{dx^2} + \lambda_x X = 0 \tag{11.52}$$

$$\frac{d^2 Y}{dy^2} + \lambda_y Y = 0 \tag{11.53}$$

$$\frac{d^2 Z}{dz^2} + \lambda_z Z = 0, \tag{11.54}$$

where

$$\lambda_x + \lambda_y + \lambda_z = k^2. \tag{11.55}$$

We see that $X(x)$, $Y(y)$ and $Z(z)$ are solutions to three ordinary differential equations, and (11.55) must be satisfied as well.

We now advance to the inhomogeneous case. To solve the inhomogeneous partial differential equation,

$$\left[\frac{\partial^2}{\partial x^2} + \frac{\partial^2}{\partial y^2} + \frac{\partial^2}{\partial z^2} + k^2 \right] G(x, y, z; x', y', z') = -\delta(x - x')\delta(y - y')\delta(z - z') \tag{11.56}$$

the *ansatz*[2] for the GSV is to use a product solution in conjunction with a linear operator $\mathcal{K}$

$$G(x, y, z; x', y', z') = \mathcal{K}[G_x(x, x')G_y(y, y')G_z(z, z')], \tag{11.57}$$

and $\mathcal{K}$ will be defined later on. Substituting (11.57) into (11.56) gives

$$\mathcal{K}\left[G_y G_z \frac{d^2 G_x}{dx^2} + G_x G_z \frac{d^2 G_y}{dy^2} + G_x G_y \frac{d^2 G_z}{dz^2} + k^2 G_x G_y G_z \right] = -\delta(x - x')\delta(y - y')\delta(z - z'). \tag{11.58}$$

Based on (11.52)–(11.54) we are tempted to try

$$\frac{d^2 G_x}{dx^2} + \lambda_x G_x = -\delta(x - x') \tag{11.59}$$

$$\frac{d^2 G_y}{dy^2} + \lambda_y G_y = -\delta(y - y') \tag{11.60}$$

$$\frac{d^2 G_z}{dz^2} + \lambda_z G_z = -\delta(z - z'). \tag{11.61}$$

---

[2] The German word *ansatz* is often used in physics and mathematics. It means 'an educated guess that is later verified by the results'.

The 1D Green's functions are called 'characteristic Green's functions'. Using (11.59)–(11.61) to replace the derivatives in (11.58) gives us

$$\mathcal{K}[(-\lambda_x - \lambda_y - \lambda_z + k^2)G_x G_y G_z - G_y G_z \delta(x - x') - G_x G_z \delta(y - y') - G_x G_y \delta(z - z')]$$
$$= -\delta(x - x')\delta(y - y')\delta(z - z'). \tag{11.62}$$

If we impose the condition (11.55) for the generalized case, then (11.62) simplifies to

$$\mathcal{K}[G_y G_z \delta(x - x') + G_x G_z \delta(y - y') + G_x G_y \delta(z - z')] = \delta(x - x')\delta(y - y')\delta(z - z'). \tag{11.63}$$

The solution G_x for (11.59) depends on x and x' and the parameter λ_x. From the theory of 1D Green's functions, we know that

$$\frac{1}{j2\pi} \oint_{C_x} G_x(x, x'; \lambda_x) \, d\lambda_x = -\delta(x - x'), \tag{11.64}$$

where C_x encircles all the singularities of G_x in a counterclockwise sense, in the complex λ_x plane. The singularities associated with G_y and G_z are excluded from C_x. G_x also satisfies the properties

$$\frac{1}{j2\pi} \oint_{C_y} G_x(x, x'; \lambda_x = k^2 - \lambda_y - \lambda_z) \, d\lambda_y = 0 \tag{11.65}$$

and

$$\frac{1}{j2\pi} \oint_{C_z} G_x(x, x'; \lambda_x = k^2 - \lambda_y - \lambda_z) \, d\lambda_z = 0, \tag{11.66}$$

where C_y encloses only the singularities of G_y in the complex λ_y plane and C_z encloses only the singularities of G_z in the complex λ_z plane. Relations similar to (11.64)–(11.66) hold for G_y and G_z, that is,

$$\frac{1}{j2\pi} \oint_{C_y} G_y(y, y'; \lambda_y) \, d\lambda_y = -\delta(y - y') \tag{11.67}$$

$$\frac{1}{j2\pi} \oint_{C_z} G_z(z, z'; \lambda_z) \, d\lambda_z = -\delta(z - z') \tag{11.68}$$

$$\frac{1}{j2\pi} \oint_{C_x} G_y(y, y'; \lambda_y = k^2 - \lambda_x - \lambda_z) \, d\lambda_x = 0 \tag{11.69}$$

$$\frac{1}{j2\pi} \oint_{C_z} G_y(y, y'; \lambda_y = k^2 - \lambda_x - \lambda_z) \, d\lambda_z = 0 \tag{11.70}$$

$$\frac{1}{j2\pi} \oint_{C_x} G_z(z, z'; \lambda_z = k^2 - \lambda_x - \lambda_y) \, d\lambda_x = 0 \tag{11.71}$$

$$\frac{1}{j2\pi} \oint_{C_y} G_z(z, z'; \lambda_z = k^2 - \lambda_x - \lambda_y) \, d\lambda_y = 0. \tag{11.72}$$

We can use (11.64)–(11.72) to motivate the definition for the linear operator $\mathcal{K}$:

$$\mathcal{K}[\cdot] = K \oint_{C_x} \oint_{C_y} [\cdot] \, d\lambda_x \, d\lambda_y, \tag{11.73}$$

where K is a constant to be determined. Applying this operator to (11.63) gives us

$$K \oint_{C_x} \oint_{C_y} G_y(\lambda_y) G_z(k^2 - \lambda_x - \lambda_y)\delta(x - x')\, d\lambda_x\, d\lambda_y$$

$$+ K \oint_{C_x} \oint_{C_y} G_x(\lambda_x) G_z(k^2 - \lambda_x - \lambda_y)\delta(y - y')\, d\lambda_x\, d\lambda_y$$

$$+ K \oint_{C_x} \oint_{C_y} G_x(\lambda_x) G_y(\lambda_y)\delta(z - z')\, d\lambda_x\, d\lambda_y$$

$$= \delta(x - x')\delta(y - y')\delta(z - z'). \tag{11.74}$$

The first two integrals are zero, because G_z has no singularities inside C_x or C_y. The third integral can be evaluated with the aid of (11.64) and (11.67):

$$K \oint_{C_x} G_x(\lambda_x) \oint_{C_y} G_y(\lambda_y)\, d\lambda_y\, d\lambda_x\, \delta(z - z') = K \oint_{C_x} G_x(\lambda_x)\, d\lambda_x (-j2\pi\delta(y - y'))\delta(z - z')$$

$$= K(-j2\pi\delta(x - x'))(-j2\pi\delta(y - y'))\delta(z - z')$$

and this must equal the right-hand side of (11.74), $\delta(x - x')\delta(y - y')\delta(z - z')$, which leads us to the conclusion that we must have

$$K = \frac{-1}{4\pi^2}.$$

The ansatz (11.57) and operator (11.73) along with K now provide us with the formal solution of the inhomogeneous partial differential equation (11.56) as

$$G(x, y, z; x', y', z') = \frac{-1}{4\pi^2} \oint_{C_x} \oint_{C_y} G_x(x, x'; \lambda_x)$$

$$\times G_y(y, y'; \lambda_y) G_z(z, z'; \lambda_z = k^2 - \lambda_x - \lambda_y)\, d\lambda_x\, d\lambda_y, \tag{11.75}$$

where G_x, G_y and G_z are 1D solutions to the ordinary differential equations (11.59)–(11.61).

Equation (11.75) is the final form of the GSV. It is important to note that in G_z the variable λ_z is replaced by $\lambda_z = k^2 - \lambda_x - \lambda_y$ prior to the evaluation of the contour integrals in the λ_x and λ_y planes.[3]

It is very useful to note that the constructed solution in (11.75) can be expressed in other ways and that similar logic can be used to construct the following alternative representations of G:

$$G(x, y, z; x', y', z') = \frac{-1}{4\pi^2} \oint_{C_x} \oint_{C_z} G_x(x, x'; \lambda_x)$$

$$\times G_y(y, y'; \lambda_y = k^2 - \lambda_x - \lambda_z) G_z(z, z'; \lambda_z)\, d\lambda_x\, d\lambda_z \tag{11.76}$$

or even

$$G(x, y, z; x', y', z') = \frac{-1}{4\pi^2} \oint_{C_y} \oint_{C_z} G_x(x, x'; \lambda_x = k^2 - \lambda_y - \lambda_z)$$

$$\times G_y(y, y'; \lambda_y) G_z(z, z'; \lambda_z)\, d\lambda_y\, d\lambda_z. \tag{11.77}$$

[3] Felsen and Marcuvitz (1994, Section 3.3c) obtained (11.75) by a different procedure, involving the manipulation of completeness relations.

The integrations in (11.75)–(11.77) look difficult, but they can sometimes be carried out in a simple way. Consider, for example, (11.75). If the problem has finite boundaries along the x and y coordinates, then we can find the orthonormal eigenfunctions $\{\phi_n(x)\}$ by solving

$$\left[\frac{d^2}{dx^2} + \lambda_n\right]\phi_n(x) = 0 \tag{11.78}$$

with its boundary conditions to obtain

$$G_x(x, x'; \lambda_x) = \sum_n \frac{\phi_n^*(x')\phi_n(x)}{\lambda_n - \lambda_x} \tag{11.79}$$

and also find the orthonormal eigenfunctions $\{\psi_m(y)\}$ by solving

$$\left[\frac{d^2}{dy^2} + \lambda_m\right]\psi_m(y) = 0 \tag{11.80}$$

with its boundary conditions to obtain

$$G_x(y, y'; \lambda_y) = \sum_m \frac{\psi_m^*(y')\psi_m(y)}{\lambda_m - \lambda_y}. \tag{11.81}$$

Using (11.79) and (11.81) in (11.75) and applying the residue theorem at the poles of G_x in the λ_x plane and the poles of G_y in the λ_y plane gives

$$G(x, y, z; x', y', z') = \sum_n \phi_n^*(x')\phi_n(x) \sum_m \psi_m^*(y')\psi_m(y)G_z(z, z'; \lambda_n, \lambda_m), \tag{11.82}$$

where G_z satisfies (11.61) and some specific boundary conditions in the z direction.

11.8.1 Reduction to 2D

The same techniques can be used to solve the 2D wave equation

$$\left[\frac{\partial^2}{\partial x^2} + \frac{\partial^2}{\partial y^2} + k^2\right]G(x, y; x', y') = -\delta(x - x')\delta(y - y') \tag{11.83}$$

with the result that

$$G(x, y; x', y') = \frac{j}{2\pi}\oint_{C_x} G_x(x, x'; \lambda_x)G_y(y, y'; k^2 - \lambda_x)\, d\lambda_x \tag{11.84}$$

or alternatively,

$$G(x, y; x', y') = \frac{j}{2\pi}\oint_{C_y} G_x(x, x'; k^2 - \lambda_y)G_y(y, y'; \lambda_y)\, d\lambda_y, \tag{11.85}$$

where $\lambda_x + \lambda_y = k^2$ is used to eliminate λ_x or λ_y from the integrand.

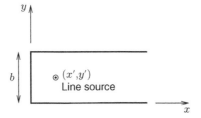

Figure 11.3 2D metallic waveguide with width b and a line source at (x', y').

Example 11.4 (2D Waveguide) Use the GSV to find the Green's function for the 2D metallic waveguide in Figure 11.3. The Green's function satisfies the wave equation (11.83), the Dirichlet boundary conditions

$$G(0, y; x', y') = 0; \quad 0 \le y \le b$$

$$G(x, 0; x', y') = 0, G(x, b; x', y') = 0; \quad 0 \le x < \infty$$

and the radiation condition for $e^{j\omega t}$ time dependence as $x \to +\infty$. Use a construction in the λ_y plane. Also find E_z for electric line source excitation.

Solution: G_x vanishes at $x = 0$ and obeys the radiation condition as $x \to \infty$. Using the UT method gives

$$G_x(x, x'; \lambda_x) = \frac{\sin \sqrt{\lambda_x} x_< \, e^{-j\sqrt{\lambda_x} x_>}}{\sqrt{\lambda_x}}.$$

In order to ensure decay as $x \to \infty$, the branch cut for the square root should be chosen so that $\text{Im} \sqrt{\lambda_x} \le 0$.

G_y vanishes at $y = 0$ and $y = b$. Using the eigenfunction series (11.41) (with x replaced by y) gives

$$G_y(y, y'; \lambda_y) = \frac{2}{b} \sum_{n=1}^{\infty} \frac{\sin(n\pi y'/b) \sin(n\pi y/b)}{(n\pi/b)^2 - \lambda_y}.$$

Substituting G_x and G_y into the λ_y-plane GSV in (11.85),

$$G(x, y; x', y') = \frac{j}{\pi b} \sum_{n=1}^{\infty} \oint_{C_y} \frac{\sin \sqrt{k^2 - \lambda_y} x_< \, e^{-j\sqrt{k^2-\lambda_y} x_>}}{\sqrt{k^2 - \lambda_y}} \frac{\sin(n\pi y'/b) \sin(n\pi y/b)}{(n\pi/b)^2 - \lambda_y} d\lambda_y.$$

The branch cut of G_x is shown in Figure 11.4(a)–(c). Since $\lambda_y = k^2 - \lambda_x$, case (c) is in fact the λ_y plane. The poles of G_y are shown in Figure 11.4(d). The singularities of G_x and G_y are shown in (e). The contour C_y must enclose the poles of G_y while excluding the branch cut of G_x, so C_y is deformed as shown in (e).

The integral is equal to $j2\pi$ times the counterclockwise residues at the poles $\lambda_y = (n\pi/b)^2$, so that

$$G(x, y; x', y') = \frac{2}{b} \sum_{n=1}^{\infty} \frac{\sin k_x x_<}{k_x} e^{-jk_x x_>} \sin \frac{n\pi y'}{b} \sin \frac{n\pi y}{b},$$

where $k_x = \pm\sqrt{k^2 - (n\pi/b)^2}$.

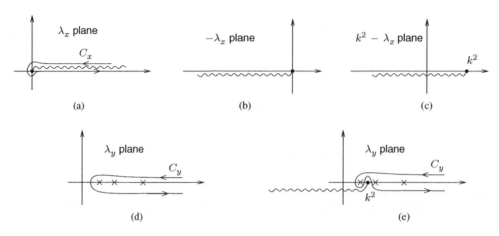

Figure 11.4 (a) G_x branch cut in the complex λ_x plane; (b) in the $-\lambda_x$ plane; (c) in the $k^2 - \lambda_x$ plane. (d) Poles of G_y in the λ_y plane. (e) Singularities of G_x and G_y in the λ_y plane.

When $k > n\pi/b$, $k_x = +\sqrt{k^2 - (n\pi/b)^2}$ gives the required $+x$ propagation, and when $k < n\pi/b$ (the waveguide is below cutoff), $k_x = -j\sqrt{(n\pi/b)^2 - k^2}$ gives the required decay as $x \to \infty$.

The field due to an electric line source I_0 at (x', y') follows from Equation (10.32) and is $E_z = -j\omega\mu I_0 G$. ∎

Example 11.5 (Wedge) Use the GSV to find the Green's function for a 2D metallic wedge in Figure 11.5. The Green's function satisfies the wave equation (11.83), the Dirichlet boundary conditions $G = 0$ at $\phi = 0$ and $\phi = n\pi$ and the radiation condition for $e^{j\omega t}$ time dependence as $x \to +\infty$. Also find E_z for electric line source excitation.

Solution: The polar version of the GSV can be found and is

$$G(\rho, \phi; \rho', \phi') = \frac{j}{2\pi} \oint_{C_\phi} G_\rho(\rho, \rho'; \lambda_\rho) G_\phi(\phi, \phi'; \lambda_\phi) \, d\lambda_\phi.$$

The ϕ problem satisfies

$$\frac{d^2 G_\phi}{d\phi^2} + \lambda_\phi G_\phi = -\delta(\phi - \phi')$$

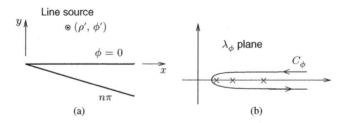

Figure 11.5 (a) Metallic wedge with faces at $\phi = 0$ and $\phi = n\pi$. (b) Singularities in the λ_ϕ plane.

and $G_\phi = 0$ at $\phi = 0$ and $\phi = n\pi$. The UT method can be used to obtain G_ϕ. Then, the completeness relation provides orthonormal eigenfunctions from which

$$G_\phi(\phi, \phi'; \lambda_\phi) = \frac{2}{n\pi} \sum_{m=1}^{\infty} \frac{\sin(m\pi\phi'/n) \sin(m\phi/n)}{(m/n)^2 - \lambda_\phi}.$$

The ρ problem satisfies

$$\frac{1}{\rho} \frac{d}{d\rho}\left(\rho \frac{dG_\rho}{d\rho}\right) + \left(k^2 - \frac{\lambda_\phi}{\rho^2}\right) G_\rho = -\frac{\delta(\rho - \rho')}{\rho}$$

with finiteness at $\rho = 0$ and the radiation condition at $\rho \to \infty$. From the UT method, the solution is

$$G_\rho = \frac{\pi}{j2} J_{\sqrt{\lambda_\phi}}(k\rho_<) H^{(2)}_{\sqrt{\lambda_\phi}}(k\rho_>).$$

The Bessel and Hankel functions are of order $\sqrt{\lambda_\phi}$. The J Bessel function is finite at $\rho = 0$, and the Hankel function obeys the radiation condition at $\rho \to \infty$.

Using G_ρ and G_ϕ in the GSV,

$$G(\rho, \phi; \rho', \phi') = \frac{1}{2n\pi} \oint_{C_\phi} J_{\sqrt{\lambda_\phi}}(k\rho_<) H^{(2)}_{\sqrt{\lambda_\phi}}(k\rho_>) \sum_{m=1}^{\infty} \frac{\sin(m\pi\phi'/n) \sin(m\phi/n)}{(m/n)^2 - \lambda_\phi} d\lambda_\phi.$$

G_ρ has no singularities in the λ_ϕ plane. The integral is equal to $j2\pi$ times the counterclockwise residues at the poles $\lambda_\phi = (m/n)^2$ so that

$$G(\rho, \phi; \rho', \phi') = \frac{1}{jn} \sum_{m=1}^{\infty} J_{m/n}(k\rho_<) H^{(2)}_{m/n}(k\rho_>) \sin\frac{m\phi'}{n} \sin\frac{m\phi}{n}.$$

$E_z(\rho, \phi)$ due to an electric line source I_0 at (ρ', ϕ') follows from Equation (10.32) and is $E_z = -j\omega\mu I_0 G$. ∎

11.8.2   Relation Between 2D and 3D

A 2D Green's function is given by

$$G_{2D}(\rho, \rho'; k_t) = \frac{j}{2\pi} \oint_{C_y} G_x(x, x'; k_t^2 - \lambda_y) G_y(y, y'; \lambda_y) d\lambda_y,$$

where $\lambda_x + \lambda_y = k_t^2$. By comparing the contour integral representation of G for 2D and 3D, it is possible to see that the 3D Green's function is obtainable from 2D via

$$G_{3D}(\mathbf{r}, \mathbf{r}') = \frac{j}{2\pi} \oint_{C_z} G_{2D}(\rho, \rho'; k_t = \sqrt{k^2 - \lambda_z}) \frac{e^{-j\sqrt{\lambda_z}|z-z'|}}{j2\sqrt{\lambda_z}} d\lambda_z. \tag{11.86}$$

Knowing a 2D Green's function in cylindrical coordinates, this can be used to find the point source field from the line source field.

Usually, G_{2D} involves a Hankel function, which can be replaced by its large argument approximation in the far field. The integral can then be carried out using the method of stationary phase.

11.9   Further Reading

The construction of 1D Green's functions is described in Arfken and Weber (2005, Chapter 10). Techniques for constructing 2D and 3D Green's functions are attributed to Titchmarsh (1946), Marcuvitz (1951) and Felsen (1957). These techniques are described in the books by Collin (1991, Chapter 2) and Felsen and Marcuvitz (1994, Chapters 3–6); both of these are highly recommended. These two books also provide Green's functions for many other cases that have not been covered in this chapter.

The very useful notes by Kouyoumjian (1980) are gratefully acknowledged in the preparation of this chapter.

References

Arfken GB and Weber HJ (2005) *Mathematical Methods for Physicists*. Elsevier.
Collin RE (1991) *Field Theory of Guided Waves*. Oxford University Press.
Erdelyi A, Magnus W, Oberhettinger F and Tricomi FG (1953) *Higher Transcendental Functions*. McGraw-Hill.
Felsen LB (1957) Alternative field representations in regions bounded by spheres, cones, and planes. *IEEE Trans. Antennas Propag.* **AP-5**(1), 109–121.
Felsen LB and Marcuvitz N (1994) *Radiation and Scattering of Waves*. IEEE Press.
Kouyoumjian RG (1980) Advanced electromagnetic theory EE 817. Ohio State University course notes.
Marcuvitz N (1951) Field representations in spherically stratified regions. *Commun. Pure Appl. Math.* **4**, 263–315.
Titchmarsh EC (1946) *Eigenfunction Expansions Associated with Second Order Differential Equations*. Clarendon Press.

Problems

11.1   Find the Green's function on the interval $0 \leq x \leq 1$ for

$$\left[\frac{d}{dx} \left(x \frac{d}{dx} \right) - \frac{\nu^2}{x} \right] G(x, x') = -\delta(x - x')$$

$$G(0, x') < \infty, \frac{dG}{dx}(1, x') = 0.$$

Here, ν is real and positive. To obtain the two linearly independent solutions of the homogeneous differential equation, try a solution x^k.

11.2   Find the Green's function on the interval $0 \leq x \leq 1$ for

$$\left[\frac{d}{dx} \left(\frac{1}{1 - cx} \frac{d}{dx} \right) \right] G(x, x') = -\delta(x - x')$$

$$G(0, x') = 0, G(1, x') = 0.$$

To solve the homogeneous problem, note that one of the two solutions is a constant. Then find the other solution by integration.

11.3   Find the Green's function on the interval $-\infty < x < \infty$ for

$$\left[\frac{d^2}{dx^2} + k_0^2 n^2(x) \right] G(x, x') = -\delta(x - x').$$

For $x \leq 0$, we have $n(x) = n_1$, and for $x \geq 0$, we have $n(x) = 1$. For convenience, we may denote $k_1 = n_1 k_0$ when $x < 0$. Assume that the source is in the region $x' > 0$. The solution should satisfy the radiation condition for $e^{j\omega t}$ time dependence, as $x \to \pm\infty$. To solve the problem, impose the continuity of G and dG/dx at $x = 0$.

11.4 Find the Green's function on the interval $-d \leq x < \infty$ for

$$\left[\frac{d}{dx}\left(p(x)\frac{d}{dx} \right) + k^2 \right] G(x, x') = -\delta(x - x'),$$

where d is a positive number. Assume that the source point is restricted to $x' \geq 0$. Also, $p(x) = 1$ for $x \geq 0$ and $p(x) = m$ for $x < 0$. k and m are constants. (It is convenient to define $k_1 = k/\sqrt{m}$.) The boundary conditions are $G(-d, x') = 0$ and the radiation condition for $e^{j\omega t}$ time dependence as $x \to \infty$. At $x = 0$, both G and $p\, dG/dx$ should be continuous.

11.5 Solve the differential equation

$$\frac{d^2 y}{dx^2} + k^2 y = \sin 3x; \quad 0 \leq x \leq 1$$

subject to the boundary conditions $y(0) = y(1) = 0$. First, find a Green's function for the related problem

$$\left[\frac{d^2}{dx^2} + k^2 \right] G(x, x') = -\delta(x - x')$$

that satisfies $G(0, x') = G(1, x') = 0$. Then, use G in a convolution integral

$$y(x) = -\int_0^1 \sin 3x'\, G(x', x)\, dx'$$

to find $y(x)$. Note that G has two different forms, one for $x > x'$ and another for $x < x'$ so the integral should be split into two parts.

11.6 Find the complete orthonormal set of eigenfunctions $\{\phi_n(x)\}$ that satisfy

$$\left[\frac{d^2}{dx^2} + \lambda_n \right] \phi_n(x) = 0; \quad -b/2 \leq x \leq b/2$$

$$\phi_n(-b/2) = 0, \phi_n(b/2) = 0.$$

Do this by first finding the related Green's function $G(x, x'; \lambda)$ and then applying the completeness relation

$$\frac{1}{j2\pi} \oint G(x, x'; \lambda)\, d\lambda = -\sum_n \phi_n^*(x')\phi_n(x)$$

to deduce the $\{\phi_n\}$. You should obtain two sets of $\{\phi_n\}$, one for odd n and one for even n.

11.7 Next, consider what happens if $b \to \infty$. We wish to show that

$$\sum_n \phi_n^*(x')\phi_n(x) \to \int \phi^*(x'; \beta)\phi(x; \beta)\, d\beta$$

in which $\phi(x; \beta)$ can then be found.

To accomplish this, write the odd and even terms under a common summation. For the two types of terms, define $\beta_n = (2n-1)\pi/b$ and $\beta_n = 2n\pi/b$. Note that in the limiting process, b is large and the two types of β_n become the same, making it unnecessary to have two definitions. Then use $\beta_{n+1} - \beta_n = 2\pi/b = \Delta\beta_n \to d\beta$ to obtain the final result.

11.8 Solve Example 11.4, but this time impose the Neumann boundary condition $\partial G/\partial n = 0$ on the waveguide walls. Use G to find $H_z(x,y)$ due to a magnetic line source M_0 at (x',y').

11.9 Solve Example 11.4, except this time develop the solution using a Green's function that encircles the singularities of G_x:

$$G(x,y;x',y') = \frac{j}{2\pi}\oint_{C_x} G_x(x,x';\lambda_x)G_y(y,y';\lambda_y)\,d\lambda_x.$$

(a) Show the singularities of G_x in the λ_x plane and G_y in the λ_y plane.
(b) Show how the singularities of G_y appear in the λ_x plane.
(c) Show the integration contour in the λ_x plane.
(d) Map this into the ν plane by using the relation $\nu = \sqrt{\lambda_x}$, and express the integral in terms of ν.
(e) Find a suitable way to close the contour so that the ν plane integral can be evaluated.

11.10 Write a program to evaluate G for the 2D waveguide in Example 11.4. The source point is at $(x',y') = (0.25\text{ m}, b/2)$. Calculate $|G|$ along the line $y = b/2$. Investigate the cases of $b = 0.3$ m, 0.5 m, 1.0 m, and 1.5 m. Assume that $\lambda = 1$ m. Notice that the waveguide cutoff is when $b = \lambda/2$. Also, $|G|$ is sometimes constant, and other times it is not. Discuss your numerical results (two pages or less).

11.11 Find the Green's function for a metallic wedge as in Example 11.5, but this time impose the Neumann boundary condition $\partial G/\partial n = 0$ on the wedge faces. Use G to find $H_z(\rho,\phi)$ due to a magnetic line source M_0 at (ρ',ϕ').

11.12 Use the GSV to find the Green's function for the 2D cylindrically tipped metallic wedge in Figure 11.6(a). This is similar to Example 11.5. G_ϕ is unchanged and vanishes on the metal wedge faces. The G_ρ function should vanish at $\rho = a$ and obey the radiation condition at $\rho \to \infty$. Use the UT method with

$$U(\rho) = J_{\sqrt{\lambda_\phi}}(k\rho) + cH^{(2)}_{\sqrt{\lambda_\phi}}(k\rho)$$

$$T(\rho) = H^{(2)}_{\sqrt{\lambda_\phi}}(k\rho)$$

subject to $U(a) = 0$ and obtain c.

It turns out that G_ρ has pole singularities which occur at the complex roots ν_p of $H^{(2)}_{\nu_p}(ka) = 0$. These are the 'creeping wave' poles and are located at $\lambda_\phi = \nu_p^2$ in the λ_ϕ plane. The precise values of ν_p are not needed here but appear as in Figure 11.6(b). The contour C_ϕ captures the G_ϕ poles and easily avoids the G_ρ poles so that G can be obtained from the residues at the poles of G_ϕ.

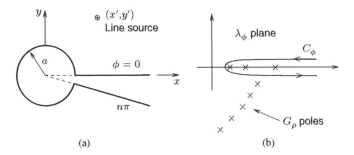

Figure 11.6 (a) Cylindrically tipped metallic wedge with faces at $\phi = 0$ and $\phi = n\pi$. The cylinder radius is a. (b) Singularities of G_ρ and G_ϕ in the λ_ϕ plane.

11.13 Derive the 2D–3D Green's function transformation formula in Equation (11.86).

11.14 Use the 2D–3D Green's function transformation in Equation (11.86) to find the 3D Green's function for a metallic wedge that is illuminated by a unit strength point source located at $(\rho', \phi', 0)$ in cylindrical coordinates. Assume Dirichlet boundary conditions $G = 0$ at $\phi = 0$ and $\phi = n\pi$.

(a) Using G_{2D} for a unit strength line source, find G_{3D} for a point source. This will be in terms of an an integral in the λ_z plane.

(b) Replace the Hankel function $H^{(2)}_{m/n}$ with its large argument approximation and use the method of stationary phase to carry out the integration. A change of variables $k_z = \sqrt{\lambda_z}$ is useful.

(c) If the source is a z-directed electric dipole of strength p_e, find $E_z(\rho, \phi, z)$. The far-field component E_θ can be obtained from $E_z = -\sin\theta E_\theta$. Show that

$$E_\theta(r, \theta, \phi) = f(r, \theta) \sum_{m=1}^{\infty} j^{(m/n)+1} J_{m/n}(k\rho' \sin\theta) \sin\frac{m\phi'}{n} \sin\frac{m\phi}{n}$$

and find $f(r, \theta)$.

11.15 Using the 2D–3D Green's function transformation, prove (4.112), repeated here

$$\frac{1}{4\pi} \int_{z'=-\infty}^{\infty} \frac{e^{-jk\sqrt{D^2+z'^2}}}{\sqrt{D^2 + z'^2}} \, dz' = \frac{1}{j4} H_0^{(2)}(kD).$$

11.16 Show that for the wave equation in spherical coordinates

$$(\nabla^2 + k^2)G(\mathbf{r}, \mathbf{r}') = -\frac{\delta(r - r')\delta(\theta - \theta')\delta(\phi - \phi')}{r^2 \sin\theta}$$

the Green's function is

$$G(\mathbf{r}, \mathbf{r}') = \frac{-1}{4\pi^2} \oint_{C_\theta} \oint_{C_\phi} G_r(r, r'; \lambda_\theta) G_\theta(\theta, \theta'; \lambda_\phi, \lambda_\theta) G_\phi(\phi, \phi'; \lambda_\phi) \, d\lambda_\phi \, d\lambda_\theta,$$

where G_r, G_θ and G_ϕ are 1D solutions to the ordinary differential equations

$$\left[\frac{d}{dr} \left(r^2 \frac{d}{dr} \right) + k^2 r^2 - \lambda_\theta \right] G_r(r, r'; \lambda_\theta) = -\delta(r - r')$$

$$\left[\frac{d}{d\theta} \left(\sin\theta \frac{d}{d\theta} \right) - \frac{\lambda_\phi}{\sin\theta} + \lambda_\theta \sin\theta \right] G_\theta(\theta, \theta'; \lambda_\phi, \lambda_\theta) = -\delta(\theta - \theta')$$

$$\left[\frac{d^2}{d\phi^2} + \lambda_\phi \right] G_\phi(\phi, \phi'; \lambda_\phi) = -\delta(\phi - \phi').$$

12

Green's Function Construction II

In Chapter 11 a waveguide and wedge were used to illustrate the usage of the generalized separation of variables (GSV) for Green's function construction. In this chapter, these techniques will be used to obtain Green's functions for several problems: a grounded dielectric slab, a dielectric half space and a bare- and dielectric-coated cylinder. A metal strip grating on a grounded dielectric slab will be used to illustrate the use of the slab Green's function, to demonstrate the application of Floquet's theorem and to obtain the complex propagation constant for this structure.

The Green's function for any open boundary problem (having a boundary at infinity) is in terms of a complex plane line integral having its endpoints at infinity and branch points in the integrand. These are the so-called Sommerfeld integrals. The analytical treatment involves a Riemann surface. The concept will be explained using the simplest possible case: the free-space Green's function. The technique will then be used in solving the aforementioned boundary value problems.

This chapter will also present the saddle point method, which is a useful approximate asymptotic technique for evaluating Sommerfeld-type integrals.

12.1 Sommerfeld Integrals

It is known that the solution to the 2D wave equation

$$\left[\frac{\partial^2}{\partial x^2} + \frac{\partial^2}{\partial y^2} + k^2\right] G(x, y; x', y') = -\delta(x - x')\delta(y - y') \tag{12.1}$$

satisfying the radiation condition at infinity is

$$G_f(x, y; x', y') = \frac{1}{j4} H_0^{(2)}(k\sqrt{(x - x')^2 + (y - y')^2}). \tag{12.2}$$

A less obvious solution to (12.1) can be found via the GSV

$$G(x, y; x', y') = \frac{j}{2\pi} \oint_{C_x} G_x(x, x'; \lambda_x) G_y(y, y'; \lambda_y = k^2 - \lambda_x) d\lambda_x. \tag{12.3}$$

G_x satisfies the radiation condition as $|x| \to \infty$. Using the UT method, G_x is found to be

$$G_x(x, x'; \lambda_x) = \frac{e^{-j\sqrt{\lambda_x}|x - x'|}}{j2\sqrt{\lambda_x}}.$$

Applied Frequency-Domain Electromagnetics, First Edition. Robert Paknys.
© 2016 John Wiley & Sons, Ltd. Published 2016 by John Wiley & Sons, Ltd.
Companion Website: www.wiley.com/go/paknys9981

Similarly, G_y satisfies the radiation condition as $|y| \to \infty$ and is

$$G_y(y, y'; \lambda_y) = \frac{e^{-j\sqrt{\lambda_y}|y-y'|}}{j2\sqrt{\lambda_y}}.$$

Denoting the free-space case as G_f, (12.3) becomes

$$G_f(\boldsymbol{\rho}, \boldsymbol{\rho}') = \frac{j}{2\pi} \oint_{C_x} \frac{e^{-j\sqrt{\lambda_x}|x-x'|}}{j2\sqrt{\lambda_x}} \frac{e^{-j\sqrt{k^2-\lambda_x}|y-y'|}}{j2\sqrt{k^2-\lambda_x}} d\lambda_x. \tag{12.4}$$

Using the substitutions

$$\nu = \sqrt{\lambda_x} \tag{12.5}$$

and

$$\kappa(\nu) = \sqrt{k^2 - \nu^2} \tag{12.6}$$

(12.4) can be rewritten as

$$G_f(\boldsymbol{\rho}, \boldsymbol{\rho}') = \frac{1}{j4\pi} \int_{-\infty}^{\infty} e^{-j\nu|x-x'|} \frac{e^{-j\kappa|y-y'|}}{\kappa} d\nu. \tag{12.7}$$

The branch cuts of G_x and G_y and the integration path C_x are shown in Figure 12.1. As chosen, they ensure that $\text{Im} \sqrt{\lambda_x} \le 0$ and $\text{Im} \sqrt{\lambda_y} \le 0$ for all values of λ_x and λ_y.

Comparing the free-space Green's function (12.2) with (12.7) tells us that

$$H_0^{(2)}(k|\boldsymbol{\rho} - \boldsymbol{\rho}'|) = \frac{1}{\pi} \int_{-\infty}^{\infty} e^{-j\nu|x-x'|} \frac{e^{-j\kappa|y-y'|}}{\kappa} d\nu. \tag{12.8}$$

This is a remarkable result, expressing the Hankel function as a spectrum of plane waves. Integral representations of Bessel functions were developed by Arnold Sommerfeld in 1896, and (12.8) is known as a *Sommerfeld integral*.

If a dipole or line source is above a half space or a dielectric slab, the scattered field will involve a Sommerfeld integral of the form

$$G_s(\boldsymbol{\rho}, \boldsymbol{\rho}') = \frac{1}{j4\pi} \int_{-\infty}^{\infty} e^{-j\nu|x-x'|} \frac{e^{-j\kappa|y-y'|}}{\kappa} \Gamma(\nu) d\nu. \tag{12.9}$$

This is similar to (12.7) except that it involves a reflection coefficient $\Gamma(\nu)$. The reflection coefficient can contain singularities such as poles and branch cuts that will have to be considered.

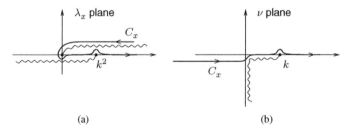

(a)   (b)

Figure 12.1 (a) Branch points $\lambda_x = 0$ for G_x and $\lambda_x = k^2$ for G_y. (b) With $\nu = \sqrt{\lambda_x}$ the λ_x plane maps to the lower ν plane.

12.2 The Function $\kappa(\nu) = \sqrt{k^2 - \nu^2}$

When developing the Sommerfeld integral, the branch cut for G_x was chosen so that $\mathrm{Im}\,\sqrt{\lambda_x} \leq 0$. Considering the mapping (12.5), it explains why nothing is mapped into the upper ν plane in Figure 12.1(b). However, requiring $\mathrm{Im}\,\sqrt{\lambda_x} \leq 0$ over the entire integration path is overly restrictive and can be relaxed. This is permissible because deforming portions of the path into regions having $\mathrm{Im}\,\sqrt{\lambda_x} > 0$ will *not* change the value of the integral – provided that no singularities were crossed in the course of the deformation.

The branch cut for G_x can be thought of as a 'gateway' or 'slot' in the λ_x plane. Passing through the slot provides access to 'new' values of λ_x such that $\mathrm{Im}\,\sqrt{\lambda_x} > 0$. (Similar remarks apply to G_y.) This involves the concept of a two-sheet Riemann surface. It is permitted that an integration path pass into and back out of the slot, along its way.

When evaluating an integral such as (12.9) with the saddle point method, it is possible that the deformed path could cross a pole of $\Gamma(\nu)$. If this happens the 'pole wave' has to be accounted for. If the pole has $\mathrm{Im}\,\kappa \leq 0$, the associated wave decays exponentially away from the reflecting surface and is said to be *proper*. The word 'proper' is customarily associated with the condition $\mathrm{Im}\,\kappa \leq 0$. Conversely, the word *improper* is associated with the condition $\mathrm{Im}\,\kappa > 0$. We will later see that surface waves on a dielectric slab can be proper or improper, whereas forward leaky waves are improper.

The function $\kappa(\nu)$ in Equation (12.6) is important because it frequently occurs in the integral representations of fields. Before such integrals can be tackled, the analytical properties of $\kappa(\nu)$ need to be understood. First, the conditions under which $\mathrm{Im}\,\kappa(\nu)$ is positive or negative must be known. Second, steps must be taken to ensure that $\kappa(\nu)$ is single valued for all points along an integration path in the complex ν plane. This will be accomplished by defining suitable branch cuts, associated with the branch points which exist at $\nu = \pm k$.

In the following discussion, $k = k_r + jk_i$ and $\nu = \nu_r + j\nu_i$. It will be assumed that a small amount of loss is present so that k_i is small and $k_i < 0$. By solving the equations $\mathrm{Im}\,\kappa = 0$ and $\mathrm{Re}\,\kappa = 0$, it can be shown (Felsen and Marcuvitz, 1994, Section 5.3b) that hyperbolas exist in the ν plane; they are defined by

$$k_r k_i = \nu_r \nu_i$$

and appear as shown in Figure 12.2. The hyperbolas possess certain important properties: (a) that $\mathrm{Im}\,\kappa = 0$ on the dashed portions of the hyperbolas and (b) that $\mathrm{Re}\,\kappa = 0$ on the dot-dashed portions. It is crucial to recognize that the properties of the hyperbolas are *not related* to how the branch cuts of κ are chosen; the latter is a separate issue and will be described next.

It is necessary to define branch cuts to ensure that $\kappa(\nu)$ is single valued for all points along an integration path $\bar{P}$ in the ν plane. The branch cuts can be defined in many ways, and one possible choice is shown in Figure 12.3, along with a typical integration path $\bar{P}$. The path can be deformed in any way, as

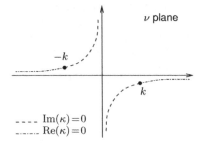

Figure 12.2 Analytical properties of the function $\kappa(\nu) = \sqrt{k^2 - \nu^2}$ in the ν plane.

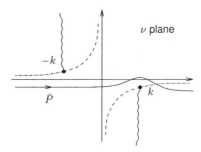

Figure 12.3   One possible choice of branch cuts for $\kappa(\nu) = \sqrt{k^2 - \nu^2}$ in the ν plane. The branch cuts define a rule that ensures the single valuedness of the function $\kappa(\nu)$ for all points along the integration path $\bar{P}$.

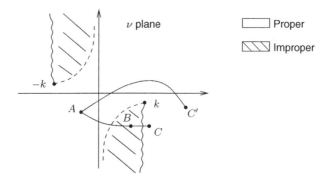

Figure 12.4   Proper and improper regions on the top sheet of the ν plane.

long as the branch cuts are not crossed in the process. If the integration path were to pass over a branch cut, then κ would abruptly change sign, making it discontinuous – which is not permissible.

The ν plane is a two-sheet Riemann surface. The top sheet of the ν plane is shown in Figure 12.4. The locations of the improper and proper regions can be explained as follows. It will be assumed that the sign of κ is chosen so that $\nu = 0$ is part of the proper region. We know that if a dotted line is crossed, then the sign of $\operatorname{Im} \kappa$ changes. We also know that if a branch cut is crossed, the sign of κ changes. Hence, as we move along a hypothetical integration path from A to B, $\operatorname{Im} \kappa$ changes sign. Moving further along from B to C, κ changes sign. The net result is that $\operatorname{Im} \kappa$ undergoes two sign changes, which implies that the signs of $\operatorname{Im} \kappa$ at point A and point C are the same. This leads to the conclusion that the unshaded region is proper, and the shaded region is improper. Following similar logic, the path AC' does not cross a dashed line or a branch cut, so all points on that path lie in the proper region.[1]

The branch cuts can take on any convenient shape. It is common to choose them so that they coincide with the dashed lines $\operatorname{Im} \kappa = 0$ as shown in Figure 12.5. Then, the improper regions vanish – that is, the entire top sheet corresponds to $\operatorname{Im} \kappa < 0$, and the entire bottom sheet corresponds to $\operatorname{Im} \kappa > 0$. This particular choice is known as the *Sommerfeld branch cuts*.[2]

---

[1] Thanks are due to Prof. D. R. Jackson who provided the description given here, in a private communication.
[2] Figure 12.5 shows two branch points; they are at $\nu = \pm k$. Figure 12.1(b) only shows one branch point, at $\nu = k$. Where is the other one? It is on the bottom sheet of the λ_x plane, where $\operatorname{Im} \sqrt{\lambda_x} > 0$.

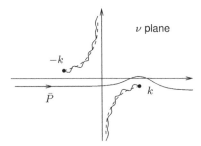

Figure 12.5   Sommerfeld's choice of branch cuts are along the dashed lines where $\operatorname{Im}\kappa(\nu)=0$.

12.3   The Transformation $\nu = k\sin w$

In carrying out integrals by the saddle point method, it is useful to introduce a change of variables so that the ν plane integration path is mapped to the w plane by the function

$$\nu = k\sin w. \tag{12.10}$$

If $w = u + jv$ then (12.10) can be rewritten as

$$\nu = k(\sin u \cosh v + j\cos u \sinh v). \tag{12.11}$$

In changing the integration variable from ν to w, (12.6) becomes

$$\kappa = \pm\sqrt{k^2 - \nu^2} = +k\cos w = k(\cos u \cosh v - j\sin u \sinh v). \tag{12.12}$$

The positive branch of the square root is a convenient choice so that κ will be positive when $-k \le \nu \le k$ and k and ν are real.

Now, consider the ν-plane points $ABCDE$ as shown in Figure 12.6(a). From Equation (12.11), every point in the ν plane (a) maps into two points in the w plane, so the mapping is not unique. If the points $ABCDE$ were on an integration path, it would be unclear how to connect the w-plane points. (Note that $\sin w = \sin(w + 2\pi)$ so the w plane repeats every 2π, and it is only necessary to show $-\pi \le u \le \pi$.)

The mapping from ν to w is useful because the branch points that were in $\kappa(\nu)$ are no longer present in $\kappa(w)$. Equation (12.12) shows us that $\kappa(w) = k\cos w$ is an entire function (that is, analytic everywhere) in the w plane, without any singularities.

Returning to the problem of the mapping's non-uniqueness, the mapping can be made unique by introducing an additional condition. A useful choice is to keep only the proper points

$$\operatorname{Im}\kappa = -\sin u \sinh v < 0 \tag{12.13}$$

and reject the improper points. Then the ν plane uniquely maps into the unshaded regions of the w plane in Figure 12.6. Conversely, we could keep only the improper points, which would map into the shaded regions.

In the ν plane, one may choose Sommerfeld branch cuts which are along $\operatorname{Im}\kappa = 0$. Then, the top sheet of the ν plane has $\operatorname{Im}\kappa(\nu) < 0$ and maps into the w-plane unshaded regions; the bottom sheet has $\operatorname{Im}\kappa(\nu) > 0$ and maps into the w-plane shaded regions. Some paths illustrating the top-sheet mapping are shown in Figure 12.7.

Integration paths are permitted to pass in and out of improper regions. As long as no singularities are crossed in the process, the integral's value will not be changed. An example of a path that is partly on the bottom sheet is shown in Figure 12.8.

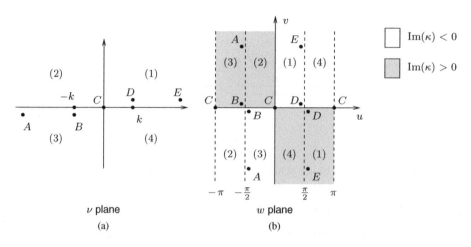

Figure 12.6   The non-unique mapping $\nu = k \sin w$.

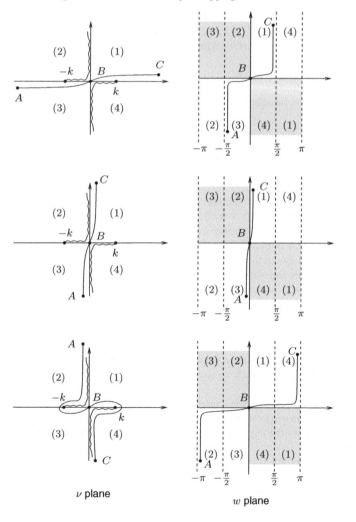

Figure 12.7   Some examples of how ν-plane integration paths map into the w plane.

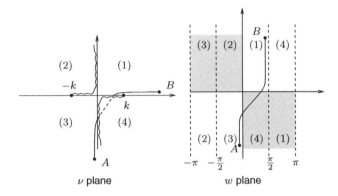

Figure 12.8 Example of how a ν-plane integration path maps into the w plane when part of the path passes onto the bottom sheet of the ν plane.

12.4 Saddle Point Method

In this section we will find a way to approximately evaluate integrals of the form

$$I(\Omega) = \int_{\bar{P}} f(z) e^{\Omega q(z)} dz \qquad (12.14)$$

where Ω is real, positive and large. $f(z)$ and $q(z)$ are complex functions of the complex variable z. $\bar{P}$ denotes a path of integration in the complex z plane.

In order to evaluate (12.14), some properties of complex functions must first be discussed. A complex function $q(z)$ that has $q'(z_s) = 0$ is said to be *stationary* at z_s. If $q''(z_s)$ and higher derivatives are non-zero, then z_s is called a first-order saddle point. At z_s the surface is locally flat, without being either a minimum or a maximum.

Let us define $z = x + jy$ and $q(z) = u(x, y) + jv(x, y)$ where u and v are real. A stationary point $q'(z_s) = 0$ implies that at z_s,

$$u_x = u_y = v_x = v_y = 0.$$

u, v must satisfy the Cauchy–Riemann equations[3]

$$u_x = v_y, u_y = -v_x$$

from which

$$u_{xx} = -u_{yy}, v_{xx} = -v_{yy}.$$

This means that the curvature of u along the x direction is opposite in sign to the curvature of u along the y direction. The same is true for v. The stationary points are therefore saddle points. The topology for u and v will be different, but the general features in either case appear as in Figure 12.9.

Next, we shall describe the ideas of *paths of constant level and constant phase*. Consider an integration path $\bar{P}$ as shown in Figure 12.10. The rate of change of $u(x, y)$ with respect to the arc length s along the path $\bar{P}$ is given by the directional derivative $\nabla u \cdot \hat{s} = du/ds$ where

$$\frac{du}{ds} = \frac{\partial u}{\partial x} \cos \alpha + \frac{\partial u}{\partial y} \sin \alpha.$$

---

[3] The Cauchy–Riemann equations are necessary conditions on $q(z)$ so that $q'(z_0)$ exists at z_0; see Churchill and Brown (2013).

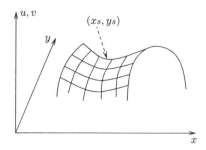

Figure 12.9   Functions $u(x,y)$ or $v(x,y)$ showing a saddle point at (x_s, y_s).

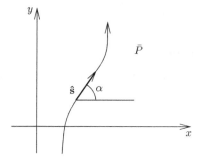

Figure 12.10   Integration path in the z plane.

The values of α that maximize this are obtained from

$$\frac{d^2u}{d\alpha\,ds} = -\frac{\partial u}{\partial x}\sin\alpha + \frac{\partial u}{\partial y}\cos\alpha = 0. \tag{12.15}$$

We also have the rate of change of $v(x,y)$ along the path as

$$\frac{dv}{ds} = \frac{\partial v}{\partial x}\cos\alpha + \frac{\partial v}{\partial y}\sin\alpha$$

and by using the Cauchy–Riemann equations, it becomes

$$\frac{dv}{ds} = -\frac{\partial u}{\partial y}\cos\alpha + \frac{\partial u}{\partial x}\sin\alpha. \tag{12.16}$$

From (12.15) we can now see that $dv/ds = 0$. By comparing (12.15) and (12.16) we can see that if we choose the direction α to maximize the change in $u(x,y)$, then we also obtain $dv/ds = 0$ or $v(x,y) =$ const. Conversely, we can show that the path that makes $u =$ const. also has the maximum rate of change in v. These ideas will be useful in the next steps.

Let us now return to the task of evaluating (12.14). It can be written as

$$I(\Omega) = \int_{\bar{P}} f(z)e^{\Omega[u(x,y)+jv(x,y)]}dz.$$

It will be assumed that (a) the integration path $\bar{P}$ can be deformed into another more suitable path P, provided that no singularities are crossed by the deformation, and that (b) the integrand is sufficiently small at the endpoints of the integration path $\bar{P}$ so that those portions of the path do not significantly contribute to the integral.

The integration path is now deformed from $\bar{P}$ into P so that it passes through the saddle point. Then, $u(x_s, y_s)$ takes on its maximum value, and the behaviour of the integrand is dominated by the term $A = |\exp(\Omega u(x, y))|$ which is large at (x_s, y_s) and decays very rapidly away from z_s. In other words, the best possible path deformation is the one that has the following properties: (a) it passes through the saddle point of $q(z)$, defined by $q'(z_s) = 0$, and (b) $\operatorname{Re} q(z) < \operatorname{Re} q(z_s)$ and $\operatorname{Im} q(z) = \text{const.}$, for all points z on the integration path. Such a path P is called the *steepest descent path* (SDP). Because of the way that $\operatorname{Re} q(z)$ behaves, we see that because Ω is large, $A = |\exp(\Omega q(z))|$ is large at z_s and decays rapidly, away from z_s. Hence the dominant contribution to $I(\Omega)$ occurs for points near z_s.

For general forms of $q(z)$, the determination of the complete SDP can be difficult. Strictly speaking, the SDP has $\operatorname{Im} q(z) = \text{const.}$ In practice, we can use the less stringent requirement that $\operatorname{Re} q(z) < \operatorname{Re} q(z_s)$ along the path; this is not the steepest path but still has an exponential decay in the integrand. The resulting error, of $O(e^{\Omega q(z) - \Omega q(z_s)})$, can be neglected (Felsen and Marcuvitz, 1994, p. 381).

12.4.1   First-Order Saddle Point

A saddle point that has $q'(z_s) = 0$ but $q''(z_s) \neq 0$ is defined as a first-order saddle point. We wish to transform the given integral (12.14) into a canonical form that can be evaluated. This is accomplished by replacing $q(z)$ by another function, a polynomial, that still represents the behaviour at z_s, though more simply. For a first-order saddle point, a good approximation near z_s turns out to be

$$q(z) = q(z_s) - s^2. \tag{12.17}$$

The stationary point z_s maps into $s = 0$ in the s plane. Figure 12.11 illustrates the relation between the z plane and s plane. From (12.17) we have that

$$q'(z)\frac{dz}{ds} = -2s \tag{12.18}$$

or

$$\frac{dz}{ds} = -\frac{2s}{q'(z)}.$$

As $z \to z_s$, we also have $s \to 0$; the indeterminate form can be evaluated by L'Hopital's rule

$$\lim_{z \to z_s} \frac{dz}{ds} = \frac{-2ds/ds}{q''(z)dz/ds} = \frac{-2}{q''(z_s)dz/ds} \tag{12.19}$$

from which

$$\frac{dz}{ds}\bigg|_{z=z_s} = \sqrt{\frac{-2}{q''(z_s)}}. \tag{12.20}$$

ds/dz is the mapping derivative, and it is finite near and at the saddle point – a desirable property.

The integral (12.14) can be rewritten in terms of s

$$I(\Omega) \sim \int_{P'} G(s)e^{\Omega[q(z_s) - s^2]}\,ds; \quad \Omega \to \infty, \tag{12.21}$$

where $G(s) = f(z)dz/ds$. Path P in the z plane maps to path P' in the s plane.

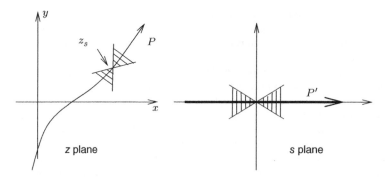

Figure 12.11 z-Plane and s-plane SDP topologies. The shaded regions represent descending directions away from the saddle point that lead to valleys.

It will be assumed that $f(z)$ is slowly varying in the vicinity of z_s. Then we can replace $f(z)$ with $f(z_s)$ and take it outside the integral so that $G(s) \approx G(0)$.

$s = 0$ maps to $z = z_s$. From (12.17) and the fact that $\mathrm{Im}\, q(z) = \mathrm{const.}$ on the SDP, we conclude that the SDP maps to the real axis of the s plane. P' in (12.21) can be replaced by $-\infty < s < \infty$. Hence we can use (12.20) and (12.21) to obtain

$$I(\Omega) \sim f(z_s)\sqrt{\frac{-2}{q''(z_s)}}\, e^{\Omega q(z_s)} \int_{-\infty}^{\infty} e^{-\Omega s^2}\, ds. \tag{12.22}$$

The last integral is equal to $\sqrt{\pi/\Omega}$. This is obtainable from the gamma function. Hence, (12.22) becomes

$$I(\Omega) \sim f(z_s)\sqrt{\frac{-2\pi}{\Omega q''(z_s)}}\, e^{\Omega q(z_s)}. \tag{12.23}$$

The symbol $\sim$ means *asymptotically equal to*. As $\Omega \to \infty$ it becomes exact. In fact, (12.23) is the leading term of an asymptotic series. Higher-order terms are available in Schafer and Kouyoumjian (1967).

One unresolved issue is how to choose the correct branch of the square root in (12.23). This term originated in (12.20) which can be rewritten as

$$\frac{dz}{ds}\bigg|_{z=z_s} = \sqrt{\frac{-2}{q''(z_s)}} = \left|\sqrt{\frac{2}{q''(z_s)}}\right| e^{j\phi_s}, \tag{12.24}$$

where ϕ_s is the phase of dz/ds at z_s. Since $dz/ds = dx/ds + j\,dy/ds$ we have

$$\mathrm{arg}\left(\frac{dz}{ds}\right) = \arctan\frac{dy/ds}{dx/ds} = \arctan\left(\frac{dy}{dx}\right)$$

so that

$$\tan\phi_s = \frac{dy}{dx}\bigg|_{(x_s, y_s)}.$$

Because dz/ds in (12.24) has a sign ambiguity, the sign of $\tan\phi_s$ must be determined by inspecting the slope of the SDP plot near z_s, as shown in Figure 12.12. The final form for the saddle point method

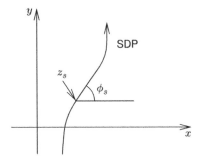

Figure 12.12 z plane, showing the angle ϕ_s at the saddle point.

(12.23) is then

$$I(\Omega) = \int_P f(z) e^{\Omega q(z)}\, dz \sim \left| \sqrt{\frac{2\pi}{\Omega q''(z_s)}} \right| e^{j\phi_s} f(z_s) e^{\Omega q(z_s)}, \qquad (12.25)$$

where the path $\bar{P}$ has been deformed to the SDP P and ϕ_s is the angle between the tangent to the SDP at $z = z_s$ and the positive real axis. We have transformed the z plane to the s plane in such a way that the descending part of the SDP is mapped to the positive real s axis. In doing so, we have removed the ambiguity associated with the mapping (12.17).

It is interesting to note that the stationary phase method is closely related to the saddle point method. If we deform the SDP into a new path where the function $e^{\Omega q(z)}$ has constant amplitude (rather than constant phase), we obtain the method of stationary phase. Furthermore, if no singularities of $f(z)$ are crossed in the deformation process, then both methods give the same result.

Example 12.1 Use the saddle point method to evaluate the following integral, which represents the zero-order Hankel function of the second kind

$$H_0^{(2)}(\Omega) = \frac{1}{\pi} \int_{\bar{P}} e^{-j\Omega \sin z}\, dz,$$

where the integration path $\bar{P}$ is defined by the straight-line segments from $(x, y) = (0, -\infty)$ to $(0, 0)$ to $(\pi, 0)$ to (π, ∞).

Solution: In this case $f(z) = 1/\pi$, and $q(z) = -j \sin z$. From $q'(z_s) = 0$ the saddle point is at $z_s = \pi/2$. The SDP is defined by $\mathrm{Im}\, q(z) = \mathrm{const.}$ which means that $\sin x \cosh y = \mathrm{const.}$ Since the SDP passes through $(x_s, y_s) = (\pi/2, 0)$, P is along

$$\sin x \cosh y = 1.$$

The paths $\bar{P}$ and P are shown in Figure 12.13(a).

From the SDP equation, at (x_s, y_s) we obtain $dy/dx = \tan \phi_s = \pm 1$ and hence $\phi_s = \pm \pi/4$. A plot of the SDP near the saddle point indicates that $\phi_s = +\pi/4$. Using (12.25) we find that

$$H_0^{(2)}(\Omega) \sim \sqrt{\frac{2}{\pi\Omega}} e^{-j(\Omega - \pi/4)}.$$

∎

In Example 12.1 the integrand is exponentially small at the path termini, and any deformed path must respect this requirement. Therefore the path can start anywhere in the strip $-\pi \leq x \leq 0$; $y \to -\infty$ and end in the strip $0 \leq x \leq \pi$; $y \to \infty$.

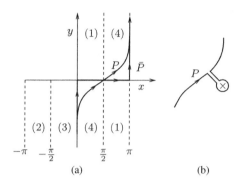

Figure 12.13 (a) Integration path $\bar{P}$ and the steepest descent path P, in the $z = x + jy$ plane. (b) SDP avoiding a pole.

Example 12.2 Use the saddle point method to evaluate an integral

$$I(\Omega) = \int_{\bar{P}} \frac{e^{-j\Omega \sin z}}{z - z_p} \, dz$$

that has a pole at $z_p = 3 + j$. The integration path $\bar{P}$ is defined by the straight-line segments from $(x, y) = (0, -\infty)$ to $(0, 0)$ to $(\pi, 0)$ to (π, ∞).

Solution: In this case $f(z) = 1/(z - z_p)$, and $q(z) = -j \sin z$. From $q'(z_s) = 0$ the saddle point is $z_s = \pi/2$. The SDP is given by $\sin x \cosh y = 1$, with $\phi_s = +\pi/4$, and is shown in Figure 12.13(a).

In deforming $\bar{P}$ into P, the pole z_p was crossed. This can be verified by noting that the point $(x, y) = (2.44, 1)$ lies on the SDP and is to the left of z_p. To avoid the pole we encircle it in the counterclockwise sense, as shown in Figure 12.13(b). Using the residue theorem, the contribution of the detour is

$$+j2\pi \frac{e^{-j\Omega \sin z_p}}{\frac{d}{dz}(z - z_p)_{z=z_p}} = +j2\pi e^{-j\Omega \sin z_p}.$$

Using (12.25) to obtain the SDP part and adding it to the residue, we obtain

$$I(\Omega) \sim \sqrt{\frac{2\pi}{\Omega}} \frac{1}{\pi/2 - (3 + j)} e^{-j(\Omega - \pi/4)} + j2\pi e^{-j\Omega \sin(3+j)}.$$

■

12.4.2   Pole Near Saddle Point

If a pole and saddle point are far apart, they can be treated separately. Otherwise, the pole will affect the saddle point topology, which requires a modified steepest descent method. Two main methods are the modified Pauli–Clemmow (Pathak and Kouyoumjian, 1970) and Van der Waerden (Felsen and Marcuvitz, 1994, Section 4.4a) approaches; the latter will be described here. If the SDP integral

$$I(\Omega) = \int_P f(z) e^{\Omega q(z)} \, dz \tag{12.26}$$

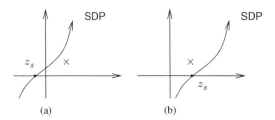

Figure 12.14 z plane, showing SDP and (a) pole not crossed, $\text{Im } b < 0$; (b) pole is crossed, $\text{Im } b > 0$.

has a saddle point at z_s and f has a simple pole at z_p, an asymptotic approximation that is valid for any $|z_s - z_p|$ is

$$I(\Omega) \sim e^{\Omega q(z_s)} \left[\pm j2a\sqrt{\pi} \, e^{-\Omega b^2} \, Q(\mp jb\sqrt{\Omega}) + \sqrt{\frac{\pi}{\Omega}} T(0) \right] ; \quad \text{Im } b \gtrless 0, \qquad (12.27)$$

where

$$a = \lim_{z \to z_p} [(z - z_p)f(z)]; \quad b = \sqrt{q(z_s) - q(z_p)}$$

$$T(0) = hf(z_s) + \frac{a}{b}; \quad h = \sqrt{\frac{-2}{q''(z_s)}}$$

$$Q(y) = \int_y^\infty e^{-x^2} \, dx; \quad Q(0) = \frac{\sqrt{\pi}}{2}.$$

The square root for h is evaluated the same way as in (12.24). The phase of b should be chosen so that $b \to (z_p - z_s)/h$ as $z_p \to z_s$. $Q(y)$ is related to the complementary error function via $Q(y) = (\sqrt{\pi}/2) \, \text{erfc}(y)$.

Figure 12.14 shows the relationship between the SDP and pole. In case (a) it is assumed that deformation to the SDP has not captured the pole, and $\text{Im } b < 0$. If the pole and saddle point are far apart, then (12.27) reduces to (12.25). In case (b) the pole has been captured, and $\text{Im } b > 0$. Equation (12.27) now includes the residue. When $\text{Im } b = 0$ the pole is on the SDP. Near $\text{Im } b = 0$ as the SDP passes over the pole, there is a jump discontinuity in (12.27) that equals the residue at the pole. There is no need to treat the pole near/far from the saddle point as separate cases, so Equation (12.27) is called a *uniform asymptotic formula*.

12.5   SDP Branch Cuts

Choosing Sommerfeld branch cuts and an SDP is an analytical convenience but not obligatory. Depending on the locations of singularities, other integration paths and branch cuts might even be preferrable. This is especially true if a Sommerfeld integral is done numerically, which becomes necessary when $\Omega = k\rho$ is not large.

The shape of branch cuts is arbitrary, and a popular alternative to Sommerfeld branch cuts is along vertical lines in the v plane, as shown in Figure 12.15. The lines are along $\text{Re } \nu = \pm k$. According to Equation (12.11), in the w plane this becomes

$$\sin u \cosh v = \pm 1. \qquad (12.28)$$

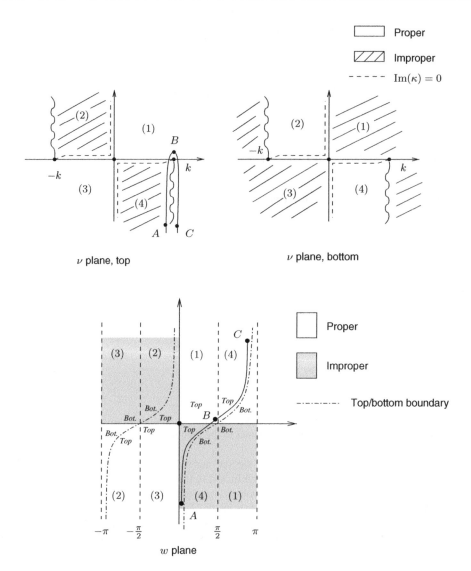

Figure 12.15 The correspondence between the ν plane and w plane. What's proper on the ν-plane top sheet is improper on the bottom sheet and vice versa.

Since these are branch cuts, (12.28) defines the top/bottom boundaries. This in turn enables us to determine how the top/bottom ν plane maps into the w plane. Coincidentally, (12.28) are also the SDP's through $w = \pm\pi/2$. The following discussion considers the branch point $\nu = k$; considerations for $\nu = -k$ are similar and omitted here.

To understand Figure 12.15 the following principles are emphasized:

- A proper/improper boundary is defined by $\operatorname{Im} \kappa = 0$.
- A top/bottom boundary is defined by the $\pi/2$ SDP.
- The top sheet of the ν plane contains both proper and improper regions, similarly for the bottom sheet.

- A proper region on the top ν sheet corresponds to an improper region on the bottom ν sheet and vice versa.
- It is not correct to associate proper/improper behaviour with top/bottom sheets.
- A given point must fall into one of four categories: top + proper, top + improper, bottom + proper and bottom + improper.
- Passage through a ν-plane branch cut from the top sheet to the bottom sheet corresponds to crossing the w-plane SDP line $\sin u \cosh v = \pm 1$.
- When the ν-plane integration path passes through a branch cut to the other sheet, points in the neighborhood of the path remain proper or improper (whatever the case may be) and do not change abruptly.

In summary: the ν-plane integration path ABC in Figure 12.15 is entirely on the top sheet. The points map into the w plane, slightly *above* the $\pi/2$ SDP, as shown. Path AB is in the improper region $\text{Im}(\kappa) > 0$, and path BC is in the proper region $\text{Im}(\kappa) < 0$.

12.6   Grounded Dielectric Slab

An electric line source of strength I_0 is at (x', y'), and the field point is at (x, y), as shown in Figure 12.16. The electric field is related to G by

$$E_z = -j\omega\mu_1 I_0 G \qquad (12.29)$$

where G satisfies the 2D wave equation

$$(\nabla^2 + k_{1,2}^2)G(\boldsymbol{\rho}, \boldsymbol{\rho}') = -\delta(|\boldsymbol{\rho} - \boldsymbol{\rho}'|). \qquad (12.30)$$

k_1 is used when $y \geq 0$ and k_2 is used when $y \leq 0$. At $y = 0$, E_z and H_x must be continuous so G must satisfy

$$G(x, y = 0^-) = G(x, y = 0^+) \qquad (12.31)$$

$$\frac{1}{\mu_2}\frac{\partial G(x,y)}{\partial y}\bigg|_{y=0^-} = \frac{1}{\mu_1}\frac{\partial G(x,y)}{\partial y}\bigg|_{y=0^+} \qquad (12.32)$$

and at the perfectly conducting ground plane

$$G(x, y = -d) = 0. \qquad (12.33)$$

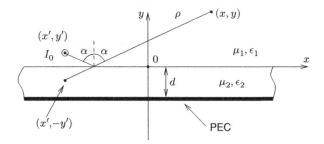

Figure 12.16   Grounded dielectric slab excited by an electric line source; image at $(x', -y')$.

The GSV allows us to construct a solution

$$G(x, y; x', y') = \frac{j}{2\pi} \oint_{C_x} G_x(x, x'; \lambda_x) G_y(y, y'; \lambda_y) \, d\lambda_x, \tag{12.34}$$

where $\lambda_x + \lambda_{y1} = k_1^2$ in region 1 and $\lambda_x + \lambda_{y2} = k_2^2$ in region 2. Continuity of the tangential fields requires that we have the same x dependence in regions 1 and 2, so we have chosen $\lambda_{x1} = \lambda_{x2} = \lambda_x$.

Using the UT method, G_x is found to be

$$G_x(x, x'; \lambda_x) = \frac{e^{-j\sqrt{\lambda_x}|x-x'|}}{j2\sqrt{\lambda_x}} \tag{12.35}$$

which is the 1D free-space Green's function. In the y direction, we need G_y for $y \geq 0$ (region 1) and $-d \leq y \leq 0$ (region 2). The basic form for region 1 is

$$G_{y1}(y, y'; \lambda_y) = \frac{(e^{j\sqrt{\lambda_{y1}}y_<} + \Gamma e^{-j\sqrt{\lambda_{y1}}y_<})e^{-j\sqrt{\lambda_{y1}}y_>}}{j2\sqrt{\lambda_{y1}}}, \tag{12.36}$$

and in region 2 a suitable form is

$$G_{y2}(y, y'; \lambda_y) = \frac{T \sin(\sqrt{\lambda_{y2}}(y_< + d)) e^{-j\sqrt{\lambda_{y1}}y_>}}{j2\sqrt{\lambda_{y1}} \sin(\sqrt{\lambda_{y2}}d)}. \tag{12.37}$$

By enforcing the field continuity conditions (12.31) and (12.32), we can obtain Γ and T. We find that

$$\Gamma(\lambda_x) = \frac{j\mu_r\kappa_1 - \kappa_2 \cot \kappa_2 d}{j\mu_r\kappa_1 + \kappa_2 \cot \kappa_2 d} \tag{12.38}$$

and

$$T = 1 + \Gamma, \tag{12.39}$$

where $\mu_r = \mu_2/\mu_1$, $\epsilon_r = \epsilon_2/\epsilon_1$, $\kappa_1 = \sqrt{k_1^2 - \lambda_x}$ and $\kappa_2 = \sqrt{k_2^2 - \lambda_x}$.

In (12.38) we note that the sign of κ_2 does not affect Γ, so there is no branch cut associated with κ_2, only with κ_1. In addition, Γ has pole singularities; these will be discussed later.

Equations (12.34)–(12.38) lead to the Green's function for $y \geq 0$

$$G(\boldsymbol{\rho}, \boldsymbol{\rho}') = \frac{j}{2\pi} \oint_{C_x} \frac{e^{-j\sqrt{\lambda_x}|x-x'|}}{j2\sqrt{\lambda_x}} \left[\frac{e^{-j\sqrt{k_1^2-\lambda_x}|y-y'|} + \Gamma(\lambda_x)e^{-j\sqrt{k_1^2-\lambda_x}(y+y')}}{j2\sqrt{k_1^2 - \lambda_x}} \right] d\lambda_x. \tag{12.40}$$

Using the substitutions $\sqrt{\lambda_x} = \nu$ and $\kappa_1 = \sqrt{k_1^2 - \nu^2}$, it can be rewritten as

$$G(\boldsymbol{\rho}, \boldsymbol{\rho}') = \frac{1}{j4\pi} \int_{-\infty}^{\infty} e^{-j\nu|x-x'|} \frac{e^{-j\kappa_1|y-y'|} + \Gamma(\nu) e^{-j\kappa_1(y+y')}}{\kappa_1} \, d\nu. \tag{12.41}$$

It is permitted to replace $|x - x'| \to (x - x')$ (see Problem 12.1). This gives an alternate form

$$G(\boldsymbol{\rho}, \boldsymbol{\rho}') = \frac{1}{j4\pi} \int_{-\infty}^{\infty} e^{-j\nu(x-x')} \frac{e^{-j\kappa_1|y-y'|} + \Gamma(\nu) e^{-j\kappa_1(y+y')}}{\kappa_1} \, d\nu. \tag{12.42}$$

The two terms represent the incident and scattered fields, so (12.41) can be written as a sum of two parts

$$G(\boldsymbol{\rho}, \boldsymbol{\rho}') = G_f(\boldsymbol{\rho}, \boldsymbol{\rho}') + G_s(\boldsymbol{\rho}, \boldsymbol{\rho}'), \tag{12.43}$$

where

$$
\begin{aligned}
G_f(\boldsymbol{\rho}, \boldsymbol{\rho}') &= \frac{1}{j4\pi} \int_{-\infty}^{\infty} e^{-j\nu|x-x'|} \frac{e^{-j\kappa_1|y-y'|}}{\kappa_1} \, d\nu \\
&= \frac{1}{j4} H_0^{(2)}\!\left(k_1 \sqrt{(x-x')^2 + (y-y')^2}\right)
\end{aligned}
\tag{12.44}
$$

$$G_s(\boldsymbol{\rho}, \boldsymbol{\rho}') = \frac{1}{j4\pi} \int_{-\infty}^{\infty} e^{-j\nu|x-x'|} \frac{e^{-j\kappa_1(y+y')}}{\kappa_1} \Gamma(\nu) \, d\nu. \tag{12.45}$$

12.6.1 Saddle Point Evaluation of G_s

To perform a saddle point evaluation of G_s in (12.45), it is expedient to make the transformation $\nu = k_1 \sin w$. In addition, we use the polar image coordinates

$$x - x' = \rho \sin \alpha \tag{12.46}$$

$$y + y' = \rho \cos \alpha. \tag{12.47}$$

Figure 12.16 shows the geometrical interpretation of (12.46) and (12.47). That is, the reflected field emanates from an image source at $(x', -y')$. With these coordinates, (12.45) can be rewritten as

$$G_s(\boldsymbol{\rho}, \boldsymbol{\rho}') = \frac{1}{j4\pi} \int_P e^{-jk_1\rho \cos(w-\alpha)} \Gamma(k_1 \sin w) \, dw. \tag{12.48}$$

The integral (12.48) for G_s has a saddle point at $w_s = \alpha$ as well as pole singularities due to the term $\Gamma(k_1 \sin w)$. For the moment, let us assume that the poles of Γ are not close to the saddle point and that no poles get crossed when the path $\bar{P}$ is deformed to the SDP P. Under these conditions the saddle point method applied to (12.48) gives

$$G_s(\boldsymbol{\rho}, \boldsymbol{\rho}') = \frac{1}{j4} \sqrt{\frac{2}{\pi k_1 \rho}} e^{-j(k_1\rho - \pi/4)} \Gamma(k_1 \sin \alpha). \tag{12.49}$$

Although this result was obtained from the saddle point method, it can also be found by using geometrical optics. The term Γ is the grounded slab reflection coefficient, evaluated at the incidence angle α.

Figure 12.17 shows the top (proper) and bottom (improper) sheets of the ν plane and the corresponding w plane. The reflection coefficient $\Gamma(\nu)$ has zeros in its denominator, giving rise to poles in G_s. There are surface wave poles on the proper sheet, and these have $\text{Im}\,\kappa_1 \leq 0$. There are also leaky wave poles on the bottom sheet with $\text{Im}\,\kappa_1 > 0$. The integration path $\bar{P}$ is entirely on the top sheet of the ν plane.

To evaluate the integral with the saddle point method, the path $\bar{P}$ has to be deformed into the SDP P, shown in Figure 12.18. The integrand is exponentially small at the path termini, and the deformed path must respect this requirement. Therefore the path can start anywhere in the strip $-\pi \leq u \leq 0; v \to -\infty$ and end in the strip $0 \leq u \leq \pi; v \to \infty$.

The SDP passes through the saddle point at $w_s = \alpha$, and $0 \leq \alpha \leq \pi/2$ is assumed. For the value of α illustrated, we see that by deforming the integration path from $\bar{P}$ to the SDP P, some surface wave poles

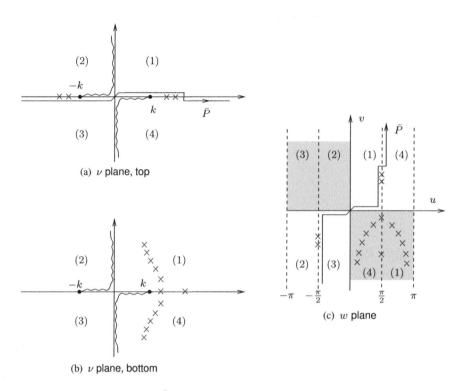

Figure 12.17   (a) Integration path $\bar{P}$ on top sheet of the ν plane and surface wave poles. (b) Bottom sheet of the ν plane, showing leaky wave poles. (c) The top and bottom sheets are mapped into the w plane.

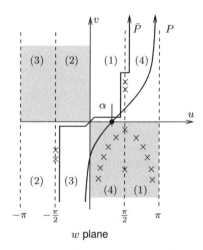

Figure 12.18   Deformation of the integration path $\bar{P}$ to the steepest descent path P, passing through the saddle point $w_s = \alpha$.

at $w_p = \pi/2 + jv_p$ were crossed. This happens at an angle $\alpha_p = \pi/2 - \arccos(\mathrm{sech}(v_p))$. For larger angles the SDP may even intercept some of the leaky wave poles in the strip $0 < u < \pi/2,\ -\infty < v < 0$. Whether or not this happens will depend on α, the dielectric material, its thickness and the frequency; for the moment this will not be considered.

To account for the surface waves, the clockwise residues must be added to G_s. Alternatively, the counterclockwise residues should be subtracted out. We will add the clockwise residues. To find the residues of the integrand in (12.48), we must find the poles associated with the reflection coefficient (12.38). It can be rewritten as

$$\Gamma = \frac{j\mu_r q - p \cot p}{j\mu_r q + p \cot p},$$
(12.50)

where

$$p = \kappa_2 d = d\sqrt{k_2^2 - k_1^2 \sin^2 w}$$

$$q = \kappa_1 d = d\sqrt{k_1^2 - k_1^2 \sin^2 w} = k_1 d \cos w$$

and $\mu_r = \mu_2/\mu_1$. The surface wave poles come from the roots of the transcendental equations

$$p \cot p = -j\mu_r q$$
(12.51)

and

$$p^2 - q^2 = (\mu_r \epsilon_r - 1)(k_1 d)^2 = \ell^2.$$
(12.52)

Equations (12.51) and (12.52) can be solved graphically or numerically for the roots p, q. For a lossless dielectric slab with $\mu_r = 1$ and real ϵ_r, there are solutions with purely imaginary q. By defining $jq = q_0$, (12.51) and (12.52) become

$$-p \cot p = q_0$$
(12.53)

$$p^2 + q_0^2 = \ell^2.$$
(12.54)

To solve (12.53) and (12.54) graphically we plot them for real p and q_0 in the $p - q_0$ plane. The function in (12.53) can be plotted as $-p \cot p$ versus q_0, and (12.54) represents a circle of radius ℓ. The intersection of the circles with the cotangents gives us the surface wave poles; see Section 3.6.2 for a more detailed discussion.

To find the residue at a pole w_p of (12.50), we denote $\Gamma = \Gamma_N/\Gamma_D$. The clockwise integration requires that we evaluate $-j2\pi$ times the residue, with the result that

$$G_p(\boldsymbol{\rho}, \boldsymbol{\rho}') = -j2\pi \frac{1}{j4\pi} e^{-jk_1\rho\cos(w_p-\alpha)} \frac{\Gamma_N}{\frac{d}{dw}\Gamma_D}\bigg|_{w=w_p}$$

$$= \frac{j\mu_r q}{k_1 d \sin w_p A_p} e^{-jk\rho\cos(w_p-\alpha)} U(\alpha - \alpha_p),$$
(12.55)

where

$$A_p = j\mu_r \left(1 - (q/p)^2\right) - q \left(1 - (\mu_r q/p)^2\right).$$
(12.56)

The unit step function $U(\alpha - \alpha_p)$ is used because a pole contribution to G does not occur unless the angle α is sufficiently large so that $\alpha > \alpha_p$. Here, α_p denotes the angle α that causes the SDP to cross the pole at w_p.

Using (12.49) and (12.55) G_s becomes

$$G_s(\boldsymbol{\rho}, \boldsymbol{\rho}') = \frac{1}{j4} \sqrt{\frac{2}{\pi k_1 \rho}} e^{-j(k_1\rho - \pi/4)} \Gamma(k_1 \sin \alpha)$$

$$+ \sum_{w_p} \frac{j\mu_r q}{k_1 d \sin w_p A_p} e^{-jk\rho \cos(w_p - \alpha)} U(\alpha - \alpha_p). \qquad (12.57)$$

The first term is the SDP contribution and represents the reflected field. The second term represents the contributions of surface waves. The summation over w_p in (12.57) is needed when more than one surface wave pole is present.

12.6.2   Surface and Leaky Waves

For a lossless dielectric slab, the surface wave poles are at $w_p = \pi/2 + jv_p$; $v_p \geq 0$. Using the polar coordinate conversions (12.46) and (12.47), the surface wave term in (12.57) can be expressed as

$$e^{-jk_1\rho\cos(w_p - \alpha)} = e^{-k_1 \sinh v_p(y+y')} e^{-jk_1 \cosh v_p(x-x')}. \qquad (12.58)$$

We see that the surface wave exhibits an exponential decay away from the surface, in the y direction. Along x, the propagation constant is $k_1 \cosh v_p$. Since $\cosh v_p \geq 1$, this is a 'slow wave' with a phase velocity that is less than the velocity of a plane wave in a homogeneous medium with $k_1 = \omega\sqrt{\mu_1 \epsilon_1}$.

In addition to surface wave poles, there exists another set of poles known as 'leaky wave' poles. By solving (12.51) and (12.52) subject to the assumption Im $\kappa > 0$, an infinite number of complex roots can be found. These poles are shown in Figure 12.19. Since Im $\kappa > 0$, the poles lie on the *improper* sheet (the bottom sheet) of the ν plane which corresponds to the shaded region of the w plane.

Depending on the source and field point positions, the SDP may cross some of these poles in the w plane. For observation angles $0 \leq \alpha \leq \pi/2$, it is possible for the SDP to intercept the leaky wave poles that are present in the w-plane region $0 \leq u \leq \pi/2$, $v \leq 0$. In Figure 12.19, the eligible poles are along the locus $A'B'C'$. The poles are most important near grazing angles when $\alpha \approx \pi/2$.

When the leaky wave poles w_ℓ are captured by the SDP, a contribution given by

$$\sum_{w_\ell} \frac{j\mu_r q}{k_1 d \sin w_\ell A_\ell} e^{-jk\rho \cos(w_\ell - \alpha)} U(\alpha - \alpha_\ell) \qquad (12.59)$$

must be added to (12.57). Here, α_ℓ denotes the angle at which the SDP crosses a leaky wave pole w_ℓ.

As the frequency ω (or d/λ_1) is increased, one set of poles in Figure 12.19 follows a locus that begins at $\nu = -j\infty$. The pole moves along from A' to B' and then to C'. The other locus ABC is the complex conjugate of $A'B'C'$.

The leaky wave poles in the w plane are symmetric about the line $u = \pi/2$. At some particular frequency, the poles meet at B, B' to form a double pole. As the frequency is further increased, the B pole moves up to C and the B' pole moves down to C'. The meeting point of the poles B and B' is called the 'split point'.

Beyond the split point, C' migrates downwards to $w = \pi/2 - j\infty$. The other pole at C migrates upwards towards $w = \pi/2 + j0$ and evolves into the next surface wave pole. A small amount of loss will keep the two trajectories separate at the split point BB'. In the lossless case it is not possible to see which locus goes where.

The leaky wave poles in the strip $0 \leq u \leq \pi/2$ are the ones that play a primary role in the observable field. The ones in the strip $\pi/2 < u \leq \pi$ will have much less of an effect, though they can still have an

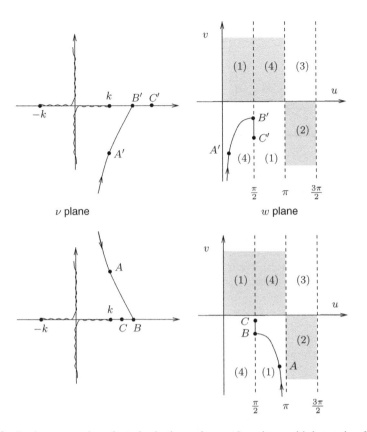

Figure 12.19 Leaky wave pole trajectories in the ν plane and w plane, with increasing frequency ω. Top figure: the poles tend to $w = \pi/2 - j\infty$ as $\omega \to \infty$. Bottom figure: as ω is increased, the leaky wave pole will move upwards from point C and pass $w = \pi/2 + j0$, evolving into a surface wave pole.

influence if they are near $w = \pi/2$. (A higher-order asymptotic analysis for a pole near a saddle point is needed to evaluate this effect.) It is a point of interest that the physically observable leaky waves are *not* the ones that evolve into observable surface waves as the frequency is increased.

The spatial behaviour of leaky waves is described by the exponential term

$$e^{-jk_1\rho\cos(w_\ell-\alpha)} = e^{-k_1 y \sin u \sinh v} \; e^{-jk_1 y \cos u \cosh v} e^{k_1 x \cos u \sinh v} \; e^{-jk_1 x \sin u \cosh v}. \qquad (12.60)$$

For leaky wave poles in the w-plane strip $0 \le u \le \pi/2$, $v \le 0$, it is seen that $\sin u \sinh v < 0$ which implies that the wave *grows* exponentially with increasing distance from the interface. It would appear that such waves violate the radiation condition. However, this is not so, because if the observer moves a sufficiently large distance away from the interface, then $\alpha \to 0$ and the SDP in Figure 12.18 will no longer capture any leaky wave poles. Hence, although the leaky wave grows away from the interface, it can only exist within a limited region of space. A geometrical interpretation of this situation is shown in Figure 12.20. This effect is accounted for by including the step function $U(\alpha - \alpha_\ell)$ in (12.59). It is important to realize that a leaky wave does not abruptly appear or disappear at the angle α_l where the SDP crosses the leaky wave pole. Applying a more detailed uniform asymptotic analysis for a pole near a saddle point reveals that the sum of the SDP and pole contributions in (12.57) remains continuous as the SDP crosses a pole.

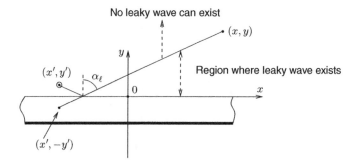

Figure 12.20   Existence region in space, for a leaky wave.

Finally it should be noted that surface waves could be proper or improper, whereas the leaky wave is improper. Poles in the strip $v < 0$, $0 < u \leq \pi/2$ could be improper surface waves (slow waves) or improper leaky waves (fast waves). The $\alpha = \pi/2$ SDP that passes through $u = \pi/2$ forms a boundary line between slow and fast behaviour.

12.6.3   TE Case

The previous results can be modified for a magnetic line source of strength M_0. In this case,

$$H_z = -j\omega\epsilon_1 M_0 G, \tag{12.61}$$

and G satisfies the same wave equation (12.30) as before.

At $y = 0$ the tangential magnetic and electric fields must be continuous, so G must satisfy

$$G(x, y = 0^-) = G(x, y = 0^+) \tag{12.62}$$

$$\frac{1}{\epsilon_2} \frac{\partial G(x, y)}{\partial y}\bigg|_{y=0^-} = \frac{1}{\epsilon_1} \frac{\partial G(x, y)}{\partial y}\bigg|_{y=0^+}. \tag{12.63}$$

E_x must vanish on the ground plane, so $\partial G/\partial y = 0$ at $y = -d$.

The resulting G is the same as in (12.41) (or (12.43)–(12.45)) except that Γ is now given by

$$\Gamma = -\frac{j\kappa_2 \tan \kappa_2 d - \epsilon_r \kappa_1}{j\kappa_2 \tan \kappa_2 d + \epsilon_r \kappa_1} \tag{12.64}$$

or equivalently

$$\Gamma = \frac{j\epsilon_r q + p \tan p}{j\epsilon_r q - p \tan p}, \tag{12.65}$$

where $\epsilon_r = \epsilon_2/\epsilon_1$. As before, $p = \kappa_2 d$ and $q = \kappa_1 d$. The surface wave poles come from the roots of the transcendental equations

$$p \tan p = j\epsilon_r q \tag{12.66}$$

and

$$p^2 - q^2 = (\mu_r \epsilon_r - 1)(k_1 d)^2 = \ell^2. \tag{12.67}$$

Analogous to (12.57), the saddle point and pole contributions are

$$G_s(\boldsymbol{\rho}, \boldsymbol{\rho}') = \frac{1}{j4}\sqrt{\frac{2}{\pi k_1 \rho}} e^{-j(k_1\rho - \pi/4)} \Gamma(k_1 \sin \alpha)$$

$$+ \sum_{w_p} \frac{j\epsilon_r q}{k_1 d \sin w_p B_p} e^{-jk\rho \cos(w_p - \alpha)} u(\alpha - \alpha_p) \qquad (12.68)$$

where

$$B_p = j\epsilon_r \left(1 - (q/p)^2\right) - q\left(1 - (\epsilon_r q/p)^2\right). \qquad (12.69)$$

As before, the summation should account for any leaky and/or surface wave poles that are crossed by the SDP.

12.6.4 Summary

The field reflected by a grounded dielectric slab was asymptotically evaluated by the saddle point method. The saddle point furnishes the reflected field, and the saddle point method gives us the same result as predicted by geometrical optics.

In applying the saddle point method, it is necessary to deform the integration path $\bar{P}$ into the SDP P. In doing so, pole singularities may get crossed. If this happens then they must be accounted for.

The poles, which come from the reflection coefficient Γ, fall into two classes: (a) surface wave poles on the proper sheet of the ν plane, having $\text{Im}\,\nu \leq 0$, and (b) leaky wave poles on the improper sheet, having $\text{Im}\,\nu > 0$.

For the TM case the incident field is the free-space Green's function (12.44), and the scattered field is (12.45). The asymptotic evaluation gives the reflected and surface wave fields (12.57). The term in (12.59) should be added to (12.57) if any leaky wave poles are crossed by the SDP. The TE case follows a similar development; G_f is unchanged, but G_s is now given by (12.68).

12.7 Half Space

A half space is shown in Figure 12.21. The $x < 0$ region is air with μ_1, ϵ_1 and the $x > 0$ region has μ_2, ϵ_2. When ϵ_2 is complex, this configuration is useful for examining radio propagation over lossy earth and seawater. We will start with electric line source excitation and include the permeabilities μ_1, μ_2 so that the field expressions for a magnetic line source can be readily obtained later on by duality.

For an electric line source of strength I_0, the electric field is $E_z = -j\omega\mu_1 I_0 G$, and E_z, H_y are continuous at $x = 0$. Constructing G_x and G_y with the UT method leads to the following Green's function when the source and field point are in region 1 ($x \leq 0$)

$$G = \frac{j}{2\pi} \oint_{C_y} \underbrace{\frac{e^{-j\kappa_1|x-x'|} + \Gamma^e e^{j\kappa_1(x+x')}}{j2\kappa_1}}_{G_x} \underbrace{\frac{e^{-j\sqrt{\lambda_y}|y-y'|}}{j2\sqrt{\lambda_y}}}_{G_y} d\lambda_y. \qquad (12.70)$$

The separation constants obey $\lambda_{x1} + \lambda_y = k_1^2 = \omega^2 \mu_1 \epsilon_1$, $\lambda_{x2} + \lambda_y = k_2^2 = \omega^2 \mu_2 \epsilon_2$ and we have defined $\kappa_1 = \sqrt{\lambda_{x1}} = \sqrt{k_1^2 - \lambda_y}$, $\kappa_2 = \sqrt{\lambda_{x2}} = \sqrt{k_2^2 - \lambda_y}$. Phase matching at $x = 0$ makes λ_y the same in both regions. The reflection coefficient is

$$\Gamma^e = \frac{\mu_2 \kappa_1 - \mu_1 \kappa_2}{\mu_2 \kappa_1 + \mu_1 \kappa_2}. \qquad (12.71)$$

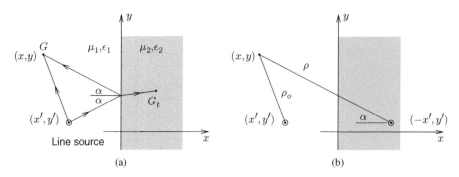

Figure 12.21 (a) Half space with air in region 1 ($x < 0$) and material in region 2 ($x > 0$). The incident and reflected angles are α. (b) Image at $(-x', y')$ and image coordinates (ρ, α).

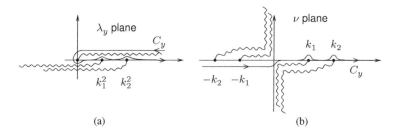

Figure 12.22 G_y has a branch point at $\lambda_y = 0$ and G_x has branch points at $\lambda_y = k_1^2, k_2^2$.

The first and second terms in G_x provide the incident and scattered fields, so G can be written as $G = G_f + G_s$. When $\Gamma^e = 0$, G reduces to the free-space Green's function G_f.

The integrand of (12.70) is sensitive to the square root sign choices for $\sqrt{\lambda_y}$, κ_1 and κ_2, so there are three branch points. G_y has a branch point at $\lambda_y = 0$, and G_x has branch points at $\lambda_y = k_1^2$, and $\lambda_y = k_2^2$. The branch cuts can be chosen as in Figure 12.22(a), making Im $\sqrt{\lambda_y} \leq 0$ as well as Im $\kappa_1 \leq 0$ and Im $\kappa_2 \leq 0$.

It is convenient to make a change of variables $\nu^2 = \lambda_y$. This removes the branch point at $\lambda_y = 0$. In the ν plane the contour C_y and branch cuts now appear as in Figure 12.22(b). Equation (12.70) becomes

$$G = \frac{1}{j4\pi} \oint_{C_y} \frac{e^{-j\kappa_1|x-x'|} + \Gamma^e e^{j\kappa_1(x+x')}}{\kappa_1} e^{-j\nu|y-y'|} d\nu \qquad (12.72)$$

with $\kappa_1 = \sqrt{k_1^2 - \nu^2}$ and $\kappa_2 = \sqrt{k_2^2 - \nu^2}$.

When applying the UT method to find G_x, it was necessary to consider E_z and H_y on both sides of the interface and their continuity; from those equations one can obtain G for $x > 0$

$$G_t = \frac{1}{j4\pi} \oint_{C_y} \frac{(1 + \Gamma^e) e^{j\kappa_1 x'} e^{-j\kappa_2 x}}{\kappa_1} e^{-j\nu|y-y'|} d\nu. \qquad (12.73)$$

For the H_z polarization, the field due to a magnetic line source of strength M_0 immediately follows from duality. The reflection coefficient is of special interest for this polarization. Applying duality to

(12.71) gives

$$\Gamma^h = \frac{\epsilon_2 \kappa_1 - \epsilon_1 \kappa_2}{\epsilon_2 \kappa_1 + \epsilon_1 \kappa_2}. \tag{12.74}$$

This polarization has a pole at $\epsilon_2 \kappa_1 + \epsilon_1 \kappa_2 = 0$. If region 2 is a dielectric with permittivity $\epsilon_r = \epsilon_2/\epsilon_1$, there is a pole ν_p at

$$\underbrace{\epsilon_r \sqrt{k_1^2 - \nu_p^2}}_{\kappa_1} = -\underbrace{\sqrt{\epsilon_r k_1^2 - \nu_p^2}}_{\kappa_2} \tag{12.75}$$

from which

$$\nu_p = \pm k_1 \sqrt{\frac{\epsilon_r}{1 + \epsilon_r}}. \tag{12.76}$$

The pole ν_p is associated with a type of surface wave that is called a *Zenneck wave*. It is bound to the interface and decays with increasing $|x|$ while propagating along the y direction. For the wave to decay in this manner, it must have $\mathrm{Im}\,\kappa_1 \leq 0$ and $\mathrm{Im}\,\kappa_2 \leq 0$. The Zenneck wave does not exist for the E_z polarization and a dielectric half space.

Because of the branch points associated with κ_1 and κ_2, there are four Riemann sheets for the complex ν plane. We can choose Sommerfeld branch cuts. Then, passing through a k_1 branch cut corresponds to a sign change in κ_1 (and hence $\mathrm{Im}\,\kappa_1$); likewise, passage through a k_2 branch cut corresponds to a sign change in κ_2 (and hence $\mathrm{Im}\,\kappa_2$). The four sheets cover all the possibilities, which are

(a) $\mathrm{Im}\ \kappa_1 < 0$ and $\mathrm{Im}\ \kappa_2 < 0$

(b) $\mathrm{Im}\ \kappa_1 > 0$ and $\mathrm{Im}\ \kappa_2 < 0$

(c) $\mathrm{Im}\ \kappa_1 < 0$ and $\mathrm{Im}\ \kappa_2 > 0$

(d) $\mathrm{Im}\ \kappa_1 > 0$ and $\mathrm{Im}\ \kappa_2 > 0$.

The Zenneck wave must lie on sheet (a); we will call it the 'top' sheet.

Choosing the correct signs for the square roots in κ_1 and κ_2 requires some care. Some numerical experiments summarized in Table 12.1 can be used to clarify the concepts.[4] To begin, suppose that there is no loss; let $\epsilon_r = 4$ and $k_1 = 1$. The pole ν_p satisfies (12.75) and is obtained from (12.76). Only ν_p^2 is needed in (12.75) so the $\pm$ sign in (12.76) does not matter. As for $\kappa_{1,2}$, Equation (12.75) shows that for real ϵ_r, both κ_1 and κ_2 are real and that a solution is only possible when $\mathrm{Re}\,\kappa_1$ and $\mathrm{Re}\,\kappa_2$ have opposite signs.

Now suppose that a small amount of loss is introduced so that $\epsilon_r = 4 - j0.4$. A small change in ϵ_r must not cause any large changes in the wave characteristics, so in the lossy case $\mathrm{Re}\,\kappa_1$ and $\mathrm{Re}\,\kappa_2$ should

Table 12.1   Calculation of κ_1 and κ_2.

ϵ_r	ν_p	$\kappa_1 = \pm\sqrt{k_1^2 - \nu_p^2}$	$\kappa_2 = \pm\sqrt{k_2^2 - \nu_p^2}$
4	± 0.894	0.447	1.79
		-0.447	-1.79
$4 - j0.4$	$\pm(0.895 - j0.00888)$	$0.446 + j0.0178$	$1.79 - j0.107$
		$-0.446 - j0.0178$	$-1.79 + j0.107$

where $k_1^2 = 1$ and $k_2^2 = k_1^2 \epsilon_r$.

[4] Thanks are due to Prof. D. R. Jackson for discussing the numerical experiment in a private communication.

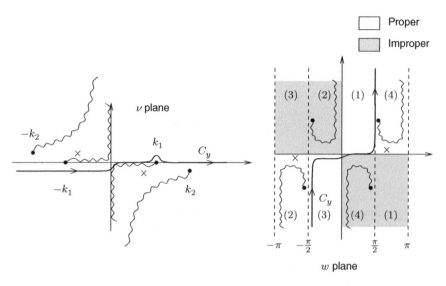

Figure 12.23   Location of the Zenneck wave poles, on the top sheet of the ν plane. k_1 is real, but the branch cuts are shown slightly displaced from the ν-plane axes for clarity. The k_2 branch cuts map to the w plane as shown.

still have opposite signs. Inspecting Table 12.1 shows that if $\operatorname{Re}\kappa_1$ and $\operatorname{Re}\kappa_2$ have *opposite* signs, then $\operatorname{Im}\kappa_1$ and $\operatorname{Im}\kappa_2$ must have the *same* signs. Therefore, in going from the lossless to the lossy case, the Zenneck pole moves from the $\operatorname{Re}\nu$ axis to the gap between the k_1 and k_2 branch cuts, as shown in Figure 12.23. The transformation $\nu = k_1 \sin w$ removes the branch points at $\nu = \pm k_1$ and maps the k_2 branch cuts to the w plane as shown in the figure. The w plane has two sheets, with $\operatorname{Im}\kappa_2 < 0$ on the top sheet and its branch cuts providing access to the lower sheet with $\operatorname{Im}\kappa_2 > 0$.

12.7.1   Asymptotic Evaluation

The contour integral expression (12.72) for G contains a saddle point, branch cuts and, for the H_z polarization, a Zenneck pole. If the source and field point are away from the interface, the saddle point method can be used to evaluate the integral. The procedure is similar to the grounded slab. With reference to Figure 12.21(b) we can define polar image coordinates

$$-(x + x') = \rho \cos \alpha \tag{12.77}$$

$$y - y' = \rho \sin \alpha. \tag{12.78}$$

The Green's function (12.72) has free space and scattered parts $G = G_f + G_s$. To evaluate G_s with the saddle point method, we employ the polar image coordinates and the change of variables $\nu = k_1 \sin w$. The path C_y is deformed into the SDP P. Then,

$$G_s(\boldsymbol{\rho}, \boldsymbol{\rho}') = \frac{1}{j4\pi} \int_P e^{-jk_1\rho\cos(w-\alpha)}\, \Gamma^{e,h}(k_1 \sin w)\, dw. \tag{12.79}$$

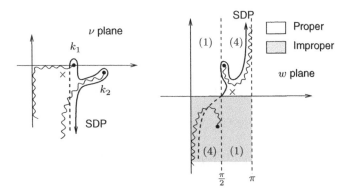

Figure 12.24   SDP in the ν plane and w plane, when $w_s = \alpha = \pi/2$. The pole is only present for the H_z polarization.

The saddle point is at $w_s = \alpha$, and as long as $\alpha \not\approx \pi/2$ the saddle point will be isolated from any other singularities. Under this condition we can apply the saddle point formula (12.25) with the result that

$$G_s(\boldsymbol{\rho}, \boldsymbol{\rho}') = \frac{1}{j4}\sqrt{\frac{2}{\pi k_1 \rho}} e^{-j(k_1 \rho - \pi/4)} \Gamma^{e,h}(k_1 \sin \alpha). \qquad (12.80)$$

This is the same result as geometrical optics, with $\Gamma^{e,h}$ being the Fresnel reflection coefficient. At grazing incidence, $\alpha \to \pi/2$; for the E_z polarization, $\Gamma^e \to 1$, so the incident and reflected rays add. For the H_z polarization, $\Gamma^h \to -1$, so the incident and reflected rays cancel. The saddle point formula only provides terms of $O(1/\sqrt{k_1 \rho})$, so the cancellation does not imply the total field is zero – as there can be higher-order terms.

When $\alpha \approx \pi/2$ the GO result is incomplete. The SDP requires a detour around a branch point at $\nu = k_2$, as shown in Figure 12.24. The depiction of the Sommerfeld branch cuts here assumes that region 1 is lossless and region 2 is lossy. As before, the saddle point provides the GO part of the field. The detour around the branch point provides an additional contribution that is called a *lateral wave*. For this type of wave, an asymptotic analysis reveals that $G \propto 1/\rho^{3/2}$ in 2D or $G \propto 1/r^2$ in 3D. The lateral wave decays faster than the GO part, and is usually negligible whenever $\epsilon_2 > \epsilon_1$ and/or losses are present. When $\epsilon_1 > \epsilon_2$, total internal reflection can occur. In this case the incident ray refracts, travels parallel to the boundary (i.e. 'laterally') and then refracts away from the interface. A lateral wave can be stronger than the GO part when region 1 is lossy and region 2 is lossless.

For $\alpha \approx \pi/2$ and the H_z polarization, there is an additional contribution from the Zenneck pole near the saddle point as shown in Figure 12.24. When deforming the integration path to the SDP, the pole is *not* crossed. Therefore there is no residue. However, the pole affects the nearby saddle point, necessitating the uniform asymptotic formula (12.27).

Equation (12.27) is valid for a function $f(z)$ that has a simple pole but is otherwise slowly varying. That is, $f(z) = g(z)/(z - z_p)$ where $g(z)$ is slowly varying. This requirement may not be met for the half space. When $\alpha \approx \pi/2$ and $|\epsilon_r|$ are large, the numerator of Γ^h can change rapidly. A solution is to substitute $\Gamma^h = 1 + \hat{\Gamma}$ where

$$\hat{\Gamma} = \Gamma^h - 1 = \frac{\epsilon_r \kappa_1 - \kappa_2}{\epsilon_r \kappa_1 + \kappa_2} - 1 = \frac{-2\kappa_2}{\epsilon_r \kappa_1 + \kappa_2}$$

which provides a slowly varying numerator. A good approximation of $\hat{\Gamma}$ for w near α is

$$\hat{\Gamma}(k_1 \sin w) \approx \hat{\Gamma}(k_1 \sin \alpha) \frac{\alpha - w_p}{w - w_p}.$$

With Equation (12.79) and these approximations, G_s can be expressed as

$$G_s = \frac{1}{j4\pi} \int_P e^{-jk_1\rho\cos(w-\alpha)} \left(1 + \hat{\Gamma}(k_1 \sin w)\right) dw$$

$$= \frac{1}{j4} H_0^{(2)}(k_1\rho) + \frac{1}{j4\pi} \hat{\Gamma}(k_1 \sin \alpha)(\alpha - w_p) \int_P \frac{e^{-jk_1\rho\cos(w-\alpha)}}{w - w_p} dw. \quad (12.81)$$

The '1' in the $1 + \hat{\Gamma}$ term provides a free-space Green's function with the source at the image point. The next step is to evaluate the integral

$$I_s = \int_P \frac{e^{-jk_1\rho\cos(w-\alpha)}}{w - w_p} dw.$$

This can be done with (12.27) where $\Omega = k_1\rho$, $q(w) = -j\cos(w - \alpha)$, $f(w) = 1/(w - w_p)$, $a = 1$, $b = e^{-j\pi/4}\sqrt{2}\sin((w_p - \alpha)/2)$ and $h = \sqrt{2}e^{j\pi/4}$. The pole is not crossed by the SDP, so $\operatorname{Im} b < 0$. The result is

$$I_s = e^{-jk_1\rho} \left[-j2\sqrt{\pi}\, e^{-k_1\rho b^2} Q(jb\sqrt{k_1\rho}) + \sqrt{\frac{\pi}{k_1\rho}} \left(\frac{\sqrt{2}e^{j\pi/4}}{\alpha - w_p} + \frac{1}{b} \right) \right]. \quad (12.82)$$

Combining $G = G_f + G_s$,

$$G = \frac{1}{j4} H_0^{(2)}(k_1\rho_0) + \frac{1}{j4} H_0^{(2)}(k_1\rho) + \frac{1}{j4\pi} \hat{\Gamma}(k_1 \sin \alpha)(\alpha - w_p) I_s. \quad (12.83)$$

The distance ρ_0 is from the line source to the field point, and ρ is from the image to the field point, as shown in Figure 12.21(b).

For the case of grazing incidence, the term

$$(\alpha - w_p)\left(\frac{\sqrt{2}e^{j\pi/4}}{\alpha - w_p} + \frac{1}{b} \right) \to 0.$$

With $\alpha = \pi/2$ and $\rho_0 = \rho$, using the large-argument forms for H_0 in (12.83) and the asymptotic approximation $Q(y) \sim 1/(2y\, e^{y^2})$ in (12.82), one finds that the two Hankel functions in (12.83) are cancelled out by the Q contribution. Therefore, at grazing, G is zero to $O(1/\sqrt{k_1\rho})$.

The Zenneck wave created a lot of excitement in the early 1900s because it is a pole wave. Considered alone, its residue provides a contribution of the form $G_{sw} = Ke^{-jk\rho\cos(w_p-\alpha)}$ (with $K = \text{const.}$) which implies propagation without a $1/\sqrt{k_1\rho}$ decay. However the pole does not exist in isolation, so such a wave cannot be physically observed. It has an indirect effect, affecting the saddle point. By including the next higher-order term in $Q(y) \sim 1/(2y\, e^{y^2})(1 - 1/(2y^2))$, one finds that $G \propto 1/(k_1\rho)^{3/2}$.

A more friendly form of (12.83) can be obtained by factoring out the asymptotic form of $H_0^{(2)}(k_1\rho)$ in the third term of (12.83) and combining the second and third terms to obtain

$$G = \frac{1}{j4} H_0^{(2)}(k_1\rho_0) + \frac{1}{j4} H_0^{(2)}(k_1\rho) \left(1 + S(\rho, \alpha)\right), \quad (12.84)$$

where

$$S(\rho, \alpha) = (\alpha - w_p)\hat{\Gamma}(k_1 \sin \alpha)\sqrt{\frac{k_1\rho}{2}} \, e^{-j\pi/4}$$

$$\times \left[-j2 \, e^{-k_1\rho b^2} Q(jb\sqrt{k_1\rho}) + \sqrt{\frac{1}{k_1\rho}} \left(\frac{\sqrt{2}e^{j\pi/4}}{\alpha - w_p} + \frac{1}{b} \right) \right]. \tag{12.85}$$

For a z-directed dipole, G can be obtained via the 2D to 3D transformation in (11.86). The result in the $z = 0$ plane is

$$G = \frac{e^{-jk_1 r_0}}{4\pi r_0} + \frac{e^{-jk_1 r}}{4\pi r}(1 + S(r, \alpha)). \tag{12.86}$$

Although r_0, r are the same as ρ_0 and ρ, they have been changed to follow the notational conventions of 3D.

To find the field of a z-directed magnetic dipole, $H_z = -j\omega\epsilon_1 p_m G$ with $\hat{\Gamma} = \Gamma^h - 1$; likewise, for a z-directed electric dipole, $E_z = -j\omega\mu_1 p_e G$ with $\hat{\Gamma} = \Gamma^e - 1$.

12.7.2 Vertical Electric Dipole

We will now find the field produced by an electric dipole that is perpendicular to a horizontal and lossy ground. This is known as 'the Sommerfeld problem'. The ground is a lossy dielectric, typically earth or seawater. The field can be obtained from the magnetic dipole solution in the previous section. Knowing **H**, we can find **E** and its component normal to the interface. Then, by placing an electric dipole normal to the interface and applying reciprocity, the radiation field can be found.

The configuration is shown in Figure 12.25. In (a), a z-directed magnetic dipole p_m^a produces $H_z^a = -j\omega\epsilon_1 p_m^a G$. Taking the curl of **H** gives **E**. The x component requires $\partial H_z/\partial y$. Considering the rectangular form of G in (12.72), one finds that

$$\frac{\partial G}{\partial y} = \mp j\nu G; \, y \gtrless y'$$

and at the saddle point, $\nu = k_1 \sin \alpha$. Therefore, $E_x^a = -j p_m^a k_1 \sin \alpha \, G$. The configurations in Figure 12.25(a) and (b) are reciprocal so that

$$p_m^a \, H_z^b = -p_e^b \, E_x^a$$

from which

$$H_z^b = j p_e^b k_1 \sin \alpha \, G.$$

Figure 12.25(c) shows the conventional 3D coordinates that are usually used for the Sommerfeld problem. With G from (12.86), the incident and scattered waves become

$$H_\phi = \frac{j p_e k_1}{4\pi} \left(\frac{e^{-jk_1 r_0}}{r_0} \sin \alpha_0 + \frac{e^{-jk_1 r}}{r}(1 + S(r, \alpha)) \sin \alpha \right) \tag{12.87}$$

and since these are spherical waves, $E_\theta = \eta_1 H_\phi$ where $\eta_1 = \sqrt{\mu_1/\epsilon_1}$.

At large distances and grazing incidence, the field is zero to $O(1/k_1 r)$. The behaviour is dominated by the next higher-order term of $Q(jb\sqrt{k_1 r})$ that is in $S(r, \alpha)$. One finds that $|H_\phi| \sim 1/(k_1 r)^2$. Therefore, the field drops off at the rate of 12 dB per double distance, in contrast to the 6 dB per double distance

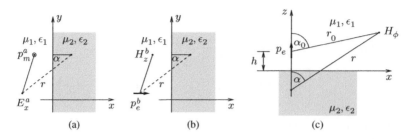

(a)   (b)   (c)

Figure 12.25 Half space with (a) magnetic dipole p_m^a. (b) Reciprocal problem with electric dipole p_e^b. (c) Conventional coordinates for vertical dipole in 3D.

for a field in free space or on a perfectly conducting ground. This $1/(k_1 r)^2$ wave is known as a *ground wave*. It is not a surface wave, though it is sometimes called a *Norton surface wave*.

Example 12.3 (VED on Lossy Earth) A vertical electric dipole has an arbitrary strength p_e and is directly on lossy earth. Calculate the ground wave along the earth's surface for a distance of 1–100 km. The earth has $\epsilon_r = 15$ and $\sigma = 0.012$ S/m. Consider two frequencies, 0.3 and 3 MHz.

Solution: The pertinent geometry is in Figure 12.25(c) and $h = 0$. The field is obtained from Equation (12.87). This has been coded in PROGRAM ved. For each frequency the magnetic field has been normalized to 1 A/m at a distance of 1 km and plotted in Figure 12.26.

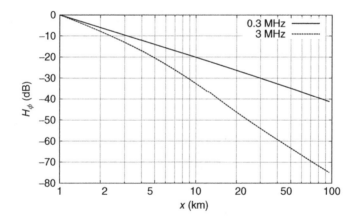

Figure 12.26 Normalized $|H_\phi|$ for a vertical electric dipole above lossy earth.

At 0.3 MHz and large distances, the field decays at the rate of 6 dB per double distance. Raising the frequency to 3 MHz, the decay rate increases to 12 dB per double distance. Therefore, lower frequencies are desirable for ground wave propagation.

GO does not provide a satisfactory result. The GO reflection coefficient is $\Gamma^h = -1$, so the incident and reflected fields cancel, giving $H_\phi = 0$. ∎

For ground wave calculations the Zenneck pole ν_p is obtained from (12.76). In the w plane this is at $w_p = \arcsin(\nu_p/k_1)$. It is required that $\operatorname{Im} w_p \geq 0$, as shown in Figure 12.24. To get this, the function

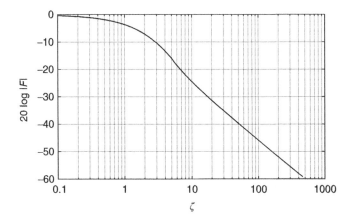

Figure 12.27   Attenuation function $|F|$ versus numerical distance ζ.

$\arcsin(z) = -j \ln(jz \pm \sqrt{1 - z^2})$ should be used with the minus sign on the square root. For large $|\epsilon_r|$ and arg $\epsilon_r \approx -\pi/2$, it can be shown that $w_p \approx \pi/2 + e^{j\pi/4}/\sqrt{|\epsilon_r|}$.

With the source and field points on the ground, $\alpha \to \pi/2$. Assuming a large and lossy ϵ_r, (12.87) can be simplified and reduces to

$$H_\phi = jp_e k_1 \frac{e^{-jkr}}{2\pi r} F(\zeta),$$ (12.88)

where

$$F(\zeta) = (1 - j2\sqrt{\zeta}\, e^{-\zeta} Q(j\sqrt{\zeta})).$$ (12.89)

F is a correction factor for lossy ground, and $F = 1$ when the ground is a perfect electric conductor. The argument of Q is approximated as $jb\sqrt{k_1 r} \approx j\sqrt{\zeta}$, where

$$\zeta = \frac{k_1 r}{2|\epsilon_r|}$$ (12.90)

is called Sommerfeld's 'numerical distance'. Figure 12.27 shows that for $\zeta = 5, 50$ and 500, the loss factor $-20 \log |F|$ is about $16, 40$ and 60 dB, respectively. This is an additional loss on top of the spherical wave $1/r$ amplitude loss for a perfect ground.

12.8   Circular Cylinder

We wish to find the Green's function for the circular cylinder, shown in Figure 12.28. A line source is at (ρ', ϕ'), and the field point is at (ρ, ϕ). The cylinder radius is a. We need to solve the 2D wave equation

$$\left[\frac{1}{\rho} \frac{\partial}{\partial \rho} \left(\rho \frac{\partial}{\partial \rho} \right) + \frac{1}{\rho^2} \frac{\partial^2}{\partial \phi^2} + k^2 \right] G(\rho, \phi; \rho', \phi') = -\frac{\delta(\rho - \rho')\delta(\phi - \phi')}{\rho}$$ (12.91)

for $a \le \rho < \infty$, $-\pi < \phi \le \pi$. The boundary conditions given are

$$G(a, \phi; \rho', \phi') = 0,$$ (12.92)

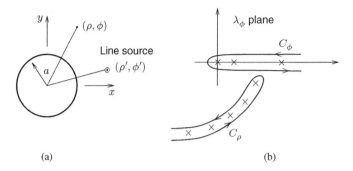

Figure 12.28 (a) Circular cylinder with radius a, source at (ρ', ϕ') and field point at (ρ, ϕ). (b) Singularities of G_ρ and G_ϕ in the λ_ϕ plane.

the radiation condition for $e^{j\omega t}$ time dependence, as well as the requirement that G should be 2π periodic in ϕ

$$G(\rho, \phi; \rho', \phi') = G(\rho, \phi + 2\pi\ell; \rho', \phi'); \quad \ell = \pm 1, \pm 2, \pm 3, \ldots . \tag{12.93}$$

Using the GSV,

$$G(\rho, \phi; \rho', \phi') = K \oint_{C_\phi} G_\rho(\rho, \rho'; \lambda_\rho) G_\phi(\phi, \phi'; \lambda_\rho) \, d\lambda_\phi. \tag{12.94}$$

Substituting (12.94) into (12.91),

$$K \oint_{C_\phi} \frac{G_\phi}{\rho} \frac{d}{d\rho} \left(\rho \frac{d}{d\rho} G_\rho \right) + \frac{G_\rho}{\rho^2} \frac{d^2 G_\phi}{d\phi^2} + k^2 G_\rho G_\phi \, d\lambda_\phi = -\frac{\delta(\rho - \rho') \delta(\phi - \phi')}{\rho}. \tag{12.95}$$

The 1D wave equations

$$\frac{1}{\rho} \frac{d}{d\rho} \left(\rho \frac{dG_\rho}{d\rho} \right) + \lambda_\rho G_\rho = -\frac{\delta(\rho - \rho')}{\rho} \tag{12.96}$$

$$\frac{d^2 G_\phi}{d\phi^2} + \lambda_\phi G_\phi = -\delta(\phi - \phi') \tag{12.97}$$

are used to replace the second derivatives in (12.95) so that

$$K \oint_{C_\phi} - \lambda_\rho G_\rho G_\phi - G_\phi \frac{\delta(\rho - \rho')}{\rho} - \lambda_\phi \frac{G_\rho G_\phi}{\rho^2} - \frac{G_\rho}{\rho^2} \delta(\phi - \phi')$$

$$+ k^2 G_\rho G_\phi \, d\lambda_\phi = -\frac{\delta(\rho - \rho') \delta(\phi - \phi')}{\rho}. \tag{12.98}$$

By imposing the condition

$$\lambda_\rho + \frac{\lambda_\phi}{\rho^2} = k^2 \tag{12.99}$$

the first, third and fifth terms in (12.98) cancel out. Furthermore, the contour C_ϕ excludes the singularities of G_ρ so that the fourth term integrates to zero. The relation

$$\oint_{C_\phi} G_\phi(\phi, \phi'; \lambda_\phi) \, d\lambda_\phi = -j2\pi\delta(\phi - \phi')$$

reduces (12.98) to

$$K j2\pi\delta(\phi - \phi') \frac{\delta(\rho - \rho')}{\rho} = -\frac{\delta(\rho - \rho')\delta(\phi - \phi')}{\rho}$$

from which

$$K = \frac{j}{2\pi}.$$

Hence (12.94) becomes

$$G(\rho, \phi; \rho', \phi') = \frac{j}{2\pi} \oint_{C_\phi} G_\rho(\rho, \rho'; \lambda_\rho) G_\phi(\phi, \phi'; \lambda_\phi) \, d\lambda_\phi. \tag{12.100}$$

Let us now return to solving the one-dimensional wave equations (12.96), (12.97). Equation (12.93) requires that G should be 2π periodic in ϕ, so we need to impose that condition on G_ϕ in (12.97). Related to (12.97) is the eigenvalue problem

$$\left[\frac{d^2}{d\phi^2} + \lambda_m \right] \psi_m(\phi) = 0$$

which has orthonormal eigenfunctions

$$\psi_m(\phi) = \frac{1}{\sqrt{2\pi}} e^{jm\phi}$$

and eigenvalues $\lambda_m = m^2$. Using the techniques in Section 11.7 we can obtain the series representation of G_ϕ

$$G_\phi(\phi, \phi'; \lambda_\phi) = \sum_{m=-\infty}^{\infty} \frac{\psi_m^*(\phi')\psi_m(\phi)}{\lambda_m - \lambda_\phi} = \sum_{m=0}^{\infty} \frac{\epsilon_{0m}}{2\pi} \frac{\cos m(\phi - \phi')}{m^2 - \lambda_\phi}, \tag{12.101}$$

where ϵ_{0m} is Neumann's number, with $\epsilon_{00} = 1$ and $\epsilon_{0m} = 2$ for $m \neq 0$. G_ϕ has simple poles at $\lambda_\phi = m^2$. Next, we need to solve (12.96) for G_ρ. Since $\lambda_\rho = k^2 - \lambda_\phi/\rho^2$, (12.96) becomes

$$\frac{1}{\rho} \frac{d}{d\rho} \left(\rho \frac{dG_\rho}{d\rho} \right) + k^2 G_\rho - \frac{\lambda_\phi}{\rho^2} G_\rho = -\frac{\delta(\rho - \rho')}{\rho}. \tag{12.102}$$

This is Bessel's differential equation, and the solutions are $Z_{\sqrt{\lambda_\phi}}(k\rho)$ where Z can be any two of J, Y, $H^{(1)}$ and $H^{(2)}$ as independent solutions. G_ρ must satisfy the radial boundary condition (12.92) so that $G_\rho(a, \rho') = 0$. Using the UT method, we can choose

$$U(\rho) = J_{\sqrt{\lambda_\phi}}(k\rho) + c H^{(2)}_{\sqrt{\lambda_\phi}}(k\rho), \tag{12.103}$$

and requiring $U(a) = 0$ gives us the constant

$$c = -\frac{J_{\sqrt{\lambda_\phi}}(ka)}{H^{(2)}_{\sqrt{\lambda_\phi}}(ka)}. \tag{12.104}$$

The radiation condition requires $e^{-jk\rho}/\sqrt{\rho}$ behaviour as $\rho \to \infty$, so we must choose

$$T(\rho) = H^{(2)}_{\sqrt{\lambda_\phi}}(k\rho). \tag{12.105}$$

The conjunct is

$$W(T,U) = W(H^{(2)}_{\sqrt{\lambda_\phi}}(k\rho), J_{\sqrt{\lambda_\phi}}(k\rho)) = \frac{j2}{\pi}. \tag{12.106}$$

It should be remembered that the derivatives used in evaluating W are with respect to ρ and not $k\rho$. From (12.103)–(12.106) we obtain

$$G_\rho(\rho, \rho') = \frac{-j\pi}{2} \left[J_{\sqrt{\lambda_\phi}}(k\rho_<) - \frac{J_{\sqrt{\lambda_\phi}}(ka)}{H^{(2)}_{\sqrt{\lambda_\phi}}(ka)} H^{(2)}_{\sqrt{\lambda_\phi}}(k\rho_<) \right] H^{(2)}_{\sqrt{\lambda_\phi}}(k\rho_>). \tag{12.107}$$

G_ρ has poles which occur at the complex roots ν_p of $H^{(2)}_{\nu_p}(ka) = 0$. These are the 'creeping wave' poles and are located at $\lambda_\phi = \nu_p^2$. G_ϕ has poles at $\lambda_\phi = m^2$. The poles of G_ρ and G_ϕ in the λ_ϕ plane are illustrated in Figure 12.28(b).

Substituting (12.101) and (12.107) in (12.100),

$$G(\rho, \phi; \rho', \phi') = \frac{1}{8\pi} \sum_{m=0}^{\infty} \oint_{C_\phi} \left[J_{\sqrt{\lambda_\phi}}(k\rho_<) - \frac{J_{\sqrt{\lambda_\phi}}(ka)}{H^{(2)}_{\sqrt{\lambda_\phi}}(ka)} H^{(2)}_{\sqrt{\lambda_\phi}}(k\rho_<) \right]$$

$$\cdot H^{(2)}_{\sqrt{\lambda_\phi}}(k\rho_>) \frac{\epsilon_{0m} \cos m(\phi - \phi')}{m^2 - \lambda_\phi} d\lambda_\phi. \tag{12.108}$$

The contour C_ϕ excludes the singularities of G_ρ. The integral is evaluated by taking $j2\pi$ times the residues at the poles $\lambda_\phi = m^2$ of G_ϕ, with the result that

$$G(\rho, \phi; \rho', \phi') = \frac{1}{j4} \sum_{m=0}^{\infty} \epsilon_{0m} \cos m(\phi - \phi')$$

$$\times \left[J_m(k\rho_<) - \frac{J_m(ka)}{H^{(2)}_m(ka)} H^{(2)}_m(k\rho_<) \right] H^{(2)}_m(k\rho_>). \tag{12.109}$$

This is the formal solution for G. The first term can be interpreted as the incident field, and the second term provides the scattered field. An interesting and useful by-product of this derivation is that the free-space Green's function can be expressed as

$$G_f = \frac{1}{j4} H^{(2)}_0(k|\boldsymbol{\rho} - \boldsymbol{\rho}'|) = \frac{1}{j4} \sum_{m=0}^{\infty} \epsilon_{0m} \cos m(\phi - \phi') J_m(k\rho_<) H^{(2)}_m(k\rho_>). \tag{12.110}$$

12.8.1 Creeping Waves

In the previous section we used an angular Green's function G_ϕ that is 2π periodic. We will now find a different Green's function G_ϕ^∞ that satisfies the radiation condition as $|\phi| \to \infty$. Contrary to the usual

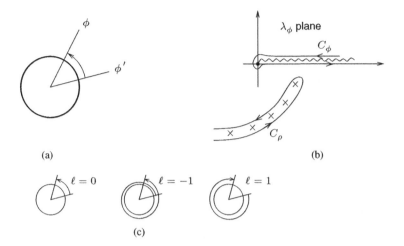

Figure 12.29 (a) Angular domain $-\infty < \phi, \phi' < \infty$ for G_ϕ^∞. (b) Singularities of G_ρ and G_ϕ^∞ in the λ_ϕ plane. (c) Interpretation of ℓ for the encirclements.

way of reckoning angles as 2π periodic, the angular domain shown in Figure 12.29(a) covers $-\infty < \phi < \infty$. This is non-physical, in that $G_\phi^\infty(\phi, \phi') \neq G_\phi^\infty(\phi + 2\pi, \phi')$. However it will provide a Green's function for circumferentially propagating waves, which are useful for representing waves travelling along a curved surface, known as creeping waves.

A solution for the angular wave equation

$$\left(\frac{d^2}{d\phi^2} + \lambda_\phi\right) G_\phi^\infty(\phi, \phi') = -\delta(\phi - \phi') \tag{12.111}$$

that satisfies the radiation condition in the infinite angular domain can be found with the UT method. The result is

$$G_\phi^\infty(\phi, \phi') = \frac{e^{-j\sqrt{\lambda_\phi}|\phi - \phi'|}}{j2\sqrt{\lambda_\phi}}. \tag{12.112}$$

Since the angular domain extends to infinity, G_ϕ^∞ has a branch point at $\lambda_\phi = 0$. The branch cut is chosen so that $-2\pi < \arg \lambda_\phi \leq 0$, giving Im $\sqrt{\lambda_\phi} \leq 0$. The singularities of G_ρ and G_ϕ^∞ appear as in Figure 12.29(b).

To validate this concept, it will be shown that summing up all of the waves that encircle the cylinder gives back our original G_ϕ in (12.101) that is 2π periodic. These encirclements appear as in Figure 12.29(c). At a fixed point ϕ, an infinite number of encirclements occur in both the counterclockwise and clockwise directions. Replacing ϕ with $\phi + 2\pi\ell$ leads to

$$G_\phi(\phi, \phi') = \sum_{\ell=-\infty}^{\infty} G_\phi^\infty(\phi + 2\pi\ell, \phi') = \sum_{\ell=-\infty}^{\infty} \frac{e^{-j\sqrt{\lambda_\phi}|\phi - \phi' + 2\pi\ell|}}{j2\sqrt{\lambda_\phi}}, \tag{12.113}$$

and the summation index ℓ is associated with encirclements. If we restrict ϕ and ϕ' to the range $(-\pi, \pi)$, then (12.113) can be rewritten without the absolute value signs as

$$G_\phi(\phi,\phi') = \sum_{\ell=0}^{\infty} \frac{e^{-j\sqrt{\lambda_\phi}(\phi-\phi')-j2\pi\ell\sqrt{\lambda_\phi}}}{j2\sqrt{\lambda_\phi}} + \sum_{\ell=-1}^{-\infty} \frac{e^{j\sqrt{\lambda_\phi}(\phi-\phi')+j2\pi\ell\sqrt{\lambda_\phi}}}{j2\sqrt{\lambda_\phi}}; \quad \phi > \phi' \qquad (12.114)$$

$$G_\phi(\phi,\phi') = \sum_{\ell=0}^{\infty} \frac{e^{j\sqrt{\lambda_\phi}(\phi-\phi')-j2\pi\ell\sqrt{\lambda_\phi}}}{j2\sqrt{\lambda_\phi}} + \sum_{\ell=-1}^{-\infty} \frac{e^{-j\sqrt{\lambda_\phi}(\phi-\phi')+j2\pi\ell\sqrt{\lambda_\phi}}}{j2\sqrt{\lambda_\phi}}; \quad \phi < \phi'. \qquad (12.115)$$

Using the geometric series

$$\sum_{\ell=0}^{\infty} e^{-j\ell x} = \frac{e^{jx/2}}{j2\sin(x/2)}; \qquad \sum_{\ell=-1}^{-\infty} e^{j\ell x} = \frac{e^{-jx/2}}{j2\sin(x/2)}$$

we can rewrite (12.114) and (12.115) as

$$G_\phi(\phi,\phi') = -\frac{\cos[\sqrt{\lambda_\phi}(\pi - |\phi-\phi'|)]}{2\sqrt{\lambda_\phi}\sin[\sqrt{\lambda_\phi}\pi]}, \qquad (12.116)$$

where $|\phi-\phi'| < 2\pi$.

It is not obvious that Equation (12.116) is 2π periodic. Even less obvious is the fact that it is equivalent to (12.101). The equivalence can be established by making use of the completeness relation

$$\frac{1}{j2\pi}\oint_{C_\phi} G_\phi(\phi,\phi';\lambda_\phi)\,d\lambda_\phi = -\sum_m \psi_m^*(\phi')\psi_m(\phi) = \sum \text{Res}\, G_\phi(\phi,\phi';\lambda_\phi). \qquad (12.117)$$

Here, C_ϕ encloses the singularities of G_ϕ in the λ_ϕ plane, as shown in Figure 12.28(b). With (12.116) in (12.117),

$$\sum \text{Res}\left[\frac{-\cos[\sqrt{\lambda_\phi}(\pi - |\phi-\phi'|)]}{2\sqrt{\lambda_\phi}\sin[\sqrt{\lambda_\phi}\pi]}\right] = -\sum_m \psi_m^*(\phi')\psi_m(\phi). \qquad (12.118)$$

The left-hand side does not depend on the sign of $\pm\sqrt{\lambda_\phi}$, so there is no branch cut for the square root. There are poles at $\sqrt{\lambda_\phi} = m$, where $m = 0,1,2,\cdots$. Taking the residue,

$$\sum_{m=0}^{\infty} \frac{-\cos[m(\pi - |\phi-\phi'|)]}{2\frac{d}{d\lambda}\{\sqrt{\lambda_\phi}\sin[\sqrt{\lambda_\phi}\pi]\}|_{\lambda_\phi=m^2}} = -\sum_m \psi_m^*(\phi')\psi_m(\phi)$$

or

$$-\frac{1}{2\pi}\sum_{m=0}^{\infty} \epsilon_{0m}\cos m(\phi-\phi') = -\sum_m \psi_m^*(\phi')\psi_m(\phi).$$

Knowing this sum permits us to obtain

$$G_\phi(\phi,\phi') = \frac{1}{2\pi}\sum_{m=0}^{\infty} \frac{\epsilon_{0m}\cos m(\phi-\phi')}{m^2 - \lambda_\phi}$$

which is identical to (12.101), the 2π periodic Green's function found earlier.

In summary, we have found a Green's function G_ϕ^∞ in (12.112) that represents a creeping wave, satisfying the radiation condition in the infinite angular domain $-\infty < \phi < \infty$. Summing up the multiple encirclements in (12.113) can be done in closed form and leads to (12.116). The expressions (12.113) and (12.116) are 2π periodic and are equivalent to G_ϕ in (12.101), found earlier.

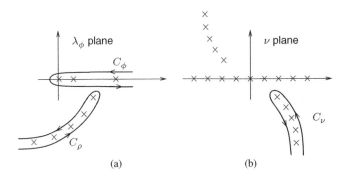

Figure 12.30 Integration paths in the λ_ϕ plane and ν plane.

12.8.2 Residue Series

We have been using (12.100) to construct the cylinder Green's function, with C_ϕ as in Figure 12.30(a). This contour can be deformed into $-C_\rho$ (the clockwise direction of C_ρ) without crossing any singularities. Therefore we can rewrite (12.100) and obtain an alternative Green's function

$$G(\rho, \phi; \rho', \phi') = \frac{1}{j2\pi} \oint_{C_\rho} G_\rho(\rho, \rho'; \lambda_\rho) G_\phi(\phi, \phi'; \lambda_\phi) \, d\lambda_\phi.$$

With a change of variables $\nu = \sqrt{\lambda_\phi}$ and using contour C_ν, this becomes

$$G(\rho, \phi; \rho', \phi') = \frac{1}{j2\pi} \oint_{C_\nu} G_\rho(\rho, \rho'; \nu) G_\phi(\phi, \phi'; \nu) 2\nu \, d\nu. \tag{12.119}$$

The contour C_ν is shown in Figure 12.30(b). Using (12.107) for G_ρ and (12.113) for G_ϕ in (12.119),

$$G(\rho, \phi; \rho', \phi') = \frac{j}{4} \sum_{\ell=-\infty}^{\infty} \oint_{C_\nu} \left[J_\nu(k\rho_<) - \frac{J_\nu(ka)}{H_\nu^{(2)}(ka)} H_\nu^{(2)}(k\rho_<) \right]$$
$$\cdot H_\nu^{(2)}(k\rho_>) e^{-j\nu|\phi-\phi'+2\pi\ell|} \, d\nu. \tag{12.120}$$

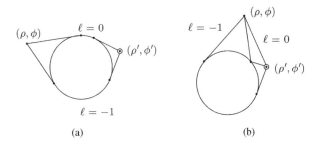

Figure 12.31 Circular cylinder, $\ell = 0$ and $\ell = -1$ rays. (a) Shadow region, (b) lit region.

The poles of G_ϕ appear at $\lambda_\phi = m^2$ in the λ_ϕ plane or at $\nu = m$ in the ν plane, where $m = 0, \pm 1, \pm 2, \ldots$. There are an infinite number of poles associated with G_ρ which occur at the zeros of $H_{\nu_p}^{(2)}(ka) = 0$, enumerated as $p = 1, 2, \ldots$. Loosely speaking, these creeping wave poles ν_p appear as in Figure 12.30. Their characteristics are summarized in Appendix C. Hankel functions of complex order can be computed using SUBROUTINE huni.

It is tempting to proceed with a residue evaluation at the poles ν_p of G_ρ. This gives the so-called residue series. It turns out that the residue series converges for the $\ell \neq 0$ terms but may or may not converge for the $\ell = 0$ term. This will be explained shortly. For this reason it is prudent to keep the $\ell = 0$ term separate, as an integral. Taking $j2\pi$ times the counterclockwise residues at the poles ν_p for the $\ell \neq 0$ terms,

$$G(\rho, \phi; \rho', \phi') = \frac{j}{4} \oint_{C_\nu} \left[J_\nu(k\rho_<) - \frac{J_\nu(ka)}{H_\nu^{(2)}(ka)} H_\nu^{(2)}(k\rho_<) \right] H_\nu^{(2)}(k\rho_>) e^{-j\nu|\phi - \phi'|} \, d\nu$$

$$+ \frac{\pi}{2} \sum_{\ell=-\infty}^{\infty}{}' \sum_{p=1}^{\infty} \frac{J_{\nu_p}(ka) H_{\nu_p}^{(2)}(k\rho_<) H_{\nu_p}^{(2)}(k\rho_>)}{\frac{\partial}{\partial \nu} H_\nu^{(2)}(ka)|_{\nu=\nu_p}} e^{-j\nu_p|\phi - \phi' + 2\pi\ell|}. \qquad (12.121)$$

The prime on the summation denotes exclusion of the $\ell = 0$ term. The term involving $J_\nu(k\rho_<)$ does not contribute to the residues.

The terms indexed by ℓ in (12.121) can be associated with rays. The interpretation of the $\ell = 0$ and $\ell = -1$ terms in is shown in Figure 12.31. The $\ell = 0$ ray is associated with the integral in (12.121). When the field point is in the shadow region, the integral can be evaluated by a residue series. Therefore the integral can be dropped, with its contribution accounted for by including the $\ell = 0$ term in the summation.

When the field point is in the lit region, the integral has two saddle points, which provide the incident and reflected rays of geometrical optics. The residue series cannot be used for the $\ell = 0$ term in the lit region because it is divergent. Since $|\phi - \phi'| \leq \pi$ the $\ell \neq 0$ rays are always in the shadow region, and their residue series are convergent. For the lit-region asymptotic analysis involving two saddle points, the reader is referred to Felsen and Marcuvitz (1994, Section 6.7) for the details.

Near a shadow boundary, a uniform asymptotic evaluation for two saddle points near a pole is needed. When the source and field points are away from the cylinder's surface, the result is in terms of Pekeris functions. If either or both points are on the surface, it is in terms of Fock functions. These special functions appear in the UTD.

12.8.3   Other Boundary Conditions

The radial Green's function (12.107) satisfies the Dirichlet boundary condition $G = 0$ at $\rho = a$. It can be writtten (with $\nu = \sqrt{\lambda_\phi}$) as

$$G_\rho = \frac{-j\pi}{2}[J_\nu(k\rho_<) - f(\nu)H_\nu^{(2)}(k\rho_<)]H_\nu^{(2)}(k\rho_>), \qquad (12.122)$$

where

$$f(\nu) = \frac{J_\nu(ka)}{H_\nu^{(2)}(ka)}. \qquad (12.123)$$

By changing $f(\nu)$, other boundary conditions can be met. For the Neumann boundary condition $\partial G/\partial \rho = 0$, we should replace $f(\nu)$ by

$$f(\nu) = \frac{J_\nu'(ka)}{H_\nu'^{(2)}(ka)}. \qquad (12.124)$$

The impedance boundary condition

$$\frac{\partial G}{\partial \rho} - jk\bar{C}G = 0 \tag{12.125}$$

is satisfied when

$$f(\nu) = \frac{J_\nu'(ka) - j\bar{C}J_\nu(ka)}{H_\nu'^{(2)}(ka) - j\bar{C}H_\nu^{(2)}(ka)}. \tag{12.126}$$

The complex constant $\bar{C}$ is related to the normalized surface impedance or admittance. For an electric line source illumination, $E_z = -j\omega\mu_0 I_0 G$ and the normal derivative gives H_ϕ. From (12.125) it then follows that $\bar{C} = \eta H_\phi/E_z$ and the surface impedance is $Z = E_z/H_\phi = \eta/\bar{C}$; equivalently $\bar{C} = \eta Y$ is the normalized surface admittance. For a magnetic line source, $H_z = -j\omega\epsilon_0 M_0 G$ and the normal derivative gives E_ϕ. It follows that $\bar{C} = -E_\phi/\eta H_z$ and the surface impedance is $Z = -E_\phi/H_z = \eta\bar{C}$; equivalently, the normalized surface impedance is $\bar{C} = Z/\eta$. The case of a perfectly conducting cylinder is recovered when $\bar{C} \to \infty$ for an electric line source and $\bar{C} = 0$ for a magnetic line source.

12.9 Strip Grating on a Dielectric Slab

A metal strip grating on a grounded dielectric slab is shown in Figure 12.32. There is an infinite number of strips, the strip width is w, and the periodicity is p. The TE case is assumed, so H_z, E_x and E_y are non-zero. Two situations will be considered. First, we will treat the structure as a waveguide and find the complex propagation constant β. Second, we will find the scattered field under plane wave illumination.

We will begin with the waveguide problem. The first step is to find E_x at $y = 0$ due to the surface current J_x on a strip. Then, Floquet's theorem (introduced in Chapter 6) is used to determine J_{xp} on the periodic array of strips. The MoM is used to enforce $E_x = 0$ on the strip at $-w/2 \leq x \leq w/2$. From Floquet's theorem we will then know the currents on all of the strips, the propagation constant β, and all of the space harmonics.

We can obtain E_x due to J_x by using reciprocity and the configuration shown in Figure 12.33. Case (a) shows a magnetic line source above a grounded dielectric slab with no strips. The source is located at $(x,y) = (x',y')$ and given by $\mathbf{M} = \hat{z}M_0\delta(x-x')\delta(y-y')$. From Section 12.6.3 the magnetic field is

$$H_z(x,y) = \frac{-\omega\epsilon_1 M_0}{4\pi}\int_{-\infty}^{\infty} e^{-j\nu(x-x')}\frac{e^{-j\kappa_1|y-y'|} + \Gamma_0^h(\nu)\,e^{-j\kappa_1(y+y')}}{\kappa_1}\,d\nu, \tag{12.127}$$

where the magnetic field reflection coefficient is

$$\Gamma_0^h(\nu) = -\frac{j\kappa_2\tan\kappa_2 d - \epsilon_r\kappa_1}{j\kappa_2\tan\kappa_2 d + \epsilon_r\kappa_1} \tag{12.128}$$

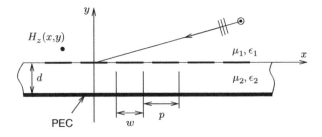

Figure 12.32 Plane wave incident on a metal strip grating, on a grounded dielectric slab.

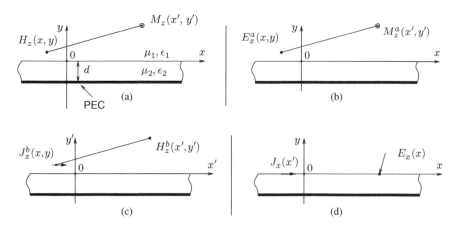

Figure 12.33   Grounded dielectric slab with no strips. (a) H_z due to magnetic line source, (b) E_x^a due to magnetic line source M_z^a and (c) H_z^b due to electric line dipole J_x^b. (d) Source and field points at $y = 0$.

and $\epsilon_r = \epsilon_2/\epsilon_1$, $\kappa_1 = \sqrt{k_1^2 - \nu^2}$, $\kappa_2 = \sqrt{k_2^2 - \nu^2}$. The notation Γ_0^h is being used to emphasize that this is the H_z polarization and that there are no strips.

We can use $\nabla \times \mathbf{H}$ to find E_x in Figure 12.33(b) so that

$$E_x = \frac{1}{j\omega\epsilon_1}\frac{\partial H_z}{\partial y}. \tag{12.129}$$

E_y could also be found from H_z, but it is not needed.

Next, consider the reciprocal configurations in Figure 12.33(b) and (c). In case (b), a z-directed magnetic current $\mathbf{M}^a = \hat{z}M_0^a\delta(x - x')\delta(y - y')$ produces $\mathbf{E}^a$, $\mathbf{H}^a$ at (x, y). We can obtain $E_x^a(x,y)$ from (12.127) and (12.129) so that

$$E_x^a(x, y) = \frac{jM_0^a}{4\pi}\frac{\partial}{\partial y}\int_{-\infty}^{\infty} e^{-j\nu(x-x')}\frac{e^{-j\kappa_1|y-y'|} + \Gamma_0^h\, e^{-j\kappa_1(y+y')}}{\kappa_1}\, d\nu. \tag{12.130}$$

In case (c), the source is an x-directed electric current element $\mathbf{J}^b = \hat{x}I_t^b\delta(x' - x)\delta(y' - y)$ at (x, y), and the field point is at (x', y'). From reciprocity,

$$E_x^a(x, y)\, I_t^b = -H_z^b(x', y')\, M_0^a.$$

Using this and (12.130) gives

$$H_z^b(x', y') = \frac{-jI_t^b}{4\pi}\frac{\partial}{\partial y}\int_{-\infty}^{\infty} e^{-j\nu(x-x')}\frac{e^{-j\kappa_1|y-y'|} + \Gamma_0^h\, e^{-j\kappa_1(y+y')}}{\kappa_1}\, d\nu. \tag{12.131}$$

Equation (12.131) and $\nabla \times \mathbf{H}$ with respect to the field point coordinates gives E_x at (x', y')

$$E_x(x', y') = \frac{1}{j\omega\epsilon_1}\frac{\partial H_z}{\partial y'}$$

so that

$$E_x^b(x',y') = \frac{-I_t^b}{4\pi\omega\epsilon_1} \frac{\partial}{\partial y} \frac{\partial}{\partial y'} \int_{-\infty}^{\infty} e^{-j\nu(x-x')} \frac{e^{-j\kappa_1|y-y'|} + \Gamma_0^h \, e^{-j\kappa_1(y+y')}}{\kappa_1} d\nu. \qquad (12.132)$$

The absolute value in the integrand can be thought of as

$$|y - y'| = \begin{cases} (y - y'); \, y > y' \\ (y' - y); \, y < y' \end{cases}$$

which can be used to show that

$$\frac{d}{dy}|y - y'| = \mathrm{sgn}(y - y'); \quad \frac{d}{dy'}|y - y'| = -\mathrm{sgn}(y - y')$$

and

$$\frac{d^2}{dydy'}|y - y'| = -1.$$

Also needed is

$$\frac{d}{dy}(y + y') = \frac{d}{dy'}(y + y') = 1$$

from which

$$\frac{d^2}{dydy'}(y + y') = 1.$$

Using these expressions in (12.131) and (12.132) gives

$$H_z^b(x',y') = \frac{-I_t^b}{4\pi} \int_{-\infty}^{\infty} e^{-j\nu(x-x')}[\mathrm{sgn}(y - y')e^{-j\kappa_1|y-y'|} + \Gamma_0^h \, e^{-j\kappa_1(y+y')}] d\nu \qquad (12.133)$$

and

$$E_x^b(x',y') = \frac{-I_t^b}{4\pi\omega\epsilon_1} \int_{-\infty}^{\infty} e^{-j\nu(x-x')}(-j\kappa_1)^2 \frac{-e^{-j\kappa_1|y-y'|} + \Gamma_0^h \, e^{-j\kappa_1(y+y')}}{\kappa_1} d\nu. \qquad (12.134)$$

From (12.134) with $y = y' = 0$

$$E_x^b(x') = \frac{-I_t^b}{2\pi\omega\epsilon_1} \int_{-\infty}^{\infty} \frac{e^{-j\nu(x-x')}}{(1/\kappa_1) - j(\epsilon_r/\kappa_2)\cot\kappa_2 d} d\nu. \qquad (12.135)$$

The final configuration needed for the strip-grating problem is shown in Figure 12.33(d). The source point has been renamed as x', and the field point is x. The integrand of (12.135) is not sensitive to the sign of $(x - x')$. Assuming a unit-strength source with $I_t^b = 1$ and dropping the 'b' superscript,

$$E_x(x) = \frac{-1}{2\pi\omega\epsilon_1} \int_{-\infty}^{\infty} \frac{e^{-j\nu(x-x')}}{(1/\kappa_1) - j(\epsilon_r/\kappa_2)\cot\kappa_2 d} d\nu. \qquad (12.136)$$

12.9.1   *Spectral Domain*

Equation (12.136) is the E_x field due to a unit-strength infinitesimal source $J_x(x) = \delta(x - x')$ – so it is an impulse response. To assist this discussion let us assume the impulse is at $x' = a$. Then we can rewrite

(12.136) as a convolution

$$E_x(x) = \int_{-\infty}^{\infty} G(x - x') J_x(x') dx' = G(x) \otimes J_x(x),$$ (12.137)

where

$$J_x(x) = \delta(x - a)$$ (12.138)

and

$$G(x) = \frac{-1}{2\pi\omega\epsilon_1} \int_{-\infty}^{\infty} \frac{e^{-j\nu x}}{(1/\kappa_1) - j(\epsilon_r/\kappa_2)\cot\kappa_2 d} d\nu.$$ (12.139)

The form of (12.139) inspires the definition of a Fourier transform and its inverse

$$\bar{f}(\nu) = \int_{-\infty}^{\infty} f(x) e^{j\nu x} dx = \mathscr{F}(f)$$ (12.140)

$$f(x) = \frac{1}{2\pi} \int_{-\infty}^{\infty} \bar{f}(\nu) e^{-j\nu x} d\nu = \mathscr{F}^{-1}(\bar{f}).$$ (12.141)

The integral in (12.141) represents a sum of plane waves $e^{-j\nu x}$ with complex wavenumbers ν. This is usually thought of as a 'spectrum' of plane waves, so the transform domain ν is called the 'spectral domain'. The x domain is the 'space domain'.

From (12.138) and (12.140)

$$\bar{J}_x(\nu) = e^{j\nu a}$$ (12.142)

and from (12.139) and (12.141)

$$\bar{G}(\nu) = \frac{-1}{\omega\epsilon_1[(1/\kappa_1) - j(\epsilon_r/\kappa_2)\cot\kappa_2 d]}.$$ (12.143)

Since the space-domain expression for E_x in (12.137) is a convolution, the corresponding spectral-domain version will be a product

$$E_x(x) = G(x) \otimes J_x(x) = \mathscr{F}^{-1}(\bar{G}(\nu)\bar{J}_x(\nu)).$$ (12.144)

Using (12.141)–(12.144) gives

$$E_x(x) = \frac{-1}{2\pi\omega\epsilon_1} \int_{-\infty}^{\infty} \frac{e^{-j\nu(x-a)}}{(1/\kappa_1) - j(\epsilon_r/\kappa_2)\cot\kappa_2 d} d\nu$$ (12.145)

which is the same as (12.136) with $x' = a$. Although this is not a new result, it does show that the spectral-domain approach has an advantage: it avoids the space-domain convolution.

12.9.2 Floquet Harmonics

Periodic structures are described as having 'cells'. If the period is p then we can define $-p/2 \le x \le p/2$ as the 'unit cell'. Let $J_x(x)$ be the current on the strip in the unit cell; the current is unknown. According to Floquet's theorem, a periodic structure with a periodic excitation (e.g. a plane wave) will have periodic strip currents.

We want to find the phase constant β for a mode propagating along the grating. In this case the form of the current on each strip will be the same, but there will be a phase shift from cell to cell. The periodic structure's current $J_{xp}(x)$ is made up of phase-shifted replicas of the unit cell's current $J_x(x)$ so that

$$J_{xp}(x) = J_x(x) \otimes \sum_{n=-\infty}^{\infty} e^{-j\beta x} \delta(x - np). \tag{12.146}$$

The cell-to-cell phase shift is determined by β which is unknown and will be found from the MoM.

The Fourier transform of the periodic current is

$$\bar{J}_{xp}(\nu) = \bar{J}_x(\nu) \cdot \mathscr{F} \left[\sum_{n=-\infty}^{\infty} e^{-j\beta x} \delta(x - np) \right]. \tag{12.147}$$

The Poisson summation formula (see Section 6.5.1) and $\delta(pt) = \delta(t)/|p|$ allow us to evaluate the Fourier transform of the sum

$$\mathscr{F} \left[\sum_{n=-\infty}^{\infty} e^{-j\beta x} \delta(x - np) \right] = \sum_{n=-\infty}^{\infty} e^{j(\nu-\beta)np} = \frac{2\pi}{p} \sum_{n=-\infty}^{\infty} \delta(\nu - \beta - 2n\pi/p), \tag{12.148}$$

so the periodic current is

$$\bar{J}_{xp}(\nu) = \bar{J}_x(\nu) \cdot \frac{2\pi}{p} \sum_{n=-\infty}^{\infty} \delta(\nu - \beta - 2n\pi/p). \tag{12.149}$$

The electric field can now be found from (12.144) with the periodic current (12.149). The inverse Fourier transform is very easy because it only requires the integration of delta functions. Therefore,

$$E_x(x) = \frac{1}{2\pi} \int_{-\infty}^{\infty} \bar{G}(\nu) \bar{J}_{xp}(\nu) e^{-j\nu x} \, d\nu = \frac{1}{p} \sum_{n=-\infty}^{\infty} \bar{G}(\nu_n) \bar{J}_x(\nu_n) e^{-j\nu_n x} \tag{12.150}$$

where the delta functions in $\bar{J}_{xp}$ have sifted out specific values of ν

$$\nu_n = \beta + \frac{2n\pi}{p}. \tag{12.151}$$

In the MoM formulation the weighted average of E_x should vanish on the strips. If the testing function is $s(x)$, then

$$\int_{-w/2}^{w/2} E_x(x) s(x) \, dx = 0. \tag{12.152}$$

Using (12.150) and (12.152), plus (12.140) for obtaining $\bar{s}$,

$$\frac{1}{p} \sum_{n=-\infty}^{\infty} \bar{G}(\nu_n) \bar{J}_x(\nu_n) \bar{s}(-\nu_n) = 0. \tag{12.153}$$

To complete the MoM solution, we must choose suitable basis and testing functions. A good approximation for the true strip current is $J_x(x) = A_0 I(x)$ where A_0 is an arbitrary constant (we can choose $A_0 = 1$) and

$$I(x) = \frac{1}{\pi} \frac{\cos \pi x/w}{\sqrt{(w/2)^2 - x^2}}. \tag{12.154}$$

The denominator term (Butler and Wilton, 1980) ensures that $J_x \to 0$ with the correct singular behaviour at the edges $x = \pm w/2$. Using the transform pair (Bateman, 1954, p. 43)

$$\frac{1/\pi}{\sqrt{a^2 - x^2}} \leftrightarrow J_0(a\nu) \tag{12.155}$$

writing the cosine in (12.154) as complex exponentials and applying the shifting theorem $\bar{f}(\nu - a) \leftrightarrow e^{-ja x} f(x)$ gives

$$\bar{I}(\nu) = \frac{1}{2}\left[J_0\left(\frac{w\nu + \pi}{2}\right) + J_0\left(\frac{w\nu - \pi}{2}\right)\right]. \tag{12.156}$$

J_0 is the zero-order Bessel function. Remembering the definition of the Fourier transform (12.139), we thus have

$$\bar{I}(\nu) = \frac{1}{\pi}\int_{-w/2}^{w/2} \frac{\cos \pi x/w}{\sqrt{(w/2)^2 - x^2}} e^{j\nu x}\, dx = \frac{1}{2}\left[J_0\left(\frac{w\nu + \pi}{2}\right) + J_0\left(\frac{w\nu - \pi}{2}\right)\right]. \tag{12.157}$$

Using Galerkin testing, $J_x(x) = s(x) = I(x)$ and (12.153) becomes

$$\frac{1}{p}\sum_{n=-\infty}^{\infty} \bar{G}(\nu_n)[\bar{I}(\nu_n)]^2 = 0. \tag{12.158}$$

Equation (12.158) with ν_n from (12.151) is a transcendental equation that can be numerically solved for the root β (which may be complex). Any root-finding algorithm needs to start with an educated guess, and a good starting point is to recognize that if w is very small, the strips won't have a much of an effect on the electric field. Then, β can be estimated from the grounded slab waveguide solution in Section 3.6.2. A contour plot over the complex-β plane showing the log magnitude of (12.158) can be very helpful. A program for finding β is discussed in Problem 12.13.

Once β is known, (12.151) gives the propagation constants for all of the space harmonics. From (12.151), ν_n and β have the same imaginary parts. If we denote $\nu_n = \beta_n - j\alpha$ and $\beta = \beta_0 - j\alpha$ then (12.151) becomes

$$\beta_n - j\alpha = \beta_0 - j\alpha + \frac{2n\pi}{p} \tag{12.159}$$

which implies that

$$\beta_n = \beta_0 + \frac{2n\pi}{p}. \tag{12.160}$$

The space harmonics propagate along x as $e^{-\alpha x}e^{-j\beta_n x}$. Once β_0 and α are known, all of the other β_n immediately follow from (12.160).

Any space harmonic could have $\beta_n > 0$ (a forward wave) or $\beta_n < 0$ (a backward wave). If $|\beta_n| < k_1$ it is a fast wave which will radiate; endfire radiation occurs when $\beta_n \to k_1$ and backward endfire when $\beta_n \to -k_1$. By choosing the strip grating's permittivity, dielectric thickness, period and strip width, the behaviour of the space harmonics can be controlled and used to make a leaky wave antenna; see, for example, Oliner and Jackson (2007).

In the y direction, the propagation behaves as $e^{-j\kappa_{1n}|y|}$ where

$$\kappa_{1n} = \sqrt{k_1^2 - \nu_n^2} = \sqrt{k_1^2 - (\beta_n - j\alpha)^2}. \tag{12.161}$$

The square root can be chosen so that $\mathrm{Im}\,\kappa_{1n} < 0$ for exponential decay in y (a proper wave) or $\mathrm{Im}\,\kappa_{1n} > 0$ for exponential growth (an improper wave). The forward fast harmonics have $0 < \beta_n < k_1$

and Im $\kappa_{1n} > 0$ (improper) is needed; the backward fast harmonics have $-k_1 < \beta_n < 0$ and Im $\kappa_{1n} < 0$ (proper) is needed. Slow space harmonics have $|\beta_n| > k_1$ and are chosen as proper.

Notice that we have found β without specifying a source. This is very much like the rectangular metal waveguide in Chapter 3, where β_{mn} for the modes were found, without regard to the excitation. In the case of the strip grating, the sum of the space harmonics is a mode. Since a grounded slab can have many surface wave modes, one can expect that other modes (other solutions of (12.158) for $\beta = \beta_0 - j\alpha$) may exist on a strip grating.

Another point of interest is that (12.158) can be thought of in MoM terms as $\mathbf{ZI} = \mathbf{V}$, where $\mathbf{V} = 0$ and $\mathbf{Z}$ is a one-by-one matrix. Therefore we have solved the problem $\mathbf{Z} = 0$. If we were to expand the current as N unknowns, the corresponding matrix equation $\mathbf{ZI} = 0$ would have a solution when det $\mathbf{Z} = 0$. This condition would then give β.

Example 12.4 (Radiation Pattern) A TE strip grating is semi-infinite on $0 \leq x < \infty$; otherwise it is similar to Figure 12.32. The frequency is 27 GHz. The grating supports a harmonic n with $\alpha/k_1 = 0.001$ and $\beta_n/k_1 = 0.5$. At $y = 0$, it is given that $H_z(x) = H_0 e^{-\alpha x} e^{-j\beta_n x}$ and $H_0 = $ const. Find the radiation pattern.

Solution: $y = 0$ is an aperture with H_z known. Using a surface equivalent for $y > 0$, $\mathbf{J}_s = \hat{\mathbf{x}} H_0 e^{-\alpha x} e^{-j\beta_n x}$. $\mathbf{E}_{tan} \neq 0$ on the aperture, but we do not need it if we put a perfect magnetic conductor at $y = 0^-$. Then we double the electric current and use the free-space Green's function (4.109), repeated here:

$$H_z = e^{j\pi/4} \sqrt{\frac{k}{8\pi}} \frac{e^{-jk\rho}}{\sqrt{\rho}} \int_{C_0} \hat{\mathbf{z}} \cdot (\hat{\boldsymbol{\ell}}' \times \hat{\boldsymbol{\rho}}) e^{jk\hat{\boldsymbol{\rho}}\cdot\boldsymbol{\rho}'} J_\ell(\boldsymbol{\rho}') d\ell'.$$

With $k = k_1$, $\hat{\boldsymbol{\rho}} = \hat{\mathbf{x}} \cos\phi + \hat{\mathbf{y}} \sin\phi$, $\boldsymbol{\rho}' = \hat{\mathbf{x}} x'$ and $\hat{\boldsymbol{\ell}}' = \hat{\mathbf{x}}$, it becomes

$$H_z = \sqrt{\frac{jk_1}{2\pi}} H_0 \sin\phi \frac{e^{-jk\rho}}{\sqrt{\rho}} I$$

where

$$I = \int_0^\infty e^{-(\alpha + j\beta_n - jk_1 \cos\phi)x'} dx' = \frac{1}{\alpha + j\beta_n - jk_1 \cos\phi} = \frac{1/k_1}{0.001 + j0.5 - j\cos\phi}$$

and $k_1 = 566$ rad/m. If the $\sin\phi$ factor is neglected, the pattern has a peak at 60°.

With a larger α the aperture field $H_z(x)$ would decay more quickly, leaving in effect a smaller aperture and a broader beam. If we had a slow wave with $\beta_n/k_1 > 1$, there would still be a beam but no peak. Only fast waves can form a beam off endfire. It is noted that the assumed aperture field is from an infinite slab, so this solution is a PO approximation. ∎

12.9.3 Reflection

The previous section treated the strip grating as a waveguide with an unknown propagation constant β. Now, the plane wave reflection coefficient for this same structure will be obtained.

The current on the strips is unknown and can be found from the MoM. An integral equation can be developed by imposing $E_x(x) = 0$ on the strips. In terms of incident and scattered fields,

$$E_x^s(x) = -E_x^{inc}(x).$$

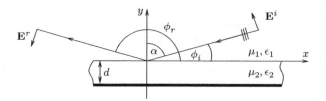

Figure 12.34 TE incident and reflected plane waves $\mathbf{E}^i$, $\mathbf{E}^r$ in the presence of a grounded dielectric slab.

However, it is important to notice that the slab Green's function is being used. Therefore, J_x radiates E_x^s in the *presence* of the slab. Likewise, the incident plane wave is in the *presence* of the slab and the *absence* of the strips. Therefore, the 'incident' field $E_x^{inc}(x)$ consists of two parts: the incident plus slab-reflected plane waves. This can be written as

$$E_x^s(x) = -(E_x^i(x) + E_x^r(x)) = -E_x^{ir}(x) \tag{12.162}$$

where E^{ir} is used to denote the incident plus slab-reflected field.

In the absence of the strips, the incident and reflected fields in Figure 12.34 are[5]

$$\mathbf{E}^i = \hat{\boldsymbol{\phi}}_i \, E_0 \, e^{jk_1(x\cos\phi_i + y\sin\phi_i)} \tag{12.163}$$

$$\mathbf{E}^r = -\hat{\boldsymbol{\phi}}_r \, E_0 \, \Gamma_0^h \, e^{-jk_1(x\cos\phi_r + y\sin\phi_r)} \tag{12.164}$$

with $\phi_r = \pi - \phi_i$, $\hat{\boldsymbol{\phi}}_i = -\hat{\mathbf{x}}\sin\phi_i + \hat{\mathbf{y}}\cos\phi_i$ and $\hat{\boldsymbol{\phi}}_r = -\hat{\mathbf{x}}\sin\phi_r + \hat{\mathbf{y}}\cos\phi_r$. The reflected field can be found from ordinary geometrical optics or, equivalently, the saddle point method. Similar to (12.49), there is a saddle point at $\nu^r = k_1 \sin\alpha$. Since $\alpha + \phi_i = \pi/2$,

$$\nu^r = k_1 \cos\phi_i. \tag{12.165}$$

From (12.128),

$$\Gamma_0^h(\nu^r) = -\frac{jk_2^r \tan\kappa_2^r d - \epsilon_r \kappa_1^r}{jk_2^r \tan\kappa_2^r d + \epsilon_r \kappa_1^r}, \tag{12.166}$$

where

$$\kappa_1^r = \kappa_1(\nu^r) = \sqrt{k_1^2 - k_1^2\cos^2\phi_i} = k_1 \sin\phi_i$$

and

$$\kappa_2^r = \kappa_2(\nu^r) = \sqrt{k_2^2 - k_1^2\cos^2\phi_i}.$$

Evaluating the incident plus reflected fields at $y = 0$ gives

$$E_x^i(x) + E_x^r(x) = E_x^{ir}(x) = \frac{-2E_0 \sin\phi_i \, e^{jk_1 x\cos\phi_i}}{1 - j\epsilon_r (\kappa_1^r/\kappa_2^r) \cot\kappa_2^r d}. \tag{12.167}$$

---

[5] Equation (12.128) is the magnetic field reflection coefficient, so the electric field reflection coefficient is $-\Gamma_0^h$.

The periodic current $\bar{J}_{xp}$ is still as in (12.146). However, unlike the case in the previous section, β is known. If the boundary conditions are to be met in all the cells, the incident, reflected and scattered fields should all have the same $e^{-j\beta x}$ dependence.[6] Therefore $\bar{J}_{xp}$ must have

$$\beta = -k_1 \cos \phi_i. \tag{12.168}$$

The unit-cell current is assumed to be of the form in (12.154) so that

$$J_x(x) = \frac{AE_0}{\pi} \frac{\cos \pi x/w}{\sqrt{(w/2)^2 - x^2}} \leftrightarrow AE_0 \bar{I}(\nu) \tag{12.169}$$

and $\bar{I}(\nu)$ is given by (12.156). The unknown constant A will be found from the MoM. Using (12.150) with this current gives the scattered field

$$E_x^s(x) = \frac{AE_0}{p} \sum_{n=-\infty}^{\infty} \bar{G}(\nu_n)\bar{I}(\nu_n)e^{-j\nu_n x}. \tag{12.170}$$

Next, Galerkin testing is applied

$$\int_{-w/2}^{w/2} E_x^s(x)s(x)\,dx = -\int_{-w/2}^{w/2} E_x^{ir}(x)s(x)\,dx \tag{12.171}$$

with a testing function

$$s(x) = \frac{1}{\pi} \frac{\cos \pi x/w}{\sqrt{(w/2)^2 - x^2}}.$$

From (12.171) with (12.167) and (12.170),

$$\frac{AE_0}{p} \sum_{n=-\infty}^{\infty} \bar{G}(\nu_n)[\bar{I}(\nu_n)]^2 = \frac{2E_0 \bar{I}(\beta) \sin \phi_i}{1 - j\epsilon_r(\kappa_1^r/\kappa_2^r) \cot \kappa_2^r d}. \tag{12.172}$$

Equation (12.172) with (12.151) and (12.168) gives A, so the strip current is now known.

The next step is to find the reflection coefficient for the slab with strips. H_z^s due to J_x can be found from (12.133). It is convenient to work with the magnetic field because it has only one component. By interchanging the primed and unprimed coordinates, putting the current at $y' = 0$ and adopting the Fourier transform notation, we obtain

$$H_z^s(x, y) = \frac{1}{2\pi} \int_{-\infty}^{\infty} \bar{G}^h(\nu, y)\bar{J}_x(\nu)e^{-j\nu x}\,d\nu \tag{12.173}$$

where

$$\bar{G}^h(\nu, y) = \frac{e^{-j\kappa_1 y}}{1 - j\epsilon_r(\kappa_1/\kappa_2) \cot \kappa_2 d}. \tag{12.174}$$

The unit-cell current is $\bar{J}_x(\nu) = AE_0 \bar{I}(\nu)$. Similar to (12.150), we obtain

$$H_z^s(x, y) = \frac{1}{2\pi} \int_{-\infty}^{\infty} \bar{G}^h(\nu, y)\bar{J}_{xp}(\nu)e^{-j\nu x}\,d\nu$$

$$= \frac{AE_0}{p} \sum_{n=-\infty}^{\infty} \bar{G}^h(\nu_n, y)\bar{I}(\nu_n)e^{-j\nu_n x}. \tag{12.175}$$

---

[6] This phase-matching requirement was also used for the wire array in Section 6.5.

The propagation characteristics of a space harmonic can be understood by looking at the (x, y) dependence of (12.175) which provides

$$H_z^s \propto \bar{G}^h(\nu_n, y)e^{-j\nu_n x} = \bar{G}^h(\nu_n, 0)e^{-j\kappa_{1n}y}e^{-j\nu_n x}$$

and using (12.151), (12.168) with $\kappa_{1n} = \sqrt{k_1^2 - \nu_n^2}$,

$$H_z^s \propto \bar{G}^h(\nu_n, 0)e^{-j\kappa_{1n}y}e^{-j\beta_n x}$$

where

$$\kappa_{1n} = \sqrt{k_1^2 - (k_1 \cos \phi_i - 2n\pi/p)^2}; \ \beta_n = -k_1 \cos \phi_i + 2n\pi/p.$$

Since κ_{1n} could be real or imaginary, the previous result reveals that some harmonics may exponentially decay in y, whereas others may propagate. All this depends on p, the incidence angle and the frequency. For the $n = 0$ harmonic, $\kappa_{10} = k_1 \sin \phi_i$, $\beta_0 = -k_1 \cos \phi_i$ and using $\phi_r = \pi - \phi_i$,

$$H_z^s \propto e^{-jk_1 y \sin \phi_r}e^{-jk_1 x \cos \phi_r}.$$

Therefore the $n = 0$ harmonic radiates in the specular direction. If there are any other $n \neq 0$ radiating harmonics, they will radiate in off-specular directions; these are grating lobes. Generally, when computing κ_{1n} in the sums of space harmonics, one must be careful to choose $\text{Im}\,\kappa_{1n} < 0$ when κ_{1n} is complex so that no harmonics grow exponentially away from the strip-grating surface.

The final step is to find the reflection coefficient. Only the $n = 0$ harmonic contributes to the reflected field, which is in the specular direction. Using (12.163), (12.164) and (12.175), we can define a magnetic field reflection coefficient for the grating as

$$\Gamma_g^h(\phi_i) = \left.\frac{H_z^r + H_z^s}{H_z^i}\right|_{y=0} = \Gamma_0^h(\beta) - \frac{\eta_1 A \bar{G}^h(\beta, 0)}{p}\bar{I}(\beta) \tag{12.176}$$

and the electric field reflection coefficient is $\Gamma_g^e = -\Gamma_g^h$. In this result we have used $\Gamma_0^h(\beta) = \Gamma_0^h(-\beta)$ and $-\nu^r = \nu_0 = \beta = -k_1 \cos \phi_i$. The first term is simply the reflection coefficient for the slab with no strips, and the second term accounts for the strip scattering.

12.9.4 Discussion

The properties of a strip-grating mode and the behaviour of the reflection coefficient are closely related; this will now be discussed. Figure 12.35(a) shows a guided mode. If there are no strips then a slow wave with $\alpha = 0$ (a surface wave) propagates in the slab; the mode decays exponentially in the y direction, so it does not radiate.

When strips are added, the mode can be expressed as an infinite sum of space harmonics. All the β_n and α can be found from the solution of (12.158). The mode will decay as $e^{-\alpha x}$ because energy is being lost to radiation as it travels – so this is called a 'leaky mode'.

The mode is in turn an infinite sum of space harmonics. Since α is the same for all of the space harmonics, the term 'leaky' is associated with the mode and not with individual space harmonics. When $\alpha \neq 0$ a mode will have one or more space harmonics that are fast; it is usually said that the 'whole mode' is leaky.

Any β_n that is fast will radiate at an angle ϕ_0 where $\cos \phi_0 = \beta_n/k_1$. (In Figure 12.35(a) it is assumed that only one harmonic is fast.) The phase velocity of a space harmonic can be positive or negative, in accordance with the sign of β_n. However, the group velocity is always positive.

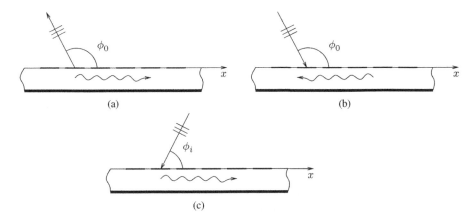

Figure 12.35   (a) A mode having a radiating space harmonic at ϕ_0. (b) An incident plane wave excites a mode. (c) By symmetry this is the same as (b); $\phi_i = \pi - \phi_0$.

Case (b) shows the reciprocal situation with an incident plane wave. Conceptually, the main difference between cases (a) and (b) is the direction of propagation; also case (b) will have $\alpha = 0$. Case (c) is the same as (b) except that symmetry in x has been used to reverse the appearance. The angles are related by $\phi_i = \pi - \phi_0$. The key point is that if we illuminate the strip grating at an angle that matches a leaky mode's radiating direction (ϕ_0 in case (a)), a propagating space harmonic will be excited inside the slab as in cases (b) and (c).

The relationships between these situations are best illustrated by making some calculations. The parameters are $p = 5$ mm, $w = 2$ mm, $d = 2.5$ mm and $\epsilon_r = 3.5$, and the frequency is 27 GHz.

First, for a grounded slab with no strips, we can use the methods in Section 3.6.2 to find the dominant (i.e. lowest) mode, having $\beta/k_1 = 1.607$ and $\alpha = 0$. Since $\alpha = 0$ there is no decay as the mode propagates along x; this is a surface wave. (This was called a TM mode in Chapter 3, but it is TE_z in the present coordinates.)

Next, the strips are added. Using PROGRAM tegrating to solve for the complex root $\nu_0 = \beta_0 - j\alpha$ of Equation (12.158), one finds that $\alpha/k_1 = 0.00579$, $\beta_0/k_1 = 1.664$ and $\beta_{-1}/k_1 = -0.557$. Since $\alpha \neq 0$ the mode decays along x, and it is a leaky mode, losing energy to radiation as it propagates.

Since $\beta_0/k_1 = 1.664$ for the strip grating is close to $\beta/k_1 = 1.607$ for the slab with no strips, the $n = 0$ space harmonic can be thought of as a 'perturbed surface wave'. However, $\alpha \neq 0$ so it is *not* a surface wave – it is a leaky wave. With $\beta_0/k_1 > 1$, the beam radiates in the forward-endfire ($\phi = 0$) direction but does not have a visible peak.

The $n = -1$ harmonic is fast and radiates at $\cos\phi_0 = \beta_{-1}/k_1$ or $\phi_0 = 123.83°$. For the parameters chosen, all the other β_n are slow, so they can only radiate in the forward-endfire ($\beta_n/k_1 \geq 1$) or backward-endfire ($\beta_n/k_1 \leq -1$) directions.

Let us now consider an incident plane wave as in case (c), with $\phi_i = 56.17°$. From (12.168) it is required that $\beta_0/k_1 = -0.557$; consequently $\beta_1/k_1 = 1.664$. Therefore, β_0 in case (a) is the same as β_1 in case (c). It shows that an incident wave at this angle causes a propagating wave inside the slab, with the same phase velocity as in case (a).

Next, we will calculate the grating's reflection coefficient. Assuming the same parameters as before and using (12.176), one finds that the magnitude of the reflection coefficient is $|\Gamma_g^h| = 1$. The phase is more interesting and is shown in Figure 12.36(a). Near $\phi_i = 56°$, the phase goes through a rapid change of almost 360°. Somewhere in that range, the incident wave and specular reflection cancel. The effect can be seen by calculating the magnetic field at $(x, y) = (0, 0)$

$$H_z = H_z^i + H_z^r + H_z^s = -\frac{E_0}{\eta_1}(1 + \Gamma_g^h)$$

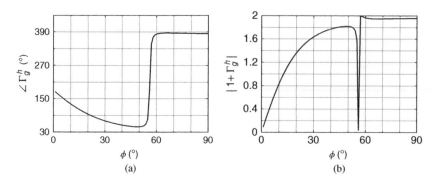

Figure 12.36   TE strip grating; the parameters are $p = 5$ mm, $w = 2$ mm, $d = 2.5$ mm and $\epsilon_r = 3.5$ and the frequency is 27 GHz. (a) Phase of the reflection coefficient Γ_g^h. (b) The field amplitude at $(x, y) = (0, 0)$ is $|H_z| \propto |1 + \Gamma_g^h|$.

from which $|H_z| \propto |1 + \Gamma_g^h|$; this is shown in Figure 12.36(b). The very sharp dip at $\phi_i = 56°$ is apparent. At this angle the incident field couples strongly into the slab. This effect is known as *Wood's anomaly*. An explanation in terms of leaky wave phenomena was found by Hessel and Oliner (1965).

12.9.5   Other Related Cases

The TE strip grating is readily extended to other situations such as a slab with no ground plane or no slab at all. For instance, if the strips are in free space, then we just need to replace the slab Green's function with the free-space Green's function. This is readily accomplished by making $\Gamma_0^h = 0$. As a result, (12.143) and (12.174) should be replaced by

$$\bar{G}(\nu) = -\frac{\kappa_1}{2\omega\epsilon_1} \qquad (12.177)$$

and

$$\bar{G}^h(\nu) = \frac{1}{2}e^{-j\kappa_1 y}. \qquad (12.178)$$

To find a solution for the TM polarization, the steps are similar to the TE case. In the TM case, the electric field has only a z component. The strips have a z-directed strip current that becomes infinite at the sharp edges. A good approximation for the current is

$$J_z(x) = \frac{1}{\pi}\frac{1}{\sqrt{(w/2)^2 - x^2}}. \qquad (12.179)$$

For strips in free space, Richmond (1980) used this 'edge mode' in conjunction with a Fourier series expansion for the current and showed that adding the edge mode improves convergence.

12.10   Further Reading

The dielectric slab Green's function is discussed in Collin (1991, Chapter 2). The general case of a multilayer planar ground is in Wait (1996). A detailed description of surface waves and leaky waves in the complex wavenumber plane is given in Jackson and Oliner (2008).

The slab, half space, cylinder and many other canonical configurations are developed in Felsen and Marcuvitz (1994, Chapters 5 and 6). Other helpful books are by Tyras (1969) and Ishimaru (1991). The book by Wait (1959) covers cylinders and wedges.

The field of a dipole above a lossy earth or sea was first studied by Sommerfeld in 1909. A good account is available in Felsen and Marcuvitz (1994) and Collin (1985, 2004).

A detailed presentation of both TE and TM strip gratings can be found in Peterson et al. (1998, Chapter 7). Strip gratings can be used to make highly directive leaky wave antennas; see, for example, Walter (1965) or Oliner and Jackson (2007).

2D periodicity in a plane is of course also possible. Periodic arrays of patches, slots, dipoles and other types of elements can be used to make *frequency selective surfaces* which have frequency-dependent reflection and transmission characteristics. These are described in Munk (2000), Cwik (2008) and other sources.

References

Bateman H (1954) *Tables of Integral Transforms* vol. 1. McGraw-Hill.

Butler CM and Wilton DR (1980) General analysis of narrow strips and slots. *IEEE Trans. Antennas Propag.* **AP-28**(1), 42–48.

Churchill RV and Brown JW (2013) *Complex Variables and Applications*. McGraw-Hill.

Collin RE (1985) *Antennas and Radiowave Propagation*. McGraw-Hill.

Collin RE (1991) *Field Theory of Guided Waves*. Oxford University Press.

Collin RE (2004) Hertzian dipole radiating over a lossy earth or sea: some early and late 20th-century controversies. *IEEE Antennas Propag. Mag.* **46**(2), 64–78.

Cwik T (2008) Frequency-selective screens. In *Modern Antenna Handbook* (ed. Balanis CA) John Wiley & Sons, Inc. pp. 779–828.

Felsen LB and Marcuvitz N (1994) *Radiation and Scattering of Waves*. IEEE Press.

Hessel A and Oliner AA (1965) A new theory of Wood's anomalies on optical gratings. *Appl. Opt.* **4**(10), 1275–1297.

Ishimaru A (1991) *Electromagnetic Wave Propagation, Radiation and Scattering*. Prentice-Hall.

Jackson DR and Oliner AA (2008) Leaky-wave antennas. In *Modern Antenna Handbook* (ed. Balanis CA) John Wiley & Sons, Inc. pp. 325–367.

Munk BA (2000) *Frequency Selective Surfaces, Theory and Design*. John Wiley & Sons, Inc.

Oliner AA and Jackson DR (2007) Leaky-wave antennas. In *Antenna Engineering Handbook* (ed. Volakis J) McGraw-Hill pp. 11.1–11.56.

Pathak PH and Kouyoumjian RG (1970) The dyadic diffraction coefficient for a perfectly conducting wedge. Technical Report 2183-4, ElectroScience Laboratory, Ohio State University, Columbus OH.

Peterson AF, Ray SL and Mittra R (1998) *Computational Methods for Electromagnetics*. IEEE Press.

Richmond JH (1980) On the edge mode in the theory of TM scattering by a strip or strip grating. *IEEE Trans. Antennas Propag.* **AP-28**, 883–887.

Schafer RH and Kouyoumjian RG (1967) Higher order terms in the saddle point approximation. *Proc. IEEE* **55**(8), 1496–1497.

Tyras G (1969) *Radiation and Propagation of Electromagnetic Waves*. Academic Press.

Wait JR (1959) *Electromagnetic Radiation from Cylindrical Structures*. Pergamon Press.

Wait JR (1996) *Electromagnetic Waves in Stratified Media*. IEEE Press.

Walter CH (1965) *Traveling Wave Antennas*. McGraw-Hill.

Problems

12.1 The Green's function $G(\rho, \rho')$ in (12.41) has a term involving $|x - x'|$. Use a change of variables $\nu \to -\nu$ and the fact that κ_1, κ_2 are even functions of ν to obtain the slightly different looking but equivalent form (12.42) where $|x - x'| \to (x - x')$.

12.2 The Green's function $G(\rho, \rho')$ in (12.41) is only valid for field points $y \geq 0$. Find $G(\rho, \rho')$ for field points inside the dielectric; $-d \leq y \leq 0$.

12.3 For the grounded slab in Figure 12.16, assume there is a magnetic line source of strength M_0 and derive the TE Green's function in (12.68) for the scattered field.

12.4 The Hankel function $H_0^{(2)}(k\rho)$ can be represented by the integral

$$H_0^{(2)}(k\rho) = \frac{1}{\pi}\int_{\bar{P}} e^{-jk\rho\cos w}\,dw.$$

Denoting $w = u + jv$, the path $\bar{P}$ runs from $(u,v) = (-\pi/2, -\infty)$ to $(-\pi/2, 0)$ to $(\pi/2, 0)$ to $(\pi/2, \infty)$.

(a) Locate the saddle point. Use the saddle point method to evaluate the integral for large $k\rho$.
(b) Write the SDP in the form $v = v(u)$, and plot the SDP as a function of u.
(c) Plot the integrand as a function of u for points on the SDP. Use $k\rho = 2$.
(d) Evaluate the integral numerically, along the linear path segments $\bar{P}$. Obtain the results for $k\rho = 0.2$ and $k\rho = 2$; compare with the asymptotic formula.
(e) The steepest descent path P can be written as $v = v(u)$. Hence, u can be eliminated. Write the integral in the form

$$H_0^{(2)}(k\rho) = \frac{1}{\pi}\int_{P}(\cdots)\,dv.$$

Plot the real and imaginary parts of the integrand as a function of v for the case $k\rho = 2$.
(f) Evaluate the integral numerically on the steepest descent path P, using $k\rho = 2$.

12.5 An impenetrable planar impedance boundary exists at $y = 0$. The region above it is free space with μ_0, ϵ_0. The excitation is a magnetic line source of strength M_0, and so $H_z = -j\omega\epsilon_0 M_0 G$. The surface impedance is $Z_s = E_x/H_z$, and the normalized surface impedance is $\bar{Z}_s = Z_s/\eta_0$, where $\eta_0 = \sqrt{\mu_0/\epsilon_0}$. The impedance boundary condition is

$$\frac{\partial G}{\partial y} = jk_0\bar{Z}_s G$$

where $k_0 = \omega\sqrt{\mu_0\epsilon_0}$.
(a) Find the Green's function that satisfies the impedance boundary condition and the 2D wave equation

$$(\nabla^2 + k_0^2)G(x,y;x',y') = -\delta(x-x')\delta(y-y').$$

Develop your formulation for G so that it is in terms of a reflection coefficient, using the w plane.
(b) Find the location of the surface wave pole in the w plane. You should assume that the surface impedance is lossless and inductive, that is, $\bar{Z}_s = \bar{R}_s + j\bar{X}_s$ has $\bar{R}_s = 0$ and $\bar{X}_s \geq 0$. An inductive impedance boundary supports a surface wave in the TE case.
(c) Use the saddle point method to obtain G in terms of three parts: the incident, reflected and surface wave field.
    You should find that at grazing incidence, the incident and reflected fields cancel, leaving only the surface wave field.

12.6 Derive the half-space Green's functions G and G_t in Equations (12.72) and (12.73).

12.7 A vertical electric dipole is above seawater, which has $\epsilon_r = 81$ and $\sigma = 4$ S/m. Calculate the horizontal magnetic field for a range of 1–100 km. You can use PROGRAM ved. Plot the result in dB and normalize to 0 dB at 1 km. Do this at 0.3 and 3 MHz.
    Compare with lossy earth, in Example 12.3. You should find that propagation on seawater is much better than on earth.

12.8 For a cylinder with the Neumann boundary condition, derive Equation (12.124). For the impedance boundary condition, derive Equation (12.126).

12.9 In the following problems, find the eigenfunction series for a cylinder of radius a. Assume a magnetic line source M_0 is at $(\rho', \phi') = (a, 0)$. In all cases simplify your result as much as possible by using Wronskians.

(a) Find $H_z(\rho, \phi)$ for a perfectly conducting cylinder.
(b) Find $E_\phi(\rho, \phi)$, for a perfectly conducting cylinder.
(c) Using E_ϕ from the previous step, apply reciprocity and find $E_z(\rho, \phi)$ due to a ϕ-directed magnetic line dipole of strength M_d at $(\rho', \phi') = (a, 0)$.
(d) Find H_z for an impedance boundary cylinder with a magnetic line source, having a normalized impedance $\bar{C} = -E_\phi / \eta H_z$ at $\rho = a$.

12.10 Make a root-finding program that calls SUBROUTINE huni and solves for the roots ν_p of $H_{\nu_p}^{\prime(2)}(x) = 0$. Compute the first three roots when $x = 5$, and compare with the roots given by Equation (C.19).

12.11 Find the Green's function for a perfectly conducting cylinder of radius a with a material coating μ_1, ϵ_1 and thickness $d = b - a$. The ambient medium is air, with μ_0, ϵ_0. The TM case is assumed, so $E_z = 0$ at $\rho = a$ and E_z, H_ϕ are continuous at the material–air interface $\rho = b$.

Assume that the source is outside the coating at $\rho' > b$ and the field point is at some point $a \le \rho < \rho'$ which could be inside the coating or in the air. Find G_ρ with the UT method, using $U(\rho) = AH_\nu^{(1)}(k_1\rho) + BH_\nu^{(2)}(k_1\rho)$ inside the material and $U(\rho) = H_\nu^{(1)}(k_0\rho) + CH_\nu^{(2)}(k_0\rho)$ outside the material, along with $T(\rho) = H_\nu^{(2)}(k_0\rho)$.

(a) Convert the boundary and continuity conditions for E_z and H_ϕ into conditions for G and $\partial G/\partial \rho$ to find A, B and C.
(b) Show that G_ρ is of the form in Equation (12.122) (with $k = k_0$) except that now

$$f(\nu) = \frac{J_\nu'(k_0 b) - j\bar{C}_\nu J_\nu(k_0 b)}{H_\nu^{\prime(2)}(k_0 b) - j\bar{C}_\nu H_\nu^{(2)}(k_0 b)},$$

where

$$\bar{C}_\nu = \bar{C}_\nu^e = -j\sqrt{\frac{\epsilon_r}{\mu_r}} \frac{H_\nu^{\prime(2)}(k_1 b)H_\nu^{(1)}(k_1 a) - H_\nu^{(2)}(k_1 a)H_\nu^{\prime(1)}(k_1 b)}{H_\nu^{(2)}(k_1 b)H_\nu^{(1)}(k_1 a) - H_\nu^{(2)}(k_1 a)H_\nu^{(1)}(k_1 b)}.$$

(c) Repeat for the TE case and show that

$$\bar{C}_\nu = \bar{C}_\nu^h = -j\sqrt{\frac{\mu_r}{\epsilon_r}} \frac{H_\nu^{\prime(2)}(k_1 b)H_\nu^{\prime(1)}(k_1 a) - H_\nu^{\prime(2)}(k_1 a)H_\nu^{\prime(1)}(k_1 b)}{H_\nu^{(2)}(k_1 b)H_\nu^{\prime(1)}(k_1 a) - H_\nu^{\prime(2)}(k_1 a)H_\nu^{(1)}(k_1 b)}.$$

12.12 Show that for the TE grounded slab in Figure 12.33(c), if $y = 0^+$ and $y' = 0^-$ then (12.133) gives the magnetic field just below the current

$$H_z^b(x') = \frac{-I_t^b \epsilon_r}{2\pi} \int_{-\infty}^{\infty} \frac{\kappa_1 e^{-j\nu(x-x')}}{j\kappa_2 \tan \kappa_2 d + \epsilon_r \kappa_1} d\nu,$$

and with $y = 0^-$ and $y' = 0^+$, the magnetic field just above the current is

$$H_z^b(x') = \frac{I_t^b}{2\pi} \int_{-\infty}^{\infty} \frac{e^{-j\nu(x-x')}}{1 - j\epsilon_r(\kappa_1/\kappa_2) \cot \kappa_2 d} d\nu.$$

12.13 Find β for a TE strip grating having $p = 5$ mm, $w = 2$ mm, $d = 2.5$ mm and $\epsilon_r = 3.5$ when the frequency is 27 GHz. Solve (12.158) with (12.151) for the complex phase constant $\beta = \beta_0 - j\alpha$. This can be done with the secant method; see PROGRAM tegrating.

Tabulate β_n/k_1 for $-5 \leq n \leq 5$. Observe that the $n = -1$ harmonic is the only one that is fast; $|\beta_{-1}| < k_1$. Therefore only the $n = -1$ harmonic is capable of radiation off endfire or backward endfire.

You should find that $\beta_0/k_1 = 1.664$ and $\alpha/k_1 = 0.00579$; the other β_n follow from this.

12.14 Calculate and plot the radiation pattern of the $n = -1$ space harmonic in Problem 12.13. Note that $\beta_{-1} < 0$ which will make the beam point backwards into the $\phi > 90°$ region. Assume a semi-infinite grating, along $0 \leq x < \infty$ as in Example 12.4.

12.15 Calculate and plot β_{-1} versus ω for the $n = -1$ space harmonic in Problem 12.13. Do this for a frequency range of 26–28 GHz. From the data at 27 GHz obtain the phase velocity u_p, and from a numerical derivative, the group velocity u_g. Since this is a backward wave, you should find that $u_p < 0$, whereas $u_g > 0$.

12.16 Find the magnetic field reflection coefficient for a TE metal strip grating in free space. First, derive the expressions (12.177) and (12.178). Then obtain new expressions similar to (12.172) and (12.176) for the current strength A and reflection coefficient Γ_g^h.

12.17 A TE strip grating has $p = 5$ mm and $w = 2$ mm and the frequency is 27 GHz. The strips are in free space. Calculate the reflection coefficient $|\Gamma_g^h|$ for $0 \leq \phi < 90°$. Check your program using a broadside-incident plane wave. You should obtain $|\Gamma_g^h| \approx 0.2$.

12.18 A TM strip grating has $p = 5$ mm and $w = 2$ mm and the frequency is 27 GHz. The strips are in free space. Calculate the reflection coefficient $|\Gamma_g^e|$ for $0 \leq \phi < 90°$. Check your program using a broadside-incident plane wave. You should obtain $|\Gamma_g^e| \approx 0.9$.

Appendix A

Constants and Formulas

A.1 Constants

Charge on an electron $e = -1.60217656535 \times 10^{-19}$ C
Speed of light $c = 1/\sqrt{\mu_0 \epsilon_0} = 2.99792458 \times 10^8$ m/s
Permittivity of free space $\epsilon_0 = 8.854187816 \times 10^{-12}$ F/m
Permeability of free space $\mu_0 = 4\pi \times 10^{-7}$ H/m
$\pi = 3.141592653589793238462643$
Euler's constant $\gamma = 0.5772156649015328606065121209$
$e^\gamma = 1.781072417990197985236504$

A.2 Definitions

$$\text{Kronecker delta} \quad \delta_{mn} = \begin{cases} 1; & m = n \\ 0; & m \neq n \end{cases}$$

$$\text{Neumann's number} \quad \epsilon_{mn} = \begin{cases} 1; & m = n \\ 2; & m \neq n \end{cases}$$

$$\text{Signum function} \quad \text{sgn}(x) = \begin{cases} 1; & x > 0 \\ -1; & x < 0 \end{cases}$$

$$\text{Sampling function} \quad \text{sa}(x) = \frac{\sin x}{x}$$

$$\text{Sinc function} \quad \text{sinc}(x) = \frac{\sin \pi x}{\pi x}$$

Applied Frequency-Domain Electromagnetics, First Edition. Robert Paknys.
© 2016 John Wiley & Sons, Ltd. Published 2016 by John Wiley & Sons, Ltd.
Companion Website: www.wiley.com/go/paknys9981

A.3 Trigonometry

$$e^{j\theta} = \cos\theta + j\sin\theta; \quad \text{Euler's identity} \tag{A.1}$$

$$\cos\theta = \frac{(e^{j\theta} + e^{-j\theta})}{2}; \quad \sin\theta = \frac{(e^{j\theta} - e^{-j\theta})}{j2} \tag{A.2}$$

$$\cosh\theta = \frac{(e^{\theta} + e^{-\theta})}{2}; \quad \sinh\theta = \frac{(e^{\theta} - e^{-\theta})}{2} \tag{A.3}$$

$$\cos j\theta = \cosh\theta; \quad \sin j\theta = j\sinh\theta \tag{A.4}$$

$$\cosh j\theta = \cos\theta; \quad \sinh j\theta = j\sin\theta \tag{A.5}$$

$$e^{j(A+B)} = e^{jA}e^{jB} = (\cos A + j\sin A)(\cos B + j\sin B) \tag{A.6}$$

and taking the real and imaginary parts gives

$$\cos(A + B) = \cos A\cos B - \sin A\sin B \tag{A.7}$$

$$\sin(A + B) = \sin A\cos B + \cos A\sin B \tag{A.8}$$

which can be used to obtain

$$\cos(A + jB) = \cos A\cosh B - j\sin A\sinh B \tag{A.9}$$

$$\sin(A + jB) = \sin A\cosh B + j\cos A\sinh B \tag{A.10}$$

$$2\cos A\cos B = \cos(A + B) + \cos(A - B) \tag{A.11}$$

$$2\sin A\sin B = -\cos(A + B) + \cos(A - B) \tag{A.12}$$

$$2\cos A\sin B = \sin(A + B) - \sin(A - B). \tag{A.13}$$

If $A = B$ then

$$2\cos^2 A = \cos 2A + 1 \tag{A.14}$$

$$2\sin^2 A = -\cos 2A + 1 \tag{A.15}$$

$$2\cos A\sin A = \sin 2A. \tag{A.16}$$

The previous relations can be used to show that

$$2\cos(A + B)\cos(A - B) = \cos 2A + \cos 2B \tag{A.17}$$

$$2\cos(A + B)\sin(A - B) = \sin 2A - \sin 2B \tag{A.18}$$

$$-2\sin(A + B)\sin(A - B) = \cos 2A - \cos 2B. \tag{A.19}$$

A.4   The Impulse Function

The impulse function is not a function in the usual sense. It is not defined directly but by its property:

$$\int_{-\infty}^{\infty} \delta(t) f(t) dt = f(0),$$
(A.20)

where $f(t)$ is an arbitrary function that is continuous at $t = 0$. Manipulations with impulse functions rely on a branch of mathematics called the theory of distributions. Some useful results are (Papoulis 1962, Appendix I)

$$\delta(at) = \frac{1}{|a|}\delta(t)$$
(A.21)

$$f(t)\delta(t) = f(0)\delta(t)$$
(A.22)

$$f(t)\delta'(t) = f(0)\delta'(t) - f'(0)\delta(t)$$
(A.23)

$$f(t)\delta''(t) = f(0)\delta''(t) - 2f'(0)\delta'(t) + f''(0)\delta(t)$$
(A.24)

$$t\delta(t) = 0$$
(A.25)

$$t\delta'(t) = -\delta(t).$$
(A.26)

Reference

Papoulis A (1962) *The Fourier Integral and its Applications.* McGraw-Hill.

Appendix B

Coordinates and Vector Calculus

In Figure B.1, the location of point P is specified by three numbers, being (x, y, z) or (ρ, ϕ, z) or (r, θ, ϕ). At P we also have three orthogonal basis vectors. The direction in which a coordinate increases defines a corresponding basis vector. As examples, the basis vector $\hat{\boldsymbol{\theta}}$ points in the direction of the increasing scalar quantity θ. The basis vector $\hat{\mathbf{x}}$ points in the direction of increasing scalar x and so on.

B.1   Coordinate Transformations

The dot products in Table B.1 are useful for 'extracting' a specific component of a vector. Suppose that $\mathbf{A}$ is a known vector. Its $\hat{\mathbf{x}}$ component would be $\hat{\mathbf{x}} \cdot \mathbf{A}$. Similarly, Its $\hat{\boldsymbol{\rho}}$ component would be

$$A_\rho = \hat{\boldsymbol{\rho}} \cdot \mathbf{A} = \hat{\boldsymbol{\rho}} \cdot (\hat{\mathbf{x}} A_x + \hat{\mathbf{y}} A_y) = A_x \cos \phi + A_y \sin \phi.$$

The table can also be used for converting a vector, for example,

$$\hat{\boldsymbol{\rho}} = \hat{\mathbf{x}} \cos \phi + \hat{\mathbf{y}} \sin \phi.$$

The table also serves as a mnemonic aid for coordinate conversions

$$x = \rho \cos \phi; \; y = \rho \sin \phi \tag{B.1}$$

$$x = r \sin \theta \cos \phi; \; y = r \sin \theta \sin \phi; \; z = r \cos \theta. \tag{B.2}$$

B.2   Volume and Surface Elements

From the cuboids in Figure B.2, it follows that

$$dV = dx \cdot dy \cdot dz \tag{B.3}$$

$$dV = d\rho \cdot \rho d\phi \cdot dz \tag{B.4}$$

$$dV = dr \cdot r \sin \theta d\phi \cdot r d\theta. \tag{B.5}$$

Likewise, the cuboid side lengths can give differential surface elements, for example, on the cylindrical face $\rho = $ const. we have $dS = \rho d\phi \cdot dz$ and $d\mathbf{S} = \hat{\boldsymbol{\rho}} \, \rho d\phi \cdot dz$.

Applied Frequency-Domain Electromagnetics, First Edition. Robert Paknys.
© 2016 John Wiley & Sons, Ltd. Published 2016 by John Wiley & Sons, Ltd.
Companion Website: www.wiley.com/go/paknys9981

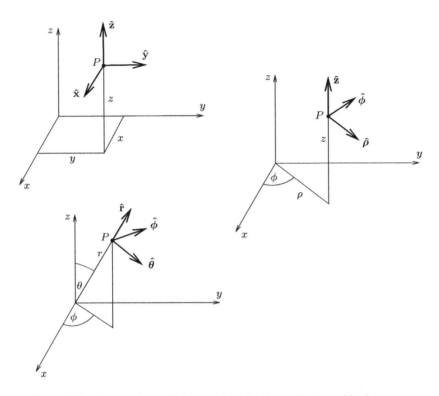

Figure B.1 Rectangular, cylindrical and spherical coordinates and basis vectors.

Table B.1 Unit vector dot products.

	$\hat{\mathbf{x}}$	$\hat{\mathbf{y}}$	$\hat{\mathbf{z}}$
$\hat{\boldsymbol{\rho}}$	$\cos\phi$	$\sin\phi$	0
$\hat{\boldsymbol{\phi}}$	$-\sin\phi$	$\cos\phi$	0
$\hat{\mathbf{z}}$	0	0	1

	$\hat{\mathbf{x}}$	$\hat{\mathbf{y}}$	$\hat{\mathbf{z}}$
$\hat{\mathbf{r}}$	$\sin\theta\cos\phi$	$\sin\theta\sin\phi$	$\cos\theta$
$\hat{\boldsymbol{\theta}}$	$\cos\theta\cos\phi$	$\cos\theta\sin\phi$	$-\sin\theta$
$\hat{\boldsymbol{\phi}}$	$-\sin\phi$	$\cos\phi$	0

	$\hat{\boldsymbol{\rho}}$	$\hat{\boldsymbol{\phi}}$	$\hat{\mathbf{z}}$
$\hat{\mathbf{r}}$	$\sin\theta$	0	$\cos\theta$
$\hat{\boldsymbol{\theta}}$	$\cos\theta$	0	$-\sin\theta$
$\hat{\boldsymbol{\phi}}$	0	1	0

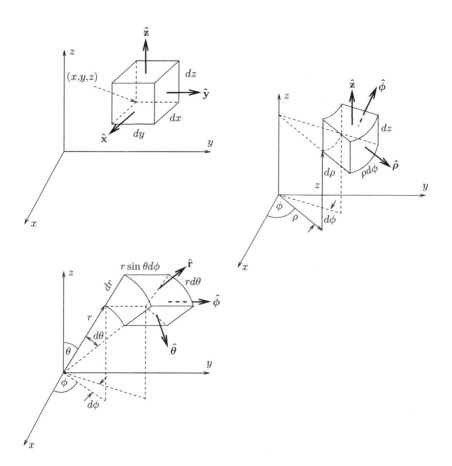

Figure B.2   Differential volumes and surfaces.

B.3   Vector Derivatives

Rectangular coordinates

$$\nabla f = \hat{\mathbf{x}}\frac{\partial f}{\partial x} + \hat{\mathbf{y}}\frac{\partial f}{\partial y} + \hat{\mathbf{z}}\frac{\partial f}{\partial z} \tag{B.6}$$

$$\nabla \cdot \mathbf{A} = \frac{\partial A_x}{\partial x} + \frac{\partial A_y}{\partial y} + \frac{\partial A_z}{\partial z} \tag{B.7}$$

$$\nabla \times \mathbf{A} = \hat{\mathbf{x}}\left(\frac{\partial A_z}{\partial y} - \frac{\partial A_y}{\partial z}\right) + \hat{\mathbf{y}}\left(\frac{\partial A_x}{\partial z} - \frac{\partial A_z}{\partial x}\right) + \hat{\mathbf{z}}\left(\frac{\partial A_y}{\partial x} - \frac{\partial A_x}{\partial y}\right) \tag{B.8}$$

$$\nabla^2 f = \frac{\partial^2 f}{\partial x^2} + \frac{\partial^2 f}{\partial y^2} + \frac{\partial^2 f}{\partial z^2} \tag{B.9}$$

Cylindrical coordinates

$$\nabla f = \hat{\boldsymbol{\rho}}\frac{\partial f}{\partial \rho} + \hat{\boldsymbol{\phi}}\frac{1}{\rho}\frac{\partial f}{\partial \phi} + \hat{\mathbf{z}}\frac{\partial f}{\partial z} \tag{B.10}$$

$$\nabla \cdot \mathbf{A} = \frac{1}{\rho} \frac{\partial(\rho A_\rho)}{\partial \rho} + \frac{1}{\rho} \frac{\partial A_\phi}{\partial \phi} + \frac{\partial A_z}{\partial z} \tag{B.11}$$

$$\nabla \times \mathbf{A} = \hat{\boldsymbol{\rho}} \left(\frac{1}{\rho} \frac{\partial A_z}{\partial \phi} - \frac{\partial A_\phi}{\partial z} \right) + \hat{\boldsymbol{\phi}} \left(\frac{\partial A_\rho}{\partial z} - \frac{\partial A_z}{\partial \rho} \right) + \hat{\mathbf{z}} \frac{1}{\rho} \left(\frac{\partial(\rho A_\phi)}{\partial \rho} - \frac{\partial A_\rho}{\partial \phi} \right) \tag{B.12}$$

$$\nabla^2 f = \frac{1}{\rho} \frac{\partial}{\partial \rho} \left(\rho \frac{\partial f}{\partial \rho} \right) + \frac{1}{\rho^2} \frac{\partial^2 f}{\partial \phi^2} + \frac{\partial^2 f}{\partial z^2} \tag{B.13}$$

Spherical coordinates

$$\nabla f = \hat{\mathbf{r}} \frac{\partial f}{\partial r} + \hat{\boldsymbol{\theta}} \frac{1}{r} \frac{\partial f}{\partial \theta} + \hat{\boldsymbol{\phi}} \frac{1}{r \sin \theta} \frac{\partial f}{\partial \phi} \tag{B.14}$$

$$\nabla \cdot \mathbf{A} = \frac{1}{r^2} \frac{\partial(r^2 A_r)}{\partial r} + \frac{1}{r \sin \theta} \frac{\partial(\sin \theta A_\theta)}{\partial \theta} + \frac{1}{r \sin \theta} \frac{\partial A_\phi}{\partial \phi} \tag{B.15}$$

$$\nabla \times \mathbf{A} = \hat{\mathbf{r}} \frac{1}{r \sin \theta} \left(\frac{\partial(\sin \theta A_\phi)}{\partial \theta} - \frac{\partial A_\theta}{\partial \phi} \right)$$
$$+ \hat{\boldsymbol{\theta}} \frac{1}{r} \left(\frac{1}{\sin \theta} \frac{\partial A_r}{\partial \phi} - \frac{\partial(r A_\phi)}{\partial r} \right) + \hat{\boldsymbol{\phi}} \frac{1}{r} \left(\frac{\partial(r A_\theta)}{\partial r} - \frac{\partial A_r}{\partial \theta} \right) \tag{B.16}$$

$$\nabla^2 f = \frac{1}{r^2} \frac{\partial}{\partial r} \left(r^2 \frac{\partial f}{\partial r} \right) + \frac{1}{r^2 \sin \theta} \frac{\partial}{\partial \theta} \left(\sin \theta \frac{\partial f}{\partial \theta} \right) + \frac{1}{r^2 \sin^2 \theta} \frac{\partial^2 f}{\partial \phi^2} \tag{B.17}$$

B.4 Vector Identities

$$\mathbf{A} \cdot \mathbf{B} \times \mathbf{C} = \mathbf{A} \times \mathbf{B} \cdot \mathbf{C} = \mathbf{C} \times \mathbf{A} \cdot \mathbf{B} \tag{B.18}$$

$$\mathbf{A} \times (\mathbf{B} \times \mathbf{C}) = \mathbf{B}(\mathbf{A} \cdot \mathbf{C}) - \mathbf{C}(\mathbf{A} \cdot \mathbf{B}) \tag{B.19}$$

$$(\mathbf{A} \times \mathbf{B}) \times \mathbf{C} = \mathbf{A} \times (\mathbf{B} \times \mathbf{C}) - \mathbf{B} \times (\mathbf{A} \times \mathbf{C}) \tag{B.20}$$

Note that $\mathbf{A} \times (\mathbf{B} \times \mathbf{C}) \neq (\mathbf{A} \times \mathbf{B}) \times \mathbf{C}$.

$$\nabla \cdot \nabla \times \mathbf{A} = 0 \tag{B.21}$$

$$\nabla \times \nabla f = 0 \tag{B.22}$$

$$\nabla^2 f = \nabla \cdot \nabla f \tag{B.23}$$

$$\nabla(fg) = f \nabla g + g \nabla f \tag{B.24}$$

$$\nabla(\mathbf{A} \cdot \mathbf{B}) = (\mathbf{A} \cdot \nabla)\mathbf{B} + (\mathbf{B} \cdot \nabla)\mathbf{A} + \mathbf{A} \times (\nabla \times \mathbf{B}) + \mathbf{B} \times (\nabla \times \mathbf{A}) \tag{B.25}$$

$$\nabla \cdot (f\mathbf{A}) = \mathbf{A} \cdot \nabla f + f \nabla \cdot \mathbf{A} \tag{B.26}$$

$$\nabla \cdot (\mathbf{A} \times \mathbf{B}) = \mathbf{B} \cdot \nabla \times \mathbf{A} - \mathbf{A} \cdot \nabla \times \mathbf{B} \tag{B.27}$$

$$\nabla \times (f\mathbf{A}) = (\nabla f) \times \mathbf{A} + f \nabla \times \mathbf{A} \tag{B.28}$$

$$\nabla \times (\mathbf{A} \times \mathbf{B}) = \mathbf{A} \nabla \cdot \mathbf{B} - \mathbf{B} \nabla \cdot \mathbf{A} + (\mathbf{B} \cdot \nabla)\mathbf{A} - (\mathbf{A} \cdot \nabla)\mathbf{B} \tag{B.29}$$

In a term such as $(\mathbf{A} \cdot \nabla)\mathbf{B}$, the part $(\mathbf{A} \cdot \nabla)$ is an operator that differentiates $\mathbf{B}$, and so, $(\mathbf{A} \cdot \nabla)\mathbf{B} \neq (\nabla \cdot \mathbf{A})\mathbf{B}$.

The scalar and vector Laplacians are different:

$$\nabla^2 f = \frac{\partial^2 f}{\partial x^2} + \frac{\partial^2 f}{\partial y^2} + \frac{\partial^2 f}{\partial z^2} \quad \text{scalar Laplacian} \tag{B.30}$$

$$\nabla^2 \mathbf{A} = \hat{\mathbf{x}} \nabla^2 A_x + \hat{\mathbf{y}} \nabla^2 A_y + \hat{\mathbf{z}} \nabla^2 A_z \quad \text{vector Laplacian.} \tag{B.31}$$

In rectangular coordinates, the vector Laplacian has a convenient expression in terms of scalar Laplacians. However, in other coordinate systems, there are no equivalent formulas, for example, $\nabla^2 \mathbf{A} \neq \hat{\rho} \nabla^2 A_\rho + \hat{\phi} \nabla^2 A_\phi + \hat{\mathbf{z}} \nabla^2 A_z$. In cylindrical and spherical coordinates, it is necessary to expand the right-hand side of

$$\nabla^2 \mathbf{A} = -\nabla \times \nabla \times \mathbf{A} + \nabla \nabla \cdot \mathbf{A}. \tag{B.32}$$

The double curl is often written without parentheses; however, one should remember that its meaning is always $\nabla \times \nabla \times \mathbf{A} = \nabla \times (\nabla \times \mathbf{A})$; note that $(\nabla \times \nabla) \times \mathbf{A} = 0$.

B.5   Integral Relations

$$\int_V \nabla f \, dV = \oint_S f \, d\mathbf{S} \tag{B.33}$$

$$\int_V \nabla \times \mathbf{A} \, dV = -\oint_S \mathbf{A} \times d\mathbf{S} \tag{B.34}$$

$$\int_S \hat{\mathbf{n}} \times \nabla f \, dS = \oint_C f \, d\boldsymbol{\ell} \tag{B.35}$$

$$\oint_S \mathbf{F} \cdot d\mathbf{S} = \int_V \nabla \cdot \mathbf{F} \, dV \quad \text{divergence theorem} \tag{B.36}$$

$$\oint_C \mathbf{F} \cdot d\boldsymbol{\ell} = \int_S \nabla \times \mathbf{F} \cdot d\mathbf{S} \quad \text{Stokes theorem.} \tag{B.37}$$

The relevant surfaces are shown in Figure B.3. For the divergence theorem, a volume V is bounded by a closed surface S having an outward-pointing surface area element $d\mathbf{S} = \hat{\mathbf{n}} \, dS$. In Stokes' theorem, the dot product of $\mathbf{F}$ with $d\boldsymbol{\ell} = \hat{\mathbf{t}} \, d\ell$ takes the part of $\mathbf{F}$ that is along the curve's tangent $\hat{\mathbf{t}}$. The direction of the line integration along C can be chosen either way and does not have to be in the same direction as $d\boldsymbol{\ell}$. However the right-hand rule must be followed. That is, the right-hand fingers follow the direction of C and the thumb points in the direction of the surface area element $d\mathbf{S} = \hat{\mathbf{n}} \, dS$.

If $\bar{\mathbf{C}}$ is a dyadic then we have

$$\oint_S \hat{\mathbf{n}} \cdot \bar{\mathbf{C}} \, dS = \int_V \nabla \cdot \bar{\mathbf{C}} \, dV \quad \text{divergence theorem} \tag{B.38}$$

$$\oint_C \hat{\mathbf{t}} \cdot \bar{\mathbf{C}} \, d\ell = \int_S \hat{\mathbf{n}} \cdot (\nabla \times \bar{\mathbf{C}}) \, dS \quad \text{Stokes theorem.} \tag{B.39}$$

2D versions in z-invariant coordinates can be obtained as in Figure B.4. For the divergence theorem we replace $dV \to dx \, dy$ and $d\mathbf{S} \to \hat{\mathbf{n}} \, d\ell$ which points out of S. For Stokes theorem, $d\mathbf{S} \to \hat{\mathbf{z}} \, dx \, dy$ and since $d\mathbf{S}$ is chosen in the $+z$ direction, C is counterclockwise following the right-hand fingers. The results are

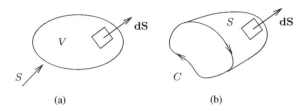

Figure B.3   3D mathematical surfaces associated with the integral formulas. (a) Closed surface S and volume V. (b) Open surface S and bounding contour C.

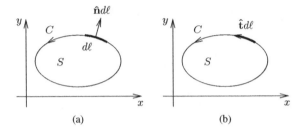

Figure B.4   2D mathematical surfaces associated with the (a) divergence theorem and (b) Stokes theorem.

$$\oint_C \mathbf{F} \cdot \hat{\mathbf{n}} \, d\ell = \int_S \nabla \cdot \mathbf{F} \, dx \, dy \quad \text{divergence theorem, 2D} \tag{B.40}$$

$$\oint_C \mathbf{F} \cdot \hat{\mathbf{t}} \, d\ell = \int_S \nabla \times \mathbf{F} \cdot \hat{\mathbf{z}} \, dx \, dy \quad \text{Stokes theorem, 2D} \tag{B.41}$$

and the unit vectors obey $\hat{\mathbf{n}} \times \hat{\mathbf{t}} = \hat{\mathbf{z}}$.

B.5.1   Green's Identities

$$\int_V (\nabla f \cdot \nabla g + g \nabla^2 f) \, dV = \oint_S g \nabla f \cdot \mathbf{dS} \quad \text{Green's first identity} \tag{B.42}$$

$$\int_V (g \nabla^2 f - f \nabla^2 g) \, dV = \oint_S (g \nabla f - f \nabla g) \cdot \mathbf{dS} \quad \text{Green's second identity} \tag{B.43}$$

$$\int_V \nabla \cdot (\mathbf{A} \times \nabla \times \mathbf{B}) \, dV = \int_V (\nabla \times \mathbf{A}) \cdot (\nabla \times \mathbf{B}) - \mathbf{A} \cdot \nabla \times \nabla \times \mathbf{B} \, dV$$

$$= \oint_S \mathbf{A} \times (\nabla \times \mathbf{B}) \cdot \mathbf{dS} \tag{B.44}$$

$$\int_V \nabla \times (\nabla \times \mathbf{A}) \cdot \mathbf{B} - \mathbf{A} \cdot \nabla \times (\nabla \times \mathbf{B}) \, dV = \oint_S [(\nabla \times \mathbf{A}) \times \mathbf{B} + \mathbf{A} \times (\nabla \times \mathbf{B})] \cdot \mathbf{dS} \tag{B.45}$$

and with a vector $\mathbf{A}$ and dyadic $\bar{\mathbf{C}}$,

$$\int_V \nabla \times (\nabla \times \mathbf{A}) \cdot \bar{\mathbf{C}} - \mathbf{A} \cdot \nabla \times (\nabla \times \bar{\mathbf{C}}) \, dV = \oint_S [(\nabla \times \mathbf{A}) \times \bar{\mathbf{C}} + \mathbf{A} \times (\nabla \times \bar{\mathbf{C}})] \cdot d\mathbf{S}. \quad (B.46)$$

2D versions of Green's identities can be obtained by replacing $dV \to dx \, dy$ and $d\mathbf{S} \to \hat{n} d\ell$.

B.5.2   Helmholtz's Theorem

Theorem 1: Helmholtz's theorem is

$$\mathbf{F}(\mathbf{r}) = -\nabla \int_V \frac{\nabla' \cdot \mathbf{F}(\mathbf{r}')}{4\pi |\mathbf{r} - \mathbf{r}'|} \, dV' + \nabla \oint_S \frac{\mathbf{F}(\mathbf{r}') \cdot d\mathbf{S}'}{4\pi |\mathbf{r} - \mathbf{r}'|}$$

$$+ \nabla \times \int_V \frac{\nabla' \times \mathbf{F}(\mathbf{r}')}{4\pi |\mathbf{r} - \mathbf{r}'|} \, dV' - \nabla \times \oint_S \frac{\mathbf{F}(\mathbf{r}') \times d\mathbf{S}'}{4\pi |\mathbf{r} - \mathbf{r}'|}. \quad (B.47)$$

A derivation can be found in Collin (1991, pp. 799–801). If S is at infinity and $\mathbf{F}$ vanishes there, the surface integrals will be zero. Then, if the divergence and curl of $\mathbf{F}$ are specified everywhere in V, $\mathbf{F}$ will also be known.

Theorem 2: Any differentiable vector function $\mathbf{F}$ can be written as a sum of irrotational and solenoidal parts

$$\mathbf{F}(\mathbf{r}) = \nabla g + \nabla \times \mathbf{G}. \quad (B.48)$$

This follows from Theorem 1 with

$$g(\mathbf{r}) = -\int_V \frac{\nabla' \cdot \mathbf{F}(\mathbf{r}')}{4\pi |\mathbf{r} - \mathbf{r}'|} \, dV; \quad \mathbf{G}(\mathbf{r}) = \int_V \frac{\nabla' \times \mathbf{F}(\mathbf{r}')}{4\pi |\mathbf{r} - \mathbf{r}'|} \, dV'.$$

Theorem 3: An irrotational vector is expressible as

$$\mathbf{F} = \nabla g.$$

This is because $\nabla \times \nabla g \equiv 0$ for any scalar function g.

Theorem 4: A solenoidal vector is expressible as

$$\mathbf{F} = \nabla \times \mathbf{G}.$$

This is because $\nabla \cdot \nabla \times \mathbf{G} \equiv 0$ for any vector function $\mathbf{G}$.

Reference

Collin RE (1991) *Field Theory of Guided Waves*. Oxford University Press.

Appendix C

Bessel's Differential Equation

Any differential equation of the form

$$y''(x) + \frac{b(x)}{x}y'(x) + \frac{c(x)}{x^2}y(x) = 0$$

with $b(x)$ and $c(x)$ analytic can be solved by the *method of Frobenius* (Arfken and Weber 2005, Chapter 9). An important special case is Bessel's differential equation

$$y''(x) + \frac{1}{x}y'(x) + \frac{x^2 - \nu^2}{x^2}y(x) = 0, \qquad (C.1)$$

where ν is a non-negative real number.

C.1 Bessel Functions

Application of the method of Frobenius leads to two series solutions. The first series is called the *Bessel function of the first kind* and is given by

$$J_\nu(x) = \sum_{m=0}^{\infty} \frac{(-1)^m}{m!(m+\nu)!}\left(\frac{x}{2}\right)^{2m+\nu} \qquad (C.2)$$

When $\nu \neq$ integer, the second solution is $J_{-\nu}(x)$. The general solution is then

$$y(x) = C_1 J_\nu(x) + C_2 J_{-\nu}(x),$$

where C_1 and C_2 are arbitrary constants. The *gamma function* is a generalized factorial, having the property $n! = \Gamma(n+1)$. It is used to extend the solution to noninteger orders ν.

When ν is an integer, the second solution has a logarithmic term. It is called the *Bessel function of the second kind*. When $\nu = 0$, it is given by

$$Y_0(x) = \frac{2}{\pi}\ln(\gamma x/2)J_0(x) + \frac{2}{\pi}\sum_{m=1}^{\infty}\frac{(-1)^{m+1}}{(m!)^2}\left(\frac{x}{2}\right)^{2m}\phi(m), \qquad (C.3)$$

Applied Frequency-Domain Electromagnetics, First Edition. Robert Paknys.
© 2016 John Wiley & Sons, Ltd. Published 2016 by John Wiley & Sons, Ltd.
Companion Website: www.wiley.com/go/paknys9981

where $\quad \phi(m) = 1 + 1/2 + 1/3 + \cdots + 1/m \quad$ and $\quad \ln \gamma = 0.57721566\cdots = \lim\limits_{n \to \infty} (1 + 1/2 + \cdots + 1/n - \ln n)$ is *Euler's constant*.

When $\nu = n$ is an integer, and $n > 0$, the second solution is

$$Y_n(x) = \frac{2}{\pi} \ln(\gamma x/2) J_n(x) - \frac{1}{\pi} \sum_{m=0}^{n-1} \frac{(n-m-1)!}{m!} \left(\frac{2}{x}\right)^{n-2m}$$

$$- \frac{1}{\pi} \sum_{m=0}^{\infty} \frac{(-1)^m}{m!(m+n)!} \left(\frac{x}{2}\right)^{n+2m} [\phi(m) + \phi(m+n)]. \tag{C.4}$$

The general solution for integer order n is given by

$$y(x) = C_1 J_n(x) + C_2 Y_n(x).$$

Figures C.1 and C.2 show $J_n(x)$ and $Y_n(x)$.
The first few roots x_n of $J_m(x_n) = 0$ and $J'_m(x_n) = 0$ are given in Table C.1.
Hankel functions of the first and second kind are defined by

$$H_\nu^{(1)}(x) = J_\nu(x) + jY_\nu(x) \tag{C.5}$$

$$H_\nu^{(2)}(x) = J_\nu(x) - jY_\nu(x). \tag{C.6}$$

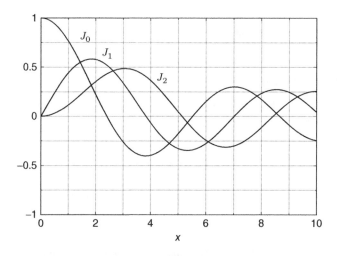

Figure C.1   Bessel functions of the first kind, $J_n(x)$.

Table C.1   First few roots of $J_m(x) = 0$ and $J'_m(x) = 0$.

	$J_m(x_n) = 0$			$J'_m(x_n) = 0$		
m	x_1	x_2	x_3	x_1	x_2	x_3
0	2.405	5.520	8.654	3.832	7.016	10.174
1	3.832	7.016	10.174	1.841	5.331	8.536
2	5.135	8.417	11.620	3.054	6.706	9.970

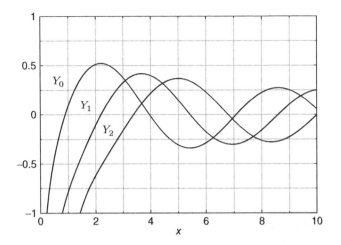

Figure C.2   Bessel functions of the second kind, $Y_n(x)$.

A Taylor series can be used to find approximations for small x,

$$J_0(x) \approx 1 - \frac{1}{4}x^2 + \frac{1}{64}x^4 + \cdots \tag{C.7}$$

$$\pi Y_0(x) \approx 2 J_0(x)\{\ln(x/2) + \gamma\} + \frac{1}{2}x^2 - \frac{3}{64}x^4 + \cdots . \tag{C.8}$$

By using $J_0' = -J_1$, $Y_0' = -Y_1$, we also have

$$J_1(x) \approx \frac{1}{2}x - \frac{1}{16}x^3 + \cdots \tag{C.9}$$

$$\pi Y_1(x) \approx x \ln(x/2) - \frac{2}{x} + \left(\gamma - \frac{1}{2}\right)x + \frac{6}{32}x^3 + \cdots . \tag{C.10}$$

Here, $\gamma = 0.577215665 \ldots$ is Euler's constant. From the leading term of the series, $H_0^{(2)}(x)$ can be approximated as

$$H_0^{(2)}(x) \approx 1 - j\frac{2}{\pi}\ln(\gamma_0 x/2) , \tag{C.11}$$

where $\gamma_0 = \exp \gamma = 1.781072416 \ldots$.

Hankel function asymptotic approximations for large x (the Debye approximation) are

$$H_\nu^{(1)}(x) \sim \sqrt{\frac{2}{\pi x}} e^{j(x - \nu \pi/2 - \pi/4)} + O(x^{-3/2}) \tag{C.12}$$

$$H_\nu^{(2)}(x) \sim \sqrt{\frac{2}{\pi x}} e^{-j(x - \nu \pi/2 - \pi/4)} + O(x^{-3/2}) \tag{C.13}$$

and the real or imaginary part of $H^{(1)}$ gives expressions for J_ν or Y_ν. Differentiating and retaining only the leading term, the derivatives are

$$H_\nu'^{(1)}(x) \sim j H_\nu^{(1)}(x) + O(x^{-3/2}) \tag{C.14}$$

$$H_\nu'^{(2)}(x) \sim -j H_\nu^{(2)}(x) + O(x^{-3/2}) . \tag{C.15}$$

Hankel function asymptotic approximations (the Watson approximation) for large x and large ν with $\nu \approx x$ are (Bowman et al., 1969), (Felsen and Marcuvitz, 1994, p. 418)

$$H_\nu^{(1)}(x) \sim \frac{-j}{m\sqrt{\pi}} W_1(t) + O(m^{-3}) \tag{C.16}$$

$$H_\nu^{(2)}(x) \sim \frac{j}{m\sqrt{\pi}} W_2(t) + O(m^{-3}), \tag{C.17}$$

where $m = (x/2)^{1/3}$, $t = (\nu - x)/m$ and $W_{1,2}(t)$ are Fock-type Airy functions that are defined in terms of Miller-type Airy functions $\text{Ai}(t)$, $\text{Bi}(t)$ as

$$W_{1,2}(t) = \sqrt{\pi}(\text{Bi}(t) \pm j\text{Ai}(t)).$$

C.2   Roots of $H_{\nu_p}^{(1,2)}(x) = 0$

The equation $H_{\nu_p}^{(2)}(x) = 0$ has an infinite number of roots ν_p, enumerated $p = 1, 2, \dots$. The first few are near $\nu_p \approx x$ and can be found from the Watson approximation (C.17); in other words $W_2(t) = 0$. This has complex roots $t_p = \alpha\, e^{-j\pi/3}$ where $\alpha_p = 2.3381, 4.0879, 5.5205, \dots$ (Felsen and Marcuvitz, 1994, Appendix 6a). Therefore, $H_{\nu_p}^{(2)}(x) = 0$ has roots

$$\nu_p = x + m\, \alpha_p\, e^{-j\pi/3}. \tag{C.18}$$

The roots of $H_{\nu_p}^{\prime(2)}(x) = 0$ follow from the derivative of (C.17) which is $H_\nu^{\prime(2)}(x) \sim (-j/m^2)\sqrt{\pi}\, W_2'(t)$. The roots of $W_2'(t) = 0$ are at $t_p = \beta\, e^{-j\pi/3}$, where $\beta_p = 1.0188, 3.3482, 4.8201, \dots$. Therefore, the roots of $H_{\nu_p}^{\prime(2)}(x) = 0$ are at

$$\nu_p = x + m\, \beta_p\, e^{-j\pi/3}. \tag{C.19}$$

The roots of $H_\nu^{(1)}$ and its derivative can be found by replacing $e^{-j\pi/3} \to e^{j\pi/3}$ in the expressions for ν_p.

The roots of $H_\nu^{(2)}$ and its derivative lie on a line in the complex ν plane that starts at $\nu = x$ and extends downwards at an angle of $-60°$ into the fourth quadrant. This is true for the first few roots; however when $|\nu| \gg x$ the Watson approximation is no longer valid, and the Debye approximation is needed. One finds that for large $|\nu|$ the roots lie along an asymptote that is at $-90°$.

Having to switch between Watson and Debye representations can be avoided by using Olver's uniform asymptotic formulas. These have been coded in SUBROUTINE huni (Paknys 1992) and permit the computation of Hankel functions and derivatives for complex argument and order. Together with a root-finding program, ν_p can be computed.

C.3   Integrals

Here, Z_n is J_n, Y_n or H_n.

$$\int^x Z_n(kx)Z_n(\ell x)x\,dx = \frac{x}{k^2 - \ell^2}[kZ_n(\ell x)Z_{n+1}(kx) - \ell Z_n(kx)Z_{n+1}(\ell x)] \tag{C.20}$$

$$\int^x Z_n^2(kx)x\,dx = \frac{x^2}{2}\left[Z_n^{\prime 2}(kx) + \left(1 - \frac{n^2}{k^2 x^2}\right)Z_n^2(kx)\right] \tag{C.21}$$

$$2\pi j^n\, J_n(x) = \int_0^{2\pi} e^{jx\cos\theta}\cos n\theta\,d\theta. \tag{C.22}$$

C.4　Orthogonality

If $x_p = \alpha_p a$, $p = 1, 2, 3, \ldots$ are the roots of

$$AJ_m(x_p) + Bx_p J'_m(x_p) = 0 \tag{C.23}$$

the aforementioned integrals can be used to show that

$$\int_0^a J_m(\alpha_p \rho) J_m(\alpha_q \rho) \rho d\rho = \begin{cases} (a^2/2)\left[J'^2_m(\alpha_p a) + \left(1 - \dfrac{m^2}{(\alpha_p a)^2}\right) J^2_m(\alpha_p a)\right]; p = q \\ 0; p \neq q \end{cases} \tag{C.24}$$

In the special case where $B = 0$, the numbers α_p satisfy $J_m(\alpha_p a) = 0$ so that

$$\int_0^a J_m(\alpha_p \rho) J_m(\alpha_q \rho) \rho d\rho = \begin{cases} (a^2/2) J'^2_m(\alpha_p a); p = q \\ 0; p \neq q \end{cases} \tag{C.25}$$

This can be used to develop a Fourier–Bessel series:

$$f(\rho) = \sum_{p=1}^{\infty} A_p J_m(\alpha_p \rho), \tag{C.26}$$

where m is arbitrary.

C.5　Recursion Relations

Here, Z_n is J_n, Y_n or H_n.

$$Z'_0(x) = -Z_1(x) \tag{C.27}$$

$$Z_{n+1}(x) = \frac{2n}{x} Z_n(x) - Z_{n-1}(x) \tag{C.28}$$

$$Z'_n(x) = \frac{1}{2}[Z_{n-1}(x) - Z_{n+1}(x)] \tag{C.29}$$

$$xZ'_n(x) = xZ_{n-1}(x) - nZ_n(x) \tag{C.30}$$

$$xZ'_n(x) = nZ_n(x) - xZ_{n+1}(x) \tag{C.31}$$

$$[x^n Z_n(x)]' = x^n Z_{n-1}(x) \tag{C.32}$$

$$[x^{-n} Z_n(x)]' = -x^{-n} Z_{n+1}(x). \tag{C.33}$$

C.6　Gamma Function

The gamma function is defined by

$$\Gamma(t) = \int_0^{\infty} x^{t-1} e^{-x}\, dx. \tag{C.34}$$

Carrying out an integration by parts, one can obtain

$$\Gamma(t+1) = t\Gamma(t). \tag{C.35}$$

Using $\Gamma(1) = 1$ and (C.35) it follows that

$$\Gamma(n) = (n-1)! \tag{C.36}$$

for a positive integer n. Since $\Gamma(t)$ can be used with noninteger values of t, it is sometimes called the 'generalized factorial function'. A few values are

$$\Gamma(1) = 1; \quad \Gamma(-1/2) = -2\sqrt{\pi}; \quad \Gamma(1/2) = \sqrt{\pi}; \quad \Gamma(3/2) = \frac{1}{2}\sqrt{\pi}. \tag{C.37}$$

From the leading terms in the power series of $J_\nu(x)$ and $Y_\nu(x)$, we can obtain the following small-argument approximations:

$$J_\nu(x) \approx \frac{(x/2)^\nu}{\Gamma(\nu+1)} \tag{C.38}$$

$$Y_\nu(x) \approx \frac{-\Gamma(\nu)}{\pi(x/2)^\nu}; \nu > 0 \tag{C.39}$$

$$Y_0(x) \approx \frac{2}{\pi}(\ln(x/2) + \gamma). \tag{C.40}$$

C.7   Wronskians

The Wronskian determinant for any two solutions T, U is defined by

$$w(T, U) = \begin{vmatrix} T & U \\ T' & U' \end{vmatrix}$$

and $w \neq 0$ implies that the solutions are independent.

Bessel's differential equation is a Sturm–Liouville problem having $p(x) = x$. Related to the Wronskian is the conjunct $W = p(x)w(T, U)$ which does not depend on x. We can choose any convenient value of x when evaluating it. It is convenient to assume that x is small, so that we can use the small-argument approximations (C.38)–(C.40). Solutions of Bessel's differential equation obey

$$Z_\nu'(x) = \frac{\nu}{x}Z_\nu(x) - Z_{\nu+1}(x)$$

so the Wronskian of $J_\nu(x)$ and $Y_\nu(x)$ for small x is

$$w(J_\nu(x), Y_\nu(x)) = J_\nu Y_\nu' - J_\nu' Y_\nu = -J_\nu Y_{\nu+1} + J_{\nu+1} Y_\nu \approx \frac{2}{\pi x}.$$

Since $p(x) = x$, the conjunct is

$$W(J_\nu(x), Y_\nu(x)) = \frac{2}{\pi}.$$

The conjunct does not depend on x, which implies that the Wronskian result previously is not restricted to small x but is valid for any x. Some Wronskian relations are

$$J_\nu(x)J'_{-\nu}(x) - J'_\nu(x)J_{-\nu}(x) = -\frac{2\sin(\nu\pi)}{\nu x} \tag{C.41}$$

$$J_\nu(x)Y'_\nu(x) - J'_\nu(x)Y_\nu(x) = \frac{2}{\pi x} \tag{C.42}$$

$$H_\nu^{(1)}(x)H_\nu^{(2)}(x) - H_\nu^{'(1)}(x)H_\nu^{(2)}(x) = -\frac{j4}{\pi x} \tag{C.43}$$

$$J_\nu(x)H_\nu^{'(2)}(x) - J'_\nu(x)H_\nu^{(2)}(x) = -\frac{j2}{\pi x} \tag{C.44}$$

$$J_\nu(x)H_\nu^{'(1)}(x) - J'_\nu(x)H_\nu^{(1)}(x) = \frac{j2}{\pi x}. \tag{C.45}$$

C.8   Spherical Bessel Functions

The radial part of the wave equation in spherical coordinates is

$$\frac{d}{dr}\left(r^2\frac{df(kr)}{dr}\right) + [(kr)^2 - n(n+1)]f(kr) = 0. \tag{C.46}$$

Its solutions are *spherical Bessel functions* $b_n(kr)$, which can be $j_n(kr)$, $y_n(kr)$, $h_n^{(1)}(kr)$ or $h_n^{(2)}(kr)$. These are related to the cylindrical Bessel functions by

$$b_n(kr) = \sqrt{\frac{\pi}{2kr}}B_{n+\frac{1}{2}}(kr). \tag{C.47}$$

The derivatives are

$$b'_0 = -b_1 \tag{C.48}$$

$$b'_n = \frac{nb_{n-1} - (n+1)b_{n+1}}{2n+1}; n \geq 1. \tag{C.49}$$

There are also Schelkunoff's spherical Bessel functions $\hat{B}_n(kr)$ defined by

$$\hat{B}_n(kr) = krb_n(kr) = \sqrt{\frac{\pi kr}{2}}B_{n+\frac{1}{2}}(kr). \tag{C.50}$$

The derivative is

$$\hat{B}'_n(kr) = b_n(kr) + krb'_n(kr). \tag{C.51}$$

$\hat{B}_n$ satisfies the differential equation

$$\frac{d^2\hat{B}_n(kr)}{dr^2} + \left(k^2 - \frac{n(n+1)}{r^2}\right)\hat{B}_n(kr) = 0 \tag{C.52}$$

which is useful for finding $\hat{B}''_n$. In the far field,

$$\hat{H}_n^{(2)}(kr) \sim j^{n+1}e^{-jkr} \tag{C.53}$$

$$\hat{H}_n^{\prime(2)}(kr) \sim (-j)\hat{H}_n^{(2)}(kr),\qquad\qquad\qquad\text{(C.54)}$$

where the prime denotes differentiation with respect to the argument; $d/d(kr)$.

References

Arfken GB and Weber HJ (2005) *Mathematical Methods for Physicists*. Elsevier.

Bowman JJ, Senior TBA and Uslenghi PLE (1969) *Electromagnetic and Acoustic Scattering by Simple Shapes*. North Holland.

Felsen LB and Marcuvitz N (1994) *Radiation and Scattering of Waves*. IEEE Press.

Paknys R (1992) Evaluation of Hankel functions with complex argument and complex order. *IEEE Trans. Antennas Propag.* **40**, 569–578.

Appendix D

Legendre's Differential Equation

Any differential equation of the form

$$y''(x) + p(x)y'(x) + q(x)y(x) = r(x)$$

with $p(x)$, $q(x)$, $r(x)$ analytic can be solved by the *power series method*. An important special case is *Legendre's differential equation of order* α:

$$(1 - x^2)y''(x) - 2xy'(x) + \alpha(\alpha + 1)y(x) = 0 \qquad\qquad (\text{D.1})$$

where α is a non-negative real number. It has singular points at $x = \pm 1$ and a power series solution that converges on $-1 \leq x \leq 1$. Proceeding with the power series method yields the two series

$$y(x) = a_0 \left(1 - \frac{(\alpha + 1)\alpha}{2!}x^2 + \frac{(\alpha + 3)(\alpha + 1)\alpha(\alpha - 2)}{4!}x^4 - \cdots\right)$$

$$+ a_1 \left(x - \frac{(\alpha + 2)(\alpha - 1)}{3!}x^3 + \frac{(\alpha + 4)(\alpha + 2)(\alpha - 1)(\alpha - 3)}{5!}x^5 - \cdots\right),$$

where a_0 and a_1 are arbitrary constants.

D.1 Legendre Functions

When $\alpha = n =$ integer, one of the series is of finite length and the other series is of infinite length. The finite length series is called a *Legendre polynomial* and is denoted as $P_n(x)$. The infinite series is denoted as $Q_n(x)$. The common way of writing the solution of Legendre's equation is

$$y(x) = C_1 P_n(x) + C_2 Q_n(x),$$

where C_1 and C_2 are arbitrary constants.

Applied Frequency-Domain Electromagnetics, First Edition. Robert Paknys.
© 2016 John Wiley & Sons, Ltd. Published 2016 by John Wiley & Sons, Ltd.
Companion Website: www.wiley.com/go/paknys9981

It is convenient to choose the coefficients a_0 and a_1 in such a way that $|P_n(\pm 1)| = 1$. Then, the first few $P_n(x)$ are given by

$$P_0(x) = 1, P_1(x) = x, P_2(x) = \frac{1}{2}(3x^2 - 1), P_3(x) = \frac{1}{2}(5x^3 - 3x), \dots . \tag{D.2}$$

It turns out that the infinite series $Q_n(x)$ can be summed in closed form. The first few Q_n are given by (Abramowitz and Stegun, 1965)

$$Q_0(x) = \frac{1}{2} \ln\left(\frac{1+x}{1-x}\right), Q_1(x) = \frac{x}{2} \ln\left(\frac{1+x}{1-x}\right) - 1,$$

$$Q_2(x) = \frac{3x^2 - 1}{4} \ln\left(\frac{1+x}{1-x}\right) - \frac{3x}{2}, \dots . \tag{D.3}$$

D.2 Associated Legendre Functions

The associated Legendre differential equation is

$$(1 - x^2)y''(x) - 2xy'(x) + \left(n(n+1) - \frac{m^2}{1-x^2}\right)y(x) = 0 \tag{D.4}$$

and the general solution is

$$y(x) = C_1 P_n^m(x) + C_2 Q_n^m(x).$$

These are called the *associated Legendre functions* of degree n and order m and are given by

$$P_n^m(x) = (-1)^m (1 - x^2)^{m/2} \frac{d^m P_n(x)}{dx^m} \tag{D.5}$$

$$Q_n^m(x) = (-1)^m (1 - x^2)^{m/2} \frac{d^m Q_n(x)}{dx^m}. \tag{D.6}$$

P_n^m and Q_n^m are zero when $m > n$. When $m = 0$, $P_n^0 = P_n$ and $Q_n^0 = Q_n$. The $Q_n^m(x)$ functions are less useful in physical problems because they are singular at $x = \pm 1$. The first few P_n^m are

$$P_1^0(x) = P_1(x); \quad P_1^1(x) = -(1 - x^2)^{1/2} \tag{D.7}$$

$$P_2^0(x) = P_2(x); \quad P_2^1(x) = -3x(1 - x^2)^{1/2}; \quad P_2^2(x) = 3(1 - x^2). \tag{D.8}$$

D.3 Orthogonality

The Legendre polynomials are orthogonal with respect to the degree n, and the normalization

$$\bar{P}_n^m(x) = (-1)^m \sqrt{\frac{(2n+1)(n-m)!}{2(n+m)!}} P_n^m(x) \tag{D.9}$$

is sometimes used to make them orthonormal, so that

$$\int_{-1}^1 \bar{P}_i^m(x) \bar{P}_j^m(x)\, dx = \begin{cases} 1; & i = j \\ 0; & i \neq j. \end{cases} \tag{D.10}$$

D.4   Recursion Relations

Here, L_n^m is P_n^m or Q_n^m:

$$(m - n - 1)L_{n+1}^m + (2n + 1)xL_n^m - (m + n)L_{n-1}^m = 0 \tag{D.11}$$

$$L_n^{m+1} + \frac{2mx}{(1 - x^2)^{1/2}}L_n^m + (n + m)(n - m + 1)L_n^{m-1} = 0 \tag{D.12}$$

$$L_n^{m'}(x) = \frac{1}{1 - x^2}[-nxL_n^m + (n + m)L_{n-1}^m] \tag{D.13}$$

$$= \frac{1}{1 - x^2}[(n + 1)xL_n^m - (n - m + 1)L_{n+1}^m] \tag{D.14}$$

$$= \frac{mx}{1 - x^2}L_n^m + \frac{(n + m)(n - m + 1)}{(1 - x^2)^{1/2}}L_n^{m-1} \tag{D.15}$$

$$= -\frac{mx}{1 - x^2}L_n^m - \frac{1}{(1 - x^2)^{1/2}}L_n^{m+1}. \tag{D.16}$$

D.5   Spherical Form

In spherical coordinates we usually encounter the Legendre function $P_n^m(\cos\theta)$. Some useful relations for this case are

$$\frac{dP_n(\cos\theta)}{d\theta} = P_n^1(\cos\theta) \tag{D.17}$$

$$\frac{d^2P_n(\cos\theta)}{d\theta^2} = \frac{dP_n^1(\cos\theta)}{d\theta} = -\sin\theta P_n'^1(\cos\theta), \tag{D.18}$$

where the prime denotes differentiation with respect to the argument: $d/d(\cos\theta)$. Since P_n obeys Legendre's differential equation

$$\sin\theta\frac{d^2P_n(\cos\theta)}{d\theta^2} + \cos\theta\frac{dP_n(\cos\theta)}{d\theta} + n(n + 1)P_n(\cos\theta)\sin\theta = 0 \tag{D.19}$$

we have that

$$\sin\theta\frac{dP_n^1(\cos\theta)}{d\theta} + \cos\theta P_n^1(\cos\theta) + n(n + 1)P_n(\cos\theta)\sin\theta = 0 \tag{D.20}$$

or

$$\frac{dP_n^1(\cos\theta)}{d\theta} = -\cot\theta P_n^1(\cos\theta) - n(n + 1)P_n(\cos\theta). \tag{D.21}$$

Another relation that is useful near $\theta = 0$ or π can be obtained from recursion

$$\frac{P_n^1(\cos\theta)}{\sin\theta} = -\frac{P_n^2(\cos\theta)}{2\cos\theta} - \frac{n(n + 1)P_n(\cos\theta)}{2\cos\theta} \tag{D.22}$$

which shows that $P_n^1(\cos\theta)/\sin\theta$ is not singular at those angles.

Reference

Abramowitz M and Stegun I (1965) *Handbook of Mathematical Functions*. Dover Publications.

Appendix E

Complex Variables

This appendix summarizes residue calculus and branch cut concepts. Recommended supplementary reading is Churchill and Brown (2013) and Arfken and Weber (2005).

E.1   Residue Calculus

If a function $f(z)$ has a singular point at z_0, it can be represented as a Laurent series

$$f(z) = \sum_{n=0}^{\infty} a_n (z - z_0)^n + \sum_{n=1}^{\infty} \frac{b_n}{(z - z_0)^n}.$$

(E.1)

The coefficient b_1 is called the *residue* and is given by

$$b_1 = \frac{1}{j2\pi} \oint_C f(z)\, dz.$$

(E.2)

The closed contour C can be any counterclockwise path enclosing the pole at $z = z_0$, and $f(z)$ is analytic everywhere else inside and on C. If b_1 can be found by some other means, (E.2) provides a powerful way to evaluate contour integrals. Since b_1 is the residue of $f(z)$ and this is so important, it is given the distinguishing notation

$$b_1 = \operatorname{Res} f(z)|_{z=z_0}.$$

For instance, the function

$$f(z) = \frac{e^{-z}}{z} = \frac{1}{z} \sum_{n=0}^{\infty} \frac{(-z)^n}{n!} = \frac{1}{z} - \frac{1}{1} + \frac{z}{2} - \frac{z^2}{6} + \cdots$$

has a pole of order 1 (called a simple pole), and the residue is $b_1 = 1$. The function

$$f(z) = \frac{e^{-z}}{z^2} = \frac{1}{z^2} \sum_{n=0}^{\infty} \frac{(-z)^n}{n!} = \frac{1}{z^2} - \frac{1}{z} + \frac{1}{2} - \frac{z}{6} + \frac{z^2}{24} - \cdots$$

Applied Frequency-Domain Electromagnetics, First Edition. Robert Paknys.
© 2016 John Wiley & Sons, Ltd. Published 2016 by John Wiley & Sons, Ltd.
Companion Website: www.wiley.com/go/paknys9981

has a pole of order 2, and the residue is $b_1 = -1$.

There are convenient formulas to find the residue. If $f(z)$ has only a simple pole at $z = z_0$ then

$$\text{Res } f(z) = \lim_{z \to z_0} (z - z_0)f(z). \tag{E.3}$$

If $f(z)$ has an nth order pole at $z = z_0$ then

$$\text{Res } f(z) = \frac{1}{(n-1)!} \lim_{z \to z_0} \frac{d^{(n-1)}}{dz^{(n-1)}}[(z - z_0)^n f(z)]. \tag{E.4}$$

Suppose that $f(z)$ has a simple pole at z_0 and can be written as a quotient $f(z) = p(z)/q(z)$ where $p(z)$ and $q(z)$ are analytic at $z = z_0$, and $q(z_0) = 0$ gives the pole. Then,

$$\text{Res } f(z) = \frac{p(z_0)}{q'(z_0)}. \tag{E.5}$$

When $f(z)$ has several poles z_p within C, their contributions can be combined. With N poles we have that the integral is equal to $j2\pi$ times the sum of the counterclockwise residues

$$\oint_C f(z)\, dz = j2\pi \sum_{p=1}^{N} \text{Res } f(z_p). \tag{E.6}$$

This is known as the residue theorem.

A proof of (E.6) will not be given, but it hinges on the result (E.2) and it is easy to show why it works. As an example, substitute $f(z) = e^{-z}/z$ into (E.2):

$$b_1 = \frac{1}{j2\pi} \oint_C \frac{e^{-z}}{z}\, dz = \frac{1}{j2\pi} \oint_C \frac{1}{z} - 1 + \frac{z}{2} - \frac{z^2}{6} + \cdots dz.$$

C has to enclose the pole of $f(z)$, which is $z_0 = 0$, so C can be a circle of radius ρ. Then, $z = \rho e^{j\phi}$ and $dz = j\rho e^{j\phi}\, d\phi$. Integrating over 2π, all of the terms in the Laurent series integrate to zero except for the one involving $1/z$ so that

$$b_1 = \frac{1}{j2\pi} \oint_C \frac{1}{z}\, dz = \frac{1}{j2\pi} \int_0^{2\pi} \frac{1}{\rho e^{j\phi}} j\rho e^{j\phi}\, d\phi = 1$$

which is the coefficient b_1 associated with the $1/z$ term in the Laurent series for $f(z)$.

E.2 Branch Cuts

The complex logarithm $f(z) = \ln z$ can be rewritten by using z in polar form, being $z = \rho e^{j\phi}$. Then,

$$f(z) = \ln z = \ln(\rho e^{j\phi}) = \ln \rho + j\phi.$$

It is clear that f is singular at $\rho = 0$. As for the angle, the domain is arbitrary. For the sake of this discussion, we can define it as $-\pi/2 \le \phi < 3\pi/2$. This selection of angles is a *branch* of the complex logarithm function and it defines a 'branch cut' as in Figure E.1(a), and $z = 0$ is the 'branch point'. Since

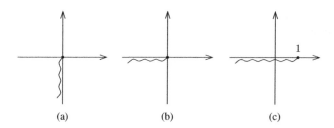

Figure E.1 Some branch cuts in the complex z plane. (a) Cut for $-\pi/2 \le \arg z < 3\pi/2$. (b) Cut for $-\pi < \arg z \le \pi$. (c) Cut for $-\pi < \arg (z - 1) \le \pi$.

$z = \rho\, e^{j(\phi + 2n\pi)}$ with $n = 0, \pm 1, \pm 2, \ldots$ are all the same z, there are an infinite number of branches of the logarithm according to

$$\ln(\rho\, e^{j\phi}) = \ln \rho + j(\phi + 2n\pi).$$

In Figure E.1(a), for points that are immediately adjacent to the branch cut, the imaginary part of $\ln z$ changes abruptly from $-\pi/2$ on the right side of the branch cut to $3\pi/2$ on the left side of the branch cut. Therefore, $\ln z$ is not analytic across the negative imaginary axis.

The branch cut definition is arbitrary and can be defined as in Figure E.1(b) whereby $-\pi < \phi \le \pi$. This is called the 'principal branch' and is what's used by the intrinsic function f=LOG(z) in a Fortran 90 computer program.

The complex square root $f(z) = \sqrt{z}$ has a branch point at $z = 0$, and a branch cut can be defined in any convenient way. Choosing it as in Figure E.1(b) and employing the polar form $z = \rho\, e^{j\phi}$, it becomes

$$f(z) = \sqrt{z} = \sqrt{\rho\, e^{j\phi}} = \sqrt{\rho}\, e^{j\phi/2}$$

and since $-\pi < \phi \le \pi$, the phase of f is limited to $-\pi/2 < \arg f \le \pi/2$. This choice is the principal branch, as used by the intrinsic function f=SQRT(z) in a Fortran 90 computer program. As before, defining other branches is permissible. The principal branch will never give Re $f(z) < 0$ and to get that requires the choice f=-SQRT(z). There are only two branches for the complex square root.

A minor variation on the square root theme worth understanding is the function

$$f(z) = \sqrt{z - 1}$$

which has a branch point at $z = 1$. The branch cut can appear as in Figure E.1(c). In this case the principal branch restricts the phase to $-\pi < \arg (z - 1) \le \pi$. The idea can be extended to a function $\sqrt{z^2 - 1} = \sqrt{z - 1}\sqrt{z + 1}$ which has branch points at $z = \pm 1$. This form appears in Sommerfeld integrals.

References

Arfken GB and Weber HJ (2005) *Mathematical Methods for Physicists*. Elsevier.
Churchill RV and Brown JW (2013) *Complex Variables and Applications*. McGraw-Hill.

Appendix F

Compilers and Programming

This appendix gives some suggestions for obtaining and using a Fortran 90 compiler. A summary of how to work from a command-line terminal is given. The operating system can be Linux/UNIX, Windows or Mac OS X. Selected features of Fortran 90 are discussed and summarized. It is assumed that the reader is familiar with programming but not necessarily Fortran 90. The gnuplot plot program is also described.

It is important to note that Fortran's syntax is not case sensitive. For readability and no other reason, capital letters will be used for the language's keywords. A mix of capital and lower-case letters will be used for variable names. Variables must start with a letter, can use an underscore and can have up to 31 characters.

Programs can be written in the fixed or free format. When using the fixed format, the program's file name has a `.f` extension, whereas for the free format it is `.f90`.

We will use the fixed format, which adheres to Fortran's original conventions. It uses an 80-character line. Columns 7–72 are for the program statements. Columns 1–5 are for numeric statement labels. A comment is made with a 'C' in column 1 or an exclamation mark '!' anywhere. Long statements can be split over several lines by putting any character in column 6; it turns that line into a continuation of the previous line. Columns 73–80 are not used.

The free format allows for lines of up to 132 characters. Column positions are not important. Comments are made with an exclamation mark '!', but the 'C' comment is not allowed. If a line ends with an ampersand '&', the next line is treated as a continuation.

F.1 Getting Started

If you have access to a centrally administered computer at your workplace or school, you might already have a Fortran 90 compiler installed. Check with your system administrator to see if it is available.

You can also install a compiler on your own computer. There is public-domain software and versions are available for Linux, Windows and OS X operating systems. You can use either g95 or gfortran; both are satisfactory. The relevant websites are (September 2015)

```
www.g95.org          The g95 compiler
gcc.gnu.org/fortran  The gfortran compiler
```

There are also commercial Fortran 90 compilers. The advantage of commercial software is that it typically includes technical support, comprehensive documentation and tools for debugging programs.

Applied Frequency-Domain Electromagnetics, First Edition. Robert Paknys.
© 2016 John Wiley & Sons, Ltd. Published 2016 by John Wiley & Sons, Ltd.
Companion Website: www.wiley.com/go/paknys9981

Table F.1   Some Linux commands. It is assumed that gfortran has been installed.

ls	;list your directory
apropos fortran	;see what the system has on this topic
man gfortran	;show the manual page for gfortran
f, b, q	;go forward, back, or quit the man page
gedit x1.f	;a popular editor; edit/create x1.f
pico x1.f	;an alternate editor
cp x1.f x2.f	;copy a file
rm x2.f	;remove the file
exit	;log out

F.1.1   Running Linux

The following discussion assumes Linux is being used. However, from a user point of view, the commands are the same on any UNIX system and its variants, for example, AIX, HP-UX, Linux, Minix and Solaris. Mac OS X is also UNIX based.

Exercise F.1 *Start a terminal and try the commands in Table F.1. Then type in, compile and run the program shown in the following. Print your program and the output. Describe the software and computer that you used.*

- *Save the program as* x1.f
- *Compile it with* gfortran x1.f *or* g95 x1.f
- *Run it with* ./a.out
- *Run it with* ./a.out >op.dat.

```
      PROGRAM example1
C234567890 program statements start in column 7.
      IMPLICIT NONE
      INTEGER:: i
      REAL:: x,y,z,pi
      WRITE(6,*)' Hello World '
      pi=4*ATAN(1.)
      DO i=0,360,30   !another way to put comments
      x=i
      y=SIN(x*pi/180.)
      z=COS(x*pi/180.)
      WRITE(6,*)x,y,z
      ENDDO
      STOP
      END PROGRAM example1
```

In the first run, the program's output goes to the screen. In the second run, the output is redirected by the > to a file called op.dat.

You can print your program by using lp x1.f. If you are working on a shared machine with many users and printers, this probably won't work. You will have to ask your system administrator what the printer's name is. Then you would print to the correct destination with something like lp -d myprinter x1.f.

Table F.2 Some Windows terminal commands.

dir	;show your directory
type x1.f	;show x1.f on the screen
help dir	;help for the dir command
cd Desktop	;change to the directory called Desktop
cd ..	;go back to the parent directory
copy x1.f x2.f	;copy a file
del x2.f	;delete a file

An alternative printing method is to make a screen dump by using the `Print Scrn` key. This will write a `png` file that can be printed or sent to other people.

F.1.2 Running Windows

If you are using Windows, you can type in programs by using the Notepad editor. Compiling and running is done from a command-line terminal, found at the following: All Programs → Accessories → Command Prompt.

Exercise F.2 *Start a terminal and try the commands in Table F.2. Then type in, compile and run* PROGRAM example1 *shown in Exercise F.1. Print your program and the output. Describe the software and computer that you used.*

- *Save the program as* x1.f
- *Compile it with* gfortran x1.f *or* g95 x1.f
- *Run it with* a.exe
- *Run it with* a.exe >op.dat.

As in Linux, the first run sends the program's output to the screen. In the second run, the output is redirected by the > to a file called op.dat.

F.1.3 Running OS X

If you are using Mac OS X, the procedures are essentially the same as with Linux. The main thing is to find the 'X11 terminal'. Once that is done, the Linux commands described previously will work in that terminal.

F.2 Fortran 90

The Fortran 90 language is widely used in computational science and in particular, high-performance computing (HPC). It is a modern object-oriented language that is highly suitable for the development and maintenance of large programs. Fortran 90 is easy to learn. It compiles into highly optimized machine code, so programs run fast. It handles arrays efficiently. We have at our disposal an enormous amount of Fortran code that has been developed and carefully tested over 30+ years. Much of it is in the public domain; see, for example, the Netlib repository (Browne et al. 1995).

The following sections describe selected features of Fortran 90 so that anyone familiar with programming, but not necessarily Fortran 90, can easily get started.[1]

F.2.1   External Subprograms

Subprograms are helpful for organizing computational tasks in a program. There are two types of subprograms: functions and subroutines. The only difference between the two is that a function can only return *one* variable. A subroutine can return many.

Examples 2 and 3 show an external function and an external subroutine. The function calculates $z = x + y$, whereas the subroutine calculates both $z_1 = x + y$ and $z_2 = x - y$.

```
C Using an external function; Example 2
      FUNCTION zxy(x1,y1)
      IMPLICIT NONE
      REAL:: zxy
      REAL:: x1,y1
      zxy=x1+y1
      RETURN
      END FUNCTION zxy
C
      PROGRAM testfn
      IMPLICIT NONE
      REAL:: x,y,z,zxy
      x=1.
      y=2.
      z=zxy(x,y)
      WRITE(6,*) x,y,z
      STOP
      END PROGRAM testfn


C Using an external subroutine; Example 3
      SUBROUTINE zxy(x1,y1,sum,diff)
      IMPLICIT NONE
      REAL:: x1,y1,sum,diff
      sum = x1 + y1
      diff= x1 - y1
      RETURN
      END SUBROUTINE zxy
C
      PROGRAM testsub
      IMPLICIT NONE
      REAL:: x,y,z1,z2
      x=1.
      y=2.
      CALL zxy(x,y,z1,z2)
      WRITE(6,*) x,y,z1,z2
      STOP
      END PROGRAM testsub
```

---

[1] Fortran 90, Fortran 95 and Fortran 2000 have subtle differences. For our purposes, these distinctions will not matter, and assuming Fortran 90 compliance will be sufficient.

The adjective *external* is very important. The variables declared in an external subprogram are always private (i.e. local). This is good for general-purpose subprograms, because the end user does not have to know about the variables inside the subprogram.

Exercise F.3 *Rewrite the program in Exercise F.1 so that y and z are calculated in an external subroutine.*

F.2.2   Internal Subprograms

A unique feature of an *internal* subprogram is that it has access to the main program's variables. Other types of subprograms do not. Internal subprograms are intended for small jobs where access to the main program's variables can be very convenient. Example 4 shows an internal function.

```
C Using an internal function; Example 4
      PROGRAM testsinc
      IMPLICIT NONE
      REAL:: x,y,pi
      pi=4*ATAN(1.)
      x=0.4
      y=sinc(x)
      WRITE(6,*) x,y
      STOP
C
      CONTAINS
      FUNCTION sinc(t)
      REAL:: sinc, t
      sinc=SIN(pi*t)/(pi*t)
      RETURN
      END FUNCTION sinc
C
      END PROGRAM testsinc
```

Note that the internal function is CONTAIN'ed within the main program. The variables declared in the internal function are *private* (i.e. local). However, the main program's variables are *public* and available inside the function. In this case, it makes pi available for use in the function.

Note that if the declaration REAL::pi were added to the function, pi would become a private variable that is local to the function. Since the local pi doesn't have a numerical value, the program would not compile.

Exercise F.4 *Rewrite the program in Exercise F.1 so that y and z are calculated in an internal subroutine. Define π in the main program, but use it in the subroutine.*

F.2.3   Modules

Example 5 is a modified version of Example 3, with the subroutine CONTAIN'ed inside a MODULE. The main program USE's the module.

```
C Subroutine inside a module; Example 5
      MODULE mymod_mod
      CONTAINS
      SUBROUTINE zxy(x1,y1,sum,diff)
```

```
          IMPLICIT NONE
          REAL:: x1,y1,sum,diff
          sum = x1 + y1
          diff= x1 - y1
          RETURN
          END SUBROUTINE zxy
          END MODULE mymod_mod
C
          PROGRAM testsub
          USE mymod_mod
          IMPLICIT NONE
          REAL:: x,y,z1,z2
          x=1.
          y=2.
          CALL zxy(x,y,z1,z2)
          WRITE(6,*) x,y,z1,z2
          STOP
          END PROGRAM testsub
```

A module allows the compiler to do more sophisticated error checking. For example, it enables the compiler to compare the variables (x,y,z1,z2) in the CALL and the variables (x1,y1,sum,diff) in the SUBROUTINE. If a mismatch is found in the number of variables or the variable type (integer, real, etc.), the compiler will flag the error. Modules also offer other advantages that will become apparent later on.

In contrast, if Example 3 had a mismatch error like CALL zxy(x,y,z1), it would go unnoticed by the compiler and produce a runtime error.

Exercise F.5 *Rewrite the program in Exercise F.1 so that y and z are calculated in an external subroutine that is inside a module. USE the module in the main program.*

Exercise F.6 *Reading assignment: other uses for modules: Consult a reference book and find out about OPTIONAL variables in subprograms. Summarize your findings in 1/2 page or less.*

F.2.4   Shared Data

A module can be USE'd to share data:

```
          MODULE commondata_mod
          IMPLICIT NONE
          REAL, PARAMETER:: pi=3.1415926535, tpi=2*pi
          END MODULE commondata_mod
```

Any program with USE commondata_mod will have access to pi,tpi.
    Another way to share data is to put it above a CONTAINS in a module:

```
          MODULE somesubs_mod
          REAL, PRIVATE:: ka, omega
          CONTAINS
...  some subprograms
          END MODULE somesubs_mod
```

In this case, `ka` and `omega` will be global to everything in the module after the `CONTAINS`. However, because of the `PRIVATE` attribute, any program USE'ing the module will *not* have access to `ka` and `omega`.

If the `PRIVATE` attribute is omitted, or if `PUBLIC` is used, then any program USE'ing the module will get access to `ka` and `omega`.

The important rule is that *any statements* above a `CONTAINS` become global. This was the case in Example 4 with the internal subprogram. It is also the case in `MODULE somesubs_mod`.

Exercise F.7 *Rewrite the program in Exercise F.1 so that y and z are calculated in an external subroutine that is contained in a module. Define the value of π in the main program, but use data sharing so that π is available inside the subroutine.*

F.2.5   Integer, Real and Complex Numbers

The statement `i=5.6` becomes `i=5` if `i` is an integer. Similarly, `i=5/2` becomes `i=2`. Generally, real and complex numbers have a decimal, but integers do not.

The `COMPLEX` variable type exists in Fortran. For instance, $z = 2 + j3$ is coded as `z=(2.0,3.0)`.

Exercise F.8 *Comment on these calculations. Are some more efficient than others? Might some of them crash? Try them, as some of the results will be a surprise. (Hint: the computer uses $a^x = e^{x \ln a}$.)*

```
REAL:: y
INTEGER:: i,j
y=2**3
y=2**3.
y=(-2)**3.
y=(-2)**3.01
i=5
j=i/2*2
j=i*2/2
```

Exercise F.9 *Complex numbers: use this code to calculate $z = 2e^{j\pi/4}$.*

```
COMPLEX:: cj, z
cj=(0.,1.)
z=2*EXP(cj*pi/4)
WRITE(6,*)z
```

F.2.6   Arrays

Suppose we want to define a matrix A and vector b

$$A = \begin{bmatrix} 2 & -1 & 1 \\ 3 & 3 & 9 \\ 3 & 3 & 5 \end{bmatrix} ; \quad b = \begin{bmatrix} -1 \\ 0 \\ 4 \end{bmatrix}.$$

We could input the elements explicitly

```
      REAL, DIMENSION(3,3):: a
      REAL, DIMENSION(3):: b
      a(1,1)=2.
      a(1,2)=-1.
etc...
      b(1)=-1.
etc...
```

It is more efficient to enter the **b** data as a vector, and for **A**, RESHAPE it into a 3×3 matrix.

```
      a=RESHAPE( (/2.,3.,3.,-1.,3.,3.,1.,9.,5./),(/3,3/) )
      b=(/-1., 0., 4./)
```

We can write out the entire matrix or just part of it:

```
      WRITE(6,*)a !The entire A, in storage order
      WRITE(6,*)(a(1,j),j=1,3) !The first row, an implied-DO
```

It is important to note that in Fortran, the matrix elements are stored sequentially in the memory, in column order

$$a_{11}, a_{21}, a_{31}, a_{12}, \cdots a_{33}.$$

Therefore, the a_{ij} elements are in column order before the reshaping.

Arrays with a zero or negative index are also possible:

```
      REAL, DIMENSION(-3:2):: u
      REAL, DIMENSION(0:5):: v
```

which defines $u(-3), u(-2), \cdots, u(2)$ and $v(0), v(1), \cdots, v(5)$.

Exercise F.10 *Fortran 90 has an intrinsic function* MATMUL *for matrix multiplication. Using the earlier* **A** *and* **b**, *evaluate the matrix product* **Ab** *by using* ans=MATMUL(a,b).

Exercise F.11 *Fortran 90 has an intrinsic dot product function for real or complex vectors*

$$\mathbf{A} \cdot \mathbf{B}^* = \text{DOT\_PRODUCT}(\text{A}, \text{B})$$

Use it to evaluate $c = \mathbf{A} \cdot \mathbf{B}^*$ *where* $\mathbf{A} = 2\hat{\mathbf{x}} + j3\hat{\mathbf{y}} + \hat{\mathbf{z}}5$ *and* $\mathbf{B} = 2\hat{\mathbf{x}} - 7\hat{\mathbf{y}} + j6\hat{\mathbf{z}}$ *and check your answer. You can define the unit vectors* hatx=(/1.,0.,0./), haty=(/0.,1.,0./), hatz=(/0.,0.,1./).

F.2.7   *Input/Output*

The input/output unit numbers 5, 6 and 0 have special meanings for reading, writing and error messages, as described in Table F.3. The names stdin, stdout and stderr are Linux terminology. For example,

```
      READ(5,*)x,y,z            ! reads from keyboard
      WRITE(6,*)a,b,c           ! writes to screen
      WRITE(0,*)'Error, j2=',j2 ! writes to screen
```

Table F.3 Input/output unit numbers.

Unit #	Name	Purpose
5	stdin	Standard input, the keyboard
6	stdout	Standard output, the screen
0	stderr	Standard error, the screen

The statements OPEN and CLOSE are used for file manipulation. Suppose we want to read some data from an existing file in.dat and write some data to a new file op.dat. We can use

```
C Open the file 'in.dat' but only if it exists
      OPEN(UNIT=15, FILE='in.dat', STATUS='OLD')
      READ(15,*) x
C Open a new file 'op.dat' and write some data
      OPEN(UNIT=16, FILE='op.dat', STATUS='NEW')
      WRITE(16,*) y
C... You can close the files (this is optional)
      CLOSE(UNIT=15)
      CLOSE(UNIT=16)
```

If the file op.dat already exists then STATUS='NEW' won't work. STATUS='UNKNOWN' is usually more useful because it will open a new file whether or not an old one exists.

Each time a READ or WRITE is executed, the current position pointer in the file moves down to a new line. Consider two versions of an input data file, one with

```
12.7   5.8   3.3
```

and another with

```
12.7
5.8
3.3
```

Then,

```
      READ(15,*) x,y,z
```

can read either file, whereas

```
      READ(15,*) x
      READ(15,*) y
      READ(15,*) z
```

cannot read the first file because the second READ statement expects to see further data on the next line, and there isn't any.

Nothing forbids us from opening/closing files on units 5, 6 or 0; however we then lose their original use as stdin, stdout and stderr.

Exercise F.12 *A data file has the following numbers and words:*

```
12.45    frequency in GHz
-1.5    x1 coordinate
1.5    x2 coordinate
```

How would you get a program to read in the three numbers and ignore the comments?

F.2.8   Format Statement

In any reading or writing, the '*' means free format. Formatted input/output is also possible and sometimes desirable. For example, a formatted write could use

```
     j=2
     pi=4*ATAN(1.)
     WRITE(6,1200)j,pi,pi,pi
1200  FORMAT(' ','the ans is=', 2X, I5, F5.2, E14.7, G14.7)
```

which produces

```
 the ans is=      2 3.14 0.3141593E+01  3.141593
```

The ' ' writes a space. Then, the character string 'the ans is =' is written out. The 2X writes two spaces. Next, I5 means that the integer j is given a space (or 'field width') of five characters. F5.2 writes pi in the 'Fixed' format in a width of five characters, with two figures after the decimal. E14.7 gives the 'Exponential' format in a width of 14 characters with 7 figures after the decimal. G14.7 is the 'General' format with a width of 14 characters and 7 significant figures. The general format uses the floating style and automatically switches to exponential if needed.

Note that the field width must also allow for signs and exponents. If it's too small the write-out will show the error as a bunch of '****'. Also, in single precision, seven or eight figures of accuracy can be expected, so writing out numbers using excessive field widths can lead to false impressions about precision. A full description of formats can be found in the references, for example, Chapman (1998).

F.3   More on the OS

There are many operating system tools, tricks and shortcuts that can make usage easier. The following remarks apply to Linux, Windows and OS X, unless otherwise noted.

F.3.1   Redirection and Pipes

A useful alternative to opening/closing a file is to use *redirection*. This can be done on the command line. By typing

```
     a.out > op.dat
```

anything that normally goes to the screen will be redirected to the file op.dat. Redirection can also be used with < for reading files, so we could have

```
     a.out < in.dat > op.dat
```

Finally, the output of one program can be fed as input to another program with | which is called a *pipe*. Suppose a1.out writes to the screen and a2.out reads from the keyboard. Then we can use

```
a1.out | a2.out
```

F.3.2   Crash Messages

When a program encounters a runtime error (a 'floating-point exception signal' or 'sigfpe') it keeps running and generates nonsense numerical results. It is better to have the program halt and report the line number with the problem. This is done by using

```
gfortran -ffpe-trap=invalid,zero,overflow -g x1.f
```

This will halt the program for errors like LOG(-1.), divide by zero, and overflow. If you leave off the -g flag, the compiler's optimizer will produce slightly faster code. But in a crash you will not get the line number.

The -ffpe-trap flag is only for gfortran, and is described in the manual pages. For other compilers the user documentation should be consulted.

F.3.3   Object Files

A compiler turns source code (.f files) into object code (.o files). A 'linker' combines your object files with system libraries that are responsible for controlling the screen and keyboard and doing many other things. The result is a binary executable, called a.out or a.exe by default.[2]

Often you will work on a main program that uses several utility subroutines in a separate file. It's a waste of time to recompile the entire program and subroutines after every minor revision. It is better to compile the utilities just once and then reuse the object files. Then you just have to recompile the smaller part that you are working on.

For example, you may compile a file mysubs.f just once. Then, you can keep changing and recompiling mypgm.f while linking with mysubs.o:

```
gfortran  -c mysubs.f            ;makes mysubs.o
gfortran  -c mypgm.f             ;makes mypgm.o
gfortran  mysubs.o mypgm.o       ;links, makes a.out
./a.out                          ;run the program
```

This also works:

```
gfortran  -c mysubs.f
gfortran  -o mypgm mysubs.o mypgm.f
./mypgm
```

A tip: The compiler stores module information in *.mod files. After many edits and recompiles, you can sometimes end up with obsolete and incorrect information in these files. It can cause the compiler to produce confusing and incorrect error messages. If in doubt, clean up with rm *.mod *.o and start with a fresh recompile.

---

[2] Once a source code is compiled, the resulting object code is in machine language that a CPU can understand. There is no longer any connection with the source's language. The linker can combine object files from Fortran, c or a mix of any other languages and create an executable.

F.3.4   *Libraries*

An even more sophisticated approach (in Linux) is to make an *archive*. The `ar` command is for the creation and maintenance of a random access library. The `ranlib` command adds a directory so that the linker can efficiently find the files. In the following example, some object files are used to create a library of very popular subroutines called `libvps.a`.

```
ar r  libvps.a  *.o          ;new library, all *.o files
ar ru libvps.a ../extra/*.o  ;update, add files
ranlib libvps.a              ;make random access library
ar -t libvps.a               ;show contents
nm -s libvps.a               ;show contents, detailed
```

The linker is instructed to use this library with the command

```
gfortran -o mypgm mypgm.f   -L/home/myacct/lib -lvps
```

The `-L` option specifies the 'library path', which tells the linker where to find `libvps.a`. The `-l` option tells the linker to use `libvps.a`. Library names always have the prefix `lib`, but it is not used in the `-l`. Further details can be found on the manual page: `man ar`.

F.3.5   *Paths and Dots*

Question: Why do I have to type `./a.out` to run my program? Why can't I type `a.out`?
Answer: Linux has an 'environment variable' called the PATH. If you type `printenv PATH` you will see something like

    `/usr/local/bin:/bin:/usr/bin:/usr/X11R6/bin`

Every time you type a command like `ls`, the system will look for a file called `ls` in these directories, will find `/bin/ls` and then run it.

    If you created an executable called `a.out`, its location would probably be something like `/home/myname/a.out`. If you typed `a.out` it would not run, because `/home/myname` is not in the path and the system cannot find your `a.out`. On the other hand, if you typed `/home/myname/a.out` or `./a.out` it would run.

    A dot (.) means *my current working directory*, and a double dot (..) means *the parent directory*. Type `ls .` and it will show your directory. Typing `ls ..` will show the directory above it. So, typing `./a.out` is telling the system to run the `a.out` that is in your current working directory.

    In Linux you can append the current directory to your search path. The method depends on what kind of 'shell' you are using. The command is `setenv PATH ${PATH}:.` for the c-shell. Further details about paths and shells can be found in the references.

    Windows also uses (.) and (..) for the current and parent directories and has a path. If you type PATH on the command line, you will see it. The path is normally set up during the compiler installation, so `a.exe` usually runs without any difficulties.

F.4   Plotting

The public-domain program gnuplot can be used to make graphs. It is available for Linux, Windows and OS X. It can be found at (September 2015)

```
www.gnuplot.info
```

As an example, suppose we want to plot the data in file op.dat from Exercise F.1. The file has three columns of numbers. We want to plot column 1 as the x axis and column 2 as the y axis. The session on a machine called yagi would look like this:

```
yagi~>  gnuplot
gnuplot> plot 'op.dat' using 1:2 with lines
gnuplot> quit
yagi~>
```

It's useful to make a script file, for example, op.gnu, which might contain

```
set terminal postscript eps; set output 'op.eps'
set xlabel 'my x label'; set xrange[0:90]; set xtics 0,5
set ylabel 'my y label'; set autoscale y
set title 'My results'
plot 'op.dat' u 1:2 w l, 'op.dat' u 1:3 w l
set terminal x11; replot
```

The script is executed with

```
yagi~>  gnuplot
gnuplot> load 'op.gnu'
gnuplot> quit
yagi~>
```

The script writes the plot to an encapsulated postscript (eps) file called op.eps. Then it writes the plot to the x11 terminal (the screen). The eps file can be viewed with *ghostview:* gv op.eps. It can also be sent to a printer or used in a document. The file can be converted to other formats by using the *convert* utility. For instance, you can get a pdf version by using convert op.eps op.pdf.

Exercise F.13 *Modify the program in Exercise F.1 to write the data every 1°. Then plot the data. Make the plotting automatic with a script file. Print the result.*

F.5   Further Reading

A quick start for Linux can be found in Ray and Ray (2014). Scientific computation in a UNIX environment is covered by Landau and Fink (1993). The publications by Sun Microsystems (1986a,b) offer very good explanations about UNIX, at both introductory and advanced levels. Loukides (1990) describes libraries, linking c and Fortran object files, and other topics in UNIX/Fortran usage. There are many sources of information on the Internet.

To learn more about programming in Fortran 90, the following books are suggested: Akin (2003), Chapman (1998), Metcalf et al. (2004), as well as the article by Decyk et al. (c2000). The book by Redwine (1995) is good for readers who are familiar with Fortran but not Fortran 90.

References

Akin E (2003) *Object-Oriented Programming via Fortran 90/95.* Cambridge University Press.
Browne S, Dongarra J, Green S, Grosse E, Moore K, Rowan T and Wade R (1995) The Netlib mathematical software repository, http://www.netlib.org. Accessed: 2014-04-08.

Chapman SJ (1998) *Fortran 90/95 for Scientists and Engineers*. McGraw-Hill.

Decyk VK, Norton CD and Szymanski BK (c2000) Introduction to object-oriented concepts using Fortran 90, http://www.cs.rpi.edu/˜szymansk/OOF90/F90_Objects.html. Accessed: 2009-03-11.

Landau RH and Fink PJ (1993) *A Scientist's and Engineer's Guide to Workstations and Supercomputers*. John Wiley & Sons, Inc.

Loukides M (1990) *Unix for Fortran Programmers*. O'Reilly, Sebastopol CA.

Metcalf M, Reid J and Cohen M (2004) *Fortran 95/2003 Explained*. Oxford University Press.

Ray DS and Ray EJ (2014) *Unix and Linux: Visual QuickStart Guide*. Peachpit Press, Berkeley CA.

Redwine C (1995) *Upgrading to Fortran 90*. Springer-Verlag.

Sun Microsystems (1986a) *Doing More with Unix: Beginner's Guide*. Mountain View CA.

Sun Microsystems (1986b) *Setting Up Your Unix Environment*. Mountain View CA.

Appendix G

Numerical Methods

This appendix summarizes some techniques for numerical integration, root finding, the LU solution of linear equations, finding matrix eigenvalues and eigenvectors and the computation of Bessel and Legendre functions. Theoretical and practical descriptions of these and other numerical methods can be found in Faires and Burden (2013), Press et al. (1992, 1996) and Abramowitz and Stegun (1965).

Routines for matrix utilities, Bessel functions and Legendre functions were obtained from the Netlib repository (Browne et al. 1995). Fortran 90 interfaces for these subroutines have been developed by the author and are described in this appendix.

G.1 Numerical Integration

In general, it is always possible to exactly fit an Nth order polynomial through $N + 1$ points. *Simpson's rule* is derived by fitting a parabola through three sample points at $x = \{x_1, x_2, x_3\}$ of a function $f(x)$. This is most easily done by using the Lagrange polynomial method

$$f(x) = f(x_1)\frac{(x - x_2)(x - x_3)}{(x_1 - x_2)(x_1 - x_3)} + f(x_2)\frac{(x - x_1)(x - x_3)}{(x_2 - x_1)(x_2 - x_3)}$$
$$+ f(x_3)\frac{(x - x_1)(x - x_2)}{(x_3 - x_1)(x_3 - x_2)}.$$

This is a quadratic polynomial, and by substituting in $x = \{x_1, x_2, x_3\}$ it is clear that it fits the sample points exactly.

The quadratic polynomial is easily integrated. By denoting the sample points x_1, x_2, x_3 as $a, a + h, b$, it leads to Simpson's formula for two slices:

$$\int_a^b f(x)dx = \frac{h}{3}\{f(a) + 4f(a + h) + f(b)\} + E. \tag{G.1}$$

The error bound is found to be

$$E = \frac{-h^5}{90}f^{(IV)}(\xi). \tag{G.2}$$

To obtain a worst-case error estimate, the parameter $a \leq \xi \leq b$ is chosen to give the largest possible value of $|E|$. The error is proportional to the fourth derivative of f. Hence, it turns out that Simpson's rule is

Applied Frequency-Domain Electromagnetics, First Edition. Robert Paknys.
© 2016 John Wiley & Sons, Ltd. Published 2016 by John Wiley & Sons, Ltd.
Companion Website: www.wiley.com/go/paknys9981

exact not only for a quadratic function but for a cubic as well. This is an interesting and surprising result, as the method is based on fitting a parabola.

For N slices, the repeated application of the two-slice result gives

$$\int_a^b f(x)\,dx = \frac{h}{3}\{f(a) + 4f(a + h) + 2f(a + 2h) + 4f(a + 3h)$$

$$+ \cdots + 4f((N - 2)h) + 2f((N - 1)h) + 4f(Nh) + f(b)\}$$

$$- \frac{(b - a)}{180}h^4 f^{(IV)}(\xi), \tag{G.3}$$

where $h = (b - a)/N$. N has to be even. The sample weights follow the pattern $1, 4, 2, 4, \ldots 4, 1$. As before, $a \leq \xi \leq b$ is chosen to maximize the error bound.

Similarly, one can develop the *trapezoid rule* by fitting straight lines to the function sample points $f(x_n)$. The result is

$$\int_a^b f(x)\,dx = \frac{h}{2}\{f(a) + 2f(a + h) + 2f(a + 2h)$$

$$= + \cdots + 2f(Nh) + f(b)\} - \frac{(b - a)}{12}h^2 f''(\xi), \tag{G.4}$$

where as before, $h = (b - a)/N$. It is less accurate than Simpson's rule, and it doesn't save on calculations, so Simpson's rule is preferable.

An implementation of Simpson's rule is shown in the following. It evaluates

$$I = \int_0^1 \sin(\pi x)\,dx.$$

```
PROGRAM testsimp
IMPLICIT NONE
REAL, PARAMETER :: pi=3.14159265359
REAL:: a,b,h,sum
INTEGER:: i,n
a=0.
b=1.
WRITE(6,*)' even n= ???'
READ(5,*)n
h=(b-a)/n
sum=f(a)+f(b)
   DO i=1, n-1, 2
   sum=sum+4*f(a+i*h)
   ENDDO
   DO i=2, n-2, 2
   sum=sum+2*f(a+i*h)
   ENDDO
sum=sum*h/3
WRITE(6,*)sum, 2/pi
STOP
c

   CONTAINS
```

```
      FUNCTION f(t)
      REAL:: f
      REAL, INTENT(IN):: t
      f = SIN(pi*t)
      RETURN
      END FUNCTION f
c
      END PROGRAM testsimp
```

For some, it might be helpful to compare this to a c program.

```
/* simp.c */
# include <stdio.h>
# include <math.h>
# define pi 3.14159265359
/* main program */
main(){
float a,b,h,sum,f(float);
int i,N;
a=0.;
b=1.;
printf(" even N =??? ");
scanf("%d",&N);
h=(b-a)/(float)N ;
sum=f(a)+f(b) ;
for(i=1;i<=N-1;i=i+2){
sum=sum+4*f(a+i*h) ;
}
for(i=2;i<=N-2;i=i+2){
sum=sum+2*f(a+i*h) ;
}
sum=sum*h/3 ;
printf("sum=%f 2/pi= %f \n",sum, 2./pi);
}
/* function f(t) */
float f (float t){
float y;
y=sin(pi*t);
return y;
}
```

Exercise G.1 *Use* PROGRAM testsimp *to evaluate*

$$I = \int_0^1 \sin(\pi x)\, dx.$$

For $N = 6$ calculate I and find the percent error with respect to the exact answer $I_e = 2/\pi$. Also compute the error bound E. Show that the estimated error is larger than the true error.

Modify the program; make a DO-*loop over $N = 2, 4, 6, 8$. Print your final answers for I, $I - I_e$ and E versus N as a table.*

The 1D Simpson's rule is easily extended to 2D:

$$\int_{y_a}^{y_b} \int_{x_a}^{x_b} f(x,y)\, dx dy = \frac{h_x}{3}\frac{h_y}{3} \sum_{j=0}^{N_y}\sum_{i=0}^{N_x} w(i,N_x)w(j,N_y)f(x_a+ih_x, y_a+jh_y). \qquad (G.5)$$

There are N_x slices and $N_x + 1$ points along x, with $h_x = (x_b - x_a)/N_x$ and similarly $h_y = (y_b - y_a)/N_y$ for $N_y + 1$ points. The weight function w takes on the value of 1 at the end points, 4 at the odd points and 2 at the even points. In general $N_x \neq N_y$ so the indices for the end points will differ; the weight functions need to account for this.

Another popular integration method is Gaussian quadrature. This uses unequally spaced points. By adjusting the spacing, the error can be reduced. Any number of points is possible on a slice h. With two points,

$$\int_0^h f(x)\, dx = \frac{h}{2}(f(x_1) + f(x_2)) + \frac{h^5}{4320}f^{(IV)}(\xi),$$

where

$$x_{1,2} = \frac{h}{2}\left(1 \mp \frac{1}{\sqrt{3}}\right).$$

In a composite scheme with N slices,

$$\int_a^b f(x)\, dx = \frac{h}{2}\sum_{n=0}^{N-1}[f(x_1 + nh) + f(x_2 + nh)] + \frac{(b-a)}{4320}h^4 f^{(IV)}(\xi), \qquad (G.6)$$

where $h = (b-a)/N$. In a fair comparison of Simpson and two-point Gauss formulas, we note that Simpson has N slices and $N+1$ points whereas Gauss has N slices and $2N$ points. The comparison should use Gauss with N points. Under this condition, the slice size is $2h$ in the Gauss error term so that

$$\frac{(b-a)}{4320}(2h)^4 f^{(IV)}(\xi) = \frac{(b-a)}{270}h^4 f^{(IV)}(\xi).$$

Therfore, two-point Gauss offers a slightly smaller error term than Simpson by a factor of 180/270.

With any numerical integration scheme, it is possible to generate a sequence of results for decreasing step sizes $h, h/2, h/4, \ldots$. By taking linear combinations of these results, it is possible to make their error terms cancel out, thereby raising the order of the error. This is called Richardson extrapolation, and when applied to numerical integration it becomes Romberg's method.

Generating the sequence for $h, h/2, h/4, \cdots$ can be done in an iterative manner. Therefore, with the Romberg method we can watch for convergence to a desired accuracy as the h is reduced and the iteration proceeds. The details of the algorithm can be found in Faires and Burden (2013) and other references.

G.2 Root Finding

A secant (from the Latin *secare*, to cut) is a straight line that intersects (cuts) a curve at two points. To solve $f(x) = 0$, the *secant method* uses two guess values x_0 and x_1 for the roots. $f(x)$ is then approximated as a straight line that passes through these points, which gives the approximate root x_n:

$$x_n = x_0 - \frac{f(x_0)(x_1 - x_0)}{f(x_1) - f(x_0)}. \qquad (G.7)$$

Then, we discard x_0 and update $x_1 \to x_0$, $x_n \to x_1$ and repeat until $f(x_n)$ is sufficiently small.

If x_0 and x_1 are close together, we can consider $\Delta x = x_1 - x_0$ so that

$$x_n = x_0 - \frac{f(x_0)}{f'(x)}\Big|_{x=x_0} \tag{G.8}$$

which is known as the *Newton–Raphson method* and can be thought of as a special case of the secant method.

The example in the following uses the secant method to solve $w(x) = x^3 - 8 = 0$. The statement

```
CALL secant(x0,x1,xn,w)
```

uses guess values x0, x1 to find xn, the root of w(xn)=0. FUNCTION w(x) is a user-defined function. Starting with x_0 and x_1, it finds $x_2, x_3, \cdots$. When $|x_n - x_{n-1}|/|x_n| < 10^{-5}$ is reached, the root x_n is returned.

```
C Secant method of root finding
C--- Function to be solved
         MODULE root_mod
         CONTAINS
         FUNCTION w(x)
         IMPLICIT NONE
         REAL:: w
         REAL, INTENT(IN):: x
         w= x**3-8.
         RETURN
         END FUNCTION w
C---
         SUBROUTINE secant(x0g,x1g,xn,f)
         IMPLICIT NONE
         REAL:: f
         REAL, INTENT(IN):: x0g,x1g
         REAL, INTENT(OUT):: xn
         REAL:: x0,x1,f0,f1
         REAL, PARAMETER:: tol=1.E-5
         INTEGER:: i
         i=0
         x0=x0g
         x1=x1g
2        f0=f(x0)
         f1=f(x1)
         xn=x0- (f0*(x1-x0))/(f1-f0)
         i=i+1
         x0=x1
         x1=xn
         IF(i .GE. 10)GO TO 3
         if(ABS((x1-x0)/x0) .GT. tol)GO TO 2
         RETURN
3        WRITE(6,*)' Secant bad guess, i=10'
         STOP
         END SUBROUTINE secant
         END MODULE root_mod
```

```
C- - - Main program
      PROGRAM testsec
      USE root_mod
      IMPLICIT NONE
      REAL:: x0,x1,xn
      WRITE(6,*)' guess values x0, x1 = ??? '
      READ(5,*) x0,x1
      CALL secant(x0,x1,xn,w)
      WRITE(6,1202)xn
 1202 FORMAT(' ','xn= ',G14.7)
      STOP
      END PROGRAM testsec
```

The preceding code introduces some new programming ideas. Notice that the subroutine refers to a real function f that is otherwise unspecified. It is the CALL that determines $f = w$. Also, the optional attribute INTENT, an error-checking feature, is being used.

Exercise G.2 *Use* PROGRAM testsec *to solve for the real root of* $w(x) = x^3 - 8$.

Exercise G.3 *Use* PROGRAM testsec *to solve for the two other complex roots of* $w(x) = x^3 - 8$. *This can be done by declaring the necessary variables as* COMPLEX.

Check your numerical results by comparing with the exact answer. In closed form, the roots of unity are found by recognizing that $1 = e^{j2n\pi}$ *so that* $x^3 = 8e^{j2n\pi}$, *from which* $x = 8^{1/3}e^{j2n\pi/3}$; $n = 0, \pm 1, \pm 2, \ldots$.

G.3   Matrix Equations

The matrix equation $Ax = b$ can be efficiently solved with the LU method. The method provides a way of finding lower and upper triangular matrices L, U such that $A = LU$. Then, $LUx = b$. Next, we define $Ux = z$ so that $Lz = b$. We then solve $Lz = b$ for z by forward substitution. As the final step, $Ux = z$ is solved for x by back substitution.

Example G.1 Let's demonstrate the LU method with a 3×3 example, where $Ax = b$ is

$$\begin{bmatrix} 2 & -1 & 1 \\ 3 & 3 & 9 \\ 3 & 3 & 5 \end{bmatrix} \begin{bmatrix} x_1 \\ x_2 \\ x_3 \end{bmatrix} = \begin{bmatrix} -1 \\ 0 \\ 4 \end{bmatrix}.$$

The matrix A is factored as

$$A = \begin{bmatrix} 2 & -1 & 1 \\ 3 & 3 & 9 \\ 3 & 3 & 5 \end{bmatrix} = \begin{bmatrix} \ell_{11} & 0 & 0 \\ \ell_{21} & \ell_{22} & 0 \\ \ell_{31} & \ell_{32} & \ell_{33} \end{bmatrix} \begin{bmatrix} 1 & u_{12} & u_{13} \\ 0 & 1 & u_{23} \\ 0 & 0 & 1 \end{bmatrix}.$$

Carrying out the matrix multiplication row by row, $a_{11} = 2 = \ell_{11}$, $a_{12} = -1 = \ell_{11}u_{12}$, and so on, leads to all of the values for L and U:

$$LU = \begin{bmatrix} 2 & 0 & 0 \\ 3 & 4.5 & 0 \\ 3 & 4.5 & -4 \end{bmatrix} \begin{bmatrix} 1 & -1/2 & 1/2 \\ 0 & 1 & 5/3 \\ 0 & 0 & 1 \end{bmatrix}.$$

Forward substitution is used to solve $Lz = b$:

$$\begin{bmatrix} 2 & 0 & 0 \\ 3 & 4.5 & 0 \\ 3 & 4.5 & -4 \end{bmatrix} \begin{bmatrix} z_1 \\ z_2 \\ z_3 \end{bmatrix} = \begin{bmatrix} -1 \\ 0 \\ 4 \end{bmatrix}$$

with the result that $z = (-1/2, 1/3, -1)^T$. Finally, back substitution is used to solve $Ux = z$ for x:

$$\begin{bmatrix} 1 & -1/2 & 1/2 \\ 0 & 1 & 5/3 \\ 0 & 0 & 1 \end{bmatrix} \begin{bmatrix} x_1 \\ x_2 \\ x_3 \end{bmatrix} = \begin{bmatrix} -1/2 \\ 1/3 \\ -1 \end{bmatrix}$$

with the result that $x = (1, 2, -1)^T$.

When $u_{ii} = 1$ this is called *Crout's method*. We could have instead assumed that $\ell_{ii} = 1$, which is called *Dolittle's method*. If $A = A^T$ then a factorization $A = LL^T$ is possible; this is *Cholesky's method*. If A is not positive definite, then the Cholesky method still works, but L could be complex, even if A is real. A is positive definite if $x^T A x > 0$ for all $x \neq 0$.

Finding the L, U factors is a lot of work, and the LU method doesn't offer much advantage over Gaussian elimination. On the other hand, if the answers for many different values of b are needed, the L, U factors can be reused. The forward and back substitutions are very fast so the method becomes advantageous. Gaussian elimination requires that we start over for each new value of b so in this situation it is much less efficient.

The Fortran 90 routine SUBROUTINE molerLU finds the L, U factors for an $n \times n$ matrix A. The elements of L and U overwrite the original matrix A. SUBROUTINE molerSolve solves the matrix equation $LUx = b$ for the $n \times 1$ vector x. It is possible to solve $LUx = b$ for many different b vectors by using just one call to molerLU followed by multiple calls to molerSolve. The subroutines are generic, so they can be used for real or complex equations, with single or double precision. They call subroutines by C. Moler that are in LAPACK (Anderson et al. 1999).

When finished, SUBROUTINE molerDeallocate() can be used to deallocate storage that was allocated for the pivot vector ipvt, an array of n integers. To solve the matrix equation $Ax = b$,

```
USE moler_mod
CALL molerLU(n,A)
CALL molerSolve(n,A,b,x)
CALL molerDeallocate() ! this is optional
```

Optional arguments can provide the reciprocal of the condition number opt_rcond and/or the determinant opt_det:

```
CALL molerLU(n,A,opt_rcond=rcond,opt_det=det)
```

Exercise G.4 *Using* A *and* b *from Exercise G.1, solve the matrix equation* $Ax = b$ *and write out* x. *Evaluate the product* $c = Ax$ *by using the intrinsic function* MATMUL. *Confirm that* $c = b$. *Follow the method in Section F.2.6, to input* A *and* b.

G.4   Matrix Eigenvalues

The matrix equation $Ax = \lambda Bx$ can be solved for the eigenvalues λ and eigenvectors x by using the Fortran 90 SUBROUTINE garbow. It is generic, allowing for single or double precision. It calls subroutines by Garbow (1978). Typical usage is as follows:

```
USE garbow_mod
...

INTEGER:: n
COMPLEX, DIMENSION(100):: lambda
REAL, DIMENSION(100,100):: A,B
COMPLEX,DIMENSION(100,100):: X
...

    CALL garbow(n,A,B,lambda,opt_X=X)
```

The matrices A and B are real, but the eigenvalues and eigenvectors can be complex.

The program finds the eigenvalues lambda(1), lambda(2)$\cdots$ lambda(n). The eigenvectors are optional. If requested, the jth column of X gives the jth eigenvector, belonging to lambda(j).

G.5   Bessel Functions

For a series solution using cylindrical modes, we need to know Bessel functions of integer order n. The most efficient method is to find just two known orders, and then obtain the rest of them with a recursion:

$$Z_{n-1}(x) + Z_{n+1}(x) = (2n/x)Z_n(x).$$

Here, Z_n can be J_n, Y_n or H_n. The recursion with respect to order has to be done in the direction that the functions increase, otherwise severe rounding errors will occur. The general procedure is to recur downward for J_n and upward for Y_n.

When $n < x$, J_n and Y_n are oscillatory, the recursion can be done either upward or downward in order. When $n > x$, J_n decays, so recursion has to be done downward from the highest order. Thus, one could start with $J_n = 0$ and $J_{n-1} = 10^{-35}$. When finished, the amplitude is found by applying the identity $J_0 + 2\sum_{n=1}^{\infty} J_{2n} = 1$. As for Y_n, when $n > x$ it grows exponentially, so recursion upward is appropriate. Y_0 can be found from a series, and a Wronskian can be used to find Y_1 from J_0, J_1 and Y_0.

As can be seen, the computation of Bessel functions can get highly involved. What's the best approach depends on the machine precision. If x is complex or n is not an integer, it gets even more complicated.

For real non-negative orders and and complex arguments, the Fortran 90 program SUBROUTINE amos calculates Hankel functions of the first and second kind, h1, h2 and their derivatives h1p, h2p. The array elements h1(0), h1(1), h1(2) ... h1(nmax) correspond to orders $0, 1, 2, \ldots$. The index nmax can be smaller than the array's dimension. Similar remarks apply to SUBROUTINE amosj which only calculates the J Bessel functions.

Normally, integer orders are computed, but the optional `opt_fnu` can be used to obtain fractional real orders n+opt_fnu where n is an integer. If `opt_fnu` is omitted, it is taken to be zero.

The subroutines are generic, allowing for single or double precision. The argument x can be real or complex. They call subroutines by Amos (1986). Typical usage is as follows:

```
      USE amos_mod
      REAL:: x
      INTEGER, PARAMETER:: ndim=100,nmax=30
      COMPLEX, DIMENSION(0:ndim):: h1,h1p,h2,h2p
      REAL, DIMENSION(0:ndim):: bj,bjp
...
C For Hankel functions
      CALL amos(nmax,x,h1,h1p,h2,h2p,opt_fnu=0.)
C For J Bessel functions
      CALL amosj(nmax,x,bj,bjp,opt_fnu=0.)
```

In this example, integer orders $0 \le n \le 30$ are calculated.

G.6   Legendre Polynomials

The normalized Legendre polynomials $\bar{P}_n^m(x)$ can be computed by using a recursion technique (Smith et al. 1981). The normalization is in accordance with Equation (D.9). A Fortran 90 program SUBROUTINE `sol` (named after Smith, Olver and Lozier) is an interface to their subroutine. The usage is straightforward, as follows:

```
      USE sol_mod
      REAL:: x
      INTEGER, PARAMETER:: ndim=100,nmax=20
      REAL, DIMENSION(0:ndim,0:ndim):: pbar
      INTEGER, PARAMETER, DIMENSION(2):: opt_m=(/0,1/)
...
      CALL sol(nmax,x,pbar,opt_m)
```

This calculates `pbar(n,m)`$=\bar{P}_n^m(x)$ for degree n and order m where n=0,1,2,$\cdots$ nmax and m=0,1,2,$\cdots$ n. $\bar{P}_n^m(x) = 0$ when $n > m$. The optional parameter `opt_m` can be used to reduce the range of orders to `opt_m(1)` $\le$ m $\le$ `opt_m(2)`. In this example, only $\bar{P}_n^0$ and $\bar{P}_n^1$ are calculated for $0 \le n \le 20$.

References

Abramowitz M and Stegun I (1965) *Handbook of Mathematical Functions*. Dover Publications.

Amos DE (1986) Algorithm 644: A portable package for Bessel functions of a complex argument and nonnegative order. *ACM Trans. Math. Softw.* **12**(3), 265–273.

Anderson E, Bai Z, Bischof C, Blackford S, Demmel J, Dongarra J, Du Croz J, Greenbaum A, Hammarling S, McKenney A and Sorensen D (1999) *LAPACK Users' Guide* third edn. Society for Industrial and Applied Mathematics, Philadelphia PA.

Browne S, Dongarra J, Grosse E and Rowan T (1995) The netlib mathematical software repository, http://www.netlib.org. Accessed: 2014-04-08.

Faires JD and Burden RL (2013) *Numerical Methods*. Brooks/Cole.

Garbow BS (1978) Algorithm 535: the QZ algorithm to solve the generalized eigenvalue problem for complex matrices [F2]. *ACM Trans. Math. Softw.* **4**(4), 404–410.

Press WH, Teukolsky SA, Vetterling WT and Flannery BP (1992) *Numerical Recipes in Fortran* vol. **1**. Cambridge University Press.

Press WH, Teukolsky SA, Vetterling WT and Flannery BP (1996) *Numerical Recipes in Fortran 90* vol. **2**. Cambridge University Press.

Smith JM, Olver FWJ and Lozier DW (1981) Extended-range arithmetic and normalized Legendre polynomials. *ACM Trans. Math. Softw.* **7**(1), 93–105.

Appendix H

Software Provided

Table H.1 Software list.

Name	Description	Reference
PROGRAM mlslab	Multilayer slab	Chapter 2
PROGRAM wgstep	Mode matching, H-plane waveguide step	Chapter 3
PROGRAM iris	Mode matching, H-plane waveguide iris	Chapter 3
PROGRAM ridgewg	Ridge waveguide k_c	Chapter 3
PROGRAM ezcyl	E_z for a cylinder	Chapter 5
PROGRAM jxjz	Half-plane surface currents	Chapter 5
PROGRAM sphrcs	Sphere RCS	Chapter 5
PROGRAM mmtmstrip	MoM, TM strip	Chapter 6
PROGRAM mmtestrip	MoM, TE strip	Chapter 6
PROGRAM wire	MoM, thin wire	Chapter 6
PROGRAM fem2d	FEM 2D Laplace's equation	Chapter 7
PROGRAM utd1	UTD half plane	Chapter 8
MODULE trfn_mod	UTD transition functions	Chapter 8
PROGRAM ved	Vertical electric dipole	Chapter 12
PROGRAM tegrating	TE strip grating	Chapter 12
PROGRAM huni	Hankel functions, complex order	Appendix C
PROGRAM testsimp	Simpson's method	Appendix G
PROGRAM testsec	Secant method	Appendix G
MODULE amos_mod	Hankel and Bessel functions	Appendix G
MODULE moler_mod	LU matrix solver	Appendix G
MODULE sol_mod	Legendre functions	Appendix G
MODULE garbow_mod	Matrix eigenvalues	Appendix G

Table H.1 lists the software provided at www.wiley.com/go/paknys9981.

Applied Frequency-Domain Electromagnetics, First Edition. Robert Paknys.
© 2016 John Wiley & Sons, Ltd. Published 2016 by John Wiley & Sons, Ltd.
Companion Website: www.wiley.com/go/paknys9981

Index

Printed and bound by CPI Group (UK) Ltd, Croydon, CR0 4YY

16/04/2025

14658398-0003